(continued from front flap)

diation sensors, shaft-angle encoders (and the digital techniques underlying their operation), transducers with integral excitation and signal-conditioning circuitry, the laser interferometer, gyros, electro-optical sensors, photodiodes, and contactless sensors.

• Appendices contain the new International System of Units, suggested drawing symbols to represent all types of transducers in electrical diagrams and schematics, and a complete glossary of terms pertaining to transducers and related measuring devices.

About the Author

HARRY N. NORTON developed the "error-band" concept and has been extremely active in the programs of the Instruments Society of America (ISA). In this organization he has been Standards Director, Vice Chairman for Planning of the 13th National ISA Aerospace Instrumentation Symposium, President of the San Diego Section, and chairman and developer of numerous conference and symposium sessions.

Handbook of Transducers
for Electronic
Measuring Systems

PRENTICE-HALL SERIES IN ELECTRONIC TECHNOLOGY

Dr. Irving L. Kosow, editor

Charles M. Thomson, Joseph J. Gershon, and Joseph A. Labok, consulting editors

Anderson, Santanelli and Kulis   Alternating Current Circuits and Measurements: A Self-Instructional Programed Manual

Anderson, Santanelli and Kulis   Direct Current Circuits and Measurements: A Self-Instructional Programed Manual

Babb   Pulse Circuits: Switching and Shaping

Barrington   High Vacuum Engineering

Benedict and Weiner   Industrial Electronic Circuits and Applications

Berkley   Laboratory Course in Pulse Circuitry

Branson   Introduction to Electronics

Cook and Liff   Frequency Modulation Receivers

Corning   Transistor Circuit Analysis and Design

Cowles   Transistor Circuits and Applications

DeFrance   Electrical Fundamentals: Part I, Direct Current; Part II, Indirect Current

Doyle   Pulse Fundamentals

Eadie   Introduction to the Basic Computer

Federal Electric Corporation   Mathematics for Electronics: A Self-Instructional Programed Manual

Federal Electric Corporation   Special Purpose Transistors: A Self-Instructional Programed Manual

Federal Electric Corporation   Transistors: A Self-Instructional Programed Manual

Flores   Computer Design

Hewlett-Packard   Microwave Theory and Measurements

Jackson   Introduction to Electric Circuits, 2nd ed.

Johnson   Servomechanisms

Kosow   Electric Machinery and Control

Lenk   Data Book for Electronic Technicians and Engineers

Lenk   Handbook of Oscilloscopes: Theory and Application

Lenk   Handbook of Practical Electronic Tests and Measurements

PRENTICE-HALL INTERNATIONAL, INC., *London*
PRENTICE-HALL OF AUSTRALIA, PTY. LTD., *Sydney*
PRENTICE-HALL OF CANADA, LTD., *Toronto*
PRENTICE-HALL OF INDIA PRIVATE LTD., *New Delhi*
PRENTICE-HALL OF JAPAN, INC., *Tokyo*

Handbook of Transducers
for Electronic
Measuring Systems

HARRY N. NORTON

Jet Propulsion Laboratory
California Institute of Technology

PRENTICE-HALL, Inc., Englewood Cliffs, N.J.

13-382242-7

Library of Congress Catalog Card Number 76-76879

Printed in the United States of America

Preface

The rapid growth of the instrumentation field during the past decade has few parallels in history. The sophisticated electronic measurement and control instrumentation systems now in routine operation in many countries would have appeared of marginal credibility as recently as the mid-1950's. Of the two subfields of instrumentation, control instrumentation—the basis of automation—has received considerably more emphasis in technical as well as popular literature than measuring instrumentation, whose end product is solely data. Yet, to the many professionals who suddenly found themselves working on or using electronic measuring systems, the new electronic measuring technology is as glamorous as a completely automated power station is to industrial engineers. It also appears surrounded by a certain aura of magic, especially when it becomes necessary to understand the operation and function of the sensing portion of the system—the transducers. Unfortunately, the "magic" nature of transducers has often led to waste of time and money and to misinterpretation of data.

One of the main objectives to be met by this book is to strip away any magic from transducers used in electronic measuring systems and to expose their true nature, which should be comprehensible without much difficulty to anyone having little more than a thorough high-school education (science major).

This book should be useful to transducer designers and to those who manufacture, inspect, market, and sell transducers. Since it is application-oriented, however, rather than development-oriented, it should be even more useful to the users of transducers, and those responsible for selecting, purchasing, and testing transducers in behalf of the users.

Transducer users are found in virtually all major industries and sciences, wherever measurements must be recorded or displayed at a point not identical to the point where the measurement is made. The metals, plastics, nuclear, aerospace, transportation, chemical, petroleum, paper, and textile industries, the biological, medical, oceanographic, atmospheric, natural, physical, and educational sciences, to name a few examples, all use measuring systems of varying complexity to varying degrees.

This book is also intended as a textbook and reference book for students in technical institutes, trade schools, and junior colleges, and for college and university students at the undergraduate, graduate, post-graduate, and even post-doctorate levels.

Chapters I, II, and III are general and introductory. The arabic-numbered chapters constitute the main portion of the book. Each chapter covers the transducers used for one measurand. The chapters are numbered in alphabetical order of measurands. Appendices following the last measurand chapter contain certain reference information including a glossary of terms.

The contents of each chapter are arranged in accordance with a format standardized for this book to the best extent possible. Introductory material, centered about basic concepts, is followed by a description of sensing elements, transducer design, and operation based mostly on transduction principle used, essential performance characteristics (as disclosed primarily in specifications) and applications, verification of transducer performance (calibration and testing), measuring devices used for ranges of the measurand and for related measurands (not treated in a chapter of their own), and a bibliography for those wanting further information on particular topics covered in the chapter. The "related laws" are intended as memory refreshers.

Explanations of basic concepts and definitions of terms are adapted from those shown in various dictionaries, textbooks, ISA and USA Standards, and other reference material. Most of the nomenclature and terminology used in this book is intended to be consistent with ISA Standard S37.1. When two or three equivalent terms are shown in the text, the author has used his prerogative to indicate such terms as *preferred* (acceptable, or commonly used in particular fields) or "colloquial," (but not preferred) in the manner just shown.

Since this book gives transducers an application-oriented treatment, emphasis is placed on designs which are commercially available and used in two or more major applications. Specific devices are described in each of the chapters to illustrate operating principles, design features, and other textual matter. They represent only a fraction of the numerous usable designs developed by the transducer industry to fill measurement needs. It should be noted that all commercial devices described are covered by one or more patents, and rights to their construction are held solely and entirely by their manufacturers. It should also be realized that mergers and acquisitions occur so frequently in the instrument industry that company names in picture captions may not reflect the current name of the transducer's manufacturer. Statements as to the relative merits of transducers or transducer types described in this book reflect, of course, only the opinion of the author.

Some sensing devices which (in the author's opinion) either have limited applications, are not yet fully developed into production hardware, or have doubtful practical value are mentioned by cursory descriptions. Many "theoretically possible" combinations of sensing and transduction elements are omitted entirely, even when their feasibility has been demonstrated in the laboratory.

The widely divergent interests of the users of this book made it impractical to include problems for student use. Instructors should not find it difficult to develop problems fitting the curriculum of a given course. When developing problems for most curricula, the following should not be overlooked: the selection of a specific transducer type (with reasons given for the selection) for a stated measurement problem; a plausible explanation for a stated data anomaly when transducer type, normal measurand levels, and normal environmental conditions are known; the selection of generic types of transducers for various measurands for a stated measuring system; prediction of a transducer's behavior during a stated environmental test; interpretation of a given set of calibration data; and a critical analysis of a list of specifications for a stated transducer type.

Finally, a word of caution must be directed to the reader regarding the use of transducers and their associated measuring systems. The prime purpose of this book is to describe transducers, not measuring techniques. Although a number of relatively straightforward measurement problems can be solved solely on the basis of the material presented here, most successful transducer applications require considerable additional training and experience in measurement engineering and a solid educational foundation in the physical sciences relating to the measurand. Practical experience, often gained by discovering one's own mistakes, is probably the best road towards obtaining valid data from transducer installations. The bibliography of each chapter contains a sampling of some general works and many specialized texts and papers offering additional information. The titles alone imply the vast amount of research effort by the many specialists in each measurement field whose work could only be intimated or treated very briefly in this book.

HARRY N. NORTON

Acknowledgments

The preparation of this book was vastly facilitated by the help and encouragement I received from a large number of my friends, all dedicated workers in the instrumentation and electronic fields.

The strongest source of encouragement has been my long-standing association with the ISA (Instrument Society of America), whose Executive Director, Herbert S. Kindler, permitted me to use all material, both published and in draft stages, developed by the SP31 and SP37 transducer standards committees operating under my direction as well as by other standards committees working in related areas. My thanks are extended to all former and current members of these committees. Preparatory studies for this book also enabled me to make more of a contribution to the drafting and editing of pertinent ISA standards. Hence, any similarities between material in this book and in ISA transducer standards are very definitely intentional.

The book was conceived, portions of it were written, and most of the reference material for it was collected during my period of employment at the Convair Division of the General Dynamics Corporation in San Diego. While at Convair, I received considerable encouragement from two instrumentation engineering managers, Charles E. Wilson and John A. Hughes, and from administrative engineering manager Dean H. McCoy, who supported my participation in ISA and sponsored my attendance at technical conferences in the USA as well as in Europe. My admiration and appreciation are extended to the International Measurements Confederation (headquartered in Budapest), their tri-annual Congress, IMEKO, its dedicated Permanent Secretary, Dr. G. Striker, and to Congress participants such as Professor N. I. Chistiakov, who presented me with several up-to-date Russian books in the transducer field.

Portions of some of the chapters in this book contain material similar to that used in a number of my technical articles which appeared in *Space/Aeronautics*, *Instruments and Control Systems*, and *Ground Support Equipment*. These articles are included in the bibliography of the applicable chapters. Notably, some of the transducer basics of Chapter II were included in an article in *Space/Aeronautics*, October 1962, Part 2.

The following have contributed to this book by reviewing the chapters indicated and giving me the benefit of their comments:

I:
II: } H. C. Keith; C. E. Wilson; C. H. Colvin; M. Brady
III:

1: R. R. Bouche; H. D. Morris; L. L. Lathrop

2: P. E. Humphrey

3: J. S. Hernandez; J. E. Elliott

4: R. L. Galley; G. J. Lyons

5: M. J. Lebow; H. W. Rosenberg; H. L. Jensen, Jr.

6: F. C. Quinn; D. Chleck

7: R. L. Bachner; M. Rome

8: C. W. Silver; M. D. Bennett

9: Mrs. R. Stephens; N. G. Anton; P. Polishuk; T. F. Heinsheimer

10: C. J. Elliott; J. E. Smith

11: D. N. Keast; J. W. Day; D. E. Risty

12: M. T. Zimmerman; J. C. Palmer; C. R. Forrester

13: A. D. Kurtz; D. R. Harting; S. P. Wnuk, Jr.

14: F. D. Werner; K. J. Knudsen; S. B. Williams

HARRY N. NORTON

Contents

Electronic Measuring Systems

1

1.1 Introduction

The science of measuring, the field of *metrology*, received its greatest impetus with the development of electronic techniques. These allowed remote display of the measured quantity, enabled many new types of measurements to be taken under a variety of adverse conditions, improved the accuracy of a large number of conventional measurements, and introduced a large measure of convenience into virtually all measurements.

Electronic measuring systems can be quite compact, to the extent of incorporating the sensing device as well as the display device in a single, small package. They can, on the other hand, be very large and complex, include a great number of subsystems, rely on radio transmission and retransmission as signal links, and utilize buildings full of data storage and display equipment.

The majority of electronic measurement systems have their data display device, or devices, located at a certain distance from the point at which the measurement is taken. Such remote-display measuring systems are generally referred to as *telemetry* systems. The verb "telemeter" is derived from Greek words—the adjective *tele* (meaning far, distant) and the Greek root *metr*, as employed in the verb *metrein* (to measure) or the noun *metron* (a measure). Bearing in mind a later consonant shift from "t" to "s," we can see that the German word *messen* (to measure) as well as the word "measure," itself, are also derived from the same Greek root, as is our word "metrology."

I.2 Basic Measuring Systems

The most elementary electronic measuring system is shown in Figure I-1. It consists of:

the *transducer*, which converts the *measurand* (measured quantity, property, or condition) into a usable electrical output.

the *signal conditioner*, which converts the transducer output into an electrical quantity suitable for proper operation of the display device.

the *power supply*, which feeds the required electrical power to the signal conditioner, provides excitation for all except "self-generating" types of transducers, and may also furnish electric power to certain types of display devices.

the *display*, or *readout device*, which displays the required information about the measurand.

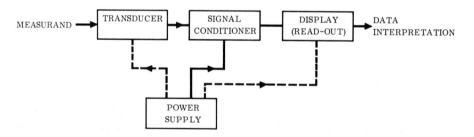

Fig. I-1. Basic electronic measuring system.

Transducers, because of their immense variety, constitute the key portion of each of the measuring systems in which they are used. Signal conditioners may vary in complexity from a simple resistance network or impedance-matching device to multistage amplifiers with or without demodulators, analog-to-digital converters, spectral-density analyzers, and other elaborate circuitry. Among commonly used display (readout) devices are circular-chart and strip-chart recorders, analog or digital meters, numerical printers, oscilloscopes (which may be equipped with a camera so that permanent records can be obtained), and discrete-level indicator lights.

In some systems the signal-conditioning and power-supply functions are incorporated in the readout unit. In others, the signal conditioner is packaged integrally with the transducers. Electronic measuring systems which appear to operate without any signal conditioning have, in virtually all instances, some signal-conditioning circuitry in the readout equipment or in the transducer.

I.3 Basic Telemetry Systems

In the remote-display electronic measuring system, the telemetry system, the function of one of the units of the basic measuring system is altered and, in a simple system as illustrated in Figure I-2, three new units are added. The function of the transducer is the same.

The *signal conditioner* converts the transducer output into an electrical quantity suitable for proper modulation of the transmitter.

The *transmitter* generates a high-frequency signal, the *carrier*, which is modulated by the properly conditioned transducer output, and feeds it into the transmission system.

The *transmission system* transmits the modulated carrier signal from the transmitter to the receiver. It can be either a radio link or a conducting link. In a radio link a transmitting antenna radiates the signal, and a receiving antenna intercepts the signal and feeds it to the receiver. In a conducting link a *coupler* feeds the signal into a conductor, and another coupler accepts the signal from the conductor and feeds it to the receiver.

The conducting link can be a special cable installed between the transmitting and receiving points solely for the purposes of the telemetry system. It can also be an electric-power line, electrical-traction wires or rails, a telephone or telegraph line, or a conductor of some other form. As in all electrical circuits, a pair of conductors is always required in a conducting link. This is accomplished in many systems by

Fig. I-2. Simple telemetry system.

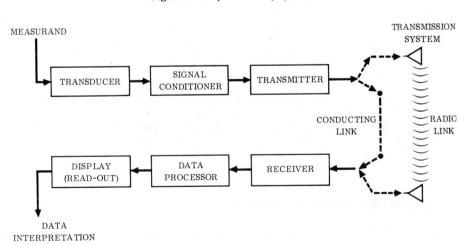

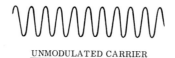

UNMODULATED CARRIER

PHASE MODULATION (PM)

AMPLITUDE MODULATION (AM)   FREQUENCY MODULATION (FM)

MODULATED CARRIER

Fig. I-3.   Principal types of carrier modulation.

using shielded, twisted, two-conductor cables or jacketed coaxial cables where either the outer shield or a special second inner shield acts as the other conductor. In other systems either the earth itself or a grounded conductor is used as one of the two required conductors in the link. Telemetry systems using electric-power lines as conducting links are often referred to as *carrier-current systems.*

The *receiver* amplifies the modulated carrier signal it receives and alters it, if necessary, to make it more suitable for demodulation. The receiver then demodulates the signal; i.e., it separates the modulation signal, which is a function of the transducer output, from the carrier signal, which was necessary to transmit this information.

The *data processor* accepts the demodulated signal from the receiver and converts it into an electrical quantity suitable for the type of display unit used.

The manner in which the transmitter's carrier signal is modulated, the type of *modulation,* deserves a more detailed description since it normally determines the nomenclature of the telemetry system (see Figure I-3). The frequency of an *amplitude-modulated* (AM) carrier remains constant while its amplitude changes with the modulating signal. The frequency and amplitude of a *phase-modulated* (PM) carrier remain constant while its phase changes with the modulating signal. The amplitude of a *frequency-modulated* (FM) carrier remains constant while its frequency changes with the modulating signal.

I.4 Multiple-Data Telemetry Systems

Most telemetry systems are designed to perform a number of different measurements rather than just one measurement. If a separate data-transmission system were used for each of a number of measurements in one system, the cost and complexity of such a multiple-link arrangement would be prohibitive.

A large variety of telemetry systems exist which incorporate *multiplexing*, the process of transmitting a number of measurements over a single data link. The two most common types of multiplexing are *frequency division multiplexing* and *time division multiplexing*. Many telemetry systems use a combination of these.

I.4.1 Frequency-division multiplexing

This method (see Figure I-4) utilizes a number of *subcarrier oscillators* (SCO), each tuned to a different frequency, generally in the audio or ultrasonic range. The subcarrier oscillator outputs are then linearly summed in voltage. The composite signal is fed to the transmitter, which modulates the carrier itself. The output of each transducer is fed to a different subcarrier oscillator which modulates the subcarrier. The modulation of both the subcarrier oscillator and the carrier oscillator in the transmitter can be of any type.

The reverse process is used to obtain a display signal for each measurement from the receiver. After removal of the carrier by demodulation, the composite signal, which now contains the frequencies of all subcarrier oscillators, is applied to a bank

Fig. I-4. Frequency-division multiplex telemetry system.

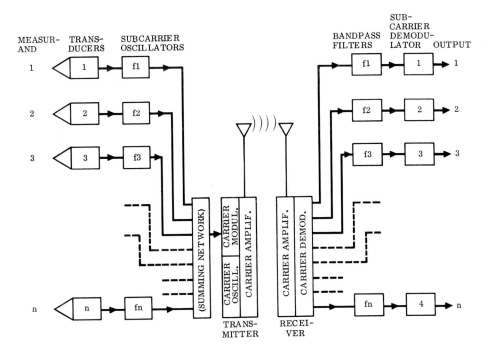

of band-pass filters. One filter is used for each of the subcarrier oscillators on the transmitter side, and each filter is tuned to the frequency of its corresponding subcarrier oscillator. The output of each band-pass filter is then demodulated to reconstitute the signal as originally applied by the transducer to the subcarrier oscillator.

The nomenclature of frequency-division multiplex telemetry systems is given by the type of subcarrier modulation followed by the type of carrier modulation. In an FM/FM system both the subcarriers and carrier are frequency-modulated. In an FM/AM system the subcarriers are frequency-modulated whereas the carrier is amplitude-modulated.

Table I-1 IRIG-Standardized Subcarrier Bands

Channel	Center frequency (KHz)	Maximum deviation (%)	Lower channel limit (KHz)	Upper channel limit (KHz)	Nominal frequency response (Hz)	Minimum sample duration (ms)†	Maximum commuta-tion rate† (samples/s)
1	0.40	±7.5	0.370	0.430	6.0	170	6.0
2	0.56	±7.5	0.518	0.602	8.4	120	8.4
3	0.73	±7.5	0.675	0.785	11	91	11
4	0.96	±7.5	0.888	1.032	14	70	14
5	1.30	±7.5	1.202	1.399	20	51	20
6	1.70	±7.5	1.572	1.828	25	39	25
7	2.30	±7.5	2.127	2.473	35	29	35
8	3.00	±7.5	2.775	3.225	45	22	45
9	3.90	±7.5	3.607	4.193	59	17	59
10	5.40	±7.5	4.995	5.805	81	12	81
11	7.35	±7.5	6.799	7.901	110	9.1	110
12	10.50	±7.5	9.712	11.288	160	6.4	160
13	14.50	±7.5	13.412	15.588	220	4.6	220
14	22.00	±7.5	20.350	23.650	330	3.0	330
15	30.00	±7.5	27.750	32.250	450	2.2	450
16	40.00	±7.5	37.000	43.000	600	1.7	600
17	52.50	±7.5	48.562	56.438	790	1.3	790
18	70.00	±7.5	64.750	75.250	1050	0.95	1050
A	22.00	±15*	18.700	25.300	660	1.5	660
B	30.00	±15*	25.500	34.500	900	1.1	900
C	40.00	±15*	34.000	46.000	1200	0.83	1200
D	52.50	±15*	44.625	60.375	1600	0.63	1600
E	70.00	±15*	59.500	80.500	2100	0.48	2100

*The "optional" channels A through E may be used in a telemetry system only if the following adjacent bands are not used:

If Channel A is used, Channels 13, 15, and B must be omitted.
If Channel B is used, Channels 14, 16, A, and C must be omitted.
If Channel C is used, Channels 15, 17, B, and D must be omitted.
If Channel D is used, Channels 16, 18, C, and E must be omitted.
If Channel E is used, Channels 17 and D must be omitted.
†For 100% duty cycle (non-return-to-zero)

The subcarrier frequencies (*channels*) of FM/FM telemetry systems have been standardized by the Inter-Range Instrumentation Group (IRIG), primarily for military airborne use of such systems. The standardized subcarrier channels and their characteristics are shown in Table I-1, together with recommended sampling duration and sampling rates for associated time-division multiplex subsystems.

I.4.2 Time-division multiplexing

In this method of multiple-data transmission the output of a number of transducers is sampled sequentially, at a rate given by the expected fluctuation of the transduced measurands. The design compromise involved in the use of such a time-sharing technique is the loss of a certain amount of information about the measurand when the latter fluctuates at a relatively high rate. However, there are comparatively few measurements in which measurand fluctuations are so rapid, and where it is so important to measure the fluctuations in their entirety, that some sort of time-division multiplexing cannot be used.

The nomenclature of time-division multiplex systems is given by the form of modulation applied to the next following building block in the transmission system, i.e., the subcarrier or the carrier oscillator. The sequential sampling of transducer outputs, which are analogs of the measurands, results in a train of pulses (*pulse train*) to be applied to this next building block. Each pulse, of course, represents the output of one transducer, sampled over a certain small period of time. The change from transducer outputs to pulse train is performed by a signal converter, or *converter*.

The simplest form of time-division multiplexing is *pulse-amplitude modulation* (PAM). Although the converter can be electromechanical or electronic, it is best illustrated by a motor-driven rotary switch (see Figure I-5). The switch sequentially

Fig. I-5. Basic PAM telemetry system.

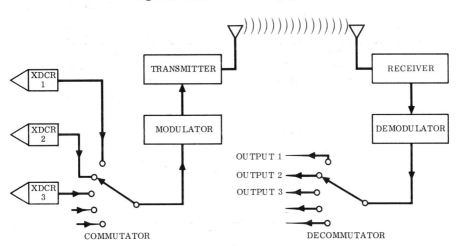

samples the outputs of a number of transducers. A similar switch, on the receiver side, is synchronized with the converter with the aid of a synchronizing signal in the pulse train. This permits the output of each transducer, as sampled by the converter, to be reconstituted for display purposes (see Figure I-6). Since the electromechanical converter used in PAM systems resembles the commutator used in d-c motors and generators, it is commonly referred to as the *commutator*. The process of sampling and conversion to a pulse train is known as *commutation*, and the process of signal separation after demodulation in the receiver is called *decommutation*.

In some commutators a signal is applied to each switch contact (commutator segment). This results in a "100% duty-cycle" or *non-return-to-zero* (NRZ) wave train. Although the NRZ method allows maximum utilization of the commutator, synchronization of the decommutator and signal-separation is sometimes more difficult than in the "50% duty-cycle" or *return-to-zero* (RZ) system in which a "reference" segment is inserted between each two "signal" segments of the commutator (Figure I-7). All reference segments are connected together, and a voltage, slightly more negative than that corresponding to zero transducer output, is applied to them. Calibration signals, representing 0% and 100%, and sometimes also 50% of full-scale standardized transducer output are usually applied to selected commutator segments so that these signals are contained in each pulse train.

PAM is the most popular analog method of time-division multiplexing. Other methods are *pulse-duration modulation* (PDM) and *pulse-position modulation* (PPM).

Fig. I-6. Signal conversion and reconversion in PAM system.

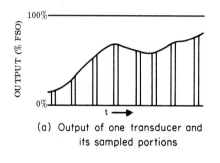

(a) Output of one transducer and
its sampled portions

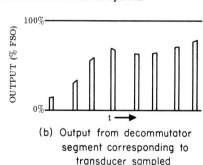

(b) Output from decommutator
segment corresponding to
transducer sampled

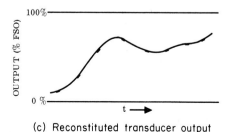

(c) Reconstituted transducer output

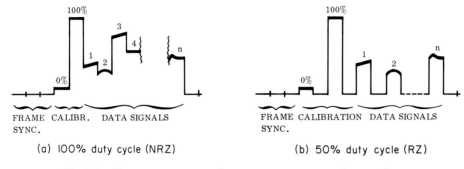

(a) 100% duty cycle (NRZ)   (b) 50% duty cycle (RZ)

Fig. I-7. Non-return-to-zero and return-to-zero commutation waveforms.

Fig. I-8. Time-division multiplex methods.

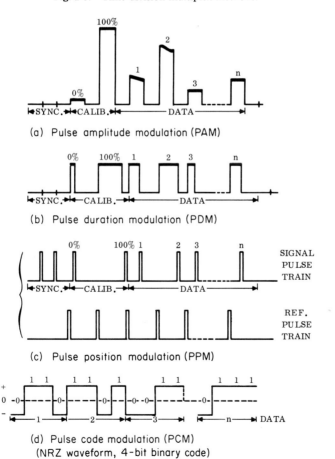

(a) Pulse amplitude modulation (PAM)

(b) Pulse duration modulation (PDM)

(c) Pulse position modulation (PPM)

(d) Pulse code modulation (PCM)
    (NRZ waveform, 4-bit binary code)

In PDM the width of each pulse is varied by the converter in proportion to the variation of the transducer output signals. In PPM each transducer output is converted into a pulse of constant amplitude and duration but is displaced in time by varying amounts from a reference pulse. The position of the signal pulse along a time axis then represents the output of one transducer. Figure I-8 illustrates the various time-division multiplex methods.

Digital time-division multiplexing by *pulse-code modulation* (PCM) is used in some of the more recently developed telemetry systems. This method utilizes a converter (*encoder*) which generates a coded-signal pulse train, usually in binary code, each time a transducer output is sampled. The code represents a certain discrete amplitude increment in the transducer output. On the receiver side the various transducer outputs are either reconstituted by a *decoder*, stored in coded form by a storage device such as a tape recorder, or fed immediately into a digital computer which manipulates the coded signal in such a manner that information about the measurand can be obtained in various forms as desired.

Alternate methods of digital encoding are *frequency-shift keying* (FSK), used in PCM/FM systems, and *phase-shift keying* (PSK) where the transducer output is converted into fixed-step changes of the phase of the modulating signal.

I.4.3 Combined multiplex telemetry systems

Telemetry systems capable of handling more than one hundred different measurements usually combine time- and frequency-division multiplexing. They can also combine two types of time-division multiplexing. Such combined multiplex systems (see Figure I-9) are typically used in pilotless aerospace vehicles where telemetry is the only means for monitoring on-board system and equipment performance.

I.4.3.1 Telemetry data processing. Provisions for performing a number of different operations on the demodulated signal from the receiver are frequently incorporated in the more advanced telemetry systems. Equipment used to perform such operations (data processing equipment) includes band-pass filters, demodulators (discriminators in the case of FM subcarriers), decoders (for PCM and other digital systems), decommutators, and various specialized signal-conditioning equipment.

Because of the multitude of data simultaneously available for display, complex telemetry systems usually incorporate data-storage equipment, such as special tape recorders. The demodulated or decommutated data can then be recorded on tape. The tape can be played back through a variety of data-conditioning equipment at any time and location desired and can be displayed in the form considered most suitable for proper analysis of the various measurements.

One or more outputs can also be fed directly to a computer which is programmed to extract the desired information from each output signal or from a combination of output signals.

Data-processing equipment can be arranged for selective or simultaneous con-

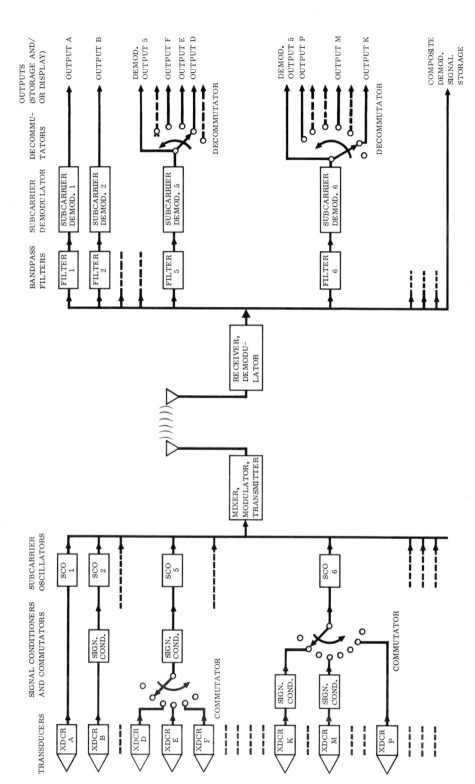

Fig. I-9. Combined multiplex telemetry system.

nection of the processed and conditioned signals to display devices such as meters, numerical printers, oscilloscopes, and circular-chart or strip-chart recorders, and to signal analyzers and computers. All these devices are used to provide or facilitate the determination of the required information about the various transduced measurands (*data reduction*). Data reduction is further facilitated by inclusion within the data-transmission system of means for providing periodic standardized calibration signals against which the data signals can be compared.

Transducers

II.1 Nomenclature

The importance of using the proper words to describe transducers, their operation, and their characteristics cannot be overemphasized. For many years the language connected with transducers was so confused that only a few engineers and scientists had a clear understanding of the wide variety of their operating principles and their behavior under different operating and environmental conditions. Not until 1960 were serious efforts made to attempt a simplified but exact classification of transducer types and their general characteristics. On May 5, 1960, the Instrument Society of America organized a committee, under the author's chairmanship, to study the need for standardization in the aerospace transducer field. Several Recommended Practice and Standard documents were developed by this committee, concurrent with the writing of this book. The basic standard, first developed, was a Recommended Practice for transducer nomenclature and terminology (RP37.1), later superseded by Standard S37.1. All portions of this book are intended to be in general agreement with applicable portions of these I.S.A. standards.

II.1.1 Basic definition of transducer

The transducer has been variously defined as a device capable of transferring energy between two or more systems, physical states, information systems, or transmission systems. These definitions are too broad for practical use. A piston or a crankshaft in an automobile engine, a valve in a steam pipe, a typewriter, a violin, and a frying pan could all be called "transducers" within these general definitions. Although the general concepts of energy transfer on which these definitions are based are of considerable academic interest, it is not expected that the term will be absorbed into the English language in such a manner that we will talk about writing our letters, making our music, or frying our eggs on a transducer.

In the field of electroacoustics, the use of "transducer" to refer to a phonograph cartridge as well as to a loudspeaker or to other sound radiators may be too firmly entrenched to undergo revision. The use of the word "transducer" in new areas should be discouraged since it is likely to result in confusion. It should be limited to the field of measuring and control instrumentation.

Transducers used for measurement are covered by the following definition:

A *transducer* is a device which provides a usable output in response to a specific measurand. (The *measurand* is the physical quantity, property, or condition which is measured.)

A number of terms have been incorrectly used in the place of "transducer." These include "end instrument" (which can also be a recorder or indicator), "gage" (which is always an indicator), "pickup" (a slang term), or "transmitter." The latter term is till in common use in the process industry although it is not only grammatically incorrect (e.g., a pressure transmitter should really transmit pressure rather than an output in response to pressure), but now it has a strong connotation connected with the radio and television communications field. The term "detector" has often been applied to radiation transducers. The only term which merits consideration as alternate to "transducer" in instrumentation work is the word "sensor."

II.1.2 Transducer description

The description of a transducer is based essentially on the following considerations:

1. What is intended to be measured by the transducer (*measurand*)?
2. In what portion of the transducer does the output originate (*transduction element*), and what is the nature of its operation (*transduction principle*)?
3. What device in the transducer responds directly to the measurand (*sensing element*)?
4. What noteworthy special features or provisions are incorporated within the transducer?
5. What are the upper and lower limits of the measurand values the transducer is intended to measure (*range*)?

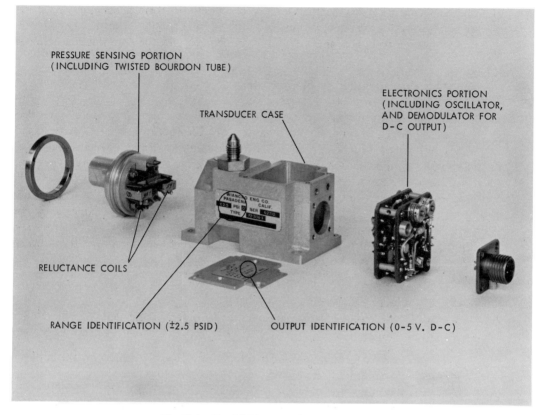

Fig. II-1. Establishing transducer nomenclature.

Applying the above to the example shown in Figure II-1 results in the following description:

1. The transducer measures *differential pressure*.
2. The transduction principle is *reluctive* (see II.2.8).
3. The sensing element is a (twisted) *Bourdon tube*.
4. The transducer contains a demodulator which converts the a-c output normally obtained from a reluctive element into a *d-c output*.
5. The transducer is intended to measure over the range *±2.5 psid*.

The transducer can therefore be described as a "±2.5 psid, d-c output, Bourdon-tube, reluctive, differential-pressure transducer."

Other transducers should be described in a similar manner, although it may not always be as easy to do so as for the example shown. Some transducers (e.g., most temperature transducers) contain a sensing element which also acts as the transduction element. Others, such as some nuclear radiation transducers, may have two or more energy conversions (secondary, tertiary, etc., sensing elements) between the element which responds directly to the measurand and the transduction element.

II.1.3 Classification by nomenclature

Several attempts have been made to classify transducers into useful categories such as: by transduction principle, by absence or presence of mechanical sensing elements, or by need, or lack of need, of external excitation. While the inclusion of such characteristics in a transducer description is quite important, an application-oriented classification of transducers, based primarily on measurands, has been found to be most useful in applied instrumentation work.

This general classification establishes the nomenclature of transducers for use in drawing and specification titles, catalogs, indices, contracts, technical papers and reports, etc. The nomenclature consists of modifiers following the word "transducer," denoting the following:

1. Measurand.
2. Transduction principle.
3. (Optional:) special feature, special provision, sensing element.
4. Range and units.

The measurand may be modified into a compound noun, if required, by the addition of a restrictive noun or adjective.

Typical modifiers are shown on Table II-1 for the purpose of clarifying the manner of transducer nomenclature assembly. It should be noted that the listing under each modifier contains only a limited number of examples. Specific descriptions and classifications of transducers can be based on one or more of these modifier headings or modifiers. Thus, though we may wish to discuss, specify, or subclassify "temperature transducers," "mass-flow-rate transducers," or "semiconductor-strain-gage pressure transducers," we may be concerned with a number of different "0 to 0.1 in. displacement transducers" utilizing various transduction principles, or we may be required to survey the general characteristics of "potentiometric transducers" used for various measurands. More than one "Modifier 4" may have to be used in the nomenclature of some transducer types (e.g., "servo, oil-damped," "d-c output, suppressed-zero," "platinum-wire, weldable," or "transistorized, triaxial").

II.2 Transduction Elements

The operation of typical transducers is covered in the sections of this book which are devoted to specific measurands. A number of basic transduction elements, common to transducers for several measurands, are described below.

Table II-1 Assembly of Typical Transducer Nomenclature

1. *Index and Title Nomenclature:* Transducer, (Modifier 1), (if required: Modifier 2), (Modifier 3), [optional: Modifier 4(a), (b), or (c)], Range and Units

2. *Text Nomenclature:* Range and Units [optional: Modifier 4(a), (b), or (c)], (Modifier 3), (if required: Modifier 2), (Modifier 1) Transducer

Modifier 1	Modifier 2	Modifier 3	Modifier 4			Range	Units of range
Measurand	(Restriction on measurand)	Transduction principle	(a) Sensing element	(b) Special feature	(c) Special provision		
Acceleration	Absolute	Capacitive	Bellows	Bonded	Amplifying	0 to 0.05	°C
Attitude	Angular	Electromagnetic	Bourdon-tube	Enclosed-element	Cementable		deg
Displacement	Differential	Inductive	Capsule	Exposed-element	D-C output	0 to 10,000	°F
Flow rate	Gage	Reluctive	Diaphragm	Oil-damped	Digital-output		cm/s
Humidity	Infrared	Photovoltaic	Float	Self-generating	Dual-output	±20	g
Light intensity	Linear	Piezoelectric	Gyro	Servo	Frequency-output		gpm
Liquid level	Mass	Potentiometric	Hot-wire	Transistorized	Integrating	−2 to +10	in.
Pressure	Surface	Resistive	Platinum-wire	Ultrasonic	Triaxial		psia
Speed	Ultraviolet	Strain-gage	Semiconductor	Unbonded	Suppressed-zero	−430 to −415	rad/s²
Temperature	Volumetric	Thermoelectric	Turbine	Vibrating-wire	Weldable		rpm

Examples of use: 1. Transducer, acceleration, piezoelectric, triaxial, ±20 *g*
2. −435 to −400°F, Germanium, Resistive, Temperature Transducer

II.2.1 Capacitive transduction

In a *capacitive* transducer the measurand is converted into a change of capacitance. Since a capacitor consists essentially of two conductors (*plates*) separated by an insulator (*dielectric*), the change of capacitance occurs typically when a displacement of the sensing element causes a moving conductive surface to move toward or away from a stationary conductive surface [see Figure II-2(a)]. In many cases the moving plate is a sensing element, such as a diaphragm or seismic mass. In other versions both plates are stationary and the change occurs in the dielectric, such as in the relative proportions of a liquid and a gas in the space between the surfaces [see Figure II-2(b)].

II.2.2 Electromagnetic transduction

In an *electromagnetic* transducer the measurand is converted into an electro-motive force (output voltage) induced in a conductor by a change in magnetic flux in the absence of excitation. This transducer is therefore self-generating. The change in magnetic flux is normally accomplished (see Figure II-3) by relative motion between a magnet or a piece of magnetic material and an electromagnet (coil with ferrous core).

II.2.3 Inductive transduction

In an *inductive* transducer the measurand is converted into a change of the self-inductance of a single coil. This is usually effected (see Figure II-4) by a displacement of the coil's core, which is linked or attached to a mechanical sensing element.

II.2.4 Photoconductive transduction

In a *photoconductive* transducer the measurand is converted into a change in resistance (conductance) of a semiconductive material by a change in the amount of illumination incident on the material (see Figure II-5). In the case of light-intensity transducers the resistance change is a direct result of a change in illumination intensity. In some other types of transducers the change in incident illumination is effected by a moving shutter between a light source and the photoresistive material. The shutter is mechanically linked to a sensing element such as a pressure capsule or a seismic mass.

II.2.5 Photovoltaic transduction

In a *photovoltaic* transducer the measurand is converted into a change in the voltage generated when a junction between certain dissimilar materials is illuminated (see Figure II-6). This principle is primarily utilized for direct measurement of light intensity. It can also be used in other transducers provided with mechanical means to change the intensity of incident illumination originating in an integral light source.

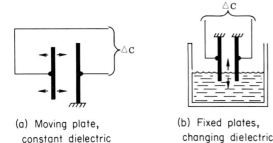

(a) Moving plate,
constant dielectric

(b) Fixed plates,
changing dielectric

Fig. II-2. Capacitive transduction.

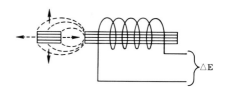

Fig. II-3. Electromagnetic transduction.

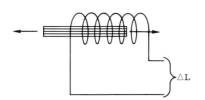

Fig. II-4. Inductive transduction.

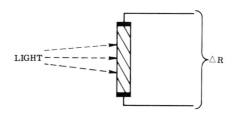

Fig. II-5. Photoconductive transduction.

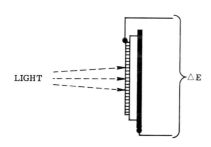

Fig. II-6. Photovoltaic transduction.

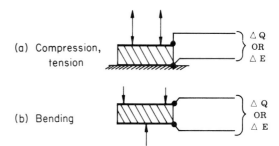

(a) Compression,
tension

(b) Bending

Fig. II-7. Piezoelectric transduction.

II.2.6 Piezoelectric transduction

In a *piezoelectric* transducer the measurand is converted into a change in the electrostatic charge (Q) or voltage (E) generated by certain crystals when mechanically stressed. The stress is developed by compression or tension forces, [Figure II-7(a)], or by bending forces [Figure II-7(b)] exerted upon the crystal directly by a sensing element or by a mechanical member linked to a sensing element.

II.2.7 Potentiometric transduction

In a *potentiometric* transducer the measurand is converted into a change in the position of a movable contact (*wiper arm*) on a resistance element. The wiper arm displacement causes a change in the ratio (*resistance ratio*) between the resistance from one element end to the wiper arm and the total element resistance. In its usual application the potentiometer element is excited by a source of a-c or d-c voltage (E_x), and the output is the *voltage ratio* (see Figure II-8).

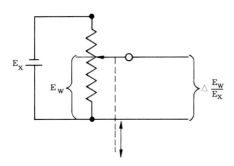

Fig. II-8. Potentiometric transduction.

II.2.8 Reluctive transduction

In a *reluctive* transducer the measurand is converted into an a-c voltage change by a change in the reluctance path between two or more coils, with a-c excitation applied to the coil system. The category of reluctive transduction elements includes those circuits frequently referred to as "variable-reluctance," "inductance-bridge," and "differential-transformer" circuits since their outputs are a-c voltage changes (see Figure II-9). The change in reluctance path is usually accomplished by the displacement of a magnetic core (sometimes referred to as the *armature*).

II.2.9 Resistive transduction

In a *resistive* transducer the measurand is converted into a change of resistance. This resistance change in a conductor or semiconductor is effected by various means, including heating or cooling, applying mechanical stresses, drying or wetting of certain materials such as electrolytic salts, or sliding a wiper arm along a rheostat-connected resistance element (see Figure II-10).

Fig. II-9. Reluctive transduction.

(a) Differential transformer   (b) Inductance bridges (variable–reluctance)

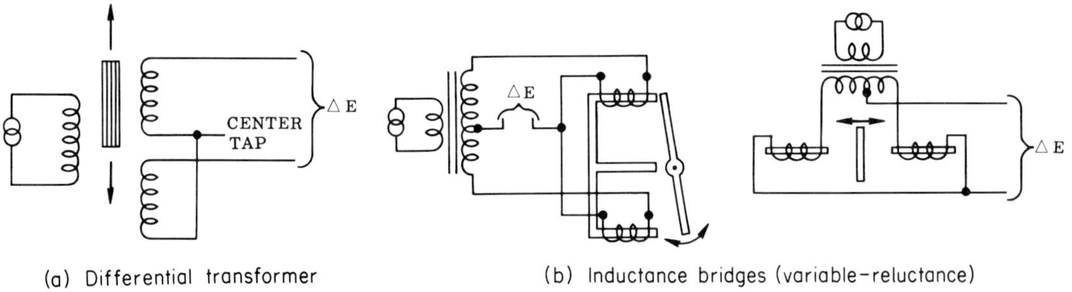

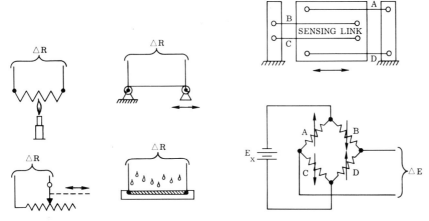

Fig. II-10. Resistive transduction.   **Fig. II-11.** Strain-gage transduction.

II.2.10 Strain-gage transduction

In a *strain-gage* transducer, a special version of the resistive transducer, the measurand is converted into a resistance change, due to strain, usually in two or four arms of a Wheatstone bridge. The bridge output is a voltage change, proportional also to the excitation voltage used. A-C or d-c excitation can be used. Figure II-11 illustrates the operation of a four-active-element, unbonded, strain-gage transduction element. Upward arrows in the schematic indicate increasing resistances, and downward arrows indicate decreasing resistances in the bridge arms for sensing-link motion toward the left.

II.3 General Characteristics

II.3.1 Measurand characteristics

A transducer is normally designed to sense a specific measurand and to respond only to this measurand. In some cases, other measurands may be calculated by their known relationship to the measurand sensed by transducer (e.g., since velocity is a change of displacement per unit time, velocity can be calculated from a measurement of displacement and time). For every measurement, then, the measurand is stated in terms of the transducer. Pressure transducers measure pressure, displacement transducers measure displacement, and acceleration transducers ("accelerometers") measure acceleration. However, pressure transducers are frequently used to measure altitude, displacement transducers can measure position or motion, and acceleration transducers are commonly used for the measurement of vibration and, by time integration, of velocity.

II.3.1.1 Range. The measurand values over which the transducer is intended to measure, denoted by their upper and lower limits, form the *range* of the transducer. The transducer range can be *unidirectional* (e.g., 0 to 10 psid, 0 to 50 *g*, 0 to 2.50 in.), *bidirectional* (e.g., ±3 *g*, ±15 psid, ±3 deg.), *asymmetrically bidirectional* (e.g., −2 to +10 *g*, −10 to +5 psig, −0.5 to −3.5 deg.), or *expanded* or "zero-suppressed" (e.g., 80 to 100 psia, +15 to +25 *g*, 3200 to 3800 rpm).

The algebraic difference between the limits of the range is the *span* of the transducer. The span of a −2 to +10 *g* acceleration transducer is 12 *g*. The span of an 80 to 100 psia pressure transducer is 20 psia.

II.3.2 Electrical design characteristics

II.3.2.1 Excitation. With the exception of self-generating types (e.g., piezo-electric, photovoltaic, and electromagnetic), transducers require external electrical *excitation* for their operation. Excitation is usually stated in terms of d-c or a-c voltage or of a current applied to the transducer.

II.3.2.2 Output. *Output* is the electrical quantity, produced by a transducer, which is a function of the applied measurand. The output is usually a continuous function of the measurand (*analog output*) in the form of current, voltage ratio, or voltage amplitude (e.g., 0 to 5 V d-c, 0 to 100% VR, 0 to 500 μA) or a variation of other parameters such as capacitance or inductance. Analog output can also be in the form of frequency (*frequency output;* e.g., 0 to 1800 Hz) or in the form of a frequency deviation from a center frequency (*frequency-modulated output;* e.g., 5400 ±400 Hz).

Some transducers produce an output that represents the magnitude of the measurand in the form of discrete quantities coded in a system of notation (*digital output;* e.g., 0 to 1056 in binary code). Digital output facilitates preserving the accuracy of the transducer throughout the measuring system.

II.3.2.3 End points. *End points* are the output values at the upper and lower limits of the transducer's range. They can be determined by a single calibration cycle, or they may be the average (mean) of end point readings obtained from two or more consecutive calibration cycles. When end points are specified, a tolerance is usually applied to them (e.g., 0.00 ±0.01 V d-c and 5.00 ±0.002 V d-c).

Theoretical end points are the specified points between which the theoretical curve is established and to which no end point tolerances apply (e.g., 5.0% VR and 95.0% VR).

II.3.2.4 Impedances. Output as well as excitation are affected by transducer and load impedances (see Figure II-12). The impedance (Z_{in}) measured across the

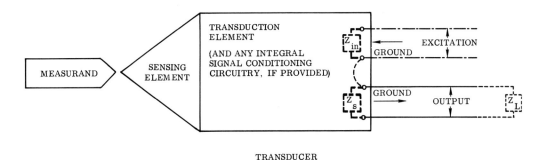

TRANSDUCER

Fig. II-12.   Basic "black-box" electrical characteristics.

excitation terminals is the *input impedance;* across the output terminals (Z_s), the *output impedance* of the transducer. The impedance (Z_L) presented to the transducer's output by the associated external circuitry and transmission line is the *load impedance.*

Mismatching of output impedances can cause an error (*loading error*) in the output. This error increases with the ratio of output impedance to load impedance.

II.3.2.5 Demodulator and amplifier characteristics. The d-c output of a transducer containing an integral output demodulator (converting a-c to d-c) may contain a measurable a-c component (*ripple*) which is undesirable and must be minimized.

Transducers which incorporate an integral amplifier share some of the latter's characteristics. These include random disturbances originating in the amplifier components and modifying the output (*noise*), changes in the characteristics of the components which result in output variations (*gain instability*), and the time which elapses between the removal of an overload applied to the transducer (causing the amplifier to limit and distort) and the restoration of the output to a value which is again within specified limits (*recovery time;* cf II.3.6.1).

The distortion in a transducer's sinusoidal (a-c) output, in the form of harmonics other than the fundamental component, is the *harmonic content* of the output. It is usually expressed as a percentage of the root mean square (rms) output.

II.3.2.6 Grounding and insulation. The excitation ground and output ground may be isolated from each other, as in bridge-type transducers (strain-gage transduction). They can also be connected together as a single ground (*common ground*), as indicated by the dotted jumper between the ground terminals in Figure II-12. One of the two ground wires can then be deleted. Both grounds are usually isolated from the transducer's case (*floating ground*).

When two or more portions of a transducer are electrically insulated from each other, the resistance between the portions, measured while a specified d-c voltage is applied, is the *insulation resistance.* The degree of insulation can also be expressed in terms of the voltage applied across the mutually insulated portions which causes arcing or conduction between them (*breakdown voltage*). It should be noted that the *breakdown voltage rating* refers to the maximum voltage which can be applied *without* causing arcing or conduction between specified insulated portions of a transducer.

II.3.3 Static performance characteristics

Transducer characteristics which are usually determined during a calibration performed at room conditions by applying the measurand in discrete amplitude intervals (see II.3.3.2) are the *static characteristics* of a transducer's performance.

II.3.3.1 Room conditions. The combination of an ambient temperature of $25 \pm 10°C$ ($77 \pm 18°F$), an ambient relative humidity of 90% or less, and an ambient (barometric) pressure of 26 to 32 in. Hg constitutes the environment known as *room conditions.* Controlling the environment within these limits is adequate for static tests on most transducers. Some types, however, are so sensitive to environmental changes, because of their range or their design, that environmental control within narrower limits may be necessary.

II.3.3.2 Calibration. It is customary to describe error characteristics in terms of their effect on output. The characteristics can be determined by a test during which known values of measurand are applied to the transducer and corresponding output readings are recorded (*calibration*).

A single performance of this test over the entire range of the transducer, once in an ascending mode and once in a descending mode, is a *calibration cycle.* A calibration usually comprises two or more calibration cycles, which may be referred to as "Run 1," "Run 2," etc.

A *static calibration* is a calibration performed under room conditions and in the absence of acceleration, shock, or vibration (unless one of these is the measurand) by application of the measurand to the transducer in discrete amplitude intervals.

A *dynamic calibration* is a calibration during which the measurand is made to vary with time in some specific manner, and the output is recorded as a function of time. Since dynamic calibrations are infrequently performed, the term "calibration" implies a static calibration unless "dynamic" is specified.

The test record obtained from a calibration is the *calibration record.* It is usually in tabular form, but may also be a graphical representation (*calibration curve*).

II.3.3.3 Error. An ideal or theoretical output/measurand relationship exists for every transducer. If the transducer were ideally designed and made from ideal materials by using ideal methods and workmanship, the output of this ideal transducer would continuously indicate the true value of the measurand. It would follow exactly the prescribed or known *theoretical curve* which specifies the relationship of the output to the applied measurand over the transducer's range. Such a relationship can be stated in the form of a mathematical equation, a graph, or a table of values.

This ideal output would be obtained regardless of the ambient environmental conditions and the operating conditions to which the transducer may be exposed. The ideal transducer would respond only to the measurand.

> **Example :** An ideal 0 to 1000 psia pressure transducer is designed to provide an output of 0 to 5 V d-c. Its output/measurand relationship is linear (see Figure II-13). An output of 2.50 V would, therefore, always indicate a pressure of 500 psia; an output of 1.00 V, a pressure of 200 psia.

The output of an actual transducer, however, is affected by the nonideal behavior of the transducer which causes the indicated measurand value to deviate from the true value. The algebraic difference between the indicated value and the true (or theoretical) value of the measurand is the transducer's *error*. Error is usually expressed in percent of full-scale output (% FSO). It is sometimes expressed in percent of the

Fig. II-13. Output/measurand relationship of ideal linear-output transducer (including application of general example to d-c output pressure transducer).

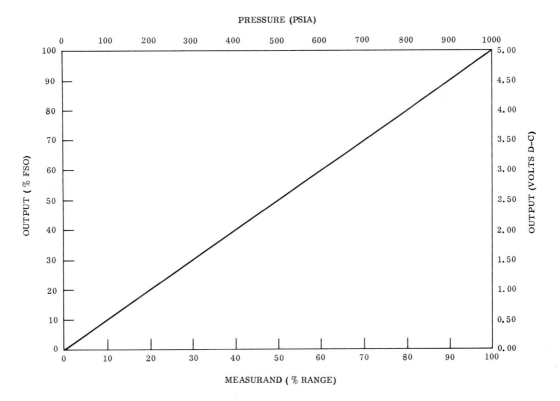

output reading of the transducer or in terms of units of the measurand. An ideal transducer would have zero error.

Accuracy is defined as the ratio of the error to the full-scale output. It can be expressed "within $\pm$ . . . % FSO" but can also be stated in terms of units of measurand or in percent of the ratio of error to output.

Although the simplest way to consider transducer errors is in terms of maximum deviations over the range from a known reference line (*error-band concept*), the existence of individual errors such as nonlinearity, nonrepeatability, zero and sensitivity shifts, and hysteresis must be recognized, and the effects of these errors on transducer behavior and data obtained should be known. Knowledge of individual errors can often be used to correct final data and thus increase their overall accuracy.

II.3.3.4 Hysteresis. When a given measurand level is approached, first with increasing values of measurand, then with decreasing measurand, the two output readings obtained usually differ from each other, primarily because of a certain amount of lag in the action of the sensing element. The maximum difference in any pair of output readings so obtained during any one calibration cycle is the *hysteresis* of the transducer (see Figure II-14). Hysteresis is usually expressed in percent of full-

Fig. II-14.   Hysteresis (scale of errors 10: 1).

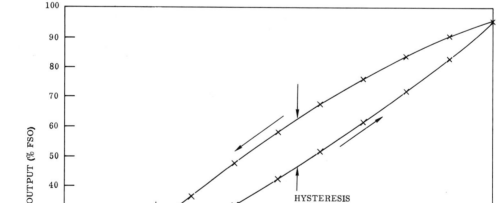

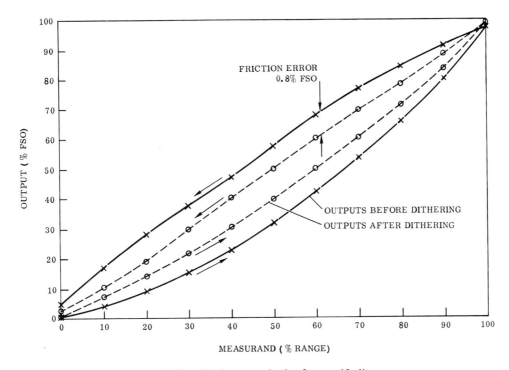

Fig. II-15. Friction error (scale of errors 10:1).

scale output. The hysteresis obtained when only a portion of the range, e.g., 0 to 30%, is traversed (*partial-range hysteresis*) is always less than the total hysteresis. Friction error (see below) is included with hysteresis unless dithering of the transducer during the calibration is specified.

II.3.3.5 Friction error. Certain types of transducers exhibit additional errors due to internal sliding friction. The friction effects can be minimized by *dithering* the transducer—applying intermittent or oscillatory acceleration forces to it—during use and during calibration. Typically, the case of the transducer can be tapped gently with a small rubber mallet, a buzzer can be mounted against the transducer, or it can be gently vibrated by means of a small vibration exciter (shaker). *Friction error* is the maximum change in output, at any measurand value within the range, before and after minimizing friction within the transducer (see Figure II-15). A calibration during which friction effects are minimized by dithering is a *friction-free calibration*. The data obtained from such a calibration should be considered valid for applied measurements and data reduction only if the environment in which the transducer operates provides the same amount of dithering used during calibration.

II.3.3.6 Repeatability. The ability of a transducer to reproduce output readings when the same measurand value is applied to it consecutively, under the same conditions, and in the same direction, is the *repeatability* of the transducer (see Figure II-16). It is normally expressed as the maximum difference between output readings at any measurand value within the range unless a specific measurand value is the sole area of interest. It is usually stated "within . . . % FSO." At least two calibration cycles must be used to determine repeatability.

In practice, two, three, and even five calibration cycles are used to determine repeatability as defined above. If the sampling is increased by performing a much larger number of calibration cycles, a statistical measure of repeatability may be obtained from the probability curves representing all output readings at each measurand level.

II.3.3.7 Linearity. The majority of transducer types in common use are designed to provide a linear output-vs-measurand relationship, primarily because this tends to facilitate data reduction. The closeness of a transducer's calibration curve to a

Fig. II-16.   Repeatability (scale of errors 10: 1).

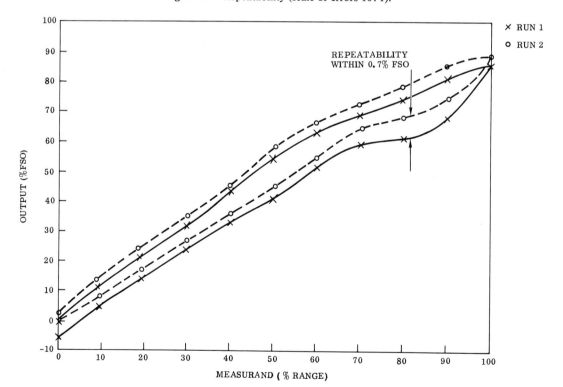

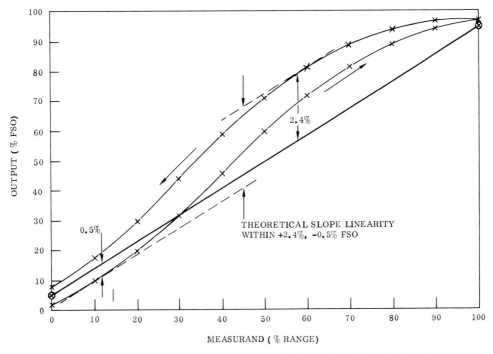

Fig. II-17. Theoretical-slope linearity; example shown for theoretical end points of 5% FSO @ 0% range, 95% FSO @ 100% range (scale of errors 10:1).

specified straight line is the *linearity* of the transducer. Linearity is expressed in the form "within ± . . . % FSO" as the maximum deviation of any calibration point from the corresponding point on the specified straight line during any one calibration cycle.

The exact nature of the straight line to which the calibration curve is compared (to which *linearity is referred*) for a given transducer must always be clearly defined. The term "linearity," by itself, means very little. For example, a transducer may have an "independent linearity" within ±0.5% FSO while its "terminal linearity" is within ±3% FSO. The statement that its "linearity is within ±0.5% FSO" may, therefore, be misleading.

Theoretical-Slope Linearity is referred to a straight line between the theoretical end points (see Figure II-17). Since no tolerances apply to theoretical end points this straight line can always be drawn without referring to any measured values.

Terminal Linearity is a special form of theoretical slope linearity for which the theoretical end points are exactly 0% and 100% of both range and full-scale output (see Figure II-18).

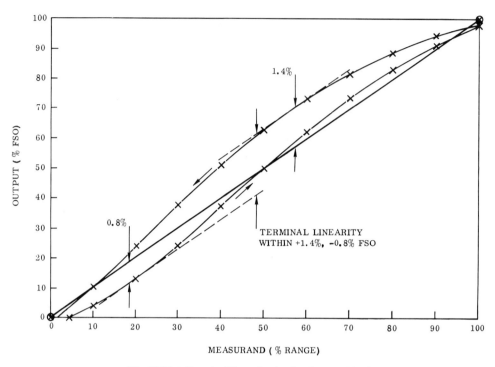

Fig. II-18.   Terminal linearity (scale of errors 10:1).

End-Point Linearity is referred to a straight line between the end points (see Figure II-19). Since end-point specifications usually carry a tolerance, this straight line cannot be drawn until end-point readings have been obtained. For the purpose of establishing end-point linearity, the end points may be specified as those obtained during any one calibration cycle, a specific calibration cycle, or as averaged readings during two or more consecutive calibration cycles.

Independent Linearity is referred to the *"best straight line,"* a line midway between the two parallel straight lines closest together and enclosing all output values obtained during one calibration cycle (see Figure II-20). It should be noted that the "best straight line" cannot be drawn until not only the end point readings but all output readings over the transducer's range have been obtained.

Least-Squares Linearity is referred to the straight line for which the sum of the squares of the residuals is minimized. The term "residuals" refers to the deviations of output readings from their corresponding points on the straight line calculated. Least-squares linearity is usually determined with the aid of a computer. The following calculations are required (derivations of equations are not shown here; they are included in most standard textbooks on statistics):

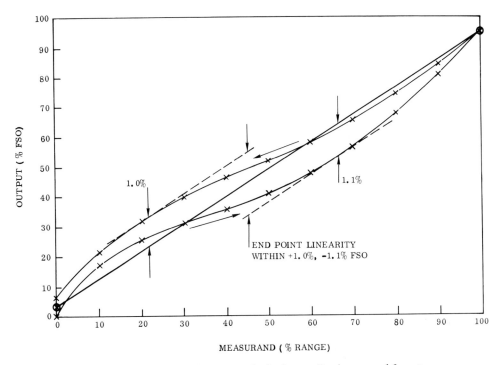

Fig. II-19. End-point linearity; zero end of reference line is averaged from two zero readings (scale of errors 10: 1).

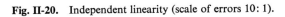

Fig. II-20. Independent linearity (scale of errors 10: 1).

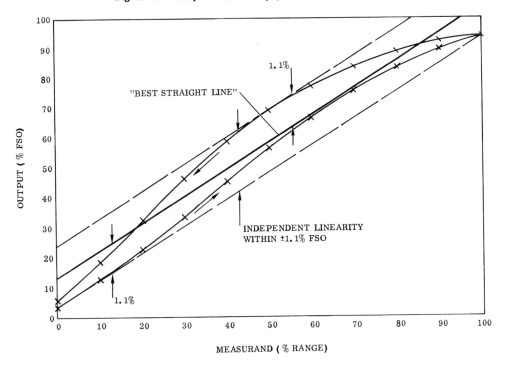

1. The line to be obtained is given by the straight-line equation

$$y = mx + b$$

where: y = output reading
x = measurand value
m = slope of line
b = intercept of line (with y-axis)

2. The slope (m) and intercept (y) of the "least-squares" straight line (least-squares line) are given by the two equations:

$$nb + m \sum_{n=1}^{k} y = \sum_{n=1}^{k} x \qquad b \sum_{n=1}^{k} y + m \sum_{n=1}^{k} y^2 = \sum_{n=1}^{k} xy$$

where n = total number of data points (or "pairs" of output-vs-measurand values)
k = total number of output or measurand values used

3. From the above two equations the slope and intercept are given by:

$$m = \frac{n \sum_{n=1}^{k} xy - \sum_{n=1}^{k} x \sum_{n=1}^{k} y}{n \sum_{n=1}^{k} y^2 - \left(\sum_{n=1}^{k} y\right)^2} \qquad b = \frac{\sum_{n=1}^{k} x - m \sum_{n=1}^{k} y}{n}$$

Example: A pressure transducer is calibrated by applying the measurand (pressure) in ten equal-increment levels, first in an increasing direction and then in a decreasing direction. As shown in the chart below, k for measurand (x) values is 11, for output (y) values 21; the total number of data points (n) is 21. By using actual values from the calibration records, each summation becomes a constant when its calculation is set up as shown below the table.

Measurand	Output	
	Increasing	Decreasing
x_1	y_1	y_{21}
x_2	y_2	y_{20}
x_3	y_3	y_{19}
x_4	y_4	y_{18}
x_5	y_5	y_{17}
x_6	y_6	y_{16}
x_7	y_7	y_{15}
x_8	y_8	y_{14}
x_9	y_9	y_{13}
x_{10}	y_{10}	y_{12}
x_{11}	y_{11}	

$$\sum_{n=1}^{k} x = 2x_1 + 2x_2 + \cdots + 2x_{10} + 2x_{11}$$

$$\sum_{n=1}^{k} y = y_1 + y_2 + \cdots + y_{21}$$

$$\sum_{n=1}^{k} y^2 = y_1{}^2 + y_2{}^2 + \cdots + y_{21}{}^2$$

$$\sum_{n=1}^{k} xy = x_1 y_1 + x_2 y_2 + \cdots + x_{11} y_{11} + x_{10} y_{12} + x_9 y_{13} + \cdots + x_1 y_{21}$$

Linearity for Special Requirements. One of the above types of linearity should be suitable for almost any application. Several additional types (listed below) have been used occasionally. Their use should be restricted to only those applications where data requirements make them absolutely necessary.

Independent Linearity with Forced Zero is determined in the same manner as independent linearity with the additional requirement that the two parallel straight lines closest together and enclosing all output values must be drawn in such a manner that the resulting "best straight line" passes through the "0% Range, 0% FSO" point.

Independent Linearity with Fixed y-intercept is similar to "independent linearity with forced zero." The only difference is that the "best straight line" passes through the "0% Range, y% FSO" point, with the value of y specified, instead of the "0% Range, 0% FSO" point.

Point-Based Linearity is referred to that "best straight line" which also passes through a specified "% Range, % FSO" point other than a point at 0% Range.

Least-Average Linearity is similar to "least-squares linearity"; it is referred to a straight line for which the average residuals (instead of the sum of the squares of the residuals) are minimized.

The determination, or specification, of linearity for a transducer with bidirectional range may require special considerations. The reference line may have to pass through the "zero measurand, y% FSO" point (with the value of y at a specific output level except when output is also bidirectional, in which case y is at "zero output"). In this case, point-based linearity or, preferably, terminal linearity should be used. Examples of reference lines for bidirectional range transducers are illustrated under "static error band" (see II.3.3.12).

The specification and determination of linearity on the basis of a full calibration cycle (with increasing and decreasing measurand values) has been a commonly accepted practice. It is justified by the necessity and desire to define deviations from an assumed linear transducer behavior regardless of the direction of measurand change.

It has been proposed by some workers in the field, however, that such definitions of linearity are not "pure," since hysteresis is included in the bidirectional calibration cycle. They suggest that linearity should be specified and determined on the basis of the ascending half of a calibration cycle only. Others feel that the only meaningful linearity specification is one which defines the linearity of a calibration cycle's mean-output curve. In some commercial literature the term "combined linearity and hysteresis" has even been applied to one or another type of linearity as described in this chapter. An effective means of circumventing such controversy is given by the alternative use of error-band specifications (see II.3.3.12).

II.3.3.8 Conformance. When the transducer is intended to provide a nonlinear output-vs-measurand relationship, the term "linearity" is not applicable. The closeness of a calibration curve to a specified curve is the *conformance* (sometimes called "conformity") of the transducer. Although determination of "least-average" or "least-squares" and other types of conformance are common, conformance is easiest to state as *theoretical-curve conformance*, referred to a theoretical curve (see Figure II-21). The theoretical curve can be defined by a table, a graph, or an equation.

Fig. II-21. Conformance, referenced to theoretical curve (scale of errors 10:1).

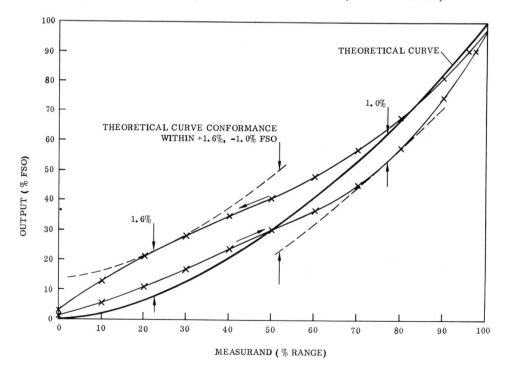

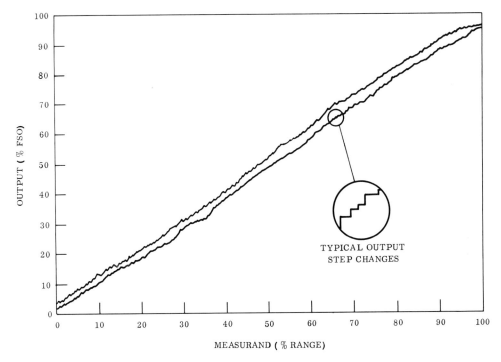

Fig. II-22. Resolution.

II.3.3.9 Curve shifts. The ability of a transducer to retain its performance throughout its specified operating and storage life (*stability*) can be affected by overloads, aging, environmental conditions, and frequent use. Such instabilities are often stated as a change in the slope of the calibration curve (*sensitivity shift*) and a parallel displacement of the entire calibration curve (*zero shift*—cf. Figure II-32 for thermal shifts).

II.3.3.10 Resolution and threshold. When the measurand is continuously varied over the range, the output changes obtained from certain types of transducers may not be perfectly smooth. Instead, the output will change in small but measurable steps (see Figure II-22). This is particularly noticeable in the output of potentiometric transducers which use wirewound elements. The magnitude of these output step changes, expressed in percent of full-scale output, is the *resolution* of the transducer. Resolution is not equal throughout the transducer's range but varies slightly from step to step.

Resolution has been commonly expressed either as maximum resolution (the greatest of all steps) or as maximum resolution of a certain percentage of all steps, with a larger maximum value specified for the remaining steps. For example, it may be specified (or it may have been determined by a test) that 95% of the steps are within 0.25% FSO, 5% of the steps are within 0.5% FSO, and no steps are greater than 0.5% FSO.

The concept of average resolution was formally introduced in ISA Standard S37.6–1967 to provide a simpler alternative for the specification and verification of resolution. *Average resolution* is defined as the reciprocal of the total number of transducer output steps over the measuring range, multiplied by 100 and expressed in percent of full-scale output (% FSO). Average resolution should be specified together with maximum resolution with smaller tolerances assigned to the former than to the latter.

A measurand change of finite magnitude is required to cause a change in the output of any transducer. The smallest change in the measurand that will result in a measurable change in output is the *threshold* of the transducer. It is usually stated in terms of units of measurand and may have different values in different portions of the range.

II.3.3.11 Error band. It is often convenient to determine or specify transducer errors in terms of the band of maximum or allowable deviations of output values from a specified reference line or curve (*error band*), instead of considering individual errors such as linearity, friction, hysteresis, and curve shifts.

The error band concept was originally developed in 1957 by G. Chasinowitz, M. R. Barlow, and the author, at General Dynamics/Convair to simplify specifications and performance verification of transducers used in the *Atlas* missile instrumentation program. Most transducers were used in applications where the measurand would vary frequently, over wide portions of the range, in increasing and decreasing directions, and with or without sufficient vibration to relieve friction in the transducer. Linear output was assumed during data reduction.

This led to the conclusion that tolerances for repeatability, hysteresis, friction error, and linearity can be replaced by a single set of tolerances on the maximum allowable deviation of any data point from a specified reference line which represents the ideal calibration curve of the transducer.

Introduction of error band specifications resulted in higher production yields by transducer manufacturers, and lower unit cost, since the manufacturer now had more leeway in his individual transducer tolerances. For example, if a particular instrument design had inherently low hysteresis and good repeatability, the manufacturer could afford to be less careful in adjusting the transducer for linearity and still meet error band requirements. The previously necessary calculations of individual errors after calibration were also eliminated. This reduced the total calibration time by 20 to 40 minutes per transducer, with the same time savings incurred during each recalibration of the transducer.

A secondary benefit gained from the use of error band specifications was the creation of *interchangeable calibration records*. If the error band specified for a given transducer model is sufficiently narrow for overall data accuracy requirements, the calibration records of all transducers bearing the same part number can be considered interchangeable within error band tolerances. The data obtained from any of these transducers can then be reduced on the basis of the ideal calibration curve without referring to the individual calibration record for each transducer. If data of greater accuracy are required for a particular measurement at a later date, the individual record can be used for a data reduction rerun. If necessary, individual errors can then still be extracted from the calibration record and used to correct the reduced data even further.

Specifications in terms of error bands have since come into widespread use. Many transducer users have analyzed the nature of their measurements, their measurement accuracy requirements, and the overall accuracy of their measurement system, and have concluded that error band specifications would best serve their purposes in a majority of applications.

Error bands can include two or more transducer characteristics depending on the conditions imposed on calibration. These conditions must always be established. In order to include repeatability, the error band is determined after the completion of at least two (but often three or more) calibration cycles. Effects of measurand overloads can be included in the error band if application of this overload is required during calibration. One or more types of environments can be adjusted to their extremes during calibration when an *environmental error band* is to be determined. Friction error can be excluded from the error band by dithering the transducer during calibration (*friction-free error band*) when it is known that the transducer will always be operating in the presence of sufficient vibration to minimize friction.

The specified error band is usually stated with equal positive and negative error tolerances (e.g., "$\pm 1.25\%$ FSO"). The error band determined by a calibration is defined by maximum positive and negative errors obtained at any point within the range. These errors are usually not equal (e.g., "$+0.8$, -0.4% FSO").

An error band can be referred to a large variety of reference lines or curves ranging from a straight line between the "0% Range, 0% FSO" and "100% Range, 100% FSO" points ("terminal" end points) to a curve empirically determined only during an actual calibration. The error band may be made applicable over only a portion of the range, with no tolerances existing over the remaining portion or portions. It can also be "stepped," so that different tolerances apply over different portions of the range. Examples of various kinds of error bands are given in the next section.

II.3.3.12 Static error band. The error band applicable at room conditions, in the absence of any shock, vibration, or acceleration (unless one of these is the measurand), is the *static error band*. The band of allowable deviations of output values from the reference line or curve under these *static conditions*, and the line or curve itself, define the shape of the static error band. The output values are always measured over at least two consecutive calibration cycles to include a minimum determination

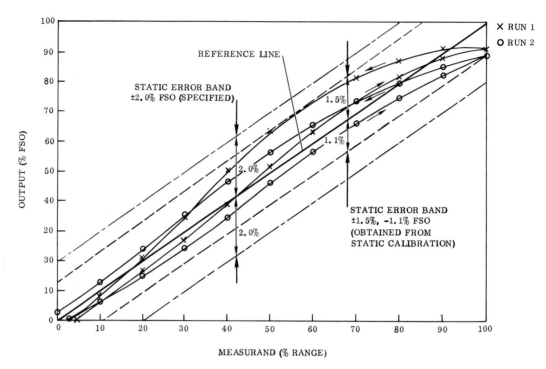

Fig. II-23. Static error band, referred to line between "terminal" end points (error scale 10: 1).

of repeatability. Only when repeatability is known to be extremely close should a single calibration cycle be used.

The static error band of a linear-output transducer is commonly referred to a straight line between either theoretical end points (as in theoretical-slope linearity) or end points set at "0% Range, 0% FSO" and "100% Range, 100% FSO" (as in terminal linearity). A static error band referred to a terminal line, as specified and as it may result from a static calibration, is illustrated in Figure II-23.

When the output-vs-measurand relationship of a transducer follows a known nonlinear relationship, the static error band follows the theoretical curve to which it is referred (see example, Figure II-24).

A static error band may show different tolerances applicable to different portions of the range (see example, Figure II-25). This results in a "stepped" error band.

Narrower static error bands can be obtained when linearity, or conformance to any prescribed curve, is not required. The absence of such a requirement is usually connected with an ability to reduce final data from a complex curve, different for each transducer. This type of data reduction is greatly facilitated by use of a computer. An example of such an error band is shown in Figure II-26. This static error band is referred to the curve through mean output readings obtained during consecutive

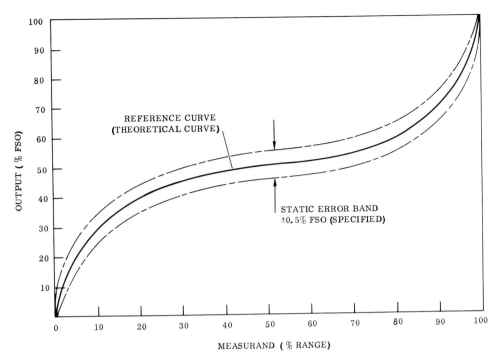

Fig. II-24. Static error band referred to theoretical curve (error scale 10: 1).

Fig. II-25. "Stepped" static error band referred to straight line between theoretical end points 10% FSO @ 0% range, 95% FSO @ 100% range (error scale 10: 1).

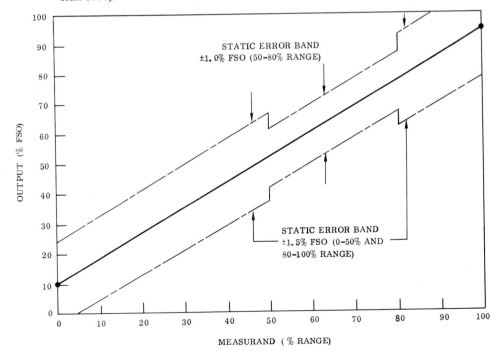

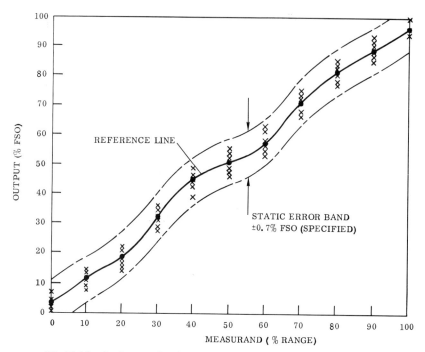

Fig. II-26. Static error band referred to mean-output curve (error scale 10: 1).

Fig. II-27. Static error band for ± 5 *g* acceleration transducer, referred to straight line between terminal end points (error scale 10: 1).

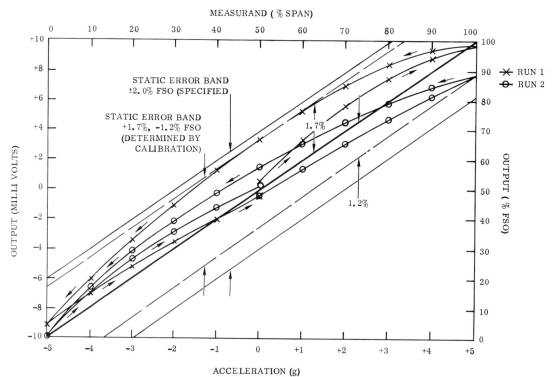

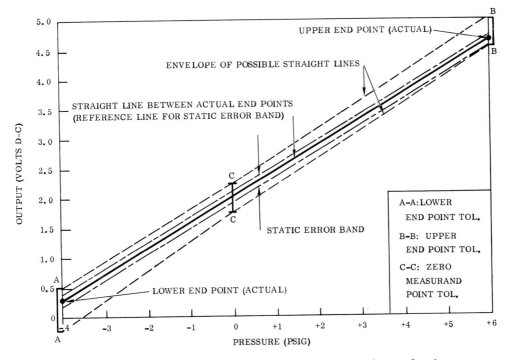

Fig. II-28. Static error band for −4 to +6 psig pressure transducer, referred to straight line between (actual) end points (scale of errors and tolerances exaggerated).

static-calibration cycles (*mean-output curve*). It includes hysteresis, repeatability, and friction error.

The selection of a reference line for a static error band of a transducer with a bidirectional range frequently involves the consideration of its intercept with the zero-measurand level. The simplest reference line is a theoretical slope or a terminal line, or a theoretical curve if the output is not intended to be linear. Figure II-27 illustrates an actual example of a static error band referred to a "terminal" straight line. If a straight line between actual end points is used, tolerances must be established for the output at zero measurand as well as for the end points. Any reference line can be used equally well for transducers with bidirectional output and with unidirectional output (see Figure II-28). A stepped static error band, narrower near zero measurand than near the end points, has been found useful for some of the more specialized measurement requirements.

II.3.4 Dynamic performance characteristics

When a transducer is used for a measurement where rapid measurand variations occur, or where step changes in measurand level have to be monitored, the transducer's

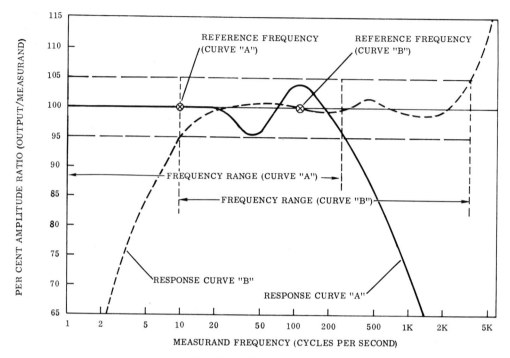

Fig. II-29. Frequency response.

dynamic characteristics must be established. If a pressure transducer, for example, is connected to a cylinder in which gas is compressed by means of a piston, the transducer's output will always follow the slow pressure rise caused by slowly pushing the piston into the cylinder. However, if the piston is linked to a rotating crankshaft, the transducer will reproduce the pressure changes in the piston less faithfully as the crankshaft speed increases. A speed is finally reached at which the transducer has almost no output since it cannot follow the pressure variations to any measurable degree. In a phonograph, the ability of the system to sense and reproduce high-frequency groove variations on a record is known as "fidelity."

II.3.4.1 Frequency response. When the measurand applied to the transducer varies sinusoidally over a stated frequency range, the change with frequency of the output/measurand amplitude ratio is the *frequency response* of the transducer. It is usually specified as "within ± ... percent from ... to ... Hz" and should be referred to a specific frequency within the range (*reference frequency*) and to a specific amplitude level (*reference amplitude*).

The examples shown in Figure II-29 represent the typical response curves of a transducer for static and dynamic measurements (Curve *A*) and a transducer for dynamic measurements only (Curve *B*). The former has a frequency response within $\pm 5\%$ from 0 to 800 Hz, referred to 10 Hz. The frequency response of the latter is within $\pm 5\%$ from 10 to 3500 Hz, referred to 100 Hz. Reference amplitude is not stated for these general examples.

A time difference always exists between output and measurand variations. The output lags behind the measurand. If this phase shift is significant to the measurement, frequency response can additionally be stated in terms of the phase difference between output and measurand.

II.3.4.2 Response time. When a step change of measurand is applied to a transducer, its output will change in the direction of the measurand change until a final output value is reached. Examples of measurand step changes are found in the sudden opening of a furnace door (temperature), the closing of a solenoid valve in an oil line (pressure), a golf club impacting with a golf ball (acceleration), a motor starting under load (torque in the shaft), etc.

The time required for the corresponding output change to reach 63% of its final (steady) value is the *time constant* (τ) of a transducer. The time required to reach a different specified percentage of this final value (e.g., 90, 98, or 99%) is the *response*

Fig. II-30. Response of overdamped transducer to step change in measurand.

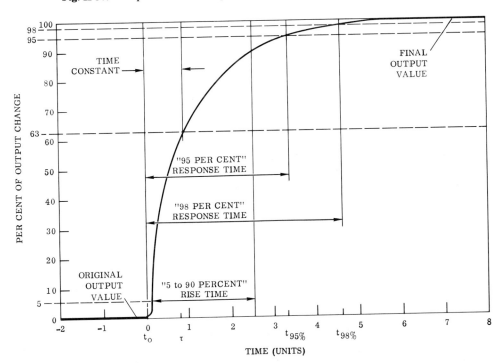

time. The time in which the output changes from a small to a large specified percentage of the final value is the *rise time* (see Figure II-30).

The measurand step change may be in an increasing or decreasing direction. The original value of the measurand can be at one of the transducer's range limits or at some other point within the range. The amplitude of the measurand change may be of any value.

When time constant, response time, or rise time has to be determined by a test, the ramp of the step change itself (rate of change of measurand) should be negligible when compared to the expected rate of change in output. When this ramp is insufficiently steep, suitable corrections must be made to the time value obtained.

It is usually simpler to perform a response-time test than to determine frequency response by application of a sinusoidally varying measurand. Data analysts, however, frequently prefer the frequency-response approach. It has been suggested that a coarse value ($\pm 10\%$) for the frequency response of an overdamped or critically damped transducer, with less than 10% overshoot, can be determined from response-time data since

$$f_{3 \text{ db}} = \frac{1}{\tau}$$

where $f_{3 \text{ db}}$ is the frequency at which the output amplitude is reduced by 50% (3 db) and τ is the time constant of the transducer as measured during a test.

While this relationship can be applied to an ideal transducer under ideal test conditions, it appears that a more realistic conversion is obtained by replacing τ, in the above equation, with $t_{95\%}$, the 95% response time (see Figure II-30).

II.3.4.3 Damping. The upper limit of frequency response and the response-time characteristics of a transducer are controlled by its energy-dissipating properties (*damping*).

If a transducer is *underdamped*, the output will rise beyond the final steady value (*overshoot*) and may then oscillate about the final value with decreasing amplitudes until the oscillations come to rest at this value (see Figure II-31). In an *overdamped* transducer the step-change-originated output change will reach its final value without overshoot or oscillations. A *critically damped* transducer operates at the point of change between the underdamped and overdamped conditions.

When a transducer's sensing element is set into free oscillation, it will oscillate at its *damped natural frequency*, or if no damping is present, at its *undamped natural frequency*.

The viscosity of a fluid (*viscous damping*), the current induced in conductors by changes in magnetic flux (*magnetic damping*), or other means can be used to effect damping in a transducer.

The ratio of the degree of actual damping to the degree of damping required to attain the state of critical damping is the *damping ratio*. A damping ratio of 1.0, therefore, signifies critical damping, damping ratios larger than 1.0 signify overdamping, and damping ratios less than 1.0 signify underdamping.

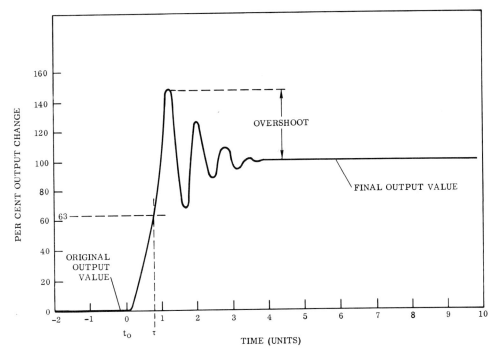

Fig. II-31. Response of underdamped transducer to step change in measurement.

II.3.5 Environmental characteristics

When the conditions under which a transducer operates differ from the previously described room conditions and other static conditions, a number of other errors can appear in the transducer's output. These errors can be considered separately, as output shifts additional to static errors, or as increased deviations of output values from an originally specified or known reference line due to environmental effects (*environmental error bands*). The specified external conditions (temperature, shock, vibration, etc.) to which a transducer may be exposed during shipping, storage, handling, and operation constitute the *environmental conditions* for any one transducer.

Environmental effects on a transducer output are normally temporary, lasting only for the duration of the exposure to a particular environment. Additional considerations must be given to possible damage during exposure, and to a permanent change in static characteristics of a transducer after exposure.

Environmental conditions are generally grouped in four categories: Storage, shipping, handling, and operating conditions.

Storage conditions include environments encountered by a transducer while stored in a carton, on a shelf, and while installed on a vehicle or on equipment awaiting tests or operational use.

Shipping conditions apply while the transducer or the equipment on which it is installed is in transit, provided the transducer is in a protective container whenever it is being handled.

Handling conditions exist for an unpackaged transducer, when it is transported, handled, and in the process of being installed.

Operating conditions comprise all environmental conditions under which the transducer must operate within specified performance tolerances.

II.3.5.1 Temperature effects.

Exposure of a transducer to ambient temperatures substantially above or below room temperature usually causes changes in the calibration curve and affects viscous damping due to changes in the damping fluid's viscosity. Excessively high temperatures can damage a transducer when uneven thermal expansion of parts causes cracking or rupture, or when solder melts. Somewhat lower temperatures may still cause permanent calibration shifts, e.g., by annealing of metal parts and ensuing changes in their spring constants. Very low temperatures can also cause rupture and similar failures by unequal thermal contraction or by freezing of entrapped moisture. They may render a transducer temporarily nonoperative, e.g., by causing solidification of damping oil or freezing of bearings.

The *operating temperature range* of a transducer is the range of environmental temperatures over which it must perform within specified error limits. When returned to and stabilized at room temperature after exposure to any temperature in this range, the transducer must again perform within original static error limits. Temperature errors are always determined after stabilization at the new temperature, usually at one of the extremes of the operating temperature range so that maximum errors can be measured.

In many types of transducers temperature effects will cause primarily a parallel displacement of the calibration curve (*thermal zero shift*) and a change in the slope of this curve (*thermal sensitivity shift*). These shifts (shown separately in Figure II-32) are usually stated in percent of full-scale output per degree Fahrenheit or Celsius ("% FSO/°F" or "% FSO/°C"). Knowledge of these individual errors is useful when the environmental temperature is known while a measurement is being made and when suitable corrections can be made to final reduced data.

When this environmental temperature is not known (except for the fact that it is within a known operating temperature range) and when changes in other characteristics such as hysteresis, repeatability, and friction error must also be considered, it is more useful to state total thermal effects as temperature error or as the temperature error band. *Temperature error* is the maximum change in output, at any measurand level within the range, when the transducer's temperature is changed from room temperature to either of the two extremes of the operating temperature range (see Figure II-33).

The *temperature error band* is the error band (see II.3.3.11) applicable over the operating temperature range. It includes all static errors, their increases or decreases

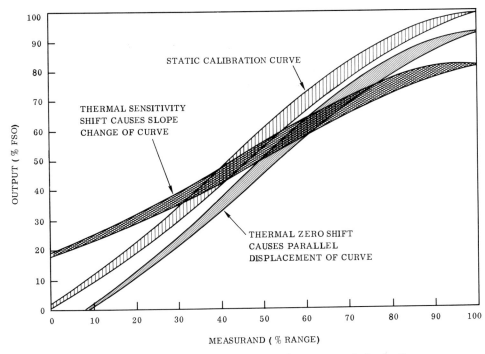

Fig. II-32. Thermal zero and sensitivity shifts, shown separately (scale of errors exaggerated).

Fig. II-33. Temperature error (at operating temperature range limits) (scale of curve shifts exaggerated).

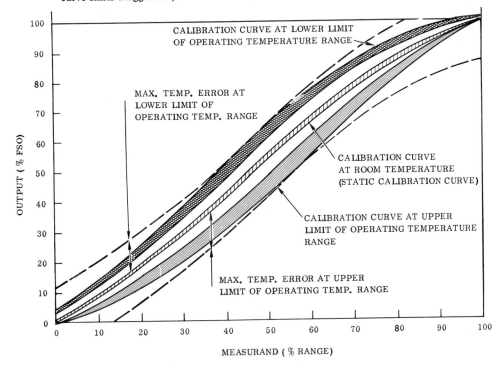

due to thermal effects, and all calibration curve shifts (see Figure II-34). All errors are referred to the same reference line used to establish the static error band.

When a transducer is subjected to a step change in its environmental temperature, a transient error (*temperature gradient error*) can appear in its output (see Figure II-35). Examples of such temperature step changes are found in sudden flow of a cryogenic or a very hot liquid through a previously empty duct or ignition of a jet or rocket engine. Temperature gradient errors are negligible in some transducer types but can exceed 30% of full scale output in other designs. The rate of change of temperature, and the temperature levels between which the change occurs, should be stated in a specification.

Due to changes in the viscosity of a damping fluid, the damping ratio can be affected by temperature changes. If a fairly constant damping is required from a transducer over the operating temperature range, tolerances should be imposed on the damping ratio (e.g., "0.7 ±0.2 from −30 to +160°F") or on the frequency response (e.g., "within ±10% from 20 to 4000 Hz at any temperature from −65 to +185 °F").

Fig. II-34.   Temperature error band referred to straight line between "terminal" end points (error scale 2: 1).

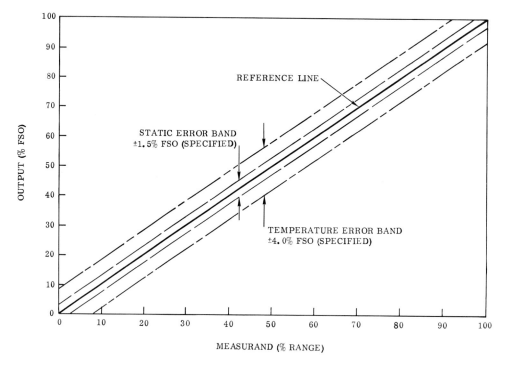

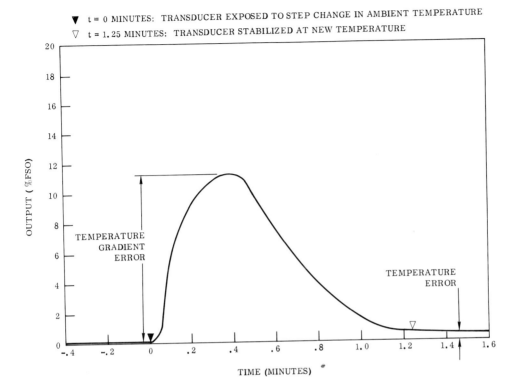

▼ t = 0 MINUTES: TRANSDUCER EXPOSED TO STEP CHANGE IN AMBIENT TEMPERATURE

▽ t = 1.25 MINUTES: TRANSDUCER STABILIZED AT NEW TEMPERATURE

Fig. II-35. Temperature gradient error (typical example shown for output at zero measurand).

II.3.5.2 Acceleration effects. Constant acceleration forces can act upon the internal components of a transducer so as to cause additional error in its output The error is usually more severe in one axis of the transducer than in other axes, depending on its construction. Acceleration may act directly on a mechanical sensing element or a mechanical linkage, causing a change of displacement; it can act upon structural supports, causing distortion which may lead to a failure; it can act upon bearing-supported rotating members, causing eccentric loading which increases friction; and it can cause mass shifts, deformations, and distortions in other ways.

The maximum difference between output readings, at any measurand value within the range, taken at room conditions with and without the application of specified constant acceleration along specified axes and directions is the *acceleration error*. This error can also be expressed as error "per *g*" (*acceleration sensitivity*) within a stated range of acceleration values (e.g., "0 to 10 *g*").

The *acceleration error band* is the error band (see II.3.3.11) applicable, or obtained during a test, when constant accelerations with a stated range of amplitudes are applied

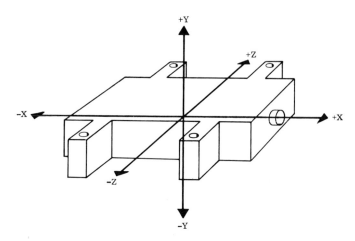

Fig. II-36.   Typical labeling of acceleration axes for a transducer.

to a transducer along specified axes, at room conditions. The acceleration error band includes maximum static and acceleration errors.

The specification or determination of acceleration errors or error bands requires reference labeling of the transducer's three orthogonal axes on a sketch of the transducer (see example, Figure II-36).

Some types of transducers are so sensitive to acceleration forces that they must be installed in such a position that effects of acceleration due to gravity are minimized. The error due to the orientation of the transducer relative to the direction in which gravity acts upon it is known as *attitude error*, although it is actually an acceleration error.

II.3.5.3 Vibration effects. Vibratory acceleration (vibration) can affect transducers in the same manner as constant acceleration. More severe effects, however, are connected with the frequencies of vibration. As vibration in a specific range of frequencies and amplitudes is applied to a transducer along a specific axis, amplified vibrations of transducer components (*resonances*) can occur within narrow frequency bands. These resonances frequently cause much larger transducer errors (see Figure II-37) than a constant acceleration of the same amplitude when applied along the same axis in either direction.

Vibration error is the maximum change in output, at any measurand value within the range, when vibration levels of specified amplitudes and range of frequencies are applied to the transducer along specified axes, at room conditions.

The *vibration error band* is the error band (see II.3.3.11) applicable, or obtained during a test, when vibration levels with a specified range of frequencies and amplitudes are applied to a transducer along specified axes at room conditions. The vibration error band, therefore, includes maximum vibration errors as well as maximum static errors except that the latter are decreased by that amount of friction error which is removed by applied vibration.

Vibration levels can be stated in terms of sinusoidal vibration or random vibration. Sinusoidal vibration is periodic vibration in a range of frequencies having either constant amplitude or a range of amplitudes following a prescribed program. Amplitudes can be stated in peak value (e.g., peak g, "$\hat{g}$"), peak-to-peak values (e.g., "g p-p") or root mean square values (e.g., "g rms"). Vibration at very low frequencies is commonly shown as single-amplitude (peak amplitude) or double-amplitude (peak-to-peak amplitude) displacement rather than as acceleration. *Random vibration* is nonperiodic vibration, described only in statistical terms. The term usually refers to vibration characterized by an amplitude distribution which essentially follows the "normal error curve" (Gaussian distribution), with amplitude and frequency limits specified. It is usually expressed in terms of g^2/Hz.

II.3.5.4 Ambient-pressure effects. Changes in ambient pressure—notably the high pressures encountered in underwater work and the very low pressures in the upper atmosphere and in space—can affect transducer performance. Such pressure changes

Fig. II-37. Typical results of vibration test on potentiometric transducer at a given program of vibration amplitudes, from 10 to 3000 Hz at one measurand value, in one transducer axis.

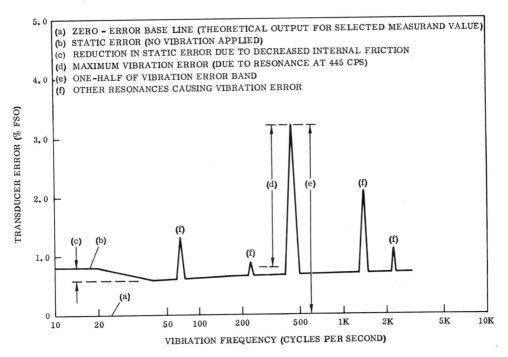

(a) ZERO – ERROR BASE LINE (THEORETICAL OUTPUT FOR SELECTED MEASURAND VALUE)
(b) STATIC ERROR (NO VIBRATION APPLIED)
(c) REDUCTION IN STATIC ERROR DUE TO DECREASED INTERNAL FRICTION
(d) MAXIMUM VIBRATION ERROR (DUE TO RESONANCE AT 445 CPS)
(e) ONE–HALF OF VIBRATION ERROR BAND
(f) OTHER RESONANCES CAUSING VIBRATION ERROR

TRANSDUCER ERROR (% FSO)

VIBRATION FREQUENCY (CYCLES PER SECOND)

can cause variations in a transducer's internal geometry and in the behavior of individual components, e.g., due to external case deformations. Outgassing of internal parts, self-heating of wires (in the absence of air as heat-transfer medium), pressures exerted by sealed cavities, and corona arcing at high-voltage terminals can affect transducer performance in a low pressure or vacuum environment if the transducer is not sufficiently sealed.

Ambient pressure error is the maximum change in output, at any measurand value within the transducer's range, when the ambient pressure is changed between specified values.

The *ambient pressure error band* is the error band (see II.3.3.11) applicable when the transducer is exposed to variations of ambient pressure within a specified range. Static errors are included in this error band.

Ambient pressure error is sometimes called "altitude error" when the range of ambient pressures is below one atmosphere and the pressures are shown in equivalent units of altitude within the earth's atmosphere (e.g., "from 0 to 30,000 ft"). This term has become less meaningful in an era where transducers may be required to operate in the atmosphere of other planets or when they are used for oceanographic measurements.

II.3.5.5 Installation effects. After a transducer has been installed and connected for its intended use, the errors in its output may be larger than those determined by a laboratory calibration. Such additional errors can be introduced when the case is not evenly machined and is deformed when all mounting screws are tightened, when the torque applied to a coupling nut on a pressure fitting causes deformation in the transducer's sensing element, or when the sensing shaft of an angular speed or displacement transducer is coupled eccentrically to a driving shaft.

Mounting error is the error resulting from mechanical deformation of the transducer caused by mounting the transducer and by making all electrical and measurand connections.

II.3.5.6 Other operating environmental effects. A number of environmental conditions, in addition to those described above in more detail, may affect the behavior of a transducer during its normal operation and can cause increased errors, erratic output, or failure. These include: the effects of high humidity (or complete immersion in liquid) on a poorly sealed transducer, particularly on its insulation resistance; the corrosive effects and insulation-reducing effects of high salt concentrations in the ambient atmosphere; various effects of measured fluids on sensing elements exposed to them; the influence of an ambient magnetic or radio-frequency field; the effects of high ambient ion density; and nuclear radiation effects.

II.3.5.7 Combined environmental effects. It may be necessary for some measurements to specify error limits for a transducer during exposure to a combination of

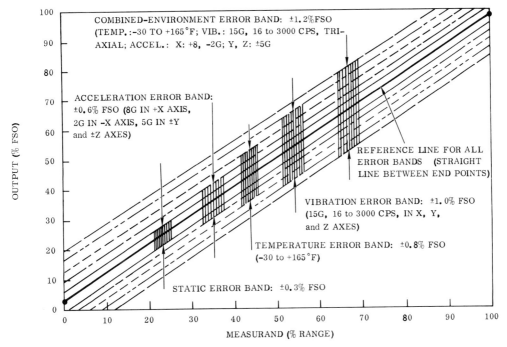

Fig. II-38. Separate and combined environmental error bands referred to straight line between (actual) end points (error scale 10: 1).

environments. This can best be achieved by specifying the error tolerances as a combined environmental error band (see Figure II-38).

This form of specification should be used only when a combined-environment test is practicable and the relatively high cost of such a test is justified by the results expected. The example illustrated would require installing the transducer in a temperature chamber which is installed on a vibration exciter, which, in turn, is mounted on a large centrifuge. If separate environmental error bands, also shown in the illustration, are specified instead, each error band can be determined by a much simpler and often readily available test set up.

It has been proposed that an approximate value for a combined environmental error band can be calculated by taking the static error band plus the square root of the sum of the squares of maximum temperature error, maximum vibration error, and maximum acceleration error. Each of these individual errors is taken as the corresponding error band less the static error band. The results of such calculations are frequently very misleading.

A more meaningful combined environmental error band is one based on those levels of each type of environment that could actually combine during transducer operation, rather than on maximum environmental limits. The assumption that, e.g., the largest acceleration in the transducer's most sensitive axis may coincide in time with the highest (or lowest) ambient temperature, lowest ambient pressure, and maximum vibration is usually quite unrealistic.

II.3.5.8 Nonoperating environmental effects. A complete transducer specification usually includes those environmental conditions to which a transducer could reasonably be exposed during storage, shipping, handling, and even use prior to the time that it yields significant data.

A transducer intended to measure accurately in a temperature environment of $+32$ to $+90°F$ may have large errors after temporary outdoor storage and exposure to solar radiation. A delicate transducer, shock-mounted and protected during use, may encounter vibration while transported by truck and shock when accidentally dropped from the height of a tailgate. A transducer may be installed in a pipe carrying liquefied gas but its output may not be recorded until some time after the pipe carries gas at a much higher temperature.

In a specification a list of nonoperating environments and their levels is usually preceded by a clause such as: "The transducer shall perform within all specified tolerances after exposure to . . . ."

If a nonoperating environment is so severe that the error band after exposure must be permitted to be larger than the static error band, a separate *post-environmental error band* may have to be specified. Examples would be a post-shock, post-low-temperature, or post-vibration error band with the exact level of the environment specified.

Temperature and relative humidity are usually shown for all phases of a transducer's life. Vibration and ambient pressure variations such as those encountered on cargo aircraft or trucks (which may also travel at high elevations) and impact shock due to accidental drops are added to shipping conditions for a properly packaged transducer. Impact shock seen by an unpackaged transducer, due to accidental drops or other mishaps, is included in handling conditions. Operation during intended use prior to obtaining significant data may call for exposure to a multitude of environmental conditions including radiant heat, various types of nuclear radiation, strong magnetic fields, high sound levels, and immersion in corrosive liquids.

II.3.5.9 Type-limited environmental effects. The environmental errors described in II.3.5.1 through II.3.5.8 are present in most transducers to varying degrees. Some transducer types are subject to certain additional errors. The following serve as examples:

1. The error due to heat conduction between sensing element and mounting or other external connections (*conduction error*) of temperature transducers.
2. The error caused by mounting-surface strain (*strain error*) in a resistive temperature transducer cemented or welded to the surface.
3. The response of acceleration transducers to acceleration forces in axes transverse to the sensing axis (*transverse sensitivity*).

4. The performance change of a differential pressure transducer caused by changes in the pressure (*reference pressure*) relative to which differential pressure is measured.

II.3.6 Reliability characteristics

Reliability generally refers to rated failure-free performance for a specified period of time. Reliability characteristics specifically comprise life ratings and numerically definable effects of exposure to excessive operating conditions.

II.3.6.1 Overload. When the magnitude of the measurand applied to a transducer is increased above the rated range, a measurand value is reached beyond which the performance of the transducer is caused to exceed specified performance tolerances after return to levels within the range. This value is the *overload* the transducer can withstand. In the case of a pressure transducer it is the specified *proof pressure*. The transducer output at this measurand value (*overload output*, *proof-pressure output*) may be required to remain within certain tolerances.

If the measurand is increased beyond the overload rating, another value will be reached beyond which permanent failure of the transducer will occur. This value is the *ultimate-overload rating* or, for a pressure transducer, the *burst-pressure rating*.

Overpressure characteristics also apply to other transducers measuring properties of pressurized fluids, e.g., liquid-level, flow, and probe-type temperature transducers. When the pressure applied to that portion of the transducer subjected to the fluid is increased above a specified *operating pressure*, the *proof-pressure* level will be reached at which the measured fluid will be contained without any leakage from the case, or with a *leakage rate* below a specified limit. If the pressure is further increased, the specified *burst-pressure rating* is attained, i.e. the maximum pressure which will not cause the sensing element or case to rupture. If this rating is exceeded, the transducer will reach the *burst point* at which it ruptures.

In conservative specifications the burst pressure rating will usually be set considerably below the expected burst point. The minimum number of applications of any type of overload or overpressure to the transducer and the time duration of each application should be included in a specification.

Another effect of overload is observed in many types of transducers, particularly those with an integral amplifier. A certain amount of time will elapse after removal of a specified overload before the transducer again performs within its specified tolerances (*recovery time*).

II.3.6.2 Life. Transducer *life* is usually stated as the minimum length of time during which a transducer can be stored or can be required to operate without changing its performance beyond specified tolerances.

Storage life is the preoperating life when the transducer is exposed to stated storage conditions (see II.3.5).

Operating life is the life during in-use operation while exposed to all stated operating conditions (see II.3.5). Transducer operation can be specified as continuous or intermittent. In the latter case the total number and the time duration of "on" and "off" cycles should always be stated.

Cycling life is a different form of operating life. It is not expressed in terms of minimum time, but as the minimum number of full-range excursions or specified partial-range excursions over which a transducer will operate such that all performance characteristics remain within their specified tolerances.

II.3.7 Mechanical design characteristics

The external mechanical characteristics of a transducer do not usually affect its performance. The intent of specifying certain case materials and the exact nature of case construction and sealing are frequently covered by stated environmental characteristics such as limitations on effects of corrosive or conducting ambient fluids.

Some mechanical design characteristics, however, are always given for every transducer. These include the general configuration, outline dimensions, mounting provisions and dimensions, and the type, size, and location of all external electrical and mechanical or pneumatic connections.

The nameplate should be located, whenever possible, on a portion of the case which is clearly visible after installation. The nameplate can be a decal or a thin metallic or plastic plate permanently fastened to the case. The information can also be etched or engraved directly on the case.

The type of information shown on the nameplate usually includes the nomenclature of the transducer, a few of the most pertinent characteristics such as range and excitation, identification of external electrical connections (pin or lead labeling), part number, serial number, manufacturer's name and address, and additional information required by applicable military specifications, industrial codes, or special needs of the user.

II.4 General Criteria for Selection

The selection of a transducer usually involves the following basic considerations:

A. *MEASUREMENT*
1. What is the real purpose of the measurement?
2. What is the measurand?
3. What range of the measurand must be displayed in final data?
4. What measurand overloads may occur before and during the time data are required?
5. With what accuracy must the measurement be presented in final data?
6. What are the lower and upper limits of flat frequency response (or the response time) needed in final data?

7. What is the nature of the fluid to be measured?
8. Where will the transducer be installed?
9. What ambient environmental conditions will exist around a transducer for this measurement?

B. *DATA SYSTEM CAPABILITY*
1. What data transmission system is used?
2. What data processing system is used?
3. What data display system is used?
4. What are the accuracy and frequency response capabilities of the transmission, processing, and display systems?
5. What transducer output will the transmission system accept with minimum signal conditioning?
6. What transducer excitation voltage is most readily available?
7. How much current can be drawn from the excitation power supply?
8. What load will the transmission circuit present to a transducer?
9. Does the transmission system provide sufficient limiting for abnormal transducer outputs?
10. Is filtering of the transducer output required, and can this be adequately provided by the transmission or data processing system?
11. To what extent does the transmission system allow for detection of, or provide compensation for, errors in the transducer output?

C. *TRANSDUCER DESIGN*
1. What are the configurational limitations?
2. What maximum error can be tolerated during static conditions and during and after exposure to known environmental conditions?
3. What are the limitations on excitation and output?
4. What power drain can be tolerated?
5. Which transduction principle (reluctive, potentiometric, strain-gage, etc.) should be utilized?
6. What are the effects of the measured fluid on the transducer?
7. Will the transducer affect the measurand so that erroneous data are obtained?
8. What cycling or operating life is required?
9. What test methods will be used to prove performance? Are these methods adequate? Are they sufficiently simple? Are they firmly established?
10. What are the failure modes of the transducer? What hazards would a failure present to the component or system in which it is installed, to adjacent components or systems, or to other portions of the data system?
11. What is the lowest level of technical competence of all personnel expected to handle, install, and use the transducer? What human engineering requirements should affect the transducer design?

D. *TRANSDUCER AVAILABILITY*
1. Is a transducer which fulfills all requirements now available?

2. What manufacturer has successfully demonstrated his ability to produce a transducer similar to the required item?
3. What experience-history exists in dealing with a proposed manufacturer?
4. Will minor redesign of an existing transducer be sufficient or will a major development effort be required?
5. Is the cost of the transducer compatible with the necessity for the measurement?
6. Can the transducer be delivered in time to meet installation schedules?
7. What is the cost and length of time of all minimum performance verification testing?

BIBLIOGRAPHY

1. Behar, M. F., *et al.*, *The Handbook of Measurement and Control.* Pittsburgh: Instruments Publishing Co., Inc., 1951.

2. Kret, D. B., *Transducers.* Allen B. DuMont Laboratories, Inc., 1953.

3. Draper, C. S., McKay, W., and Lees, S., *Instrument Engineering*, 3 vols. New York: McGraw-Hill Book Company, 1955.

4. Considine, D. M., *Process Instruments and Controls Handbook.* New York: McGraw-Hill Book Company 1957.

5. "Glossary of Terms for Flight Test Instrumentation," *AIA/ATC Report No. ARTC-16.* Los Angeles: Aerospace Industries Association, 1957.

6. Caldwell, W. I., and Coon, G. A., *Frequency Response for Process Control.* New York: McGraw-Hill Book Company, 1959.

7. Hernandez, J. S., *Introduction to Transducers for Instrumentation.* Los Angeles: Statham Instruments, Inc., 1959.

8. Lion, K. S., *Instrumentation in Scientific Research.* New York: McGraw-Hill Book Company, 1959.

9. Sarbacher, R. I., *Encyclopedic Dictionary of Electronic and Nuclear Engineering.* Englewood Cliffs, N.J.: Prentice-Hall, Inc., 1959.

10. Bibbero, R. J., *Dictionary of Automatic Control.* New York: Reinhold Publishing Corp., 1960.

11. Borden, P. A., and Mayo-Wells, W. J., *Telemetering Systems.* New York: Reinhold Publishing Corp., 1960.

12. Hernandez, J. S., "A Guide to Transducer Selection," *Electrical Design News*, February, 1960.

13. Zuehlke, A. A., "Instrument Error Band Concepts." Paper presented at I. S. A. 6th National Flight Test Symposium, May, 1960. Reprint published by Bourns, Inc., Riverside, Calif. (1960).

14. Norton, H. N.," Instrumentation Transducers," *Space/Aeronautics Research and Development Handbook*, 1960–1961.

15. Michels, W. C., ed., *The International Dictionary of Physics and Electronics* (2nd ed.). Princeton, N.J.: D. Van Nostrand Co., Inc., 1961.

16. Norton, H. N., "Transducer Field Demands Standards," *Missiles and Rockets*, Vol. 8, May 22, 1961.

17. Norton, H. N., "Formulating Aero-Space Transducer Standards," *I. S. A. Journal*, Vol. 8, September, 1961.

18. Polishuk, P., "Preliminary Study of the State-of-the-Art in Telemetry Transducers," *ASD Tech. Note 61-82*, Aeronautical Systems Division, USAF (October, 1961).

19. Tyson, F. C., *Industrial Instrumentation*. Englewood Cliffs, N.J.: Prentice-Hall, Inc., 1961.

20. Von Koch, H., and Ljungborg, G., eds., *Instruments and Measurements*. New York: Academic Press Publishers, 1961.

21. *Telemetry Transducer Handbook*, WADD Technical Report 61-67, Vols. I and II. ASD, Air Force Systems Command, USAF, Wright-Patterson AFB, Ohio (1961).

22. Norton, H. N., "Spacecraft Telemetry Transducers," *Space/Aeronautics Research and Development Handbook*, 1961–1962.

23. Holzbock, W. G., *Instruments for Measurement and Control* (2nd ed.). New York: Reinhold Publishing Corp., 1962.

24. Norton, H. N., "Transducer Selection," *Space/Aeronautics*, Vol. 38, Part 2, October, 1962.

25. "Nomenclature and Specification Terminology for Aerospace Test Transducer with Electrical Output," *I.S.A. Tentative Recommended Practice RP37.1*, 1963.

26. Cerni, R. H., and Foster, L. E., *Instrumentation for Engineering Measurement*. New York: John Wiley & Sons, Inc., 1962.

27. Leibowitz, D., "Transducers for Control," *Product Engineering*, Vol. 38, July 23, 1962.

28. Norton, H. N., "Error Band Concept Defines Transducer Performance," *Ground Support Equipment*, January, 1963.

29. "Transducer Compendium," compiled by the Instrument Society of America. New York: Plenum Press, 1963.

30. Doebelin, E. A., *Measurement Systems: Applications and Design*. New York: McGraw-Hill Book Company, 1966.

31. Striker, Dr. Gy., "Transducers for Industrial Measurements—Problems and Trends," *Paper No. HU-6, Acta IMEKO 1967*, International Measurement Confederation, Budapest, Hungary (1967).

III

Transducer Performance Determination

This chapter deals with generally applicable transducer test philosophies and methods. Additional calibration and test methods, including procedures for applying known values of the measurand to the transducer, are covered in each of the following chapters.

Transducers are used in an endless variety of applications. A set of conditions under which the transducer is expected to operate pertains to each application. Specifications for the transducer's performance usually list the most important transducer characteristics and their allowable tolerances at room conditions. They also show tolerances on changes in these characteristics which can be induced by exposure of the transducer to various environments. Transducer specification characteristics and the tests used to verify them are strongly interrelated.

It is customary to state performance tolerances for each environment (e.g., temperature error, acceleration error, etc.) and to verify these by individual-environment tests, although the operating environmental conditions usually act simultaneously upon the transducer. However, they do so to varying degrees, and the level of each of the environments changes frequently with time. Although combined-environment tests (e.g., temperature-vibration tests) on transducers are feasible, and are not infrequently performed, they are usually not considered practical for two major reasons: It is too difficult to predict a realistic set of combined environments, and the increased cost and complexity of such a test are not warranted by the possible slight gain in knowledge about the behavior of the transducer.

Similar reasoning should be applied to all transducer characteristics to be contained in a specification. Specifying characteristics which need never be verified is virtually meaningless. One of the major constraints on specifications of transducer performance is the cost and complexity of the test necessary to determine each specified characteristic with sufficient accuracy and confidence to establish specification compliance.

It is impossible to determine a hysteresis of "less than 0.05 % of full-scale output" if the best available test equipment, operated by the best trained personnel, has a cumulative accuracy within only $\pm 0.25\%$ of the transducer's full-scale output over the transducer's range. It is not practical to specify a shock test calling for $2000\,g$ shock pulses of $20\,\mu s$ duration on each of a lot of 300 transducers, which are urgently needed, if there is only one shock-test machine in the country which can be used for such a test.

Although a knowledge of test methods is required when specifying any type of component, it is of particular importance in the preparation of transducer specifications. The state of the art of transducer development, and especially of transducer testing, is relatively new. Such testing often calls for the use of costly specialized equipment and a very high degree of skill. Since the purchase of transducers frequently represents a considerable financial investment, their rejection by the buying activity can lead to bitter disagreements between user and manufacturer, and to possible legal action and litigation when any doubts exist as to the correctness of test methods used in performance determination.

The determination of transducer performance at room conditions requires a fundamental knowledge of metrology, applied physics, and electrical and mechanical engineering. Performance determination during exposure to more severe environmental conditions requires additional knowledge in the areas of test engineering, mechanics, and thermodynamics. Analysis of test results may further call for other areas of knowledge such as inorganic and organic chemistry and metallurgy.

III.1 Transducer Output Measurement

Transducers normally provide a limited number of types of output. D-C or a-c voltage output are most common. They are typical for photovoltaic, piezoelectric, reluctive, strain-gage, and thermoelectric transducers as well as for the large variety of transducers incorporating either a secondary transduction element, as in many ionizing transducers, or an output converter. The output of a potentiometric transducer is a voltage ratio which is often used as voltage output while the excitation voltage is precisely regulated and periodically monitored. The output of photoresistive, piezoresistive, and other resistive transducers is frequently measured in terms of resistance, but can also be measured in terms of a current or voltage originating in a bridge circuit to which the transducer is connected. The change in capacitance and inductance provided, respectively, by capacitive and inductive transducers can similarly be measured as such by an impedance bridge or an alternate a-c bridge; however, it is normally used in terms of either an a-c voltage or an a-c current provided by a fixed arm bridge, a d-c voltage provided by a demodulator following an a-c bridge, or by a secondary transduction element such as an ionization tube, or a frequency (or frequency-modulated) output furnished by a tuned-circuit oscillator in which the transduction element is used as the frequency-changing element. Frequency output, such as furnished by electromagnetic and vibrating-element transducers, is best measured as a frequency but is sometimes used as a d-c voltage supplied by a frequency-to-voltage converter. This applies equally to the usually aperiodic frequency output of ionizing and a few other types of transducers whose output is considered a "count" rather than a frequency.

III.1.1 Voltage measurement

The output of a majority of transducers is a d-c or a-c voltage. D-C voltages in the microvolt and millivolt ranges, up to approximately 1500 mV, are best measured with a *voltage potentiometer*. This instrument measures the transducer output by balancing it against a known voltage. The known voltage is obtained between the wiper arm and one end of a potentiometer across which a battery is connected.

In its basic operation (see Figure III-1), the wiper is moved over the potentiometer element until a null balance is read off the null indicator connected between wiper

Fig. III-1. Output voltage measurement by voltage potentiometer.

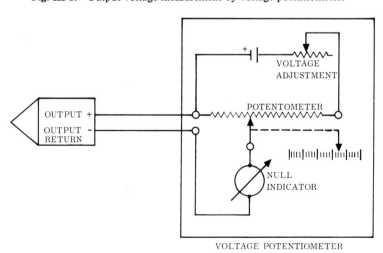

VOLTAGE POTENTIOMETER

arm and one transducer output terminal while the other output terminal is connected to the "positive" end of the potentiometer. The wiper arm is mechanically linked to a calibrated dial on which the voltage necessary to balance the unknown voltage can be read.

Voltage potentiometers are usually calibrated by connecting a known voltage across the measurement terminals. An accurate voltage (1.017 to 1.020 V) is produced by a *standard cell* such as the Weston® standard cell which has one mercury electrode and one cadmium amalgam electrode and uses a cadmium sulfate solution as electrolyte.

Commercially available voltage potentiometers usually incorporate a number of refinements such as range-multiplier resistors, devices to minimize thermoelectric potentials

Fig. III-2.   Voltage potentiometer (courtesy of Leeds & Northrup Co.).

generated by frictional heating at the contacting point of the wiper arm, optical aids to improve the accuracy of a null indication (when integral), and built-in standard cells. A typical commercial voltage potentiometer is illustrated in Figure III-2.

A voltage potentiometer always reads the open-circuit output voltage of a transducer because it presents a theoretically infinite impedance to the transducer at the precise point of null balance.

Transducer output voltages larger than 1.5 V full-scale are usually measured by means of a digital voltmeter (Figure III-3). The digital voltmeter also operates on the

Fig. III-3.   Digital voltmeter (courtesy of Electro Instruments, Inc.).

principle of the voltage potentiometer but differs primarily by using discrete resistance increments for voltage division, provided by precision resistors, instead of a continuously variable slide-wire or other continuous type of potentiometer. The unknown voltage is balanced in the voltmeter in decade steps, using either reed relays or transistor switching circuits to select the internal balancing voltage by successive approximation, i.e., until an increasingly finer null is obtained. The position of each selector switch is indicated as a numeral on the face of the digital voltmeter. For example, on a 5-digit meter, a voltage of 21.453 V would be successively approximated by the meter's automatic switching to voltage divider resistors giving first 20 V on the "Tens" switch, then 1 V on the "Units" switch, 0.4 V on the "Tenths" switch, 0.05 V on the "Hundredths" switch and, finally, 0.003 V on the "Thousandths" switch. Range changing and polarity indication is often automatic on commercially available meters.

A-C voltages can also be measured by a digital voltmeter if it is equipped with an a-c to d-c converter responding either to the average, peak, or rms value of the a-c voltage. Very accurate measurements are obtained by use of a ratio transformer in conjunction with a transfer-standard a-c voltmeter or voltammeter.

III.1.2 Current measurement

D-C currents are accurately measured by passing the current through a stable precision resistor of known value (standard resistor) and reading the resulting voltage drop across the resistor on a voltage potentiometer or digital voltmeter.

A-C currents are measured on a transfer-standard a-c voltammeter when optimum accuracy is required.

Where relatively less accuracy is required, a number of other current indicators, such as precision electronic multimeters, can be used.

III.1.3 Voltage ratio measurement

The output of potentiometric transducers is a voltage ratio. It is the ratio of the voltage between the wiper and the "ground" side of the potentiometer (resistance) element to the excitation voltage applied across the entire potentiometer element. This makes it possible to measure this form of output as a ratio, independent of excitation-supply variations.

The two types of indicators used most commonly for voltage ratio measurements are the digital electronic voltmeter with ratio-measurements provisions and the manually- or servo-balanced resistance-bridge ratiometer. The former is a digital voltmeter modified to measure and display the ratio between two voltages, one of which can be the excitation voltage; the other, the output of the transducer. The latter (see Figure III-4) is essentially a precision potentiometer of either the continuous or the decade type or a combination of these two, with digital indicator dials connected mechanically

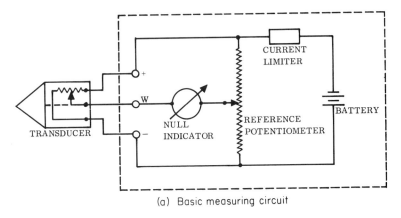

(a) Basic measuring circuit

(b) Typical ratiometer

Fig. III-4. Resistance-bridge ratiometer (courtesy of Pacific Metrology Laboratories, Inc.).

to the potentiometer shaft or (concentric) shafts so as to indicate the relative position of the potentiometer's wiper arm. A null indicator (or null-indicating galvanometer) is connected between the wiper and the external wiper terminal. A battery is connected across the potentiometer as well as across the external (transducer) excitation terminals. The transducer output is read, as "percent of voltage ratio (%VR)," directly off the mechanical indicating dials after a fine null reading has been obtained on the null indicator.

A current-limiting device, which can be a simple resistor, should always be connected in series with the battery in order to avoid burning out the transducer whenever it is incorrectly connected so as to apply the excitation between its wiper and "ground" terminal.

III.1.4 Resistance measurement

The resistance-change output of resistive transducers is almost invariably measured by means of a resistance bridge. Various commercial resistance bridges are available, all modifications of the basic bridge circuit first used by Wheatstone in 1843 and, hence, known as the *Wheatstone bridge* [see Figure III-5(a)]. The bridge

Fig. III-5.   Wheatstone bridge.

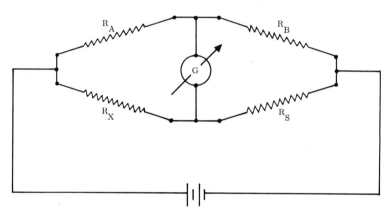

(a) Basic bridge circuit

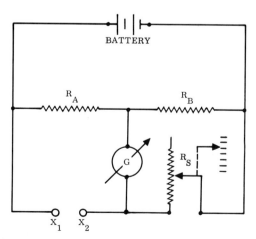

(b) Basic resistance bridge for resistance determination

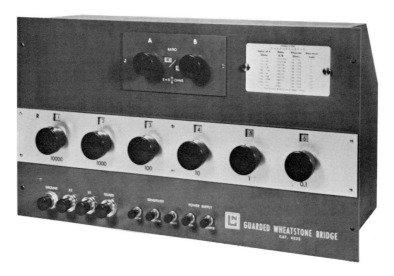

Fig. III-6. Guarded Wheatstone bridge (courtesy of Leeds & Northrup Co.).

is balanced when the current through the galvanometer (null indicator) is zero. This occurs when the ratio between resistors A and B equals the ratio between the unknown resistor X and the "standard" (reference) resistor S, or

$$\frac{R_A}{R_B} = \frac{R_X}{R_S}$$

The unknown resistance is then obtained by multiplying the value of the known resistor S by the ratio of resistances A to B, or

$$R_X = R_S \frac{R_A}{R_B}$$

Interchanging the galvanometer and the battery in their circuit position does not affect bridge balancing conditions.

The basic bridge circuit is made usable for resistance measurement by connecting a rheostat into the circuit as resistance S and connecting the unknown resistance across terminals X_1 and X_2. The electrical position of the wiper of the rheostat, which can be continuously variable or vary in predetermined discrete intervals (by decade switches), is mechanically indicated by a numerical display dial, or by a number of dials indicating decade switch positions. The basic measuring circuit is illustrated in Figure III-5(b). One of the essential refinements of this circuit is the replacement of fixed resistors A or B, or both, by adjustable resistors connected to a selector switch so as to allow ratios of R_A/R_B other than unity, e.g., 10:1, 100:1, etc., and 1:10, 1:100, etc.

Further refinements and modifications are included in such commercially available resistance bridges as the "guarded Wheatstone bridge" (Figure III-6) and the

Fig. III-7. Mueller bridge (courtesy of Leeds & Northrup Co.).

Mueller bridge which is used exclusively for the accurate determination of the resistance of platinum-wire resistive temperature transducers (Figure III-7).

III.1.5 Capacitance and inductance measurement

Although the final indication of the output of a capacitive or inductive transducer is almost never in the form of a capacitance or inductance change, it is sometimes necessary to measure such outputs because they have been specified.

A-C-excited reactance bridges (*impedance bridges*) of various types are most commonly used for measuring inductance as well as capacitance. These are similar in basic design to the Wheatstone bridge except that no more than two legs of the bridge can be purely resistive. The "standard" or "reference" leg is reactive. It is frequently a series or parallel combination of an adjustable resistor and an adjustable capacitor. The reactance (capacitive or inductive) across the transducer output terminals is then balanced against the "standard" leg of the bridge. The bridge dials are so arranged and labeled as to permit a direct reading of capacitance or inductance.

III.1.6 Frequency and pulse-count measurement

A considerable number of transducer types have frequency or pulse-count outputs. Such outputs are usually measured by means of an electronic counter or an events-per-unit-time (*EPUT*) meter. These indicators display the periodic or aperiodic transducer output in digital form, using special logic circuitry to count the "events" (cycles or pulses) repetitively over a preselected time interval (e.g., 0.1 s, 1 s, 10 s, etc.). Since time is used as reference and since it is possible to determine time with close accuracy, electronic counters and EPUT meters have inherently very small errors.

Where poorer accuracy is permissible two other types of indicators can be used. The "integrating" frequency or pulse-count meters display a voltage representing the number of "events" as integrated by suitable electronic circuitry over a fixed time interval. Somewhat better accuracy is obtained with the discriminator type of frequency meter which converts the deviation from a given center frequency into a d-c voltage over a limited range of frequencies. Discriminators are useful in measuring any rapidly fluctuating frequency-modulated transducer outputs, such as those encountered during dynamic tests of FM output transducers.

III.2 Verification of Common Design Characteristics

A number of design characteristics, common to most types of transducers, are verified before the acceptance of a transducer for its intended use. This initial inspection includes a thorough visual examination of the transducer and a number of electrical tests.

III.2.1 Voltage-breakdown test

This test, also called the high-potential, dielectric-strength, or dielectric-withstanding-voltage test, is used to verify that the transducer can safely withstand any momentary overpotentials as well as operate at normal excitation for extended periods of time. The application of a test voltage higher than rated excitation—without ensuing discharge within the transducers—is used to determine the adequacy of electrically insulating materials and of the spacing between certain mutually insulated conducting surfaces. This voltage is normally not applied across the transducer's excitation terminals, however.

The voltage-breakdown test is performed by applying a specified test voltage between mutually insulated electrical transducer connections or between ungrounded connections and the (grounded) case. The duration of the application (typically 60 s) should be specified, as well as the magnitude and other characteristics of the test voltage (e.g., 500 V a-c, rms, 60 Hz). Test failure is evidenced by surface discharge (*flashover*), air discharge (*sparkover*), or puncture discharge (*breakdown*) within the transducer. When limits are also placed on the surge current (measured at initial test voltage application) and on the maximum leakage current (over the entire duration of the test), these currents are measured and recorded.

III.2.2 Insulation resistance test

This test is used to measure the resistance offered by the insulating members of a transducer to an impressed d-c voltage tending to produce a current leakage either through or on the surface of these members. The insulation-resistance test is commonly used to determine characteristics of the insulation between the transduction element together with any integral signal-conditioning circuitry and the external case of the transducer. In some transducers, where excitation-output isolation is required, the test is additionally used to measure the insulation between excitation and output connections.

The importance of this test is given largely by the high impedance circuits frequently used in electronic measuring systems. The operation of the system can be affected severely by "ground loops"—undesirable leakage currents through structural ground planes between components of the system or between excitation-supply and signal-transmission grounds. Excessive leakage currents may also lead to further deterioration of the insulation by heating or electrolysis.

The test is performed by connecting the two voltage-carrying leads from a megohmmeter, megohm-bridge, or insulation-resistance test set to the specified connection points on the transducer (e.g., all insulated receptacle pins, in parallel, and the transducer's case). The test voltage is raised to the specified level (usually 50, 100, or 500 V d-c), and the resistance is read on the test-set meter. The reading is taken when it has stabilized at the value above the minimum specified resistance or after an electrification time of two minutes.

Insulation-resistance measurements are often repeated during and after various environmental tests in order to determine the effects of heat, moisture, dirt, oxidation, and loss of volatile materials on the required insulation characteristics of a transducer.

III.2.3 Excitation tests

These tests are performed to determine specification compliance of the transducer's excitation ratings and to verify the transducer's susceptibility to field hazards related to excitation.

Tests on transducer designs in which excitation is applied directly to the transduction element are usually limited to power-rating verification. Included here are input-impedance or transduction-element resistance measurements, determination of current drain at rated and maximum excitation voltages, measurements of power dissipation at rated and maximum excitation currents, and verification of proper transducer operation after applications of high-transient pulses of excessive excitation voltage or current.

Additional tests are performed on transducers incorporating excitation-regulation, modification, isolation, or conversion circuitry. Depending on the severity of applicable specification requirements, these may include performance determination at minimum and maximum values of excitation and after excitation polarity reversal, misapplication of excitation to output terminals, and output short-circuiting.

III.2.4 Resolution and threshold tests

Resolution tests are normally performed only on potentiometric transducers utilizing wirewound transduction elements. The purpose of this test is to verify the use of a suitable number of turns on the potentiometer winding, as well as the general quality of the winding, but primarily it verifies the absence of grossly uneven spacing between turns, of two or more turns shorting together, and of turns protruding or recessed relative to the nominal outside diameter of the wirewound resistance element.

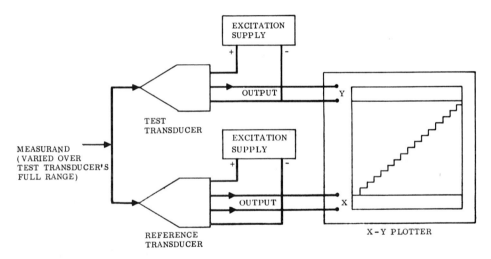

Fig. III-8.   Resolution test.

The resolution test (Figure III-8) is usually performed by applying the measurand, varied from the lower to the upper transducer range limit, simultaneously to the test transducer and to a reference transducer which has a continuous-resolution transduction element (e.g., a strain-gage type transducer). The output of the transducer under test is connected to the "*y*-axis" input of an *x-y* plotter; that of the reference transducer to the "*x*-axis" input. Typical test equipment is illustrated in Figure III-9. The resulting *x-y* plot shows the magnitude and number of all steps in the output of the test transducer, provided that the zero and gain setting of the *x-y* plotter's amplifiers were properly adjusted. Best results are obtained when the zero setting

Fig. III-9. Resolution-test console with X-Y plotter as used for pressure transducers.

is shifted several times during the test so that several full-scale plots are obtained, each showing only a selected portion of the test transducer's output but with steps magnified accordingly. The maximum magnitude of any of the output steps (maximum resolution) as well as the average magnitude of all output steps (average resolution) are then determined from examination of the x-y plot.

Threshold tests are sometimes performed on transducers having continuous-resolution transduction elements. A threshold test is simply the determination of the smallest change in the measurand that will result in a measurable change in transducer output, by use of suitable measurand and output monitoring equipment. This test may have to be performed, when required, at several measurand levels within the transducer's range.

III.2.5. Output tests

Output tests include the tests for two related characteristics, output impedance and loading error, and tests for noise in the transduction element or in the output.

The *output impedance* of capacitive, inductive, resistive, reluctive, and strain-gage transducers, whose output is not modified by active circuitry within the transducer, is simply measured with an impedance or resistance bridge.

A somewhat more elaborate method is necessary for output-impedance tests on transducers, such as d-c output transducers, which contain active circuitry between transduction element and output terminals. The substitution method (Figure III-10) is frequently used for this purpose. In this method the measurand—between 50% and 100% of the transducer's range—is applied and maintained at constant level. The output voltage is measured with a high-impedance voltmeter. A resistance decade box is then connected across the transducer's output terminals, and the resistance dials are adjusted until the output voltage is reduced to 90% of its open-circuit (no-load) value. This resistance (R_{90}) is used to calculate the output impedance which is simply $\frac{1}{9} \times R_{90}$. If saturation effects are suspected within the transducer, which may yield an incorrect output-impedance reading, the resistance can be adjusted, instead, to the value (R_{99}) required to reduce the output voltage to 99% of the open-circuit value. In this case, the output impedance is calculated as $\frac{1}{99} \times R_{99}$.

Loading-error tests are performed to determine the effect on the transducer's output of variations in the load impedance either between specified limits or between "infinity" (open circuit) and a specified value. The loading error, when expressed in percent of full-scale output (% FSO), is usually largest at the upper end point, except

Fig. III-10.   Output impedance determination by substitution method.

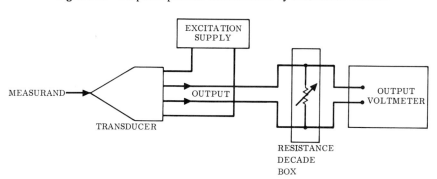

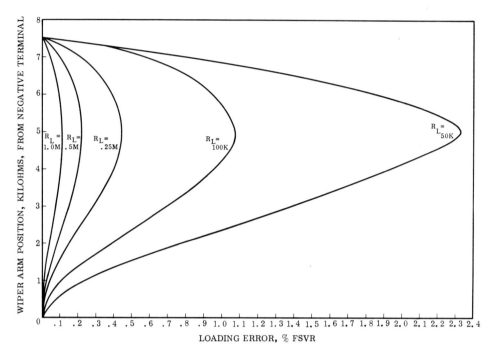

Fig. III-11. Loading error for potentiometric transducer with 7500 ohm element resistance.

for potentiometric transducers where it is maximum at 66% FSO. Figure III-11 illustrates the loading error, with various load resistances (R_L), of a potentiometric transducer having an element resistance (input impedance) of 7500 ohms. During this test the measurand is applied to the transducer and its level adjusted so that the transducer's output is at the value where maximum loading error occurs. While the measurand is held constant, the transducer's output is measured first across the maximum, then across the minimum, specified load impedance.

Since the input impedance of the read-out device froms a portion of the load impedance, it must be considered in the determination of test impedance (usually resistance) values. As illustrated in Figure III-12, the impedance Z_1 necessary to provide the maximum specified load impedance (Z_{max}) in parallel with the input impedance Z_{RO} of the readout device is

$$Z_1 = \frac{Z_{RO} \cdot Z_{max}}{Z_{RO} - Z_{max}}$$

The impedance Z_2 necessary to provide the minimum specified load impedance (Z_{min}) is

$$Z_2 = \frac{Z_{RO} \cdot Z_{min}}{Z_{RO} - Z_{min}}$$

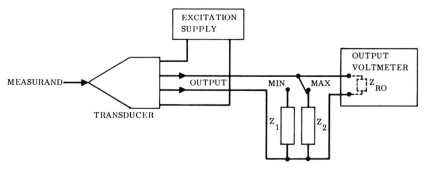

Fig. III-12.   Loading-error test.

From this test the loading error is determined as the output reading E_1 (for Z_{max}) minus the output reading E_2 (for Z_{min}). Expressed in percent of full scale output, the loading error ΔL is then

$$\Delta L = \frac{E_1 - E_2}{E_1} \times 100$$

Output noise tests. Noise in the output of potentiometric transducers frequently occurs during the stroking action of the wiper on the resistance element and is related to variations in instantaneous contact resistance. It is customary to measure such noise in terms of noise resistance using the test set-up illustrated in Figure III-13. Actuation of the "push-to-calibrate" button results in a calibration deflection on the display device for a noise resistance of 100 ohms. The transducer is then cycled over its full range by suitable variation of the measurand. The noise resistance is monitored on the display device and determined by reference to the 100 ohm calibration deflection. The usual purpose of this test is to determine the maximum noise (resistance). This value can be affected by the transducer cycling rate as well as by environmental conditions such as temperature and (unless the transducer is sealed) humidity. Transducers which have been stored for a long time frequently exhibit abnormally high noise during the first few cycles of such a test. The existence of these various effects point

Fig. III-13.   Output-noise test for potentiometric transducer.

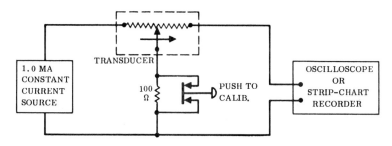

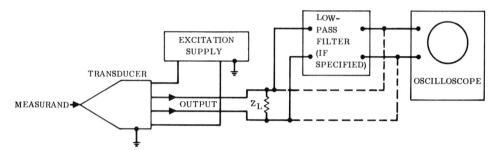

Fig. III-14.   Output-noise test for d-c output transducer.

to a need for careful definition of test conditions before starting a noise test on a potentiometric transducer.

Output noise tests on d-c output transducers and other types containing active circuitry which can modify the output are considerably simpler. Using the test setup shown in Figure III-14, the peak-to-peak output noise is read off a calibrated oscilloscope while the measurand, first at 0%, then at 100% of its range, is applied to the transducer. In some cases the use of a low-pass filter can be specified so that noise at higher frequencies is excluded from the measurement. This is specified only when noise at such higher frequencies is expected in the transducer output and the associated measuring system will in no way be adversely affected by it.

III.3 Calibration

The "calibration" of instruments, as generally defined, can include adjustments and the affixing of scale markings. The calibration of transducers having electrical output, however, is purely a test during which known values of measurand are applied to the transducer and corresponding output readings are recorded. Manufacturing processes preceding the calibration include all adjustments, stabilization, and compensation, as well as tests to verify the effectiveness of a particular process. Such tests are sometimes called "precalibrations." Only those operations normally performed during or just prior to use of a transducer can be included in its calibration. An example of such an operation is the short-time, high-temperature "bakeout" of certain vacuum transducers. Calibration methods, which differ widely depending on the measurand applied to the transducer, are discussed in more detail in each of the following chapters.

Most transducers are subjected to a *static calibration*, which is performed under room conditions by application of the measurand to the transducer in discrete amplitude intervals. Some transducer types, such as potentiometric transducers, whose output can be affected by appreciable friction within an internal linkage or the transduction element itself, may receive an additional *friction-free calibration*, during which low-level shocks or vibrations (*dither*) are applied to the unit undergoing the test so that this friction is minimized. Such a calibration is valid only when the transducer is expected to see a similar vibration or small recurrent shocks throughout its intended use. A few types of transducers, whose behavior can be accurately predicted by measuring their output at only one measurand level, may receive only a *one-point calibration*. Certain types of transducers either cannot be calibrated statically or are used primarily for dynamic measurements. They are subjected to a *dynamic calibration* (see III. 4).

III.4 Dynamic Tests

Dynamic tests are performed on transducers, generally, to determine their response to measurand fluctuations, and specifically, to measure the extent to which the output of a transducer indicates fluctuating measurand amplitudes.

Most dynamic tests take the form of a *dynamic calibration*, during which the measurand is caused to vary with time in a specified manner, and the output is recorded as a function of time. The usual purpose of a dynamic calibration is to verify the dynamic performance characteristics of a transducer (see II.3.4).

A reference transducer, having a known dynamic behavior and a flat-response frequency range greater than that of the test specimen, is connected to the test set up for dynamic calibrations of many transducer types. Hence, most such tests are comparison calibrations rather than primary dynamic calibrations.

Two test methods exist for the determination of dynamic performance characteristics. Although intended for similar purposes, the two methods differ substantially from each other. Their respective validities, where either method could be used, are sometimes controversial. The first method is the "step-function response test"; the second, the "sinusoidal response test."

The step-function response test can be performed on most types of transducers. It consists of recording the transducer output while the measurand applied to the transducer is caused to undergo a step change in its level. Step changes from 45% to 55% or from 10% to 90% of the transducer range are commonly used. The equipment used to create this step change must be capable of minimizing the rise time of the step, as seen by the transducer, so that a true step function and not a ramp function is applied to it. The equipment must also be capable of maintaining the final level of the measurand for a time long enough to yield valid output data.

The sinusoidal response test can be performed on only a limited number of transducer types, primarily because of test equipment design problems. During this test the measurand applied to the transducer is caused to undergo precisely controlled sinusoidal variations of its level. The frequency of the sinusoidal variations is increased, either continuously (*frequency sweep*) or in steps, and the transducer output is recorded continuously or for each step, respectively. The amplitude of the sinusoidal measurand fluctuation is held constant throughout the test. Typical test amplitudes are 20, 10, and 1% (the latter two are preferred) of the transducer range, peak-to-peak, within the lower or middle portion of the range.

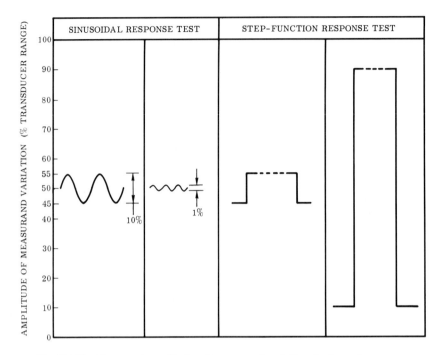

Fig. III-15. Types and amplitudes of measurand variations in dynamic response tests.

The two types of measurand changes are illustrated in Figure III-15. The amplitudes and locations within the range are those recommended by the Instrument Society of America in its Standard S26-1968.

It is generally agreed that data interpretation from sinusoidal response tests is easier and more valid than from step-function response tests when the frequency response (amplitude or phase) of a transducer is to be determined. The step-function response test can be used for determining the transducer's natural frequency, ringing period, damping ratio, time constant, rise time, response time, and overshoot.

III.5 Environmental Tests

Most transducers are used in applications where one or more environmental conditions exceed those defined as room conditions and under which their calibration record is established. The performance of a transducer in more severe environments cannot normally be expected to equal its performance under room conditions. Environmental tests must be used to determine the effects of all specified conditions on the transducer.

The two categories of environmental tests are: *operating tests*, during and after which a transducer must perform within specified tolerances; and *nonoperating tests*, during which no performance tolerances apply but after which the transducer must perform as specified.

Environmental tests are mostly performed on a qualification or sampling basis but are sometimes done as part of each transducer's acceptance test. In the latter case, care must be taken to keep environmental levels and exposure times within reasonable limits so as not to affect the durability of the transducer.

The following ground rules apply, generally, to all types of environmental tests:

1. Establish the test method to be used, in detail, prior to any test.
2. Define the test equipment and the test set up to be used, prior to any test.
3. Prevent the transducer itself and its electrical, mechanical, or pneumatic connections from affecting the environment.
4. Assure that the transducer is exposed only to the environment (or combination of environments) specified for the test.
5. Isolate the calibration equipment, especially the measurand-level sensor or indicator, from the environment.
6. Minimize any effects of the environment on the measurand.
7. Ascertain, throughout the test, the level of the environment actually seen by the transducer.
8. Record all observations completely.
9. Always perform at least a partial calibration at room conditions, shortly after the completion of an environmental test, to determine any latent or permanent effects of the environment on the transducer.

III.5.1. Temperature tests

The primary environmental equipment used for this test is the *temperature chamber*, typically a double-walled well-insulated metal box provided with a door and a number of small holes for electrical, mechanical, and pneumatic connections to the test specimen.

A mounting bracket for the test specimen is conveniently attached to the door of many chamber models (see Figure III-16). High temperatures are obtained from an electrical heating element, and low temperatures are obtained either from a refrigeration element having an external source of coolant such as liquid carbon dioxide or from a compartment filled with dry ice.

Heating as well as cooling elements are usually installed between inner and outer walls, except for the dry ice compartment used in some chambers. A small blower is installed in the chamber to improve convective heat transfer and to avoid thermal stratification. The internal chamber temperature can be preset and is then thermostatically controlled. A temperature sensor, such as a thermocouple, is attached to the transducer's case, and the case temperature is monitored until stabilization at the selected temperature is indicated on an associated temperature readout device.

Fig. III-16. Transducers prepared for temperature tests (courtesy of Pace-Wiancko Division of Whittaker Corp.).

Performance tests, e.g., a calibration, can be performed either after stabilization or after an additional exposure period (*temperature soak*), if specified.

III.5.2 Temperature shock and temperature gradient tests

Even though both of these tests involve a rapid change in the temperature seen by the transducer, the first is usually a nonoperating test and the second is usually an operating test.

The temperature shock test is used to determine whether a change in ambient temperature, between two specified levels and at a specified rapid rate (e.g., 5° per min), causes subsequently detectable damage or intolerable performance changes. This test usually simulates a shipping or other pre-use condition such as transportation in an unpressurized aircraft cargo compartment. Some temperature chambers can provide the desired rate of temperature change. Others necessitate the use of a simpler but more severe test method. The transducer, with all required connections, is first stabilized in a chamber maintained at one temperature, then removed and inserted in a second chamber precooled or preheated to a different temperature.

A performance test (including at least a partial calibration) after subsequent stabilization of the transducer at room conditions demonstrates any permanent effects of temperature shock on the transducer.

A temperature gradient test is performed on some types of transducers to determine the amount of transient deviation (*temperature gradient error*) in output, at a given measurand value, when either the ambient temperature or the measurand temperature changes at a specified (rapid) rate between two specified levels. Temperature gradient errors appear most pronounced in some pressure transducer designs in which temperature compensation is provided by temperature-sensitive resistive components located some distance away from the sensing element.

During this test the transducer output at a fixed measurand level must be recorded while a temperature difference between sensing element and transducer case is created. In a flush-diaphragm pressure transducer, for example, such a temperature gradient can be caused by rapidly immersing just the diaphragm portion into a hot or cold liquid. Other methods involve the sudden application of radiant or convected heat to the transducer's sensing element. This heat can emanate from a photographic flash bulb or similar thermal radiation source, or from a hot-air blower.

A simpler and usually adequate method calls for immersion of the case into a hot or cold liquid while the sensing element sees only the ambient room temperature. This can be accomplished on a pressure transducer by attaching a short straight section of tubing to the pressure port and immersing the transducer into a liquid, while mechanically supported by the tubing, until the liquid level is just below the pressure-port/case interface. The pressure sensed by the transducer is ambient atmospheric pressure which can be assumed to remain at a constant level during the test.

III.5.3 Acceleration tests

Acceleration tests are performed to determine the performance of a transducer during those periods of operation when it is subjected to relatively slowly varying (quasi-steady-state) acceleration. These tests are significant primarily for transducers having sensing or transduction elements in which mechanical motion occurs, or can occur, whenever such transducers are intended for use on moving vehicles or on moving portions of stationary equipment.

A rotary accelerator (*centrifuge*) is most commonly used to simulate an acceleration environment. As explained further in Section 1.5, the acceleration acting on a transducer along a specified axis can be predicted precisely from the knowledge of the radius arm (distance between centerline of transducer and center of turntable), the angular speed (in rpm) of the turntable, and the orientation of the transducer when mounted on the turntable. Unless the transducer is intended for use while mounted in only one orientation and in one location where acceleration levels acting along each axis of the transducer can be firmly established, the maximum specified acceleration is applied, during this test, in each of three orthogonal axes of the transducer and in both directions of each axis. Hence, six acceleration tests are normally performed on each transducer to determine the maximum acceleration effects that could occur during its use. Before starting the tests a sketch of the transducer is made to identify these axes (see example, Figure III-17).

During each application of acceleration the measurand is applied to the transducer, usually at one fixed level. The transducer output is recorded and compared with the output obtained prior to starting the centrifuge. The difference between these two output values is the acceleration error. Acceleration tests are quite short. Between 30 and 60 s per application is a commonly specified duration—just long enough to obtain a valid output reading.

Special fittings and fixtures are often required to apply a given measurand to the transducer while the turntable is in motion. Pressure can be applied through a rotary-seal swivel fitting in the turntable shaft. Displacement transducers can have their shaft clamped in a fixed position. Temperature transducers can be immersed in a specially designed constant-temperature bath mounted on the centrifuge turntable, with care taken to keep the acceleration from affecting the temperature of the bath during each of the short tests. Electrical connections are always made by use of slip-ring contacts in the centrifuge shaft.

Figure III-18 shows a typical test-data form for acceleration and vibration tests. The example illustrates a form used for pressure transducers with error-band specifications. It is also usable when just the errors are specified.

Fig. III-17. Identification of transducer axes for acceleration and vibration tests.

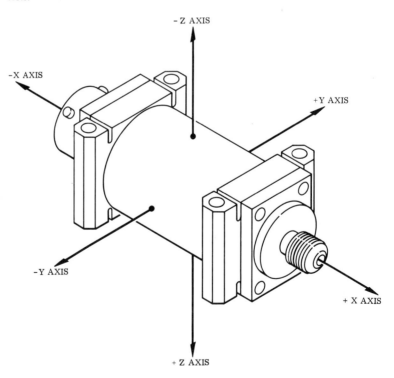

ACCELERATION TEST								
AXIS	+X	−X	+Y	−Y	+Z	−Z	**PRESSURE LEVEL USED:** _____ PSI _____	
OUTPUT BEFORE ACCEL. (mV DC)							MAX. ACCEL. ERROR: + _____ , − _____ %FSO PRE-ACCEL.	
APPLIED ACCEL. (G)							STATIC ERROR BAND:+ _____ , − _____ %FSO ACCEL. ERROR BAND:+ _____ , − _____ %FSO	
OUTPUT DURING ACCEL. (mV DC)								
ACCEL. ERROR (mV DC)							(ALLOWED ACCEL. ERROR BAND± _____ %FSO)	
COMMENTS:							TESTED BY: _____ (TECHNICIAN)	
							_____ (TEST ENGINEER)	
							DATE: _____ APPROVED BY: _____	
							WITNESSED BY: _____ (GDC INSP.)	
							WITNESSED BY: _____ (DCAS Q.A.)	

VIBRATION TEST								
AXIS	X		Y		Z		MAX. VIB. ERROR: + _____ , − _____ %FSO	
PRESSURE LEVEL USED	_____ PSI _____		_____ PSI _____		_____ PSI _____		PRE-VIB. STATIC ERROR BAND:+ _____ , − _____ %FSO	
OUTPUT BEFORE VIB.	_____ mV DC		_____ mV DC		_____ mV DC		VIB. ERROR BAND: + _____ , − _____ %FSO	
	FREQ. (CPS)	ERROR	FREQ. (CPS)	ERROR	FREQ. (CPS)	ERROR	(ALLOWED VIB. ERROR BAND: ± _____ %FSO)	
		POL.	mV DC	POL.	mV DC	POL.	mV DC	TESTED BY: _____ (TECHNICIAN)
VIBRATION ERROR							_____ (TEST ENGINEER)	
							DATE: _____ APPROVED BY: _____	
							WITNESSED BY: _____ (GDC INSP.)	
							WITNESSED BY: _____ (DCAS Q.A.)	
							COMMENTS:	

Fig. III-18. Typical data form for acceleration and vibration tests.

III.5.4 Vibration tests

These tests are performed to determine the effects of vibratory acceleration on the transducer's performance. It is customary to specify the test only for effects of linear vibration. Angular vibration is more difficult to simulate and measure, and it is assumed to exist to a lesser degree in the operating environment of a transducer.

A vibration exciter (*shaker*) is used to apply vibration to a mounting fixture on which the transducer is installed. Three types of shakers are in use for vibration testing. Motor-driven cam-equipped machines, in which the rotary motion of the electric motor shaft is converted into a linear mechanical oscillation, are now largely obsolete.

Hydraulic shakers, in which pressure to one or more hydraulic pistons is varied by means of a fast acting servo valve causing the piston shaft to oscillate, are sometimes used for very large, heavy objects and are rarely needed for transducer testing. Electromagnetic shakers (see Figure III-19) are by far the most commonly used equipment. Their operating principle is related to that of a dynamic loudspeaker such as used in many automobile radios.

The vibration fixture (see Figure III-20), which is mounted to the shaker and on which the transducer is installed so that vibration can be applied to it along any one of three orthogonal axes, requires careful design as well as a "dry run" before the actual test so as to assure the absence of mechanical resonances and of "cross talk" (induced vibration in axes other than the testing axis). Accelerometers are installed on

Fig. III-19.   Transducer vibration test using electromagnetic vibration exciter and associated control and read-out equipment (courtesy of Pace-Wiancko Division of Whittaker Corp.).

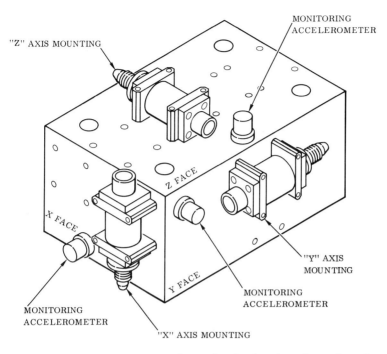

Fig. III-20. Typical vibration test fixture showing location of transducer for each test axis.

the fixture in close proximity to the transducer, for three purposes: to monitor vibration applied to the transducer, to monitor vibration in the two transverse axes (cross axes), and to control the shaker power by means of a control (*drive*) accelerometer which is connected in a feedback circuit to the shaker control console so as to maintain vibration at preselected levels. A complete test setup is illustrated in the block diagram of Figure III-21. As during acceleration tests, the measurand can be introduced from sources external to the test equipment (as illustrated), or it can be generated and maintained locally—on the transducer, on the test fixture, or in the environment immediately ambient to the shaker.

By use of appropriate control, filtering, and power-amplifying equipment, vibration is applied to the transducer in accordance with a predetermined vibration program stated in tabular or graphical form. Such a program usually describes vibration levels in terms of vertical displacement (for low frequencies) or in terms of acceleration (for higher frequencies) vs vibration frequencies, and vibration frequency vs elapsed test time (see Figure III-22).

Two types of vibration can be generated and applied to a transducer either individually or simultaneously (combined): sinusoidal vibration, whose source is usually an audio oscillator, and random vibration, whose signal emanates from a noise source

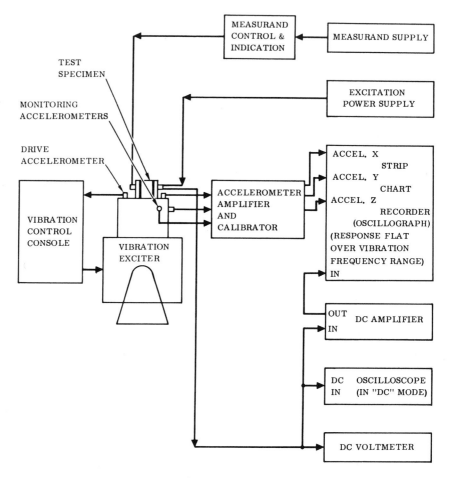

Fig. III-21. Typical transducer vibration test setup.

and passes through a number of narrow band-pass filters (equalization controls) allowing the power spectral density of the applied vibration to be adjusted for each narrow band of vibration frequencies. Sinusoidal vibration is periodic in nature. Random vibration is nonperiodic, described only in statistical terms, and is characterized by an amplitude distribution which essentially follows a Gaussian distribution ("normal error curve"). A typical random vibration program is illustrated in Figure III-23. Test duration time must be specified additionally, unless the random vibration is combined with a sinusoidal vibration program including a frequency-vs-time program.

Vibration effects on transducers, as determined during such tests, can usually be classified into three categories: permanent damage (mechanical failure), output

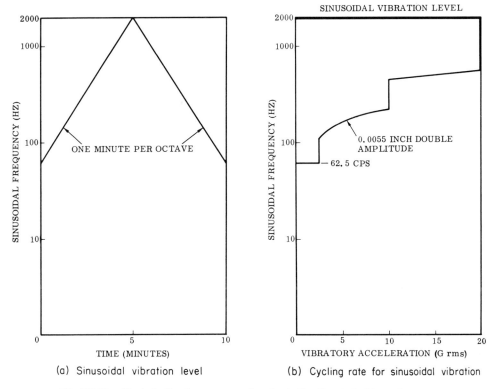

(a) Sinusoidal vibration level   (b) Cycling rate for sinusoidal vibration

Fig. III-22. Typical vibration program for short vibration test of transducers.

Fig. III-23. Typical random vibration test program.

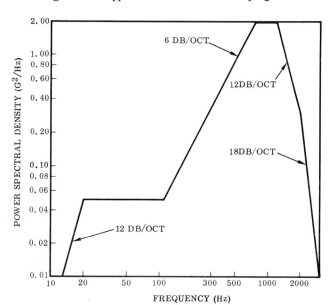

variations corresponding to applied vibration levels, and output variation due to *resonances*—amplified vibrations of internal components, within narrow frequency bands, related to the resonant frequencies of these components and excited by the applied vibration when at those frequencies.

More than most other tests, vibration tests call for the talents of an experienced test engineer, one who is capable of determining the validity and possible origins of an apparent test failure as soon as it occurs. Failures ascribed to a transducer are frequently caused by loose or improper mechanical, pneumatic, or electrical connections to the transducer, by resonances or cross talk within the fixture, by insufficient tightening of fixture or transducer mounting hardware, by inadvertent variations in the applied measurand, by incorrect settings on the control console or on an accelerometer calibrator, by improper grounding of the various interconnected pieces of test equipment, or even by unfiltered transients in the power-line voltage.

III.5.5 Shock tests

Shock tests are normally performed on transducers to determine their performance after (rather than during) exposure to mechanical shocks. Various types of shock-test machines are available for conducting such tests. They are capable of applying shocks having acceleration amplitudes over 10,000 g with time durations from less than 200 μs to over 100 ms. The shape of the shock pulse can be half-sine (upper or lower half of a sine wave), sawtooth (terminal peak sawtooth), or trapezoidal. Shock pulses can be defined either by shape, amplitude (of acceleration), and time duration, or by their frequency spectrum. Shock-test machines can be simple mechanical devices, such as a pendulous hammer swinging against a vertical surface or a weight dropping in free fall from a given height guided by vertical rails or rods. Hydraulic shock test machines are more complex but can often be programmed more accurately for a large variety of shock pulses.

Most transducer test specifications call for a total of six equal shocks to be applied to the test specimen while it is installed and connected as intended during its actual use. By mounting the transducer on different surfaces of a test fixture, shock is applied sequentially in both directions along each of the transducer's three (x-, y-, and z-) axes.

In rare cases, the performance of a transducer is also monitored during the shock applications, usually by means of an oscilloscope or oscillograph connected to its output.

III.5.6 Atmospheric environmental tests

This category includes all those tests intended to determine the effects on a transducer's operation or durability of various chemical or physical properties of the ambient fluid surrounding a transducer prior to and during its actual use. Some such tests may also verify the lack of mutual effects between ambient fluid and transducer. Most of them are nonoperating tests meant primarily for a transducer's external case or other enclosure.

Included among these tests are ambient-pressure ("altitude") tests, sunshine and ozone tests, salt-spray and salt-atmosphere tests, fungus tests, rain, humidity, and water-immersion tests, explosion (explosive-atmosphere) tests, and sand and dust tests.

Transducers having hermetically sealed, corrosion-resistant cases normally receive only two or three of these tests to provide adequate evidence of these two characteristics of their cases.

Detailed test procedures can be found in a number of civilian and military standards and specifications.

III.5.7 Special environmental tests

Certain transducer applications call for a pre-use study of maximum levels and spectra of additional environments not normally tested for and for their simulation during special tests if analyses alone cannot establish the complete absence of any effects of those environments on the reliable operation of the transducer.

This category of tests includes acoustic tests (for effects of high ambient-sound-pressure levels), nuclear radiation tests, and tests for effects of strong electromagnetic or electrostatic fields. A limited number of facilities exist where such tests can be performed adequately, using specialized equipment and highly skilled personnel.

III.5.8 Combined environmental tests

In most transducer applications two or more different types of environments exist simultaneously, each at various levels at different times. It is sometimes not possible to predict adequately the effects of a combination of environments on transducer operation from knowledge of individual environmental effects.

Combined environmental (multienvironment) tests may then have to be performed, such as temperature-vibration or temperature-altitude-humidity tests, or shock tests at cryogenic temperatures.

Since such tests typically require complex test setups their cost effectiveness deserves particular attention. Thorough analysis and expert knowledge of design, piece parts, materials, and processes used in the manufacture of a transducer can frequently obviate their necessity.

III.6 Test Documentation

Since transducer tests are usually fairly extensive and often quite costly, it is important to optimize their value by appropriate documentation.

A complete test procedure should first be written to show sequence of tests and all test methods and type of equipment to be used, illustrated by test setups in schemat-

VENDOR'S MODEL NO.	TEST FACILITY		PART NO.	
VENDOR			SERIAL NO.	
REPORT NO.	**TRANSDUCER TEST REPORT**		RANGE	
TYPE OF TEST	**POTENTIOMETER TYPE_____POSITION TRANSDUCER**			

SUMMARY OF RESULTS:

TEST	TESTED PER PROCED. NO. OR TEST WAIVED PER	PAR. NO.	PASS	FAIL				ERROR BAND	
				ERR. BD.	ELECTR.	MECHAN.	SEE COMMENTS	+ % FS	− % FS
INITIAL P.C. (PROOF CYCLE)									
STARTING FORCE									
RESOLUTION								TOTAL	WITHIN	>DOUBLE
P.C. AFTER SHIP'G SHOCK									
P.C. AFTER HANDL'G SHOCK									
P.C. AFTER SALT ATMOSPHERE									
LOW TEMP._____°F									
P.C. AFTER LOW TEMP.									
HIGH TEMP. +_____°F									
ADD'L TEMP._____°F									
P.C. AFTER HIGH TEMP.									
_____G'S VIBRATION									
P.C. AFTER_____G'S VIBR.									
ACCELERATION									
P.C. AFTER ACCEL.									
ALTITUDE									
P.C. AFTER ALTITUDE									
FREQUENCY RESPONSE								_____CPS	
P.C. AFTER IMMERSION									
P.C. AFTER SAND AND DUST									
LIFE									
FUNGUS	☐ CERTIF. ENCL'D								
FINAL PROOF CYCLE									
WEIGHT								_____OUNCES	

TESTED BY:_____ DATE TEST STARTED:_____ DATE TEST FINISHED:_____

APPROVED BY:_____ TITLE:_____ APPROVED BY:_____ TITLE:_____

INSPECTOR_____ G.S.I._____

CONVAIR/ASTRONAUTICS FORM A1921-5 (5-61) A8 _____TEST

STAMP

Fig. III-24. Transducer test report summary form.

ic diagram form. The procedure can later become a portion of the test report, unless the tests are repetitive (i.e., tests are performed frequently using the same procedure), in which case the procedure is only referenced in the report. When a test report includes the complete procedure, most narrative sections of the report become redundant and should then be omitted. Preprinted blank test data forms facilitate following a test procedure and assure data presentation in the desired sequence and the proper manner.

The test report should begin with a concise statement of purpose and results obtained, so written as to be understandable by readers having little or no technical background in this field. The report should next present a one-sheet summary of all tests performed and their individual results (see example, Figure III-24). The summary sheet should be followed by a brief, objective discussion of any anomalies observed and their actual or possible consequences. Next, individual test data sheets are included, arranged in the order of test sequence. The report ends with illustrations, such as photographs of operating test setups and special diagrams (not part of the procedure), oscillographic strip charts (appropriately identified and explained), and any substantiating information including accuracies and calibration dates of measuring devices used. Subjective interpretations or opinions should be omitted.

BIBLIOGRAPHY

1. *Test Engineering and Management.* Oakhurst, N.J.: The Mattingley Publishing Co. (monthly periodical)

2. "Resistors: Variable, Precision," *AIA-Specification NAS 810*, Washington, D.C.: Aerospace Industries Association, National Standards Association, 1955.

3. Wind, M. (ed.), *Handbook of Electronic Measurements.* Brooklyn, N.Y.: Polytechnic Press, 1956.

4. "Test Code for Industrial Control (600 Volts or Less)," *AIEE: No. 74*, New York: Institute of Electrical and Electronics Engineers, 1958.

5. McMaster, R. C. (ed.), *Nondestructive Testing Handbook.* New York: The Ronald Press Co., 1959.

6. "Environmental Testing, Aeronautical and Associated Equipment, General Specification for," *MIL-E-5272C (ASG)*, Alexandria, Virginia: Defense Supply Agency, Cameron Station, 1959.

7. Von Vick, G., "Transducer Evaluation, A Space Age Tool," *Proceedings of the National Telemetering Conference*, Santa Monica, California, May 1960.

8. "Dynamic Response Testing," *ISA Recommended Practices RP26.1 through RP26.4*, Pittsburgh: Instrument Society of America, 1960, 1961.

9. "Standard Test Methods for Electronic Parts," *EIA: RS-186-B*, New York: Electronic Industries Association, 1961.

10. "Shock and Vibration Test Equipment," *Electro-Technology*, February 1962.

11. "Environmental Test Methods for Aerospace and Ground Equipment," *MIL-STD-810A*, Washington, D.C.: Superintendent of Documents, 1964.

12. Zurstadt, H. J., "U.S. Navy Hi-Shock Tests," *Preprint No. 16.18-2-66*, Pittsburgh: Instrument Society of America, 1966.

13. "Specifications and Tests of Potentiometric Pressure Transducers," *ISA: S37.6*, Pittsburgh: Instrument Society of America, 1967.

1 Acceleration

1.1 Basic Concepts

1.1.1 Basic definitions

Acceleration is the time rate of change of velocity with respect to a reference system. It is a vector quantity.

Oscillation is the variation, usually with time, of the magnitude of a quantity with respect to a reference system when this variation is characterized by a number of reversals of direction.

Vibration (mechanical vibration) is an oscillation wherein the quantity is mechanical in nature (e.g., force, stress, displacement, velocity, acceleration). *Note:* In measurement terminology, the term "vibration" usually denotes *vibratory acceleration* and sometimes *vibratory velocity*.

A seismic system consists of a mass suspended from a (reference) base by a spring.

Damping in a seismic system is the energy-dissipating characteristic which tends to bring the system to rest when the stimulus is removed.

93

Harmonic motion is a vibration whose instantaneous amplitude varies sinusoidally with time.

Shock (mechanical shock) is a sudden nonperiodic or transient excitation of a mechanical system.

Jerk is the time rate of change of acceleration with respect to a reference system.

Mechanical impedance is the complex ratio of force to velocity during simple harmonic motion. It is a quantitative measure of the ability of a structure to resist a vibratory force.

The number of **degrees of freedom** of a mechanical system is the minimum number of independent coordinates required to define completely the position of all parts of the system at any instant of time.

Random vibration is nonperiodic vibration whose magnitude at any given time can be described only in statistical terms. *Note:* It is usually taken to mean Gaussian random vibration, whose instantaneous amplitude distribution follows a Gaussian ("normal error curve") distribution.

Periodic vibration is vibration having a waveform which repeats itself in all its particulars at certain equal time increments.

A **period** is the smallest increment of time for which the waveform of a periodic vibration repeats itself in all its particulars.

A **cycle** is the complete sequence of magnitudes of a periodic vibration that occur during one period.

The **frequency** of a periodic vibration is the reciprocal of its period.

The **phase** of a periodic vibration is the fractional part of a period through which the vibration has advanced, as measured from an arbitrary reference.

The **frequency spectrum** of a vibratory quantity is a description of its instantaneous content of components, each of different frequency and usually of different amplitude and phase.

Power density of random vibration is the mean square magnitude per unit bandwidth of the output of an ideal filter having unity gain, responding to this vibration.

A **power density spectrum** is a graphical representation of values of power density displayed as a function of frequency so as to represent the distribution of vibration energy with frequency.

1.1.2 Related laws

Displacement, velocity, acceleration

Linear (Translational)   Angular (Rotational)

$$v = \frac{dx}{dt} \qquad\qquad \omega = \frac{d\theta}{dt}$$

$$a = \frac{d^2x}{dt^2} \qquad\qquad \alpha = \frac{d^2\theta}{dt^2}$$

where: x = linear displacement
θ = angular displacement
v = linear velocity
ω = angular velocity
a = linear acceleration
α = angular acceleration
t = time

Angular frequency

$$\omega = 2\pi f$$

where: ω = angular frequency (in rad/s)
f = frequency

Newton's second law

$$F = ma$$

where: F = force (resultant of all forces acting on mass m)
m = mass
a = acceleration (of the mass)

Phase angle

$$\phi = \omega\,\Delta t = 2\pi f\,\Delta t$$

where: ϕ = phase angle (between two quantities of the same frequency varying
sinusoidally with time)
f = frequency
t = time difference (between positive maxima for the two quantities)

Simple harmonic motion

$$x = A \sin \omega t$$
$$v = \omega A \cos \omega t$$
$$a = -\omega^2 A \sin \omega t$$

where: x = displacement (instantaneous, at time t)
v = velocity (instantaneous, at time t)
a = acceleration (instantaneous, at time t)
A = maximum displacement from zero position ("zero-to-peak")
$\omega = 2\pi f$
f = frequency of oscillation

Notes: 1. "Peak-to-peak" displacement = $2A$
2. Direction of acceleration is always opposite to direction of displacement.

Mean and RMS magnitudes

$$\bar{A} = \frac{1}{t} \int_0^t A(t)\,dt \qquad A_{\mathrm{rms}} = \sqrt{\frac{1}{t} \int_0^t [A(t)]^2\,dt}$$

where: $\bar{A}$ = mean magnitude (of a vibratory quantity)
  $A(t)$ = vibratory quantity
    t = time (over which the averaging is done)
  A_{rms} = rms magnitude

Power density

$$W(f) = \frac{A(f)_{\mathrm{rms}}^2}{\Delta f}$$

where: $W(f)$ = power density
  $A(f)_{\mathrm{rms}}^2$ = mean square magnitude
   Δf = bandwidth (usually chosen as 1 Hz)

1.1.3 Units of measurement

The unit of linear (rectilinear, translational) acceleration is **g**. The quantity g is the acceleration produced by the force of earth gravity. Gravity varies with the latitude and elevation of the point of observation. The value of g has been standardized by international agreement as follows

1 (standard) $g = 980.665 \text{ cm/s}^2 = 386.087 \text{ in./s}^2 = 32.1739 \text{ ft/s}^2$

The above can be simplified, where allowable, to

1 $g = 981 \text{ cm/s}^2 = 386 \text{ in./s}^2 = 32.2 \text{ ft/s}^2$

Shock and vibration amplitudes are shown in g units.

Angular (rotational) acceleration is expressed in **radians per second squared** (*rad/s²*). One degree is equal to 0.01745 radians.

The unit for frequency is the **hertz** (*Hz*). The formerly used unit, cycles-per-second (cps) is still in common use; however, its use is discouraged by governmental authorities. The conversion is on a one-to-one basis: 1 Hz = 1 cps.

Power density is usually expressed in g^2/**Hz**.

1.2 Sensing Elements

The sensing element common to all acceleration transducers is the *seismic mass* (proof mass), which is restrained by a spring, and whose motion is usually damped in a spring-mass system (Figure 1-1). When an acceleration is applied to the transducer case, the mass moves relative to the case. When the acceleration stops, the spring

returns the mass to its original position (Figure 1-2). The small
black/white circular symbol is commonly used to denote the
location of the gravitational center of seismic mass. If acceleration were applied to the transducer case in the opposite
direction, the spring would be compressed.

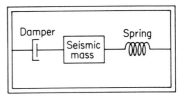

Fig. 1-1. Basic spring-mass system
of an acceleration transducer.

Under steady-state acceleration conditions the displacement y (in cm) of the mass is given by the acceleration a
(in cm/s²) multiplied by the ratio of the mass M (in gm) to the
spring constant k (in dynes/cm), or $y = aM/k$. Under dynamic
(varying) acceleration conditions the damping constant be-
comes a factor in a modified version of this relationship. The response characteristics
of spring-mass systems are adequately described in available technical literature, e.g.,
in textbooks dealing with mechanical vibrations.

The seismic mass in a linear accelerometer is usually of circular (solid or annular)
or rectangular cross section. It can be linked to the case by flexures made to slide
along a bar, or it can be otherwise restrained from motion in any but the sensing axis.

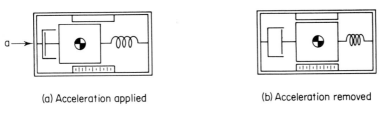

(a) Acceleration applied   (b) Acceleration removed

Fig. 1-2. Displacement of seismic mass.

The seismic mass of an angular accelerometer, which can be a disc pivoted at its
center and restrained by a spiral spring, responds to angular acceleration with an
angular displacement.

1.3 Design and Operation

1.3.1 Capacitive acceleration transducers

A change of capacitance in response to acceleration is obtained in transducers
using a fixed stator plate and a diaphragm of matching configuration to which a
circular (e.g., disc-shaped) seismic mass is attached. The diaphragm acts as restraining spring as well as the moving electrode (rotor) of the capacitor. Acceleration
acting on the mass causes the diaphragm to deflect and its capacitance to the stator
to change proportionally. If the capacitor is connected into an a-c bridge, an a-c

voltage of varying amplitude can be produced as transducer output. A frequency-modulated output can be generated by using the capacitor in an oscillator tank circuit. Transducers of this type have not been produced commercially to any significant extent. However, the capacitive principle has been applied in certain servo-accelerometer designs (see 1.3.7).

1.3.2 Piezoelectric acceleration transducers

The *piezoelectric effect* is utilized in a wide variety of transducer designs intended primarily for the measurement of vibratory acceleration and shock. This effect was discovered in 1880 by Pierre and Jaques Curie when they observed that electrical charges were produced by a quartz crystal when weights were placed on it. Subsequent development by various researchers showed that about forty types of crystalline materials generate an electrical charge when pressed or squeezed (the Greek word *piezein* means "to squeeze") so that they undergo minute changes in length, width, or thickness. The inverse effect, dimensional change when an electrical charge is applied to the crystal, was also discovered by the brothers Curie.

1.3.2.1 Crystal materials are of two basic types, natural crystals and synthetic crystals. The latter are almost invariably ceramic mixtures.

Natural piezoelectric crystals, with the exception of quartz and tourmaline, are grown artificially from aqueous solutions of the material under closely controlled conditions. To form the transduction element a slab is sliced from the crystal along a carefully chosen crystallographic axis, and plates of the desired shape are cut from the slab. Quartz (silicon dioxide) and tourmaline (a complex silicate of aluminum and boron) crystals are usually selected from those found in their natural state. Quartz crystals have also been cultured in autoclaves under controlled temperatures and pressures.

Quartz plates, when stressed in thickness-shear, thickness-expansion, or length-expansion modes, are characterized by a high and stable natural frequency and can be used over a wide range of temperatures. However, their sensitivity is quite low. Tourmaline provides a higher sensitivity than quartz but is partially hygroscopic and has a narrower operating-temperature range. Rochelle salt (sodium potassium tartrate) has a higher sensitivity than tourmaline when stressed in the face-shear or length-expansion modes but suffers from the same disadvantages. Both materials can be considered obsolete as transducer elements. ADP (ammonium dihydrogen phosphate), KDP (potassium dihydrogen phosphate), and LH (lithium sulphate) crystals are used more frequently in sonar-systems components (see 11.6.1) than in measuring transducers. DKT (dipotassium tartrate) plates, stressed in face- or thickness-shear or length-expansion modes, are found in special instrument applications. Quartz is the only natural crystal material to have found widespread usage in acceleration transducers.

Ceramic "crystals" are used much more frequently than natural crystals. The first ceramic material employed in commercial transducers was barium titanate ($BaTiO_3$). The addition of controlled impurities, such as calcium titanate, was found to improve some of its characteristics. Continuing development of piezoelectric ceramics resulted in further performance improvements when materials such as lead zirconate, lead niobate, lead zirconate/barium zirconate, and, later, lead titanate/lead zirconate mixtures, and lead metaniobate were studied and applied to transducers, frequently mixed with controlled small amounts of one or more additives. The composition of virtually all ceramics used in modern acceleration transducers is considered proprietary by their manufacturers. They are often designated by such trade names as "Piezite" (Endevco Corporation), "Glennite" (Gulton Industries), and "PZT" (Clevite Corporation).

The manufacture of a ceramic element begins with mixing of known amounts of the various compounds in a ball mill. A binder is added to the mixture which is then molded into the desired shape in a hydraulic press. The soft *bisque* is taken from the mold and sintered by firing in a kiln at a very high temperature. During subsequent cooling, the element—now a hard ceramic—is polarized by exposing it to an orienting electric field. The polarizing operation causes the ceramic to exhibit the required piezoelectric qualities. Temperature cycling, aging, and cleaning complete the process. Electrodes can be metal films made integral with the surface by high-temperature firing.

An important characteristic of piezoelectric ceramic materials is their *Curie point*, the temperature at which polarization of an element is lost when it is heated. Although most elements can be repolarized after such an occurrence, the Curie point determines the upper limit of the operating temperature range of the transducer, always specified well below this point to provide a margin of safety. Curie points vary from 120°C (barium titanate) to 570°C (lead metaniobate).

1.3.2.2 Designs are typified by those shown schematically in Figure 1-3. The acceleration sensing axis is perpendicular to the base. Crystals are so polarized as to minimize any output due to acceleration along any other axes. In all design versions at least a portion of the spring action of the spring-mass system is provided by the crystal (K) itself. The case of the basic and isolated compression designs also contributes to the elastic component of the system. A curved spring in the isolated-mass compression design adds a third elastic member. The operation of both transducer types is affected by any forces acting on the case, which tend to alter its elastic characteristics.

Case sensitivity is minimized in the following three designs; however, strains in the base of the five types can cause undesirable effects on transducer performance to various extents. In the compression design the mass is preloaded against the crystal so that an output of alternating polarity is produced when compression is alternately intensified and relieved due to vibratory acceleration. In the bender design the

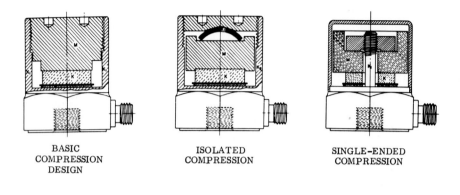

BASIC
COMPRESSION
DESIGN

ISOLATED
COMPRESSION

SINGLE-ENDED
COMPRESSION

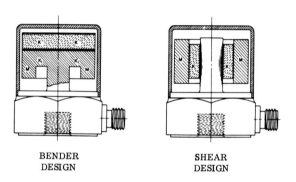

BENDER
DESIGN

SHEAR
DESIGN

Fig. 1-3. Typical piezoelectric acceleration transducer designs (courtesy of Endevco Corp.).

crystal is bonded to a mushroom-shaped mass which, in this arrangement, also acts as elastic member. The rim of the assembly deflects upward and downward with respect to its supported center. The shear design uses an annular crystal bonded to a center post on its inside surface and to an annular mass on its outside surface. Upward and downward deflection of the mass causes shear stresses across the thickness of the crystal.

Details of one type of compression design are shown in Figure 1-4. In order to increase the inherently low output of quartz crystals utilized in this transducer, seven crystals are stacked and connected for output multiplication as indicated in the schematic diagram. The miniature coaxial connector is typical for all piezoelectric acceleration transducers.

The base and case of many piezoelectric accelerometers are electrically connected to one crystal electrode or set of electrodes. The case is hermetically sealed to prevent any entry of moisture. The impedance across the crystal electrodes is very high—in

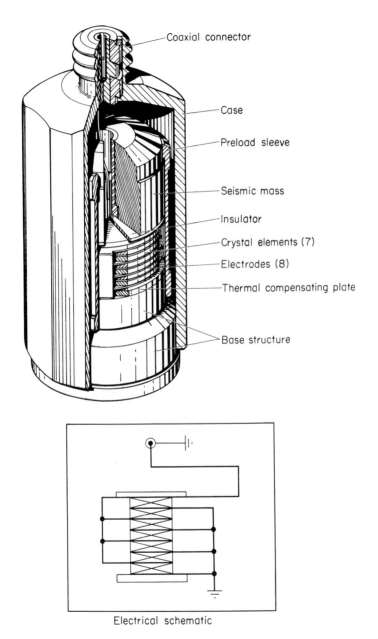

Coaxial connector

Case

Preload sleeve

Seismic mass

Insulator

Crystal elements (7)

Electrodes (8)

Thermal compensating plate

Base structure

Electrical schematic

Fig. 1-4. Quartz-crystal piezoelectric acceleration transducer (courtesy of Kistler Instrument Corp.).

the kilomegohm range—and any electrical leakage path can produce substantial reductions in transducer output. Mounting provisions usually take the form of an internal thread in the base. A mounting stud, whose protruding thread can be different from that threaded into the base, can be attached when an external thread is required. When the case must be electrically insulated from structure ground in its installation, a special insulated mounting stud can be employed. Some transducers are provided with a mounting hole through the case, along its vertical centerline. Cases too small for mounting by means of threads can be cemented to the measured structure.

1.3.2.3 Signal-conditioning circuitry is almost always required for piezoelectric accelerometers. Their output (charge or voltage) is smaller in amplitude than that which most telemetry or readout systems require. More importantly, their output impedance is much higher than usable in virtually all electronic measuring systems. To transform this impedance into one of much lower value, a cathode follower, emitter-follower, or amplifier (Figure 1-5) is connected to the transducer with as short a cable as the installation will allow.

The cable itself requires special attention since it is on the high-impedance side

Fig. 1-5. Piezoelectric acceleration transducer system with transistorized amplifier (courtesy of Gulton Industries).

of the amplifier. It should be as free as possible from *triboelectric noise* (noise induced by friction between conducting and insulating portions of the cable); it should be thin, coaxial, of low capacitance, shielded, sufficiently flexible to avoid mass loading the transducer case, and impervious to moisture. Moisture-proofing compound can also be applied over the mated connectors after installation. Several types of miniature low-noise coaxial cables have been developed especially for use with piezoelectric transducers. Triboelectric noise has been minimized in such cables by a graphite lubricant between conductor and insulation.

Two types of amplifiers have been used with piezoelectric accelerometers. The *voltage amplifier* amplifies the signal from its internal emitter follower or source follower which can provide the required high input impedance (e.g., 500 megohms) to the transducer by utilizing several cascaded stages. Changing the length of the coaxial cable from the transducer causes a corresponding change in amplifier input due to a change of the capacitive-reactance component of the input impedance. This problem is minimized in the *charge amplifier*, an operational amplifier with capacitive feedback which acts as compensation for changes in capacitance (e.g., due to variations in cable length) seen by the input terminals of the amplifier.

Several transducer designs incorporate signal-conditioning circuitry within their case. Piezoelectric accelerometers were designed with integral emitter or source followers and are only slightly larger than their corresponding "transducer-only" versions. However, at least one additional wire is needed for the power supply, and the resulting new cable and connector designs have sometimes been troublesome. The availability of microelectronic "integrated circuits" has enabled some transducer manufacturers to place a complete amplifier into the transducer case with only a slight penalty in case length. A circuit was later developed to allow the use of a simple coaxial cable for excitation as well as for output. The excitation to either an integral source follower or amplifier is supplied through a load resistor in the power supply. The output signal can then be coupled off the transducer-cable end of this resistor. The cable shield acts as common ground for output signal and excitation. An accelerometer utilizing this circuit concept (Figure 1-6) is the Piezotron®. Cable resistance and capacitance and load capacitance can affect the operation of such a transducer system. Transducers with integral circuitry are limited in their applications by the relatively narrow range of temperatures over which semiconductors can operate.

1.3.2.4 Sources of error in the output of piezoelectric accelerometer systems differ sufficiently from those of most other types of acceleration transducers to warrant a brief explanation. Errors can be produced within the transducer by mechanical deformation of the case, e.g., during installation (*strain effects*), by stresses in the case of some designs due to uneven thermal expansion (*thermal strain effects*), and by spurious electrical signals generated across the crystal electrodes due to charge separation in the crystal in response to a step change in temperature (*pyroelectric effect*). These effects can be minimized by proper transducer design. Most of the pyroelectric effects are

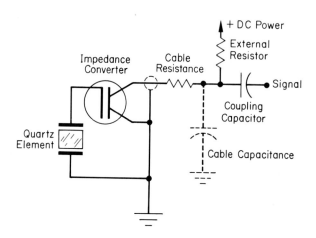

Fig. 1-6. Piezotron® circuit schematic (courtesy of Kistler Instrument Corp.).

filtered out by the low-frequency characteristics of the associated charge and voltage amplifiers. The normally encountered temperature errors—variation of transducer sensitivity from room-temperature sensitivity due to slowly changing ambient temperature—can be brought within usually acceptable tolerances by the proper choice and thermal aging of the crystal material. Sensitivity to accelerations in axes other than the sensing axis (*transverse sensitivity*) is a function of the crystal materials used, element polarization, internal mechanical design and machining, and processing of parts. The transverse sensitivity due to imperfect alignment of crystal domains in polycrystalline elements is usually quite small.

Error sources in the other portions of the transducer system include stray currents due to ground loops—caused either by poor amplifier design, lack of signal isolation, or incorrect grounding—leakage paths created by incomplete moisture proofing, cable-capacitance changes due to temperature variations, and various changes of electronic-component characteristics within the amplifier. The most frequent cause of transducer-system failure, however, is breakage of the coaxial-cable conductor at either connector due to improper handling.

1.3.2.5 Special versions of piezoelectric acceleration transducers include the triaxial accelerometer and the transducers with integral calibration provisions. A triaxial piezoelectric accelerometer (Figure 1-7) is essentially an assembly of three accelerometers in a single block with mounting provisions. The individual units are precisely aligned with respect to each other and with respect to the three axes of the block along which they are mounted. An amplifier assembly consisting of three amplifiers in one package can be used with this special transducer.

Integral calibration provisions can be of two types. One is a simple additional connector on either the transducer or the input side of the amplifier which permits the injection of a signal from an audio oscillator through a series resistor. The other employs a separate crystal mechanically linked to the transduction-element crystal. The inverse piezoelectric effect is utilized in the calibration crystal. When an a-c voltage (at audio frequencies) is applied to it through a separate connector, the separate crystal is set into mechanical oscillation and a corresponding output is obtained from the transduction crystal of the accelerometer. A transducer system with integral calibration provisions can be connected to a repeat-cycle timer switch and an audio oscillator so that the normal transducer output is periodically replaced, for a short period of time, with a calibration signal of known amplitude.

Fig. 1-7. Triaxial piezoelectric accelerometer (courtesy of Gulton Industries).

1.3.3 Photoelectric acceleration transducers

Transducers for the measurement of relatively slowly varying acceleration, using photoelectric transduction, have been produced at various times but cannot be considered as generally available models. One design uses the motion of the seismic mass to vary the position of a shutter between a light source and one or two photoconductive light sensors so that the output of the sensor, or of two sensors constituting two arms of a bridge network, is proportional to acceleration. Approaches toward development of digital-output photoelectric accelerometers include detection of the rotation modulation, due to acceleration, of an unbalanced cylinder which rotates at a constant average speed but whose instantaneous speed varies when acceleration acts on the center of seismic mass to increase or retard it over one revolution.

1.3.4 Potentiometric acceleration transducers

Acceleration transducers using potentiometric transduction are often used where a device of reasonable cost is required for moderately accurate measurements of slowly varying acceleration. In these transducers a seismic mass is mechanically linked to a wiper arm which moves over the active portion of a resistance element with full-span deflection of the mass.

The undamped natural frequency of the spring-mass system in a potentiometric accelerometer increases with the measuring span of the transducer from about 20 Hz for a span of 4 g (± 2 g range) to slightly over 60 Hz for a ± 20 g range (40 g span). Damping is almost invariably introduced to minimize vibration-induced noise in the transducer output due to wiper whipping and large changes in instantaneous contact resistance between wiper and resistance element. Magnetic, viscous, or gas damping used in these transducers reduces the upper limit of flat frequency response somewhat from that given by the undamped natural frequency which is normally taken as 0.3 times that frequency for a flatness of response within $\pm 5\%$. The permanent magnet used for magnetic (eddy-current) damping adds to transducer weight, but temperature effects on damping are small. Gas damping is also affected by temperature to only a small degree, but potential compressibility problems require a more complex design and adherence to very close machining tolerances. Viscous damping, such as attained by silicone oil having a kinematic viscosity of 300 to 500 centistokes at room temperature, is most frequently used, although temperature effects on damping due to viscosity changes are more pronounced than in magnetic or gas damping. The advantages of viscous damping are design simplicity and a relatively low transducer weight. Volumetric temperature compensation, necessitated by thermally caused volume changes in the damping liquid, can be attained by use of bimetallic members in the damping system, by placing a gas-filled sealed bellows in the liquid, or by introducing an air bubble into the liquid. The air bubble, however, should only be used when the internal design keeps it from moving into contact with any potentially vibrating component in any mounting orientation of the transducer.

Although some sort of amplification linkage is frequently used between seismic mass and wiper arm, the total stroke of the wiper contact over the potentiometric element is quite small, and the total number of wire turns is usually between 200 and 300. Platinum alloy is a typical material for wire as well as wiper contact. The enamelled wire is wound over the mandrel under constant tension. The enamel is later cleaned off the area contacted by the wiper.

Fig. 1-8. Potentiometric acceleration transducer with flexure-supported mass (courtesy of Bourns, Inc.).

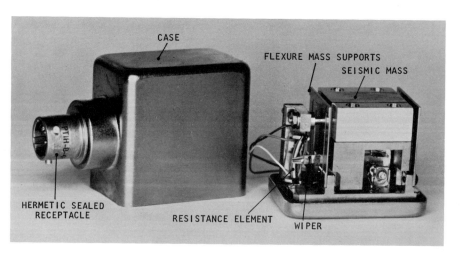

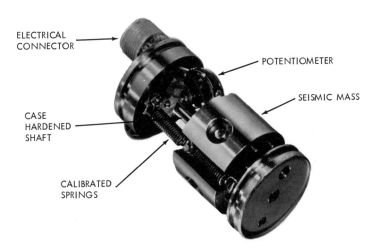

Fig. 1-9. Potentiometric accelerometer design with mass sliding on coaxial shaft (courtesy of Conrac Corp.).

An important design characteristic is the relative freedom from effects of transverse accelerations. In an ideal accelerometer the mass, together with all other components of the total spring-mass system, responds only to acceleration along the sensing axis. Transverse sensitivity can be minimized by the use of flexural springs as in the transducer shown in Figure 1-8. A different approach is used in the axial design illustrated in Figure 1-9, where the seismic mass can move only along a coaxial shaft. The springs, in conjunction with the mass, determine the measuring range. The spring-mass system of this transducer is fashioned closely after the classical concept of such a system. Motion of the mass is translated into wiper rotation by a mechanical linkage.

Both designs are of the commonly used bidirectional-range configuration; i.e., the mass can move in both directions along the sensing axis, and the wiper is at the center of the potentiometric element when no acceleration is applied. The output of a potentiometric accelerometer having a symmetric bidirectional range is 50% VR at zero g acceleration. Among other types of design the cantilevered mass, using the cantilever support as the spring, is quite popular. Most potentiometric accelerometers incorporate overload stops to limit wiper travel when accelerations exceed the transducer range.

1.3.5 Reluctive acceleration transducers

Reluctive transduction elements, of the inductance-bridge type as well as of the differential-transformer type, are used in a number of commercial accelerometers. The frequency response of these transducers, especially the inductance-bridge version with its small armature deflection, is substantially higher than that of the potenti-

Fig. 1-10. Reluctive acceleration transducer (case removed) (courtesy of Wiancko Engineering Co.).

ometric designs. The types of internal active components required in their construction—mass, spring, coils, magnetic armature—can be made from materials usable at fairly high temperatures. Excitation must be a-c and of a frequency substantially higher than the upper limit of the expected frequency response. The transducer's a-c voltage output, whose amplitude and phase are proportional to acceleration, can be demodulated into a varying d-c voltage. Use of an oscillator as excitation supply permits operation from a d-c power source.

A typical inductance-bridge type of reluctive accelerometer is shown in Figure 1-10. The two coils are mounted over the outer legs of a laminated E-core whose center leg is visible in the illustration. A spring structure with triangular end supports restrains the see saw motion of the rectangular ferromagnetic armature plate to one end of which the small cylindrical seismic mass is attached with a machine screw. The sensing axis is perpendicular to the base. When acceleration acts on the mass, the armature deflects so that the inductance of one coil increases while that of the other coil decreases. The two coils are connected as a half-bridge (the other half can be a matched pair of resistors), making the two inductance changes additive in creating a bridge unbalance. When an a-c voltage is connected across the bridge, an a-c output is produced by unbalance of the bridge. Equal downward and upward motions of the mass produce output changes of equal amplitude but of opposite phase.

A phase-sensitive demodulator can convert this output into a d-c voltage whose polarity indicates the direction of acceleration. Figure 1-11 illustrates a reluctive accelerometer which operates from a d-c excitation supply and provides d-c output. Excitation-conversion and output-conditioning circuitry is integrally packaged. The potted electronics portion is mounted to the receptacle end cap of the transducer which also holds external trimming potentiometers for "zero" and "gain" adjustments. The base of the sensing/transduction unit, the actual reluctive accelerometer in the package, becomes the opposite end cap of the case.

A different operating mode is utilized in another inductance bridge design. Its mass is a nonmagnetic diaphragm sandwiched between two air-core coils which form a half-bridge excited at radio frequencies (typically 1 mHz). Eddy currents induced in the diaphragm act to change coil inductances when motion of the diaphragm due to acceleration changes its proximity to each of the two coils.

Differential-transformer types of reluctive accelerometers, whose frequency response is inherently lower than that of most inductance-bridge designs because of the larger armature travel required, include a design using an inertial disc with an unbalanced mass. The mass subtends an arc of about 30 degrees at one portion of the circumference of the disc which is supported by cross flexures near its center of rotation. A very low natural frequency is provided by such a design.

Synchro and microsyn transduction are used in a number of linear- and angular-acceleration transducers for which a relatively low natural frequency is typical. Synchro-type accelerometers can be connected to synchro receivers for direct readout. Typical applications of such transmitter-receiver systems are found in aircraft.

Fig. 1-11.   D-C-output, reluctive acceleration transducer (courtesy of Wiancko Engineering Co.).

Reluctive transducers generally use viscous damping. Springs of special design have been employed to minimize transverse sensitivity.

1.3.6 Strain-gage acceleration transducers

Transducers converting acceleration into a change of resistance due to strain in two or, more commonly, four arms of a Wheatstone bridge exist in a variety of designs. Unbonded-metal-wire, bonded-metal-wire and metal-foil, and bonded-semiconductor gages have been utilized in this transducer type. The gages are mounted to the spring in the spring-mass system, to a separate stress member additional to the spring, or between the seismic mass and a stationary frame.

The last mentioned method is employed in the unbonded-strain-gage accelerometer shown in Figure 1-12. Four strain-gage windings are wound between insulating studs on the mass and on the frame. The wires themselves supply spring force in the spring-mass system. The rectangular mass is mechanically restrained from motion in transverse axes. The illustration shows only the upper pair of windings. The schematic diagram indicates how the lower pair is attached. Similar operation is obtained in the accelerometer shown in Figure 1-13 using a cylindrical mass. This design, however,

Fig. 1-12. Unbonded-strain-gage acceleration transducer with rectangular seismic mass (courtesy of Statham Instruments, Inc.).

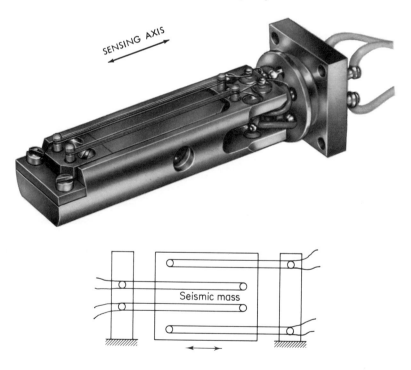

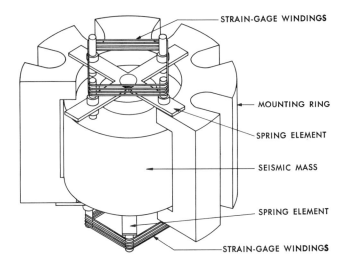

STRAIN-GAGE WINDINGS

MOUNTING RING

SPRING ELEMENT

SEISMIC MASS

SPRING ELEMENT

STRAIN-GAGE WINDINGS

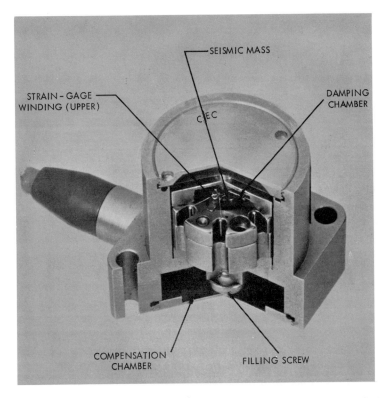

SEISMIC MASS

STRAIN-GAGE WINDING (UPPER)

DAMPING CHAMBER

COMPENSATION CHAMBER

FILLING SCREW

Fig. 1-13. Unbonded-strain-gage acceleration transducer with cylindrical seismic mass (courtesy of Consolidated Electrodynamics Corp.).

has two strain windings wound between four insulated posts attached to each of two cross-shaped springs, one above and one below the mass. The windings contribute to the spring action. The acceleration sensing axis is perpendicular to the base. In both designs the motion of the mass in one direction causes the tension in one pair of windings to be reduced while the tension in the oppositely mounted winding pair is increased. The four windings are connected as a four-arm (four-active-element) Wheatstone bridge whose output, when a-c or d-c excitation is applied across it, is proportional to acceleration and whose polarity indicates direction of acceleration. Viscous damping, using silicone oil, is usually employed to provide a damping ratio of about 0.7 at room temperature. Compensation resistors are mounted within the case.

Gas-damped unbonded-strain-gage accelerometers followed the viscous-damped types in development. Both versions are usable in a variety of applications. The advantage of gas-damped designs lies in their much smaller variation of damping ratio with temperature. A typical gas-damped accelerometer has a flat or corrugated diaphragm linked to the seismic mass. The rim of the diaphragm is sealed to the inside of the case so that it encloses a coupling chamber of small volume. This chamber is vented through a small orifice (e.g., a porous plug) to either ambient air or air (or a different gas) contained within the accelerometer. Motion of the mass produces pressure through the diaphragm within the coupling chamber. The orifice controls flow of compressed air from the chamber. The diaphragm contributes part or all of the spring action in the system. Since gas is compressible the spring-mass system has two degrees of freedom, whereas viscous-damped systems have only one degree of freedom. Damping is controlled primarily by the flow resistance of the orifice and secondarily by the stiffness of the coupling chamber.

Bonded-strain-gage accelerometers are typified by a design whose seismic mass is cantilevered by two leaf springs with gages bonded to both surfaces of both springs to enable their connection as a four-arm Wheatstone bridge. Their full-scale (unidirectional) output is generally less than that of unbonded-strain-gage designs (30 mV as compared to 40 mV at 10 V excitation). A considerably larger output (300 to 600 mV at 10 V excitation) can be obtained by use of semiconductor strain gages bonded to a cantilever beam or a compression member with mass preloading (similar to the isolated-compression piezoelectric accelerometer). Two-active-element bridge configurations are used in some semiconductor ("piezoresistive") designs. A purely piezoresistive accelerometer, using a piezoresistive two-terminal element in the same manner as piezoelectric elements are applied, is feasible but would be less sensitive than semiconductor-strain-gage types. A specialized development of a piezotransistor, on whose junction a seismic mass acts, has shown some promising characteristics.

Angular accelerometers using strain-gage transduction have been built in a few designs, using either a radially symmetric rigid rotor pivoted at its center of gravity and elastically restrained or a restrained centrally pivoted paddle which responds to the inertial forces of a liquid (filling the round case) when it rotates due to angular acceleration.

1.3.7 Servo acceleration transducers

Accelerometers using closed-loop servo systems of the force-balance, torque-balance, or null-balance type for their operation are widely used where requirements for close accuracy and high-level output warrant their relatively high cost. Their principle of operation is illustrated by the diagram shown for the transducer illustrated in Figure 1-14. The seismic mass (1) tends to move when acceleration is applied. The motion is detected by a reluctive sensor (2) and transduced into an error signal in the servo system. The error signal is amplified by the servo-nulling amplifier (3) which causes a feedback current to flow through the restoring coil (4) and the output resistor (5). The restoring coil is located in a permanent magnetic field. The current from the servo amplifier causes a torque to be produced by the coil. This torque is equal and opposite to the acceleration-generated torque acting on the mass-coil pendulum. The electromagnetic torquer, (the restoring coil) attached to a mechanical link to one end of which the mass is attached, returns the seismic system to its original position causing the error signal to be reduced to zero. At this point the servo system is in a torque-balance (generalized as *force-balance*) condition. The servo current required to achieve this balance is directly proportional to the applied acceleration. The IR drop through the output resistor, R_L, due to the same servo current, then produces the output voltage of the transducer, E_o, which is also proportional to acceleration since R_L has a fixed value.

A slightly different operation, but equally typical for a force-balance servo accelerometer, is obtained in the transducer shown in Figure 1-15. The seismic mass is integral with the restoring coil of the electromagnetic forcer as well as with the moving plate (rotor) of a capacitive motion sensor whose stator plate is fixed in the internal structure. The permanent magnet, the soft-iron piece, and the pole piece constitute the magnetic circuit of the forcer. Since an axial mass, rather than a pendulous mass, is used in this transducer, servo balance is attained by force balance instead of torque balance. The mass assembly is restrained from motion in any but the longitudinal axis (up-down in the illustration), which is the transducer's sensing axis, by specially designed and attached suspension arms. A second set of arms is used at the bottom edge of the restoring coil. Motion of the mass is detected by the capacitive sensor whose output is the error signal in the servo system. The signal is amplified by the electronic circuitry. The amplifier output current flows through the restoring coil and an output resistor. The voltage across the resistor is given by the servo current needed to retain the system in balance against the force due to acceleration. Hence, the output voltage is proportional to acceleration. A radio-frequency-interference (RFI) filter circuit is incorporated within a shielding can at the electrical-terminal end of the case. A hermetically sealed header provides the terminals.

The electronic circuitry is not only used to provide the force-rebalance current to the forcer or torquer. It is also used to set the acceleration range producing a given transducer output, to furnish any zero offset required in the output, to provide excitation conditioning, and to control the dynamic characteristics of the transducer.

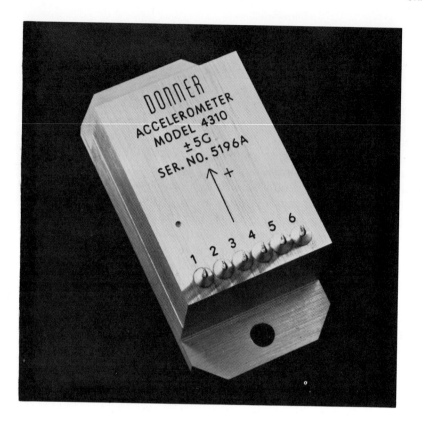

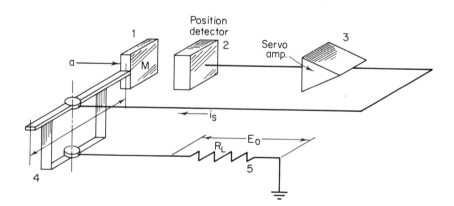

Fig. 1-14. Servo acceleration transducer with pendulous mass and reluctive motion sensor (courtesy of Systron-Donner Corp.).

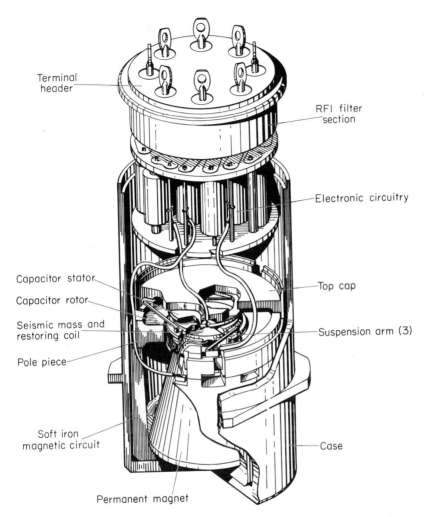

Fig. 1-15. Servo acceleration transducer with axial mass and capacitive motion sensor (courtesy of Kistler Instrument Corp.).

Design variations of these transducers have been built by a number of manufacturers and research facilities. One design uses a photoelectric motion sensor. Developments in the area of an all solid-state transducer, using a semiconductor-strain-gage motion sensor and an inverse-piezoelectric-effect forcer have shown promise. Servo accelerometers can provide one or more discrete outputs, a frequency-modulated output or a digital output. Angular accelerometers have also been built in a limited number of designs.

1.3.8 Vibrating-element acceleration transducers

The vibrating-wire principle has been applied in an acceleration transducer used, in the U.S.A., exclusively for space booster acceleration telemetry where an overall measurement-system accuracy within ±0.25% of full scale was required. Designs by two American manufacturers (Borg-Warner Electronics and Fairchild Controls) resulted from the original developmental effort of P.I. Holmes at Byron-Jackson Electronics and by G. Senseny, J. Hunt, and R. White at White Avionics Corporation. This type of accelerometer is difficult to design and manufacture and is therefore quite costly. However, its frequency-modulated output can be transmitted, received, and displayed without additional errors being introduced by the measurement system. The transducer itself is capable of representing, by its output, the applied acceleration with an error of less than ±0.1% of full scale.

The operating principle is quite simple. A tungsten wire, 0.001 in. in diameter, is attached at one end to a seismic mass and at the other end to a fixed part of the transducer case. Motion of the mass causes a change in the tension of the wire and, hence, in its resonant frequency (as in a violin string). The wire is in a permanent magnetic field established by a pair of magnets parallel to the wire, one on either side. A current passed through the wire causes it to vibrate at its resonant frequency. This oscillation is maintained in a feedback circuit whose frequency-controlling element is constituted by the wire-magnet assembly. The signal in this circuit is then coupled to an amplifier from which the transducer output is obtained. The output amplitude is essentially constant in the measuring range. The output frequency, however, varies with acceleration. It can be adjusted to vary, for a given acceleration range, over the same band of frequencies provided by an IRIG standard subcarrier oscillator in an FM/FM telemetry system.

The frequency-vs-acceleration curve is parabolic; therefore, accuracy specifications must omit linearity. Hysteresis is negligible since the spring-mass system is very stiff. Repeatability with respect to a given reference curve remains as the sole performance characteristic to which errors in the data can be ascribed. However, this design is very sensitive to temperature changes, and temperature errors can easily exceed 20 times the repeatability at the temperature at which the reference curve was established. Also, long-term drifts can be of considerable magnitudes.

Figure 1-16 shows the sensing/transduction assembly of a vibrating-wire accelerometer. This assembly is mounted within a heater jacket in a larger case which also contains the oscillator and amplifier as well as the heater-control circuitry. The use of two wires, each passing through its permanent magnets, not only makes for a symmetrical design but also allows the manufacturer to choose the wire with the better characteristics for actual use in the transducer. The travel of the mass is limited by threaded stop bushings. A small amount of damping fluid is contained in the gap around the mass. Operating theory and further descriptive details relating to vibrating-wire transducers are covered in 10.3.9.1.

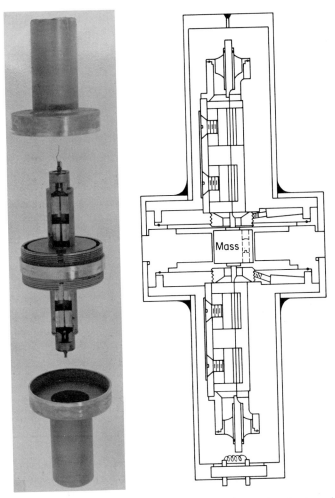

Fig. 1-16. Sensing/transduction module of vibrating-wire acceleration transducer (General Dynamics Convair photo of a design produced by Byron-Jackson Electronics, now Borg-Warner Electronics; drawing courtesy of Borg-Warner Electronics).

A similar accelerometer using a vibrating ribbon has been used in aerospace automatic control systems. Vibrating-wire ("vibrating-string") transducers in general, and the accelerometers in particular, have seen more development and applications in some other countries, notably the U.S.S.R. where, for example, various approaches have been used to linearize the output-vs-acceleration curve and to reduce long-term drift.

1.4 Performance and Applications

1.4.1 Specification characteristics

The characteristics to be considered for inclusion in an acceleration-transducer specification differ for the two major categories of transducers, based on their dynamic characteristics. The first category, which can be referred to as "steady-state acceleration transducers," includes all types normally capable of a frequency response starting at zero hertz, even when the upper limit of their frequency range extends to several hundred hertz for flat response. Potentiometric, reluctive, strain-gage, and servo accelerometers are included in this category. Capacitive, photoelectric, and vibrating-wire accelerometers also fall into this category but will not be considered further because of their relative rarity. The second category ("dynamic acceleration transducers") consists entirely of piezoelectric accelerometers.

Specification characteristics of piezoelectric acceleration transducers ("pickups") have been described in U.S.A. Standard Z24.21-1957; those of auxiliary equipment, in U.S.A. Standard S2.4–1960. Specifications and tests of piezoelectric accelerometers have also been covered in I.S.A. Recommended Practice RP37.2 (1964) which was developed under the author's direction. Similar I.S.A. documents for steady-state (strain-gage and servo) accelerometers are being developed by parallel committees.

1.4.1.1 Mechanical and electrical design characteristics of both transducer categories are listed in Table 1-1. A complete specification should include all listed characteristics, except those shown to be "supplemental" in Tables 1-1 and 1-2. Supplemental characteristics are shown in some specifications when special needs arise.

Case dimensions and mounting details should be shown on a drawing. Internal mounting threads or threaded mounting studs should be described by thread size and class. Piezoelectric accelerometers can be obtained with either integral or separate mounting studs, except for certain very small transducers for which the types of adhesive usable for mounting should be stated. Identification (nameplate) markings should include nomenclature (e.g., "Potentiometric Acceleration Transducer" or ". . . Accelerometer"), manufacturer's name, model number, serial number, nominal sensitivity for piezoelectric transducers, and range as well as identification of electrical connections for steady-state acceleration transducers. Another marking which should be on the case of all transducers is the symbol for direction of sensitivity, an arrow which indicates the sensing axis. A "+" sign placed next to the arrow indicates the direction in which the case must be accelerated to obtain an output of positive polarity. The crystal material is only specified by manufacturers, frequently by a proprietary name. Examples of sensing mode are compression, bending, and shear. Mechanical isolation is sometimes described as a feature of internal transducer construction which allows forces (e.g., bending forces and pressures) to be applied to the case with

Table 1-1 Mechanical and Electrical Design Characteristics
Specified for Acceleration Transducers

Steady-state acceleration transducers	Piezoelectric acceleration transducers
Configuration and case dimensions. Mounting provisions, mounting method, mounting dimensions. Identification and other markings. Case material and type of case sealing. Weight of transducer (and of any integral cable of specified length). Allowable mounting force and/or torque.	
Case alignment relative to actual sensing axis. Location of center of seismic mass. Method of damping and type of damping fluid (if any). Type of transduction element and descriptive details.	Crystal material and sensing mode. Mechanical isolation of crystal (supplemental).
Type of electrical connector or integral cable. Integral calibration provisions (supplemental). Allowable load impedance (range or minimum value).	
Insulation resistance at room temperature. Breakdown-voltage rating at room temperature. Input impedance (when not furnished with integral active excitation-conditioning circuitry). Output impedance. Excitation voltage. Excitation current or power drain. Frequency of a-c excitation (if used). Maximum excitation. Connector-pin, terminal, or color-coded-wire functions.	Grounded or ungrounded crystal. Output capacitance at room temperature and (supplemental), its variations over the operating temperature range. Output resistance at room temperature and (supplemental) at maximum operating temperature. Insulation resistance of ungrounded crystal at room temperature and (supplemental) at maximum operating temperature. Characteristics of cable supplied with transducer: type, length, connectors, temperature range, noise; capacitance at room temperature and (supplemental) its variation over the temperature range.

negligible resulting forces on the transduction element. Descriptive details of steady-state type transduction elements include type and number of strain-gages, wirewound or film-type potentiometric element, and type of motion sensor and forcer.

Electrical connectors are best described by their part number and number of pins. Acceptable mating connectors can also be listed. Multiconductor cable can be shielded or unshielded; the insulation and gage of each conductor and the overall insulation covering should be stated. Crystals are designated as ungrounded when the portion of the transducer intended to be in contact with the measured structure is not electrically connected to the "low" side of the crystal. Such insulation can be attained by internal construction or by use of a separate insulating stud. The output impedance

Table 1-2   PERFORMANCE CHARACTERISTICS SPECIFIED FOR ACCELERATION TRANSDUCERS

Steady-State acceleration transducers	Piezoelectric acceleration transducers
	Acceleration range.
	Acceleration overload.
Static error band, referred to a theoretical slope or a terminal line.	Reference sensitivity (at room temperature).
or	Amplitude linearity (at reference frequency).
Static error band, referred to an end-point line, a best straight line, a least-squares line, or a mean-output curve or other specified curve. *and* End points (with tolerances).	
or	
Linearity (of a specified type), hysteresis, and repeatability. *and* End points (with tolerances), *or* Lower end point (with tolerances) and full-scale output (with tolerances), *or* Zero-acceleration output (with tolerances) and sensitivity (with tolerances).	
Creep (supplemental).	Transverse sensitivity at reference frequency and (supplemental) over specified band of frequencies.
Transverse sensitivity.	
Mounting error.	Mounting error (mounting effects on sensitivity).
Overload output (supplemental).	
Stability (long-term repeatability).	
	Strain sensitivity (supplemental).
Additionally, for potentiometric transducers: Resolution.	Sensitivity stability (long-term).
Friction error (if not included in error band).	

of a piezoelectric accelerometer is specified as two separate quantities, output capacitance and output resistance ("shunting resistance").

Additional design characteristics must be shown for any separately packaged signal-conditioning equipment when such equipment is furnished as a part of a "matched" transducer system.

1.4.1.2 Performance characteristics are listed in Table 1-2. The (acceleration) range is always specified as symmetrically bidirectional for piezoelectric transducers but can be unidirectional (e.g., "0 to 10 g"), unsymmetrically bidirectional (e.g., "-2 to $+6$ g"), or symmetrically bidirectional (e.g., "± 10 g") for steady-state acceleration transducers. Acceleration beyond the specified range which can be applied to the transducer along the sensing axis without subsequent change in performance

Table 1-2   PERFORMANCE CHARACTERISTICS (Cont'd)

Steady-State acceleration transducers	Piezoelectric acceleration transducers
Frequency response (at room conditions). *or* Damped natural frequency.⎫ *and* ⎬ Damping ratio. ⎭	Frequency response (at room temperature). Mounted resonant frequency (nominal value with tolerances, or minimum value). Amplification factor at resonant frequency (supplemental).
Operating temperature range. Temperature-gradient error (supplemental). Frequency response at limits of operating range (supplemental).	
Temperature error band. *or* Temperature error. *or* Thermal zero shift. ⎫ *and* ⎬ Thermal sensitivity shift.⎭ Variation of damping ratio with temperature.	Temperature sensitivity error.
Allowable effects of: Ambient pressure variations. Humid, corrosive, and contaminating atmospheres. Ambient fluids other than air (e.g., salt water). Magnetic fields. High sound-pressure levels. Nuclear radiation. Shock. Vibration.	
Electromagnetic interference limits.	Electromagnetic interference limits (on complete transducer systems only).

characteristics is shown as overload. Performance characteristics, such as error band, linearity, friction error, and frequency response are explained in Chapter II. Error-band and linearity tolerances may be different in different portions of the range if the full range is not intended to be used in the transducer's application. The reference sensitivity of piezoelectric accelerometers, expressed as either voltage or charge per unit of acceleration, should be specified as applicable under a prescribed set of conditions including (primarily) frequency, also amplitude, mounting torque, and load impedance. Other errors are then referred to this value of sensitivity (hence the term "reference sensitivity").

Transverse sensitivity is the maximum sensitivity of a uniaxial transducer to accelerations along specified (or any) axes transverse to the sensing axis. It is usually expressed in percent of the reference sensitivity for piezoelectric accelerometers and in percent of sensitivity or of full-scale output for steady-state accelerometers. It has also been expressed as the ratio of apparent to actual acceleration (in the sensing axis).

Allowable overload output can be specified for steady-state acceleration transducers to indicate the desired effects of mechanical stops or of an electronic limiter or both. Manufacturing difficulties will be encountered when the overload output is allowed to exceed the end points by only a small amount.

The resonant frequency (always the lowest peak exceeding 1.3 times the reference amplitude when several peaks exist) of a piezoelectric accelerometer can be substantially different in the unmounted and mounted conditions. The "mounted" value is the only meaningful one. It is best stated as applicable when the transducer has been mounted to a specific type of (simple) structure with a narrow range of levels of mounting torque.

When specifying allowable effects of shock and vibration, any conflicts with range, overload, frequency-response, and transverse sensitivity specifications must be avoided.

Storage life is sometimes specified. Operating or cycling life is rarely stated for accelerometers.

Additional characteristics must be specified for separately packaged emitter followers, amplifiers, active filters, etc., which may be part of a "matched" transducer system and sold or procured as such. They include environmental characteristics, gain-adjustment range, recovery time (the time taken by an amplifier to produce normal output after removal of an overloading input signal), output impedance, excitation regulation and filtering, and common-mode rejection.

1.4.2 Consideration for selection

Primary selection criteria are frequency response and range. Applications can usually be categorized by the frequency range over which a flatness of response within $\pm 5\%$ from a (low) reference frequency must be obtained. Generally, piezoelectric accelerometers must be used when the frequency range extends above 400 Hz; semiconductor-strain-gage accelerometers can be used up to 750 Hz, several metal strain-gage and servo accelerometer designs up to 400 Hz, more versions of these two types up to 200 Hz, reluctive accelerometers up to about 80 Hz, and potentiometric accelerometers up to 10 to 20 Hz. The natural frequency and, hence, the usable frequency response of most accelerometers increases with their range. A typical potentiometric accelerometer design, for example, has a natural frequency of 20 Hz for a $\pm 2\,g$ range which rises to 65 Hz for a $\pm 20\,g$ range. The usable frequency range also depends on the damping ratio used. Figure 1-17 shows relative response curves for various damping ratios as a function of the ratio of frequency (F) and natural frequency (F_N). The damping ratio can vary considerably with temperature, especially when viscous damping is used. When piezoelectric accelerometers are used, the usefulness

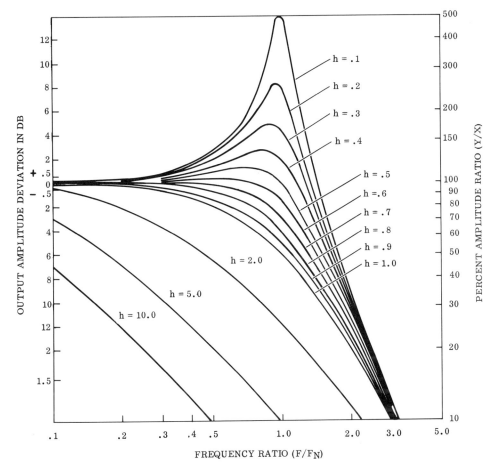

Fig. 1-17. Response of a spring-mass system to a sinusoidal acceleration for damping ratios (h) between 0.1 and 10.0.

of data at frequencies below 5 Hz must be sacrificed in the trade-off for obtaining high-frequency data.

Special filters can be connected into piezoelectric transducer systems if data above a certain frequency, normally provided by such a system, must be eliminated, e.g., because of bandwidth limitations of a telemetry system.

Acceleration ranges between ± 1 and $\pm 100\,g$ can be provided by all types of acceleration transducers. For very low-range measurement (± 100 milli-g or less), servo accelerometers provide better performance than other types, whereas piezoelectric accelerometers are most usable for high-range vibration and shock measurements.

Accuracy and cost are next in importance as factors in accelerometer selection.

Servo and vibrating-wire acceleration transducers provide the closest accuracy characteristics, and potentiometric accelerometers provide the loosest accuracy characteristics. Not surprisingly, cost is lowest for the latter types, highest for the former types. The other accelerometers occupy the region between those two extremes in both respects. The cost of any required signal-conditioning equipment is not included in these comparisons. A complete piezoelectric accelerometer system, e.g., often equals or exceeds a good quality servo accelerometer in cost. The "inertial quality" accelerometers commonly used for guidance and control rather than for measurement have not been specifically considered in this book.

Among environmental conditions the operating temperature range is the most critical factor in selection. Piezoelectric accelerometers have been designed for use over wider temperature ranges than any other types. Transducers with integral semiconductor circuits have the inherent temperature limitations associated with most semiconductor devices.

Another important consideration is the weight of the transducer, especially in medium- and high-frequency vibration measurements on relatively thin structural members where the addition of the transducer (and associated cable) mass can cause substantial changes in the dynamic behavior of the structure.

1.5 Calibration and Tests

1.5.1 Calibration

The accurate measurement of the output of accelerometers in response to acceleration, especially transient and rapidly fluctuating accelerations, has taxed the ingenuity of researchers and technicians concerned with this problem for several decades. A considerable number of reports and papers have been published to explain various methods and equipment (see 1.7). Among the most comprehensive of these documents is U.S.A. Standard S2.2–1959, "Calibration of Shock and Vibration Pickups." Calibration and test requirements are also covered in the I.S.A. RP37 series of recommended practices. Equipment and methods are described below to the extent that they are used for the calibration of production accelerometers.

Static calibrations are performed only on steady-state acceleration transducers. Dynamic calibrations are performed on piezoelectric accelerometers and, as frequency-response tests additional to calibration, on steady-state accelerometers. The most commonly used methods for calibrating linear-acceleration transducers are the static methods using gravitational and centripetal acceleration and the dynamic method using vibratory acceleration.

During gravitational and centrifuge calibrations, discrete levels of acceleration are applied to the transducer in increasing and decreasing directions. Between six and eleven levels of acceleration, including the range limits, are normally applied. At least two calibration cycles are performed at room conditions.

1.5.1.1 Gravitational calibrations use the earth's gravitational field to apply acceleration levels between 0 and $\pm 1\ g$ to the transducer. A tilt fixture (*tilting support*) is used to apply known fractions of $1\ g$ to a transducer mounted on it. One version of a tilting support is shown in Figure 1-18. Another version uses an arm which rotates about the center of a vertically mounted protractor. Depending on mounting provisions and the orientation of the sensing axis, the transducer is mounted directly to the rotating member or to an angle bracket or platform on this member. The tilt fixture is usually equipped with a vernier (e.g., worm-gear) drive and with vernier angle-reading provisions. The transducer's center of seismic mass need not be located at the center of rotation. Its sensing axis, however, must be aligned precisely with the reference axis marked on the fixture's arm or disc.

When the "positive acceleration" arrow marked on the transducer's case points vertically upward, an acceleration of $+1\ g$ is seen by the transducer which causes its seismic mass to be forced downward. When the disc is rotated so that the arrow on the case points vertically downward, a negative acceleration, $-1\ g$, is applied to the transducer. When the arrrow on the case is exactly horizontal (pointing either left or right), the acceleration along the sensing axis is zero.

For intermediate positions this acceleration (a) is then a fraction of the gravitational acceleration (g) equal to the cosine of the angular displacement from vertical (θ), or $a = g \cos \theta$. For example, if $+1\ g$ is denoted by the 0-degree and $-1\ g$ by the 180-degree marking, an acceleration of $+0.5\ g$ will be applied at the 60-degree marking and an acceleration of $-0.07\ g$ at the 94-degree marking. The calibration uncertainty

Fig. 1-18.   Calibration on tilting support.

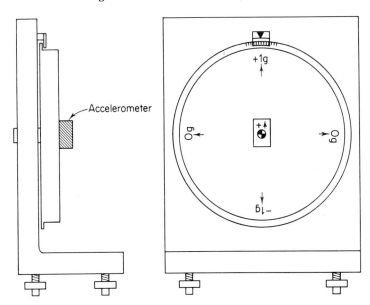

of this method is given entirely by the uncertainty of the angular-displacement measurement, provided that any difference between the local g and standard g has been
taken into consideration and that the fixture was properly leveled.

Transverse-acceleration effects can cause errors since the transverse component
of the gravitational acceleration in the same plane (a_t) seen by the transducer is equal
to the sine of the angle from vertical, or $a_t = g \sin \theta$. This effect is negligible only when
the transverse sensitivity of the transducer is negligible.

1.5.1.2 Centrifuge calibrations use the centrifugal force experienced by an object
on a rotating member to apply known levels of acceleration to a transducer. A centrifuge, as used for accelerometer calibration and acceleration testing of other transducers, consists essentially of a balanced horizontal turntable or arm whose central
shaft is driven by a motor through a continuously adjustable variable-speed drive.
The shaft is provided with slip rings and brushes to provide electrical connections to
components installed on the rotating member. A hinged protective hood is placed
over the turntable for safety purposes and an angular-speed transducer is usually
connected to its drive shaft. A frequency-output (electromagnetic or photoelectric)
tachometer is most suitable for this purpose when an electronic counter is available
for readout.

The accelerometer is installed on the turntable (Figure 1-19) at a known radial
distance (r) between its center of seismic mass and the center of the drive shaft.
The transducer's sensing axis is carefully aligned to a radial line on the arm or disc.
When the table rotates at a constant angular velocity (ω), the centripetal acceleration
acting on the seismic mass of the transducer equals the product of the radial distance
and the square of the angular velocity or $a = \omega^2 r$.

Since $\omega^2 = (2\pi f)^2 = 4\pi^2 f^2$, where f is expressed in revolutions (cycles) per
second, the acceleration, expressed in g, can be determined from the radius, expressed in inches, and the angular speed (N) in rpm from the relationship

$$a = 2.84 \times 10^{-5} N^2 r$$

While the transducer is subjected to the centripetal acceleration, the corresponding
centrifugal force on the seismic mass causes it to move in an outward direction. Unless
the radius r is much larger than the displacement (x_m) of the seismic mass, the acceleration must be calculated at each increment of angular velocity as $a = \omega^2 (r + x_m)$.
The seismic-mass deflection is sufficiently small in most transducers, and a large
enough radius can be selected on most centrifuges so that x_m can be neglected.

The location of the center of seismic mass should be stated precisely by the transducer manufacturer, preferably dimensioned with reference to the transducer's mounting holes or similar mounting provisions. If this information is not available, or is
not stated with sufficient accuracy, the gravitational center of seismic mass must
first be determined by a separate test. One such test consists of measuring the transducer output, e_1, at a radial distance, r_1 (taken to a case reference mark on the trans-

ducer), then moving the transducer slightly outward on the turntable, remounting it and measuring the new output, e_2, at the new distance, r_2 (to the same reference mark). Both measurements must be made at the same angular velocity. If the transducer's output-vs-acceleration relationship is essentially linear in the portion of the range equivalent to the output values bracketed by e_1 and e_2, the true radial distance to the center of seismic mass corresponding to the first mounting position, r, can be determined as

$$r = \frac{e_1(r_2 - r_1)}{(e_2 - e_1)}$$

The distance between case reference mark and center of seismic mass is then equal to the difference between r and r_1.

An accelerometer having a span not greater than 10 g, a small mass deflection, and good linearity can be calibrated on a centrifuge without regard to the radial distance (r) and the exact location of the center of seismic mass by a different method. The transducer is installed on a tilting support which is slowly rotated until $+1$ g is reached,

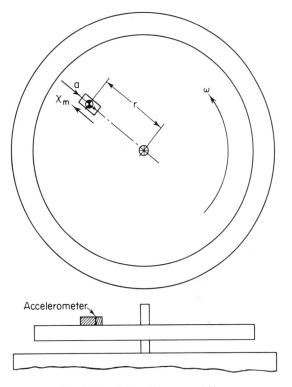

Fig. 1-19. Calibration on centrifuge.

at which point the output is recorded. The transducer is then mounted on the centrifuge at an arbitrarily chosen radial distance. The rotational speed of the turntable is gradually increased until the previously recorded transducer output is again obtained. The angular velocity at this point (ω_1) is recorded and used as reference velocity for all other levels of acceleration (a) for which the required angular velocity (ω) can then be calculated from the relationship

$$a = \frac{\omega^2}{\omega_1^2}$$

Errors in centrifuge calibrations are introduced primarily by uncertainties in the determination of angular speed since acceleration varies as the square of this speed. They can also be caused by dimensional errors in the measurement of radial distance, misalignment of the sensing axis to the radius, and a dynamic unbalance of the turntable or rotating arm.

1.5.1.3 Dynamic calibrations of piezoelectric accelerometers are usually performed as comparison calibrations using an electromagnetic vibration exciter (vibrator, *shaker*) and a piezoelectric reference accelerometer. The operation of the shaker is similar to that of a loudspeaker in that an alternating current through a movable coil in a magnetic field causes oscillatory motion of this coil. Since the coil of a shaker has to move a heavy platform whereas a loudspeaker coil moves only a paper cone, the coil and electromagnet assemblies of a shaker are much larger and stronger than those of even a good hi-fi speaker. The heavy coil also requires a special mechanical suspension, e.g., a cantilever spring arrangement, to keep the motion of the platform, attached to the coil, as purely rectilinear as possible. The acceleration (a) provided by a shaker can be calculated from the frequency (f, in Hz) and the platform displacement (x, in in.) as

$$a = 4\pi^2 f^2 x$$

where a, in in./s^2, is the peak-to-peak acceleration if x is the peak-to-peak displacement.

A comparison calibration requires the reference transducer to be mounted back-to-back with the transducer being calibrated (Figure 1-20). The characteristics of the reference accelerometer and its associated signal conditioner (usually an amplifier) must be known over the frequency and amplitude ranges used during the calibration within an uncertainty substantially smaller than the applicable performance tolerances specified for the test transducer.

One of the primary characteristics by which shaker performance is judged is the relative absence of any transverse movement of the platform ("cross talk"). When the cross-talk characteristics of a shaker used in a given calibration setup are not sufficiently well established, they can be determined by mounting a three-axis accelerometer assembly, with two dummy units added for balance, to the platform (Figure

Fig. 1-20. Dynamic comparison calibration on vibration exciter.

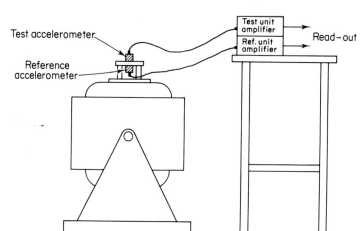

1-21) and exercising the shaker through the planned amplitude and frequency ranges while monitoring the three outputs. Transverse motion in a good calibration shaker must be as low as possible. It has been suggested that it should not exceed 1% of the axial motion. Special shakers equipped with air bearings for support of the armature have been built primarily for accelerometer calibration purposes.

Fig. 1-21. Accelerometer assembly for performance evaluation of electrodynamic vibration exciter (courtesy of National Bureau of Standards).

The mounting fixture itself should also be analyzed for possible decoupling at high frequencies, such as by interchanging the test and reference accelerometers (or two reference accelerometers), or by evaluation of relative motion using reference accelerometers of different total mass, and by prior experimental determination of frequency characteristics. Such special tests must be performed before the actual calibration. The platform should be thick enough to prevent appreciable flexure during tests and thin enough to provide the same motion to both transducers. Each transducer should be installed in the fixture with the specified mounting torque. The fixture surface to which each accelerometer is installed should be clean and flat with a smooth finish. Additional surface smoothing can be obtained by applying a thin film of oil or grease to the surface before installation.

When the test transducer is a part of a "matched" transducer system, the other parts of the system are connected to the transducer during all calibrations and tests. When it is not a part of such a system, the characteristics of the cable, amplifier, etc., connected to the transducer must be known and recorded on the test data sheet.

The output of the velocity transducer with which most shakers are furnished is used only for approximate indications of vibration amplitude. The output of the reference-transducer system is used for valid data. Similar considerations apply to the determination of vibration frequency. Rather than read the dial of the audio oscillator, whose amplified output drives the shaker, to determine frequency, we use a separate electronic frequency counter for this purpose. The drive-power waveshape is also monitored on a distortion meter or similar analyzer to assure that distortion is within acceptable limits (typically less than 3% distortion).

The calibration usually begins with the determination of reference sensitivity at a frequency chosen as reference frequency, between 40 and 100 Hz, and at an amplitude chosen as reference amplitude, large enough to provide a transducer (system) output well above the noise level. A signal-to-noise ratio of at least 40 db is desirable. The acceleration amplitude is then increased at discrete increments to determine amplitude linearity, whenever this characteristic must be demonstrated. The frequency-response test is a part of the calibration of a piezoelectric accelerometer. The reference amplitude is maintained while the vibration frequency is changed over a frequency range slightly in excess (at both ends, if feasible) of the specified range over which frequency-response tolerances apply. The output is usually recorded at ten different frequencies.

This comparison calibration method is usable at test frequencies between about 5 and 10,000 Hz. Electrodynamic vibration exciters can also be used for the determination of resonant frequencies up to 50 kHz. Optical methods are used at very low frequencies to measure platform displacement from which acceleration amplitude can be calculated when the frequency is known. Testing at very high frequencies by the comparison method can be performed on a piezoelectric vibration exciter in which the inverse piezoelectric effect is used to obtain small displacements at the ends of a stack of ceramic crystal elements. High-frequency response tests can also be performed by impact and other methods.

1.5.1.4 Other calibration methods for linear-acceleration transducers include the following:

The *pendulum calibrator* uses the centripetal acceleration along its arm while the pendulum is swinging. An accelerometer installed on a bracket at the end of the arm, with its sensing axis aligned to the centerline of the arm, is subjected to this acceleration which varies sinusoidally at a low frequency with decreasing amplitude. The acceleration acting on the transducer is related to gravitational acceleration, dimensions along the pendulum arm including the transducer, the angle over which the pendulum swings, and the radius of gyration of the pendulum.

The *chatter method* has been used to determine when the level of acceleration produced by a vibration exciter or other dynamic calibrator is $-1\,g$. It involves the use of a "chatter accelerometer" in place of a reference accelerometer. This special device consists typically of a bronze ball resting loosely against a spacer in contact with a piezoelectric crystal. When the device is mounted on a shaker platform, the ball separates upward from the spacer whenever the downward acceleration exceeds $1\,g$ and falls back to the spacer with a characteristic bounce (chatter) later in the cycle when the acceleration passes the $-1\,g$ point on its swing to positive acceleration. The chatter is often audible but is best observed in the output from the piezoelectric element.

The *reciprocity method* has been used primarily for the calibration of the velocity transducer in a shaker and of piezoelectric reference accelerometers. The principles upon which this method is based are explained in 11.5.1.1.

Some servo accelerometers have been calibrated by the *current-injection method* in which a known amount of current is caused to flow through the servo loop to simulate the servo action normally caused by acceleration forces acting on the seismic mass. The validity of this method is controversial.

A *"roulette-wheel"* vibration calibrator has been used at the Hungarian Central Measurement Research Laboratory to apply a sinusoidally varying displacement to a transducer mounted radially to a circular device furnished with an internally circumferential V-groove. The transducer is balanced by a dummy mass at the opposite end of the device which is suspended at its center by a string. A ball is set spinning along the groove by applying compressed air to it. The ball passes a proximity detector once per revolution so that its angular speed (frequency) can be determined. The circular motion causes a radial rectilinear displacement to be applied to the transducer due to the continuously varying location of a dynamic unbalance of the entire device. The amplitude of this displacement can be calculated from knowledge of the orbital radius of the ball and the masses of the ball and the entire device.

1.5.1.5 Shock calibrators are used to determine the response of accelerometers to transient accelerations with peak levels to about 15,000 g, with an uncertainty of 5 to 10% of reading. The *ballistic pendulum* calibrator contains two elongated masses, each suspended near both of its ends by strings, wires, or ribbons so that their principal axes are horizontally aligned to each other and to the direction of motion. One mass is used as hammer; the other, to the far end of which the accelerometer is attached, as the anvil. The acceleration of the anvil upon impact can be determined from its velocity change and from a record of transducer output vs time or by electrically integrating this output. A *drop-test machine*, which lets a hammer drop on a base-plate anvil, has also been used for shock-calibrating accelerometers mounted to the top of the hammer. A *drop-ball machine* lets a steel ball drop on an anvil to whose bottom surface the transducer is mounted. Upon impact the anvil support is sheared so that the anvil accelerates downward freely into a cushioned box. This machine is commonly used to perform comparison shock calibrations up to 10,000 g. The acceleration wave shape obtained in impact calibration is usually close to a half-sine wave.

1.5.1.6 Angular accelerometers can be calibrated by use of a torsional pendulum. The horizontally-deflecting torsional pendulum oscillates sinusoidally with decreasing amplitudes after it is first moved over an initial arc (θ). The angular acceleration (α) suddenly applied to a transducer mounted on the pendulum, whose undamped period of oscillation is T, when it is set free, is

$$\alpha = \frac{4\pi^2\theta}{T^2}$$

where x is expressed in rad/s^2 if θ is expressed in radians. The natural frequency of the accelerometer must be high compared to the pendulum frequency.

1.5.2 Performance verification at room temperature

1.5.2.1 Initial tests comprise mechanical and electrical checks and the calibration from which most nonenvironmental performance characteristics should be determinable. A visual inspection of case and mounting dimensions, case seals, finish, identification markings, and electrical connector is common to all accelerometers. Electrical inspection requirements depend on the type of transducer and applicable specifications. Insulation resistance is measured on all transducers except those having one side of their transduction element grounded to the case. Additional tests include: an *x-y* plot and measurement of element resistance for potentiometric accelerometers, input and output resistance determination for strain-gage accelerometers, output impedance and output noise measurement for servo accelerometers and transducers with integral amplifiers, and checks of transducer and cable capacity, cable integrity, and cable length for piezoelectric accelerometers.

Performance characteristics are determined from a record of calibration data. An example of a data sheet used for steady-state acceleration transducers is shown in Figure 1-22. This form allows the static error band, referred to a straight line between the end points, to be calculated for a strain-gage accelerometer. The end points (and their specification compliance) are first determined so that "theoretical" output values can be inserted in the appropriate column. The full-scale output, necessary to convert output deviations into "% FSO," can also be calculated from the end points. The form makes allowance for one calibration and calculation of the temperature error band at either high or low temperature.

The test data sheet shown in Figure 1-23 has been used to record acceptance-test and calibration data for piezoelectric accelerometer systems, including, in this example, temperature-test data (see 1.5.3.1). It allows for 14 frequency readings at a reference amplitude and 14 amplitude readings at a reference frequency. When less points are taken, some lines are left blank. Spaces are provided for recording the excitation (power input), gain adjustment range, insulation resistance, and output noise (with no acceleration applied) of the amplifier. Charge sensitivity, in picocoulombs per *g*, is determined by performing calibrations with charge amplifiers. It can also be calculated from voltage sensitivity by multiplying the sensitivity in volts per *g* by the total capacitance (of transducer, cable, and amplifier input) in picofarads.

1.5.2.2 Transverse-sensitivity tests are performed on all types of accelerometers, usually on a sampling basis. A steady-state acceleration transducer is tested on a centrifuge. The transducer is first installed so that acceleration is applied to it along the *y*-axis, then remounted so that acceleration acts in its *z*-axis (where the *x*-axis is the sensing axis of the transducer). A special mounting fixture is required for this purpose. In each orientation between one and three different levels of acceleration are applied to the accelerometer, and the output is measured. The output readings at

VENDOR'S MODEL NO.		PART NO.	
PURCHASE ORDER NO.		SERIAL NO.	
	INDIVIDUAL ACCEPTANCE RECORD & CALIBRATION FOR STRAIN GAGE ACCELERATION TRANSDUCER	RANGE	

1. VISUAL: Mechanical ☐   Finish ☐   Nameplate ☐   Receptacle ☐

2. ELECTRICAL: Input Resistance _____ ohms   Insulation Resistance _____ Megohms at _____ VDC

    Output Resistance_____ ohms

    EXCITATION VOLTAGE (Used During Calibration) _____ V C

3. CALIBRATION: End Points: _____ mv at _____ g, _____ mv at _____ g

    Allowed: _____ ± _____ mv at _____ g, _____ ± _____ mv at _____ g

ACCELERATION		THEORETICAL OUTPUT, mv	OUTPUT, mv (RUN 1)		OUTPUT, mv (RUN 2)		MAXIMUM ERROR		OUTPUT, mv @ _____ °F:	
±	Gs		INCREASE	DECREASE	INCREASE	DECREASE	±	% FSO	INCREASE	DECREASE

ERROR BAND: + _____ % − _____ % FSO (Referred to straight line between end points)   ALLOWED: ± _____ % FSO

TEMPERATURE ERROR BAND: + _____ % − _____ % FSO (_____ to _____ °F)   ALLOWED: ± _____ % FSO

BY: _____ DATE _____ APPROVED _____

Fig. 1-22. Test record form for accelerometer whose error bands are referred to end-point line.

VENDOR'S MODEL NO.		PART NO.
PURCHASE ORDER NO.		SERIAL NO.

INDIVIDUAL ACCEPTANCE RECORD & CALIBRATION
FOR PIEZOELECTRIC
ACCELERATION TRANSDUCER (SYSTEM)

RANGE

1. VISUAL: Mechanical ☐   Cable ☐   Finish ☐   Sealing ☐   Nameplate ☐   Receptacle ☐

2. ELECTRICAL: Pick-up Capacity_____$\mu\mu$f   Cable Capacity_____$\mu\mu$f   Cable Length_____ft.
   Insulation Resistance_____Megohms at_____VDC (min., any pin to case ground)

3. NATURAL FREQUENCY: _____Kc

4. GAIN ADJUSTMENT RANGE: _____to_____mv/g (at_____g)

5. CALIBRATION: (Static Noise Level_____mv   Power Input_____VDC, _____ma, max.)

FREQUENCY RESPONSE at_____g   LINEARITY at_____cps   TEMPERATURE EFFECTS

FREQ. CPS	OUTPUT mv	SENSITIVITY mv/g	AMPLITUDE g's	OUTPUT mv	SENSITIVITY mv/g	
						☐ PICK-UP ONLY TEMP._____°F
						☐ ACCEL. SYSTEM FREQ. RESPONSE AT_____g
						FREQ. cps
						OUTPUT mv
						SENSITIVITY mv/g
						LINEARITY at_____cps
						AMPLITUDE g's
						OUTPUT mv
						SENSITIVITY mv/g
						TEMP. ERROR:_____% ALLOWED: ±_____%
						☐ PICK-UP ONLY TEMP._____°F
						☐ ACCEL. SYSTEM FREQ. RESPONSE AT_____g
						FREQ. cps
						OUTPUT mv
						SENSITIVITY mv/g
						LINEARITY at_____cps
						AMPLITUDE g's
						OUTPUT mv
						SENSITIVITY mv/g
						TEMP. ERROR:_____% ALLOWED: ±_____%

FREQ. RESPONSE: _____%, MAX. (Ref. to_____cps); ALLOWED ±_____% SENSITIVITY: _____mv/g; ALLOWED_____mv/g

BY_____DATE_____APPROVED_____

Fig. 1-23.   Test data form for piezoelectric accelerometer system.

each acceleration level are compared to the nominal sensitivity of the transducer when transverse sensitivity is to be expressed in g/g.

A piezoelectric accelerometer is tested for transverse sensitivity on a special fixture attached to a vibration exciter which must be as free as possible of cross talk. Since any inherent cross talk varies with frequency, it can be further reduced by testing only at those frequencies where cross talk is minimized. The transducer is installed on the fixture so that acceleration is applied to it in the transverse plane. A reference accelerometer is installed on the same fixture in its normal orientation permitting the acceleration to be applied to it along its sensing axis. The test is usually performed at one frequency and is sometimes repeated at one or two other frequencies. While vibratory acceleration is applied to the fixture, the test accelerometer is rotated in 15, 30, or 45 degree increments through 360 degrees, and the output is recorded at each angular position. The maximum transverse sensitivity obtained is checked for specification compliance. A polar plot can also be drawn from all output readings to show the accelerometer's complete transverse-sensitivity characteristics.

1.5.2.3 Frequency-response tests of piezoelectric accelerometers are an essential part of their calibration, as discussed in 1.5.1.3. They are performed on steady-state acceleration transducers in addition to their static calibrations, usually on a sampling basis. Equipment usable for frequency-response tests on these transducers include the shaker, the dual centrifuge, and the gravitational dynamic calibrator. The electro-dynamic shaker (vibration exciter) is sometimes not large enough to provide the necessary amplitudes at frequencies below about 10 Hz. Other shakers, using a hydraulic or mechanical (cam-operated) drive, and other mechanical devices can be used for dynamic response tests at very low frequencies.

The *dual centrifuge* is especially useful for frequency-response tests on unidirectional-range accelerometers where shakers, with their bidirectional motion, cannot be used. Such transducers require a sinusoidal variation of acceleration about a fixed acceleration level. The equipment (Figure 1-24) consists of a small turntable, on which the accelerometer is installed with its center of seismic mass at a radial distance r_s, mounted on a large centrifuge at a radial distance r_l from the center of the large turntable. The small turntable rotates at an angular velocity ω_s either by a separate drive or by a belt drive from a pulley concentric with the large turntable which rotates at a velocity ω_l. The acceleration (a) applied to the transducer along its sensing axis is then, at any time (t)

$$a = r_l \omega_l^2 \cos \omega_s t + r_s (\omega_l \pm \omega_s)^2$$

The sign in the last term is positive when the two velocities are in the same direction, negative when they are in opposite directions (one clockwise; the other, counterclockwise).

The *gravitational (earth's field) dynamic calibrator* is a small vertical centrifuge. The acceleration along the sensing axis of a transducer mounted at a radial distance

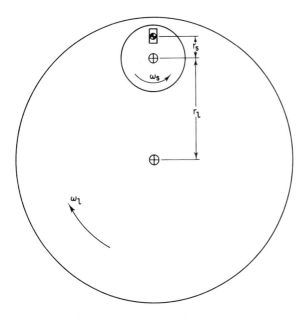

Fig. 1-24.   Dual centrifuge.

from the center of rotation has a constant centripetal component and a sinusoidally varying gravitational ($\pm 1\,g$) component. This equipment has mostly been used for tests of bidirectional accelerometers installed with their center of seismic mass at the center of the vertical disc. Accelerometers tested on a gravitational dynamic calibrator or dual centrifuge must have small transverse sensitivity since significant transverse acceleration is applied to them in the course of one revolution.

Frequency-response tests can be performed on *angular-acceleration transducers* on a *torsional-vibration calibrator*, essentially an electrically driven torsion pendulum. As designed at the National Bureau of Standards it uses a vertically mounted two-pole d-c motor with armature rotation less than one full turn. Alternating current is fed to the armature while the field is d-c excited. The accelerometer to be calibrated is mounted to a horizontal table attached to the motor shaft which is torsionally restrained by springs. The armature current is supplied by a low-frequency oscillator through a power amplifier. The angular displacement of the table is monitored by a reluctive transducer.

1.5.2.4 Mounting tests are sometimes performed, on a sampling basis, and mostly on piezoelectric accelerometers. Mounting torques equivalent to 80, 90, and 100% of the specified torque are successively used to mount the transducer to a vibration fixture while its sensitivity at the reference frequency and amplitude is determined. Insulated mounting studs are tested (on a qualification basis only) for mounting effects by applying the maximum rated torque to them several times and checking for visual evidence of mechanical damage.

1.5.3 Environmental tests

1.5.3.1 Temperature tests are so significant to the performance verification of accelerometers that abbreviated versions of such tests are frequently specified as part of the acceptance tests; i.e., each unit must be subjected to them (see Figures 1-22 and 1-23). Complete temperature tests, however, are performed on a sampling or design-qualification basis only. They comprise frequency-response and amplitude-linearity tests on piezoelectric accelerometers, calibrations at temperature extremes and tests for thermal effects on frequency response (primarily on the damping-ratio) on steady-state accelerometers, and temperature-gradient tests for both categories. The step change in temperature required for the latter tests can be obtained by sliding the accelerometer from a cooled block to a heated block, energizing a source of radiant heat close to the accelerometer for a short period of time, or immersing a well-sealed unit into hot distilled water or oil (or into an ice bath), while monitoring output changes. The other temperature tests are performed while the transducer is stabilized at the required temperature in an environmental chamber. A piezoelectric accelerometer can be vibrated by a rigid ceramic rod which passes through a hole in the chamber wall while the reference accelerometer is on the shaker platform, to which the rod is attached, so that the reference unit is well isolated from the environment within the chamber. A steady-state accelerometer can be tested on a tilt fixture, centrifuge, or double centrifuge while hooded by a small environmental chamber. The temperature of the transducer case should always be monitored by means of a thermoelectric or resistive temperature transducer.

1.5.3.2 Other environmental tests are usually performed on a qualification basis and as nonoperating tests. Brief performance tests are performed after exposure of the transducer to each specified environment (including temperatures). Ambient-pressure tests are sometimes performed as operating tests on piezoelectric acceleration transducers to determine possible performance variations due to minor case deformations. Magnetic-field tests and high sound-pressure-level tests are performed with the transducer connected but without applying acceleration to it. Immersion tests can be used to verify case sealing.

1.6 Specialized Measuring Devices

1.6.1 Special acceleration transducer designs

Besides the normally used transduction methods a number of uncommon techniques have been used in acceleration transducers with varying degrees of success. Angular accelerometers have been designed which use the liquid in a circular tube as sensing element. The flow of the liquid through the tube, due to angular acceleration, has been sensed by a deflecting vane or by use of the electrokinetic effect by which an emf is produced by the flow of certain liquids through a porous disc. A related design uses mercury in the tube, which allows flow-potential or magnetic transduction techniques to be used. Linear-acceleration transducer research has included the use of photoelastic materials, in lieu of piezoelectric elements, in conjunction with electro-optical transduction methods. Various approaches have been used in developments of omnidirectional accelerometers intended to provide an output in response to acceleration regardless of its direction.

1.6.2 Mechanical impedance heads

A variety of systems have been devised to measure the mechanical impedance of structures. Since mechanical impedance is the complex ratio of an applied force to the resulting velocity during simple harmonic motions, such systems generally require a vibration generator, a force transducer, and either a velocity transducer or an acceleration or displacement transducer whose output can be converted to velocity by electrical integration or differentiation. The force and velocity transducers must be closer to each other than the shortest wavelength of interest in the structure or object being tested. This objective can be attained in a *mechanical-impedance head* which combines a (dynamic) force and acceleration transducer in a single compact package. This transducer combination can be calibrated by methods typically used on piezoelectric force transducers.

BIBLIOGRAPHY

1. Weiss, D. E., "Design and Application of Accelerometers," *Proceedings of the Society for Experimental Stress Analysis*, Vol. IV, No. 2, pp. 89–99, 1947.

2. White, G. E., "Response Characteristics of a Simple Instrument," *Instrument Note No. 2*, Los Angeles: Statham Instruments, Inc., 1948.

3. Levy, S. and Kroll, W. D., "Response of Accelerometers to Transient Accelerations," *Research Paper 2138, Journal of Research of the National Bureau of Standards*, Vol. 45, No. 4, October 1950.

4. White, G. E. and Kerstner, S., "Basic Method for Accelerometer Calibration," *Instrument Note No. 17*, Los Angeles: Statham Instruments, Inc., 1950.

5. Smith, G. L., "Angular Accelerometer Errors Arising from a Bar-Type Design of the Seismic Mass," *Review of Scientific Instruments*, Vol. 23, No. 2, p. 97, February 1952.

6. Statham, L., "Design Parameters for Linear Accelerometers," *Instrument Note No. 9*, Los Angeles: Statham Instruments, Inc., 1952.

7. White, G. E., "Secondary Effects in Seismic System Instruments," *Instrument Note No. 23*, Los Angeles: Statham Instruments, Inc., 1952.

8. Thomson, W. T., *Mechanical Vibrations*. (2nd Ed.), Englewood Cliffs, N.J.: Prentice-Hall, Inc., 1953.

9. Perls, T. A. and Kissinger, C. W., "High-*g* Accelerometer Calibrations by Impact Methods with Ballistic Pendulum, Air Gun, and Inclined Trough," *Paper No. 54-40-2*, Pittsburgh: Instrument Society of America, 1954.

10. Rosa, G. N., "Some Design Considerations for Liquid Rotor Angular Accelerometers," *Instrument Note No. 26*, Los Angeles: Statham Instruments, Inc., 1954.

11. Wildhack, W. A. and Smith, R. O., "A Basic Method of Determining the Dynamic Characteristics of Accelerometers by Rotation," *Paper No. 54-40-3*, Pittsburgh: Instrument Society of America, 1954.

12. Edelman, S., Jones, E., and Smith, E. W., "Some Developments in Vibration Measure-

ments," *Journal of the Acoustical Society of America*, Vol. 27, No. 4, pp. 728–734, July 1955.

13. Den Hartog, J. P., *Mechanical Vibrations*. (4th ed.), New York: McGraw-Hill Book Company, 1956.

14. Levy, S. and Bouche, R. R., "Calibration of Vibration Pickups by the Reciprocity Methods," *N. B. S. Research Paper 2741, Journal of Research of the National Bureau of Standards*, Vol. 57, No. 4, October 1956.

15. "American Standard Methods for Specifying the Characteristics of Pickups for Shock and Vibration Measurement," *U. S. A. Standard Z24.21-1957*, New York: U.S.A. Standards Institute, 1957.

16. Crandal, S. H., *Random Vibration*. New York: John Wiley & Sons, Inc., 1958.

17. Stedman, C. K., "Some Characteristics of Gas-Damped Accelerometers," *Instrument Note No. 33*, Los Angeles: Statham Instruments, Inc., 1958.

18. "American Standard Methods for the Calibration of Shock and Vibration Pickups," *U.S.A. Standard S2.2-1959*, New York: U.S.A. Standards Institute, 1959.

19. Lederer, P. S., "General Characteristics of Strain-Gage Accelerometers Used in Telemetry," *N.B.S. Report 6907*, Washington, D.C.: National Bureau of Standards, July 21, 1960.

20. Thomas, I. L., Plumbly, R. W., and Pitt, R. J., "Sources of Error in Precision Force Feedback Accelerometers, and Methods of Testing," *Technical Note No. I.A.P. 1076*, Farnborough, England: Royal Aircraft Establishment, January 1960.

21. Bouche, R. R., "Improved Standard for the Calibration of Vibration Pickups," *Experimental Mechanics*, Vol. 1, No. 4, pp. 116–121, 1961.

22. Bouche, R. R., "Calibration of Shock/Vibration Pickups," *Electro-Technology*, Vol. 67, No. 5, pp. 92–93, May 1961.

23. Harris, C. M., and Crede, C. E. (eds.), *Shock and Vibration Handbook*, 3 vols. New York: McGraw-Hill Book Company, Inc., 1961.

24. Rayevskiy, N. P. and Subbotin, M. I., "Izmereniye Lineynykh Uskoreniy" (Measurement of Linear Accelerations) Moscow: Academy of Science of the U.S.S.R., 1961.

25. Sabin, H. B., "Seventeen Ways to Measure Acceleration," *Control Engineering*, pp. 106–199, February 1961.

26. Smith, R. O. and Mayo-Wells, W. J., "Methods Used at National Bureau of Standards for the Performance Testing of Spring-Mass Accelerometers," *N.B.S. Report 7092*, Washington, D.C.: National Bureau of Standards, Febraury 1961.

27. Bouche, R. R., "Characteristics of Piezoelectric Accelerometers," *Test Engineering*, p. 16, October 1962.

28. Corey, V. B., "Measuring Angular Acceleration with Linear Accelerometers," *Control Engineering*, March 1962.

29. Hilten, J. S., "Performance Tests on Two Bonded Strain Gage Accelerometers," *N.B.S. Report 7440*, Washington, D.C.: National Bureau of Standards, February 1962.

30. Hilten, J. S., "Performance Tests on Two High Sensitivity Piezoelectric Accelerometers," *N.B.S. Report 7468*, Washington, D.C.: National Bureau of Standards, March 1962.

31. Bouche, R. R. and Ensor, L. C., "Calibrators for Acceptance and Qualification Testing

of Vibration-Measuring Instruments," TP223, *Shock and Vibration Bulletin No. 33*, Washington, D.C.: The Shock and Vibration Information Center, U.S. Naval Research Labs., Dec. 1963.

32. Brown, G. W., "Accelerometer Calibration with the Hopkinson Pressure Bar," *Preprint No. 49.2.63*, Pittsburgh: Instrument Society of Amercia, 1963.

33. Hilten, J. S., "Performance Tests on an Amplifier-Accelerometer System," *N.B.S. Report 7781*, Washington, D.C.: National Bureau of Standards, January 1963.

34. "American Standard Nomenclature and Symbols for Specifying the Mechanical Impedance of Structures," *U. S. A. Standard S2.6-1963*, U.S.A. Standards Institute, New York: 1963.

35. Kemeny, T., "A Simple Method to Calibrate and Check Vibrometers" (Contribution to IMEKO 1964 Aerospace Measurements Workshop), Budapest, Hungary: Central Measurement Research Laboratories, 1964.

36. Norton, H. N., "Frontiers in Applied Aerospace Metrology (Acceleration and Pressure)," *Acta Imeko 1964*, IMEKO, Budapest, Hungary, pp. 661–673, 1964.

37. Schloss, F., "Recent Advances in Mechanical Impedance Instrumentation and Applications," *Shock and Vibration Bulletin 34*, Part 3, Washington, D.C.: The Shock and Vibration Information Center, U.S. Naval Research Lab., pp. 3–14, December 1964.

38. Vick, G. L., "A Feasibility Study of a Solid State Acceleration Sensing Technique," *Report No. FDL-TDR-64–55*, Wright-Patterson Air Force Base, Ohio: U.S. Air Force Systems Command, December 1964.

39. "Specifications and Tests for Piezoelectric Acceleration Transducers," *RP37.2*, Pittsburgh: Instrument Society of America, 1964.

40. Carver, C. F., Jr., "Acceleration Testing," *Test Engineering*, p. 34, January 1965.

41. Dunbar, L. E., "Development of an Omnidirectional Accelerometer," *Shock and Vibration Bulletin 34*, Part 4, Washington, D.C.: The Shock and Vibration Information Center, U.S. Naval Research Lab., 1965.

42. Read, J. R., "System to Calibrate Vibration Transducers at Low Displacements," *Shock and Vibration Bulletin 34*, Part 4, Washington, D.C.: The Shock and Vibration Information Center, U.S. Naval Research Lab., pp. 13–19; 1965.

43. Evans, E. J., "Piezoresistive Strain Gage Accelerometers," *Report No. FRL-TR-3374*, Dover, N.J.: Feltman Research Labs., Picatinny Arsenal, December 1966.

44. Reeds, R. T., "Piezoelectric Transducer Failure Modes and Effects," *Tech. Paper TP235*, Pasadena, California: Endevco Corp., July 1966.

45. Arlowe, H. D., Dove, R. C. and Duggin, B. W., "Faithful Transmission of Piezoelectric Transducer Data," *I.S.A. Journal*, Vol. 13, No. 1, pp. 62–66, January 1966.

46. Edelman, S., "Precision Calibration of Accelerometers," *Test Engineering*, Vol. 19, No. 5, pp. 17, November 1966.

47. Hiering, W. A. and Grisel, C. R., "Transport Aircraft Vertical Acceleration Detection," *Technical Report FAA-ADS-71*, Washington, D.C.: Federal Aviation Agency, March 1966.

48. Lederer, P. S. and Hilten, J. S., "Earth's Field Static Calibrator for Accelerometers," *NBS-TN-269*, Washington, D.C.: National Bureau of Standards, February 1966.

49. Coon, G. W. and Harrison, D. R., "Miniature Capacitive Accelerometer," *Instrumentation Technology*, Vol. 14, No. 3, p. 52 March 1967.

50. Favour, J. D., "Accelerometer Calibration by Impulse Excitation Techniques," *Preprint No. P13-1-PHYMMID-67*, Pittsburgh: Instrument Society of America, 1967.

51. Ingebritsen, O. C., "Methods for Calibrating Motion Measuring Transducers at Low Frequencies (Zero to 20 Hz)," *Preprint No. M18-4-MESTIND-67*, Pittsburgh: Instrument Society of America, 1967.

52. Lally, R. W., "Application of Integrated-Circuits to Piezoelectric Transducers," *Preprint No. P4-2-PHYMMID-67*, Pittsburgh: Instrument Society of America, 1967.

53. Ramboz, J.D., "A Proposed Method for the Measurement of Vibration Transducer Transverse Sensitivity Ratio," *Preprint No. M18-6-MESTIND-67*, Pittsburgh: Instrument Society of America, 1967.

54. Rasanen, G. K. and Wigle, B. M., "Accelerometer Mounting and Data Integrity," *Sound and Vibration*, Vol. 1, No. 11, pp. 8-15, November 1967.

55. Ruzicka, J. E., "Mechanical Vibration and Shock Terminology," *Sound and Vibration*, Vol. 1, No. 5, p. 20, May 1967.

2   Attitude

2.1 Basic Concepts

Attitude is the relative orientation of a vehicle or an object represented by its angles of inclination to three orthogonal reference axes.

Attitude transducers measure these angles of inclination with respect to one of several reference systems so that orientation can be determined.

Attitude-rate transducers measure the time rate of change of attitude, frequently referred to as "rate of rotation" or just "rate."

The **units of measurement** are angular degrees, minutes, and seconds, rarely radians, for attitude, and angular units per unit time, usually degrees per second, for attitude rate.

The attitude of a vehicle, such as a ship or an aircraft, is frequently described in terms of the orthogonal planes of the vehicle. These are designated as the **pitch, yaw,** and **roll** planes (Figure 2-1). The attitude of the vehicle can then be represented as pitch, yaw, and roll attitude.

A **bearing** is a direction at a reference point given by the angle in the horizontal plane between a reference line and the line between the reference point and the point whose bearing is specified; bearing is usually measured clockwise from the reference line. An alternate term for bearing, applied primarily to celestial navigation, is *azimuth*.

2.2 Sensing Methods

The methods of sensing attitude can be categorized on the basis of the nature of the reference system with respect to which the orientation of a vehicle or body is determined.

2.2.1 Inertial-reference sensing

The motion of a rotating body obeys Newton's First Law of Motion in an inertial frame of reference in which no forces are exerted on the body and the body is not accelerated. Unless acted upon by some unbalanced torque, a rotating body will continue turning about a fixed axis with undiminished angular speed.

Fig. 2-1. Identification of vehicle planes.

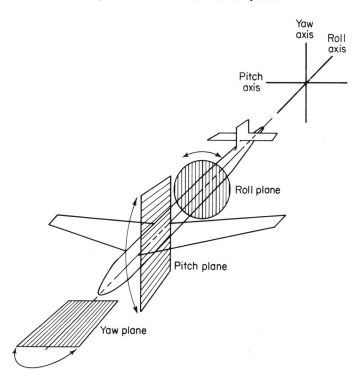

This law is applied in the *gyroscope* (*gyro*) where a wheel on an axle, supported in gimbal rings to allow the axis to assume any desired direction, is set in rotation. Since the gimbal support ideally exerts no torque on the wheel, the axis of rotation remains fixed in its attitude while the wheel is rotating even if the stand supporting the gimbal rings is moved into a different attitude. If the stand is attached to a fixed portion of the vehicle, attitude can be referred to the axis of the spinning gyro wheel.

2.2.2 Gravity-reference sensing

The force of gravity acting on a mass can be used to establish a vertical reference axis. This is exemplified by a string at whose end a plumb bob is suspended. The deflection from vertical of a fixed axis of a vehicle or body can thus be sensed by a device incorporating a mass contained or suspended in such a manner that it can always follow its attraction toward the earth.

2.2.3 Magnetic-reference sensing

If a magnetic field remains fixed in position, its poles establish a reference axis. A bar magnet aligns itself to the poles of a magnetic field, such as that of the earth. This is exemplified by the action of a compass needle. The earth's magnetic field can hence be used as reference for the attitude of a vehicle in the terrestrial horizontal plane.

2.2.4 Flow-stream-reference sensing

The direction along which a fluid flows past a vehicle moving in the fluid can be used as reference for the attitude of the vehicle, provided the vehicle itself does not alter the direction of the flow stream. This condition can be satisfied by placing a sensing element, which tends to align itself with the flow stream, ahead of the forward end of the vehicle. When devices employ this principle to sense attitude, the attitude so referenced is usually referred to as *angle of attack*.

Angle-of-attack transducers are specialized devices used almost solely on high-speed aircraft and rockets. They are briefly described in 2.6.1.

2.2.5 Optical-reference sensing

One or more reference axes for the orientation of a vehicle or body can be established by aiming an optical device on the vehicle, such as a telescope in its mount, at light-radiating celestial bodies whose position is known at the time of measurement, or at points of known position where a step change in the reflection or refraction of light occurs as seen from the vehicle. The attitude of the optical device with respect to the vehicle axes is then a measure of the vehicle's orientation to the optical targets.

A related example of these sensing devices, although used for determination of location rather than attitude, is the sextant.

Optical-reference attitude transducer systems are normally custom designed for specific aerospace missions. They are briefly covered in 2.6.2.

2.3 Design and Operation

2.3.1 Gyro attitude transducers

The operation of gyro attitude transducers is based on the space reference established by the *spin axis* of the gyro rotor. In a simple gyro (Figure 2-2), the rotor shaft is supported at each end by a bearing in a frame which in turn is free to rotate in bearings attached to a fixed structure whose attitude is to be measured with respect to the rotor spin axis. The frame (*gimbal*) rotates about an axis (*gimbal axis*) which is perpendicular to the spin axis. The fixed structure can be the case of the transducer attached to a fixed portion of the vehicle and aligned with one of the vehicle axes. Since the spatial position of the spin axis is fixed when the rotor turns, an attitude change of the case results in an angular displacement between case and gimbal shaft. This displacement can be transduced by any angular-displacement transduction element, e.g., a circular potentiometric element mounted to the case with a wiper arm attached to the gimbal shaft. The mechanism illustrated represents a *single-degree-of-freedom* gyro because the number of orthogonal axes about which the spin axis is free to rotate is one.

The essential stability characteristic of a gyro is the angular momentum (H) of the gyro rotor. The force required to deflect the spin axis from its position increases with increasing angular momentum, which is the product of the rotor angular velocity

Fig. 2-2. Basic single-degree-of-freedom gyro.

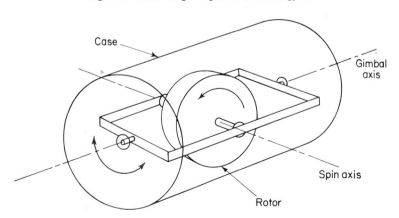

(ω) and the moment of inertia (I_r) of the rotor, or

$$H = \omega I_r$$

The moment of inertia varies with the mass and the square of the radius of the rotor. For a solid cylinder of mass M and radius r, $I_r = \frac{1}{2}Mr^2$. For a hollow cylinder of mass M and radius r and negligible wall thickness, $I_r = Mr^2$, twice that of the solid cylinder. When the wall thickness is not negligible, both inside and outside radii must be considered, and I_r of the hollow cylinder is less than twice as great but is still greater than that of the solid cylinder. Gyro rotors are therefore frequently constructed as a wheel with a thick rim and a thin web so that the shape of a hollow cylinder is approached.

All quantities in the angular-momentum relationship are vector quantities; i.e., they have magnitude as well as direction. The direction of the angular momentum vector lies along the spin axis and is arbitrarily determined by the "right-hand rule": If the fingers of the right hand are placed around the spin axis, with the ends of the fingers pointing in the direction of rotation, the extended thumb will point in the direction of the angular momentum vector.

The quantities described above are usually expressed in the absolute system of units, mass in grams (*gm*), radius in centimeters (*cm*,) moment of inertia in *gm-cm²*, angular momentum in *gm-cm²/s*. Torque, whose relationship is described below, is normally expressed in *dyne-cm* but has sometimes been expressed in *gm-cm* in the gravitational system of units.

The direction of the rotor spin axis remains fixed in inertial space until a torque is exerted on the mechanism. When the torque is applied about any axis except the spin axis, the spin axis will rotate about an axis which is perpendicular to the axis about which the torque is applied and which is also perpendicular to the spin axis. If the torque were applied in the plane of the gimbal frame as in Figure 2-2, the rotor would tend to balance the torque by deflecting angularly about the gimbal axis. This rotation of the spin axis produced by a torque (T) is known as *precession*. The precession rate (Ω) is a constant angular velocity when the torque is held constant. It is related to the torque and the rotor angular momentum (H) by

$$T = H\Omega$$

where all quantities are vectors. The mechanism will rotate about the axis of the applied torque only when it is not free to precess.

2.3.1.1 The free gyro (Figure 2-3) is a two-degree-of-freedom gyro in which the spin axis may be oriented in any specified attitude. It is commonly used as attitude transducer. The angular displacement of each of the two gimbals can be converted into an electrical output by a suitable transduction element ("pickoff"). The two outputs can represent attitude in any two of the three vehicle planes depending on the orientation of the gyro within the vehicle. The most commonly used transduction elements are potentiometric and reluctive (usually synchro-type); others, such as capacitive and photoelectric elements, have been used in special designs.

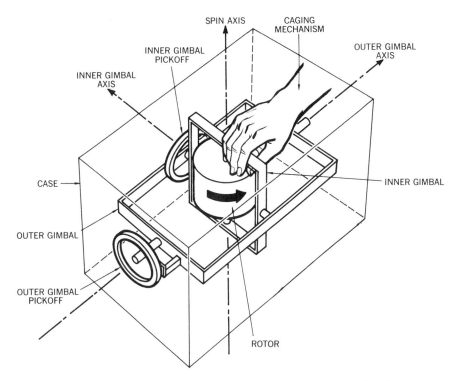

Fig. 2-3.   Two-degree-of-freedom gyro (courtesy of Conrac Corp.).

It should be noted that the inner gimbal can align itself parallel to the outer gimbal after rotating 90 degrees in either direction. This condition is known as *gimbal lock*. The inertial reference is lost when gimbal lock occurs. It can be reestablished by caging the gyro and repositioning the spin axis to the desired reference axis before uncaging the gyro. Mechanical stops are often provided to prevent inner gimbal deflection greater than ± 90 degrees.

The caging mechanism, represented by the hand in Figure 2-3 (some gyros are indeed caged manually), is used to orient and lock one or both gimbal axes to a reference position. The mechanism is unlocked (the gyro is *uncaged*) upon command, usually by an electrical solenoid, after the rotor has been set into rotation. The direction of the spin axis at the instant of uncaging determines the reference system for subsequent attitude measurements.

The gyro rotor is driven by an a-c or d-c electric motor in most designs. The rotor is integral with or closely coupled to the motor's rotating member. In some designs the rotor of the spin motor also acts as the gyro rotor. The stator can be external to a housing specially designed for this purpose. Other means of driving the rotor include a clock spring, which is mechanically wound before use of the gyro, and a

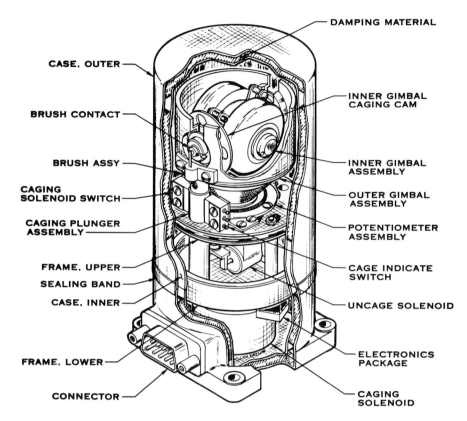

Fig. 2-4. Free-gyro potentiometric attitude transducer (courtesy of Humphrey, Inc.).

pyrotechnic charge which, during combustion, forces a stream of hot gas at small buckets or blades on the rotor surface. The latter source dictates use of the gyro only during its run-down time subsequent to the rapid exhaustion of the charge.

A typical free-gyro attitude transducer is shown in Figure 2-4. Intended for operation from a missile or aircraft battery, its d-c spin motor, integral with the inner gimbal assembly, is fed through a voltage regulator and noise filter and draws 300 ma at 28 V. The outer gimbal assembly moves a wiper arm over a potentiometric element. A similar element is usually activated by the inner gimbal as well. The gyro is electrically caged and uncaged by use of solenoids. A switch on the caging plunger assembly allows external "cage" indication. The case (2.5 in. diameter, 4.6 in. length) is of dual construction. Damping material between inner and outer case provides isolation from vibration and high-sound-pressure environments. The case is hermetically sealed.

A number of unwanted torques cause precession of the spin axis from its intended position (*drift*). These derive from internal sources such as friction in gimbal bearings, radial mass unbalance in the rotor, unbalance within the gimbal assembly, forces exerted by flexible leads, reaction in synchro elements, friction in potentiometric elements, and magnetic interaction as well as from external environmental conditions. Good gyro designs minimize sources of drift; however, it is difficult to reduce the rate of drift of motor-driven, mechanical-mass, free gyros equipped with mechanical bearings below 6 degrees per hour. The drift rate can be over 100 degrees per hour in some low-cost designs.

A number of techniques have been applied to reduce gyro drift. The use of an a-c induction or synchronous motor instead of a d-c motor usually lowers the drift rate. In the *floated gyro* the spin assembly (motor and gyro rotor), contained in a spherical sealed enclosure, is floated in a viscous liquid to give the assembly neutral buoyancy, thus eliminating purely mechanical suspension and reducing static friction in the gimbal bearings. A fluorocarbon has been commonly used as flotation liquid. During use of the gyro the temperature of the liquid should be as stable as possible and temperature gradients within the liquid should be minimized. Pressurized gas is used as flotation fluid in the *gas-bearing* gyro in which the gas can also be used to spin the rotor by acting on small buckets in its surface. Further attempts to minimize any unwanted forces on the rotor by suspending it freely have resulted in the development of an electrostatic suspension system for a spherical rotor in vacuum, spun by an external rotating magnetic field only until the desired angular speed is attained. The run-down time of the rotor in the *electrostatic gyro* is very long since virtually no retarding forces act on the spin assembly. In the *cryogenic* gyro the spherical rotor (typically made of niobium) is maintained at the temperature of liquid helium so that it is superconducting. It is freely suspended within a magnetic-field coil system by interaction between this field system and a magnetic field set up by electric currents in the superconductor. This gyro also operates during its long run-down time following a quick initial runup. Research efforts have also been applied to gyros using principles other than a spinning rotor.

Torquers are used to correct for constant drift rates due to known sources (e.g., rotation of the earth) as well as to precess the spin axis in accordance with a predetermined program. A torquer, or torque motor, is a device which exerts torque on a gimbal in response to a command signal. It is, essentially, a rotary solenoid so installed as to apply a mechanical torque to the gimbal proportional to its electrical excitation. Closed-loop servo operation can be attained by feeding an error signal, proportional to drift rate, to a servo amplifier which powers the torquer.

2.3.1.2 The vertical gyro is a pitch/roll attitude transducer. It is a special version of the two-degree-of-freedom free gyro provided with a two-axis erection system to maintain the spin axis in a vertical position (in the pitch and roll planes). Both gimbal axes are thus in a plane parallel to the earth's surface. The inner gimbal provides measurements of pitch attitude, whereas the outer gimbal displacement is

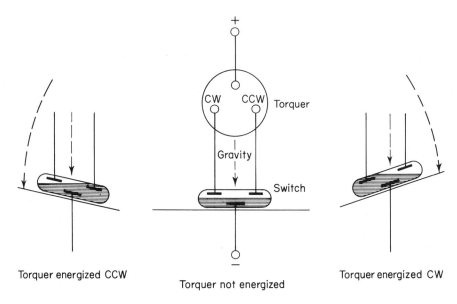

Torquer energized CCW   Torquer not energized   Torquer energized CW

Fig. 2-5. Action of gravity-sensitive switch in erection system.

a measure of roll attitude. Each of the two erection systems consists of a gravity-sensitive switch, such as a conducting-liquid switch (Figure 2-5) in which the liquid acts as common contact as well as the gravity-sensing mass, and a torquer. Mercury or an electrolyte is used as conducting liquid. An alternative erection system uses a gravity-sensitive vane to control air jets which act on the gimbal in one or the other direction. Alignment to local vertical provided by such erection systems is typically between ± 0.05 and ± 0.20 degrees.

   Since the gravity-sensing switches will respond to acceleration other than that due to gravity, the erection system is often equipped with an acceleration switch which disconnects erection power from the torquer during a vehicle maneuver which causes more than a preset level of acceleration on the gravity switch. During the time when erection is discontinued the vertical gyro operates as free gyro.

   2.3.1.3 The directional gyro is a yaw attitude transducer. It is a special version of the two-degree-of-freedom free gyro with provisions for maintaining the spin axis horizontal, i.e., parallel to the earth's surface. The inner-gimbal erection system is identical to that described for the vertical gyro except that the gravity-sensing switch is used to sense deflection of the spin axis from horizontal. The gyro is installed in the vehicle with the outer gimbal parallel to the yaw axis. The direction of the spin axis (in the horizontal plane) is normally established upon uncaging. When the direction of the spin axis is, instead, determined by reference relative to the earth's coordinates, such as by slaving the gyro to a compass synchro-transmitter, the directional gyro operates as *gyro compass*.

2.3.2 Gyro attitude-rate transducers

Attitude rate can be measured directly in terms of angular velocity by the *rate gyro*. This device (Figure 2-6) is essentially a single-degree-of-freedom gyro whose gimbal is elastically restrained and provided with damping means. The rate gyro produces an output signal due to gimbal deflection about the *output axis* by precession in response to a change in attitude rate about the gyro's measurand axis (*input axis*). The output signal is indicative of magnitude as well as direction of the vectorial attitude rate.

The dynamic response characteristics of the rate gyro can be described by a second-order linear differential equation in which the torque applied to the gimbal equals the sum of the three restraining torques, those due to the restraining spring, the damper, and the gimbal inertia. The applied torque is the product of the instantaneous attitude rate about the input axis and the angular momentum of the gyro. The gimbal deflection angle enters into each restraining torque term (as its first derivative in the damping torque term and as its second derivative in the inertial torque term). Since damping and gimbal inertia do not enter the relationship when a constant level of attitude rate is applied, the static response of the rate gyro is then essentially such that

Fig. 2-6. Basic rate gyro.

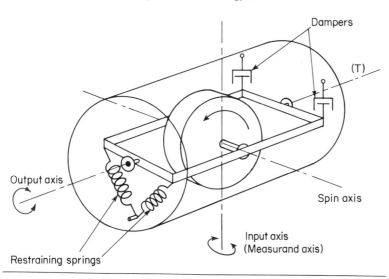

$$\theta = \frac{H\Omega}{K}$$

where: θ = gimbal deflection angle (output deflection)
         H = gyro angular momentum
         Ω = attitude rate (about input axis)
         K = spring constant (of restraining device)

The gimbal deflection is converted into an electrical output by a potentiometric or reluctive transduction element. Typical means of elastic restraint are the torsion bar, the cruciform flexure, leaf springs, the torsional spring, and the magnetic spring. Damping means are of the air-bleed-dashpot, paddle-wheel-viscous-liquid, viscous-shear, hydraulic-bleed, or eddy-current types. The rotor is driven by an a-c or d-c electric motor and sometimes by a compressed gas.

Figure 2-7 illustrates a rate-gyro mechanism which relies on a torsion bar to provide the required elastic restraint. The gimbal is of cylindrical shape and is hermetically sealed. Damping fluid fills the gap between gimbal and outer case. The rotor of the reluctive transduction element ("pickoff") is attached to the gimbal whereas its stator (inner portion) remains fixed to the case. The gyro rotor is normally driven by an a-c motor to revolve at 24,000 rpm. The actual gyro does not use the torsion bar

Fig. 2-7.   Physical arrangement (simplified) of torsion-bar restrained rate gyro (courtesy of Fairchild Controls).

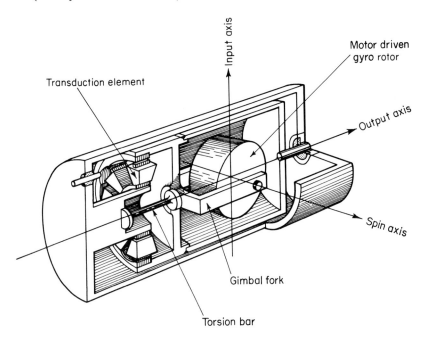

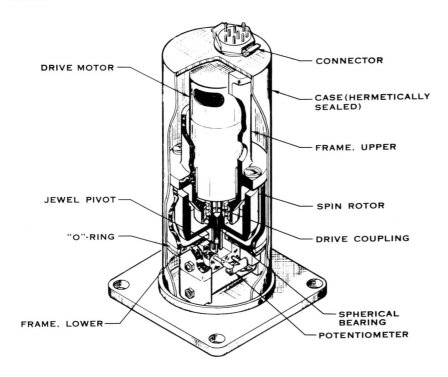

DRIVE MOTOR

CONNECTOR

CASE (HERMETICALLY SEALED)

FRAME, UPPER

JEWEL PIVOT

"O"-RING

SPIN ROTOR

DRIVE COUPLING

SPHERICAL BEARING

FRAME, LOWER

POTENTIOMETER

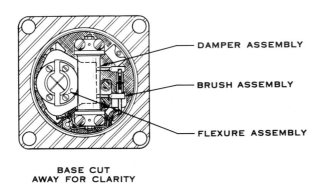

DAMPER ASSEMBLY

BRUSH ASSEMBLY

FLEXURE ASSEMBLY

BASE CUT
AWAY FOR CLARITY

Fig. 2-8. Gimbal-less potentiometric rate gyro (courtesy of Humphrey, Inc.).

as supporting member for the gimbal. The supporting function is provided by a mechanical "spider" network which circumvents the torsion bar.

In the rate gyro shown in Figure 2-8 the function of the gimbal and its supports is provided by a self-aligning spherical bearing. The inner race of the bearing and the thick-rimmed gyro rotor are attached to a cylindrical hub which is driven by the motor through a flexible coupling. The spherical outer race of the bearing, fixed to the frame which also supports the motor, allows the axis of rotation of the gyro rotor to deflect angularly from the rotational axis of the motor. The deflection of the rotor axis, a few degrees to either side of the null position as is typical for rate gyros, actuates the wiper arm in a potentiometric transduction element. The cruciform flexure assembly provides the required elastic restraint. A cylinder-piston assembly with air-bleed ports supplies damping. The damping effect can be adjusted by means of a metering screw in each of the two bleed ports.

A different design uses electrical restraint, rather than mechanical restraint, in a closed-loop servo system. Gimbal deflection from a null position is sensed by a transducing element whose output is amplified and fed to a torquer. The torquer acts on the gimbal so as to return it to the null position. The greater the deflection torque, the greater the restoring torque must be in this torque-balance system. The current supplied to the torquer is therefore proportional to the deflection torque. If a resistor is inserted in series with the torquer, the *IR* drop across the resistor becomes the output voltage of the rate gyro.

A high angular momentum and minimization of random drift, as discussed for free gyros, are also important considerations for rate gyros. Dynamic characteristics of rate gyros deserve particular attention since attitude rate measurements usually involve rapid rate fluctuations. As in all spring-mass-damper systems these include primarily the natural frequency and the damping ratio as well as the constancy of the damping ratio over the required operating temperature range. Liquid damping, usually with silicone fluid, tends to vary with temperature more than gas (air) damping. Most liquid-damped rate gyros provide for reasonably constant damping ratios in the presence of varying temperatures either by compensating devices such as expansion bellows or by use of a heater jacket or other integral temperature control.

2.3.3 Gyro attitude-rate-integrating transducers

The *rate-integrating gyro* (Figure 2-9) differs from the rate gyro primarily by the absence of a spring restraint. It is a single-degree-of-freedom gyro having viscous restraint of the spin axis about the output axis. Precession of the gimbal produces an output signal which is proportional to the integral of the attitude rate about the input axis. The case is filled with a damping liquid. The viscous shear in the liquid within the gap between the sealed cylindrical spin-assembly case and the transducer case provides the integration effect.

The rate-integrating gyro is commonly provided with a torquer on the output shaft. Various modes of operation can be obtained depending on the type of command signal applied to the torquer. Constant internal temperature is frequently attained by using an integral heater. Typical designs incorporate a microsyn-type (reluctive) transduction element and a torquer of the same type. Since a hermetic case seal is required to contain the flotation liquid (which acts as damping fluid) this type of device has often been referred to as "hermetic integrating gyro (HIG)."

The *double-integrating gyro* is a single-degree-of-freedom gyro having no elastic

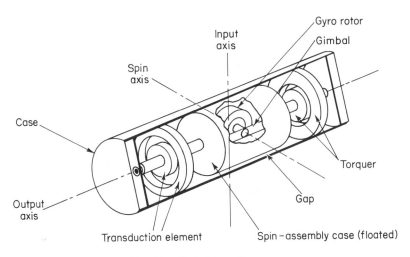

Fig. 2-9. Rate-integrating gyro.

or viscous restraint of the gimbal about the output axis. The dynamic behavior is established primarily by the inertial properties of the gimbal. The output signal, produced by gimbal angular displacement about the output axis and relative to the case, is proportional to the double integral of the attitude rate of the case about the input axis.

2.3.4 Gravity-sensing attitude transducers

Several types of transducers of relatively simple design are useful in measuring gravity-referenced attitude changes. They employ a liquid or solid mass which is either linked to a transduction element or acts as an active portion of it.

The *electrolytic-potentiometer* type transducer (Figure 2-10) resembles a carpenter's bubble level. It consists of a small curved glass tube partially filled with a conducting liquid (electrolyte). Electrodes *A* and *C* act as the end terminals, electrode *B* as the wiper terminal of a potentiometric element. As the tube is angularly deflected from a true horizontal position, the resistance to the common electrode increases from one upper electrode and decreases from the other.

To avoid polarization of the electrolyte, this device is a-c excited. The difference in a-c voltage measured from electrodes *A* and *B*, respectively, to electrode *C* is then a measure of the magnitude and direction of attitude changes from true horizontal. The transducer can be connected as two adjacent arms of a Wheatstone bridge to provide a phase-sensitive bridge output when excited from a center tapped transformer winding which acts as the two other arms of the bridge circuit. Two such transducers, installed at right angles to each other, can be used to measure attitude changes in two orthogonal planes.

Resistance at null position is between 700 and 1000 ohms for each half of the

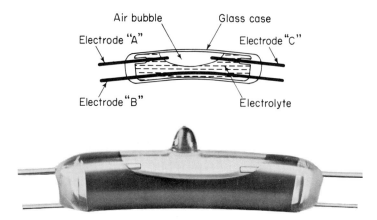

Fig. 2-10. Electrolytic-potentiometer gravity-sensing attitude transducer (courtesy of Hamlin, Inc.).

unit. The attitude range is normally ± 1 degree and the threshold is within 1 minute of arc. The electrolyte has a large negative temperature coefficient of resistance, and temperature control is required for substantial changes in ambient temperature from room temperature. Nominal excitation of the transducer, which weighs 2.3 gr, is 12.5 V a-c.

A *capacitive* transducer has been used to measure small deflections from true

Fig. 2-11. Gravity-sensing-pendulum potentiometric attitude transducer (courtesy of Humphrey, Inc.).

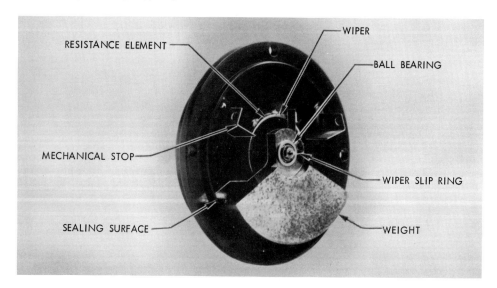

horizontal of apparatus such as seismographs. This device is partially filled with mercury and is similar in construction to the gravity-sensitive switch shown in Figure 2-5, except that the two switch contacts are replaced by capacitor plates. The surface of the mercury then acts as the common "rotor" plate of a dual-stator differential capacitor. An a-c bridge circuit, whose other two legs are resistive, is used to convert the capacity changes into changes of a-c voltage.

The *pendulum* type transducer (Figure 2-11) provides an output proportional to a change in attitude from true vertical of the object to which the transducer case is mounted. The pendulum consists of a segmented weight pivoted to rotate freely about the center of the case in a ball bearing. Rotation of the pendulum causes a wiper to travel over a resistance element. The position of the wiper on the potentiometric resistance element is such that the output is 50% voltage ratio (% VR) at true vertical and 0% and 100% VR, respectively, at the limits of the measuring range which are usually chosen as ±45 degrees.

Pendulum type transducers have also been designed with *reluctive* transduction elements. One design suspends a liquid-damped pendulum in such a manner that it can move a magnetically permeable armature relative to two pairs of inductance-bridge coils placed at right angles to each other. The output of each coil pair provides an amplitude and phase change proportional to the magnitude and direction of attitude change about the pitch axis and the roll axis, respectively.

2.3.5 Magnetic-field-sensing attitude transducers

Transducers related to the compass use the earth's magnetic field to provide a reference for an output proportional to attitude changes in the terrestrial horizontal plane. The reference is a line between the vehicle and magnetic north. A pendulous suspension can be used to stabilize the sensing mechanism in a true horizontal plane. A simple version of such a "north seeker" uses a bar-magnet assembly to actuate the wiper in a potentiometric transduction element. Other designs use reluctive (e.g., synchro) transduction elements and may operate in a closed-loop servo mode.

2.4 Performance and Applications

2.4.1 Specification characteristics

Characteristics typically specified for attitude transducers are described below primarily for the two most widely used types, free gyros and rate gyros. The title or heading of a specification must indicate the specific type of gyro described, preferably by the application-oriented nomenclature as used in this book and as prescribed in ISA Standard S37.1, e.g., "Transducer, Yaw Attitude, Potentiometric, Gyro" or "Transducer, Attitude Rate, Reluctive, Gyro" rather than the design-oriented nomenclature, "Directional Gyro" or "Rate Gyro," respectively.

2.4.1.1 Mechanical design characteristics. Essential outline and mounting dimensions are normally shown on a drawing which also shows the directions of all rotational axes (e.g., spin axis, roll axis, pitch axis, or spin, input, and output axes). The location and types of electrical connectors, cables, or terminals are included on the drawing, as are any reference surfaces or index marks to be used for proper orientation of the transducer during installations. Case sealing, case material and finish, weight, and location and contents of the nameplate are normally specified for all types of attitude transducers. Caging access and type of cage indication are described for manually caged gyros.

Additional mechanical design characteristics are frequently shown by the manufacturer. These include the angular speed of the gyro rotor, the angular freedom of rotation of gimbals, the location of mechanical stops, if any, and characteristics of attitude-rate transducers necessary to an understanding and possible calculations of their dynamic behavior, such as: type of damping means and damping fluid, nature of elastic restraint, angular momentum, spin moment of inertia, gimbal-parts moment of inertia about input and output axes, and maximum deflection of the gimbal axis.

Additional specifications may be dictated by special applications and by use of spin power other than an electric motor.

2.4.1.2 Electrical design characteristics. Excitation voltage and current (or power) and excitation frequency and phasing (for a-c excitation) are specified for the motor as well as for the transduction element(s) and any externally excited torquer(s). Starting current, as well as running current, is stated for the motor. Input impedance (element resistance) is shown for potentiometric transduction elements; input and output or load impedances, for reluctive elements. All external electrical connections should be identified, preferably by inclusion of a simplified wiring diagram. Operating voltage and current for electrical caging provisions and any integral heaters are stated. Insulation resistance is normally specified for all types of attitude transducers. Electrical cage indication, if any, is described.

2.4.1.3 Performance characteristics (at room conditions). The range is stated for all attitude transducers and for each axis (e.g., pitch and roll) in the case of two-degree-of-freedom gyros and other two-axis transducers. Full-scale output or sensitivity, zero-measurand output, linearity (with type of reference line), hysteresis and repeatability, resolution, and, sometimes, friction error of potentiometric transducers, threshold, and drift comprise the most commonly specified accuracy characteristics. Error band specifications offer an alternative to assigning tolerances to each of these characteristics. When maximum static and dynamic drift rates are not identical, tolerances are specified for both. Dynamic drift rate ("Scorsby" drift, see 2.5) is often specified as applicable under certain test conditions during which attitude is varied about specific axes in accordance with a specific program. Drift due to rotation of the earth is normally excluded from stated tolerances.

A number of time-period specifications are usually included: run-up and run-down time for the rotor, overall warm-up time, caging and uncaging time, and erection time. Tolerances for the mechanical alignment obtained by caging and erection provisions and for transverse sensitivity ("cross coupling errors") should be stated.

Dynamic performance characteristics are specified primarily for attitude-rate transducers. They can be shown in terms of frequency response but are usually described by the damping ratio and the natural frequency. The time constant ("characteristic time") is sometimes specified.

The operating life is stated on the basis of a continuous rating for either static or dynamic (Scorsby) operation or both, sometimes augmented by an intermittent rating whose operating cycles can include caging, uncaging, and erection.

2.4.1.4 Environmental characteristics. Tolerances on the effects on performance, especially on drift and dynamic characteristics, should be specified for stated types and levels of operating environmental conditions. These include primarily the operating temperature range, vibration, and acceleration. Among effects of vibratory and steady state acceleration are those due to *mass unbalance* (lack of coincidence of the center of supporting forces and the center of mass of the gyro) and *anisoelasticity* (inequality of compliance of the rotor and gimbal assembly in different directions). Shock, ambient pressure, atmospheric conditions such as humidity and salt content, and special environments such as nuclear radiation are usually specified as non-operating conditions. Magnetic fields and high sound-pressure levels, if specified, should be shown as operating conditions.

2.4.2 Applications

With the exception of gravity-sensing types, attitude and attitude-rate transducers are used primarily in aircraft and other aerospace vehicles and to a somewhat lesser extent on ships. Other applications range from automotive road testing to borehole logging. Gravity-sensing transducers find usage in a variety of measurements where small deflection from true horizontal or vertical are monitored.

Applications in vehicle attitude control (not a subject of this book) are slightly more frequent than in measurement systems. The familiar *autopilot*, for example, relies primarily on attitude sensing by a combination of such transducers. *Inertial-reference systems* and *stable platforms* are assemblies (usually attitude-controlled about at least one axis) of attitude and attitude-rate or acceleration transducers or both, with all components mechanically aligned to each other very closely. The operation and operating modes of these components can be altered by signals from a *programmer*, such as command signals to torquers, in an *inertial-guidance system*.

Measurement systems on vehicles are used for data acquisition during flight tests as well as for continuous indication of attitude in the cockpit or other control station. Two or more visual indicators can be slaved to one attitude transducer ("transmitter") provided with a synchro-type transduction element. Some transducers provide a mechanical as well as an electrical output so that they can be packaged integrally with a visual indicator and mounted on a cockpit panel for observation by crew members.

Considerations for selection of an attitude or attitude-rate transducer are based primarily on the time period (after spin-motor runup or after uncaging) during which measurements must be obtained continuously, on the allowable drift during this period and on other accuracy characteristics, on the dynamic characteristics of the measurand, and on cost. Next in importance are the characteristics and functions of the measuring system to which the transducer is connected and of the vehicle mission during which measurements are made. All other criteria enter into the final phases of hardware selection.

2.5 Calibration and Tests

2.5.1 Calibration

A calibration, as such, is performed only on gravity-sensing attitude transducers using a tilt fixture (see 1.5.1) or a vertically mounted dividing head (see 3.5.1) to vary the angular displacement of the transducer from a true vertical or horizontal reference position in discrete increments while output readings are recorded. Performance verification equivalent to that provided by a calibration is a part of the acceptance (or final production) and qualification test program for gyro-type and magnetic-field-sensing transducers. There is normally no test termed "calibration" on such transducers.

2.5.2 Tests

2.5.2.1 Gyro attitude transducers are production- and acceptance-tested on a *Scorsby table*, a test fixture capable of simulating generalized flight maneuvers. Static tests are usually performed on a sampling or qualification basis only. The Scorsby table consists essentially of a base and a tilt fixture which contains the mounting surface to which the gyro is attached. The base contains the drive motor, the drive control, control switches, and a counting device. The motor drives the tilt fixture on its horizontal rotating table in a clockwise or counterclockwise direction or in an oscillating mode as selected by actuating the appropriate control switches. The gyro mounting surface is tilted from the horizontal to a specified position about the horizontal axis after uncaging the gyro. When the table drive motor is energized, the mounting surface moves through a series of pitch, roll, and yaw maneuvers having a predetermined angular displacement. The table is normally driven to vary these displacements at a frequency of 6 cycles per minute. The counting device is actuated when the oscillating mode is selected. It causes the drive to reverse direction every 60 seconds. The *Scorsby drift* of the transducer can be established by redetermining the output after returning the mounting surface to the horizontal position following a specified period of Scorsby maneuvers.

Allowance must be made for the drift due to earth rotation, approximately 0.25 degree per minute. This is usually done by applying a correction for earth-rotation rate to the drift values determined by tests (Scorsby as well as static tests). The magnitude of the correction depends on the direction of the gyro axis relative to earth coordinates and on the latitude at which the tests are conducted. When the gyro is equipped with torquers, a command signal can be applied to these to partly or wholly correct for earth-rotation rate during tests.

Temperature tests can be performed by placing an environmental hood-type chamber over the tilt fixture while isolating the base from the thermal environment. Vibration and acceleration tests are performed without use of a Scorsby table.

Static test can be performed on the bench using a dividing head with a vertical platform or a vertical pin and indexing plate type of fixture.

2.5.2.2 Gyro attitude-rate transducers are tested for most of their characteristics on a small centrifuge (see 1.5.1) known as a *rate table*. The rate gyro is mounted at the rotational center of the rate table and is oriented so that it will be rotated about its input axis. The variable-speed drive allows applying known attitude rates in discrete increments over the transducer's range. Rate tables have been designed to operate at angular speeds between less than 0.05 and over 5000 degrees per second.

Frequency-response and other dynamic tests can be performed by mounting the transducer either to an angular-displacement arm driven by a vibration exciter or to a motor-driven vertical rotating platform attached to the rate table, with its input axis parallel to this platform so that simultaneous rotation of both test devices causes a sinusoidal variation of attitude rate.

Temperature tests can be performed by mounting an environmental chamber on the rate table and installing the transducer to the chamber floor or by using a hood-type chamber placed over the transducer. Vibration and acceleration tests are performed in the normal manner, using a test fixture which permits precise orientation of the transducer's axes during each portion of the tests.

2.6 Specialized Measuring Devices

2.6.1 Angle-of-attack transducers

Transducers measuring attitude with reference to the direction of the local fluid stream ambient to the vehicle have been used on high-speed aircraft, drones, and rockets. With the exception of one design intended for use on a submarine-launched rocket, these transducers are used exclusively in air. They exist in two basic types,

Fig. 2-12. Vane-type angle-of-attack transducer (courtesy of Conrac Corp.).

the vane type and the pressure-differential type. Although the term "angle-of-attack" has been applied to the measurand of such transducers in general, only the angle about the pitch axis is properly referred to as angle of attack. The angle about the yaw axis is more correctly termed *angle of sideslip.*

A simple vane-type transducer is illustrated in Figure 2-12. Depending on its location on the aircraft, it provides an output proportional to either angle of attack or angle of sideslip. The vane is a drag-stabilized, statically balanced airfoil, made as a wedge-shaped casting of aluminum or high-temperature-resistant stainless steel. When installed with the sharp edge of the vane pointing forward, the vane aligns itself to the direction of the local air flow ambient to the aircraft. The deflection of the vane on its lever arm causes rotation of the internal shaft to which the arm is attached. The shaft is geared to two transduction elements, either potentiometric or synchro, and to a rotational viscous-damper assembly. A deicing heater is integral with the vane.

A related device is the "ogive" transducer. Angle-of-attack sensing in the pitch and yaw axes is obtained by a projectile-shaped body with four integrally machined fins at 90 degrees from each other. The fins point aft and outward at an angle of 30 degrees with the aft portion of the body which swivels freely about the end of a supporting mast. This transducer is usually installed on a "boom" extending forward of the aircraft's nose. As the finned body aligns itself to the direction of the airstream ahead of the aircraft, its relative motions about the pitch and yaw axes actuate two internal transduction elements, usually potentiometric, one for each axis. A finless version of this transducer was also designed, intended for use on rockets during flight at hypersonic velocities.

Pressure-differential angle-of-attack transducers have been constructed in various configurations custom-designed for use on specific vehicles during specific missions. Sensing is performed by a round or pointed probe provided with two flush pressure ports. The ports are so located that the difference in pressure between them varies with the direction of air flow relative to a vehicle axis. Tubing from each port leads to a differential-pressure transducer. Two pressure-port pairs can be used for angle-of-attack measurements relative to two vehicle axes. Transducers of this type have been used on some supersonic and hypersonic aircraft and on space boosters.

2.6.2 Optical-reference attitude sensors

Determinations of position and course by observation of the sun, moon, and stars have been made on ships for centuries and on aircraft in this century. The use of such optical references for attitude measurements, however, is of much more recent origin. Optical-reference attitude sensors have been custom-designed primarily for use on certain aircraft and spacecraft and in the payloads of research balloons. A number of satellites have been guided in space flight by optical-reference attitude control systems aimed at stars (e.g., Canopus). Attitude measurements referred to a line between the vehicle and the sun, and attitude control based on such measurements, are of importance to vehicles or payloads relying for their power supply on solar-cell batteries. Solar sensors are simple inexpensive devices for monitoring a vehicle's spin rate. Electro-optical attitude sensors respond to visible or infrared radiation. Photoconductive light sensors (see 7.3.2) are used most frequently.

A related sensing principle is employed in an infrared horizon scanner which allows spacecraft attitude determination with reference to the thermal radiation discontinuity at opposite horizons of a planetary body. Infrared sensors of various designs have also found considerable military applications in aiming and missile guidance.

BIBLIOGRAPHY

1. Ferry, E. S., *Applied Gyrodynamics*. New York: John Wiley & Sons, Inc., 1932.

2. Deimel, R. F., *Mechanics of the Gyroscope*. New York: Dover Publications, Inc., 1950.

3. Lyman, J., "New Space Rate Sensing Instrument," *Aeronautical Engineering Review*, Vol. 12, pp. 24-30, 1953.

4. Morrow, C. T., "Zero Signals in Sperry Tuning Fork Gyrotron," *Journal of the Acoustical Society of America*, Vol. 27, pp. 581–585, 1955.

5. Breitwieser, C. J., "The Role of the Stable Platform in Inertial Navigation," *Interavia*, Vol. XI, No. 6, p. 447, 1956.

6. Scarborough, J. B., *The Gyroscope, Theory and Applications*. New York: Interscience Publishers, Inc., 1958.

7. Crane, S. C., "Free Gyros," *Giannini Technical Note*, Duarte, Calif.: Conrac Corp., May–June 1959.

8. Johnston, R. H., *The Rate Gyroscope*. Grand Rapids, Michigan: R. C. Allen Business Machines, Inc., Aircraft Instruments Div., 1960.

9. Savet, P. H., *Gyroscopes: Theory and Design*. New York: McGraw-Hill Book Company, 1961.

10. Pitman, G. R., Jr. (ed.), *Inertial Guidance*. New York: John Wiley & Sons, Inc., 1962.

11. Corbin, R. W., "The Application of Variable Capacitance Pickoffs to a Two-Axis Gyroscope," *Report I.R.-20*, Farnborough, England: Royal Aircraft Establishment, 1963.

12. Fischer, J. W. L., "Improved Rate Measurements with Electrically Restrained Rate Gyros," *Giannini Technical Note No. 3*, Duarte, Calif.: Conrac Corp., 1963.

13. Hamilton, A. S., "Improving Rate Tables for Gyro Testing," *Electronic Industries*, Vol. 22, No. 9, pp. 72–75, September 1963.

14. Ernst, E. H. and Tehon, S. W., "Solid State Rate-of-Turn Sensor," *Preprint No. 1.2-1-64*, Pittsburgh: Instrument Society of America, 1964.

15. Fischel, J. and Webb, L. D., "Flight-Informational Sensors, Display, and Space Control of the X-15 Airplane for Atmospheric and Near-Space Flight Missions," *NASA TN D2407*, Washington, D.C.: Office of Technical Services, 1964.

16. Volk, J. A., "Gyroscopes," *Data Systems Engineering*, Vol. 19, No. 2, pp. 28–31, February 1964.

17. "Standard Gyro Terminology" (Rev.), Washington, D.C.: Aerospace Industries Association, September 1964.

18. Hatcher, N. M., Newcomb, A. L., Jr. and Groom, N. J., "Development and Testing of a Proposed Infrared Horizon Scanner for Use in Spacecraft Attitude Determination," *NASA Technical Note TN D-2995*, Hampton, Virginia: Langley Research Center, September 1965.

19. Fraenkel, A., "Gyrodynamics," *Electro-Technology*, Vol. 78, No. 1, pp. 59–72, July 1966.

20. Hughes, W. G., "Attitude Control: Gyros as Sensors," *Report No. ESRO-TM-32*, Farnborough, England: Royal Aircraft Establishment, 1966.

21. Koppe, H., "Spirit Level Electronically Reads Angular Displacement," *Electronic Design News*, Vol. 11, No. 1, pp. 40–42, January 1966.

3 Displacement

3.1 Basic Concepts

3.1.1 Basic definitions

Position is the spatial location of a body or point with respect to a reference point.

Displacement is the vector representing a change in position of a body or point with respect to a reference point.

Linear displacement is a displacement whose instantaneous direction remains fixed.

Angular displacement is the angle between the two coplanar vectors determining a displacement.

Motion is the change in position of a body or point with respect to a reference system.

Proximity is the spatial closeness between two objects or points.

Distance is the spatial separation between two objects or points.

165

3.1.2 Related laws

Displacement, velocity, acceleration

Linear (Translational, Rectilinear)	Angular (Rotational)

$$v = \frac{dx}{dt}$$

$$a = \frac{d^2x}{dt^2}$$

$$\omega = \frac{d\theta}{dt}$$

$$\alpha = \frac{d^2\theta}{dt^2}$$

where: x = linear displacement
θ = angular displacement
v = linear velocity
ω = angular velocity
a = linear acceleration
α = angular acceleration
t = time

Static friction

$$\mu = \frac{F_T}{F_N}$$

where: μ = coefficient of static friction
F_T = force, in sliding plane, required to initiate motion
F_N = force normal to sliding plane, pressing moving surface to stationary surface

3.1.3 Units of measurement

Linear displacement is usually expressed in **inches** (*in.*) or **centimeters** (*cm*), sometimes in **feet** (*ft*) or **meters** (*m*), and very small displacements are expressed in **microinches, "mils"** (milli-inches) or **millimeters.**

The **meter** is the SI (Système Internationale d'Unités) standard unit.

Angular displacement is usually measured in **degrees** (*deg*), sometimes in **radians** (*rad*). The radian is the SI standard unit. One radian equals 57.296 degrees. This can be rounded off to 57.3 degrees where allowable. Small angular displacements can be shown in minutes or seconds (of arc).

3.2 Sensing Elements

With the exception of a few "contactless-sensing" types, displacement transducers sense displacements by means of their *sensing shaft*, which is mechanically connected to the point or object whose displacement is to be measured.

3.2.1 Sensing shafts and couplings

The displacement-sensing shafts of linear- and angular-displacement transducers, and their attachments to the point of measurement (driving point), are usually of simple mechanical design. To understand their importance, however, one must simply realize that the output of a displacement transducer indicates the position of the sensing shaft, not of the driving point. To make the two equal requires a shaft of the proper shape and strength as well as a suitable coupling device.

The coupling must be designed primarily to avoid any slippage after it is fastened. It must also be free from any unwanted play or backlash. For certain applications it is necessary to make provisions for minor misalignments between the point of measurement and the sensing shaft. Such misalignments may occur due to tolerance buildup in the mounting configuration of both transducer and measured object. They may also occur during normal operation of the measured object due to various reasons. Provisions for such minor misalignments can be designed into the coupling, into the sensing shaft itself, or into the internal attachment of the shaft within the transducer.

The shaft coupling, or one of two mating couplings, can be on the transducer-shaft end or attached to the driving point. Typical shaft ends and couplings are illustrated in Figure 3-1. The threaded end, lug, clevis, bearing type, and ball joint are used

Fig. 3-1. Typical displacement-transducer shafts and couplings.

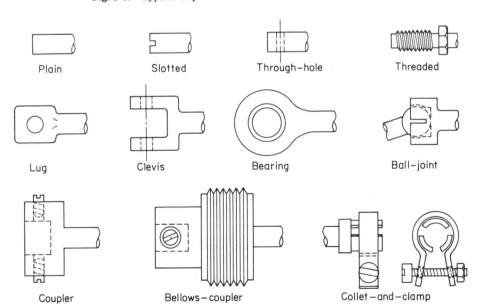

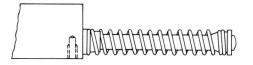

Fig. 3-2. Spring-loaded shaft.

primarily on linear-displacement transducers and the slotted end, coupler, bellows coupler, and collet are used on angular-displacement models.

Spring-loaded shafts (Figure 3-2) are required for certain applications, e.g., when an expected total displacement is larger than the transducer range and this range is intended only for measurement of that portion of the total displacement which occurs closest to the transducer.

3.2.2 Noncontacting transducer sensing elements

A number of specialized types of displacement transducers operate without use of a mechanical link between transducer and point of measurement. The sensing element of such a "contactless-sensing" or "noncontacting" transducer is more difficult to define than that of mechanically linked transducers. Depending on the design, a coherent or noncoherent light beam, or source of such light, may be considered as responding directly to the measurand, sometimes aided by a special reflecting surface at the driving point. When electromagnetic proximity transducer are used as displacement transducers, the moving object, or a portion of it, becomes essentially the transducer sensing element.

3.3 Design and Operation

Among the various types of displacement transducers, elassified primarily on the basis of their transduction principle, three groups are used most frequently: Reluctive transducers are used in a-c measuring circuits; this group includes reluctance-bridge, differential-transformer and some other types. Potentiometric transducers are used in d-c systems. Digital-output transducers are used when very close measurement accuracy is required. A number of additional types are commercially available but are used less frequently. A few designs are not yet sufficiently developed for widespread use.

3.3.1 Capacitive displacement transducers

Capacitive transducers are usually designed for linear-displacement measurements only. Three design variations exist in this category (see Figure 3-3): moving-dielectric, moving-rotor (coupled), and moving-rotor (noncontacting) transducers. In all these, displacement is converted into a change of capacitance.

3.3.1.1 Moving-dielectric type. In a typical moving-dielectric design two concentric cylindrical electrodes constituting a capacitor (rotor and stator) are fixed and station-

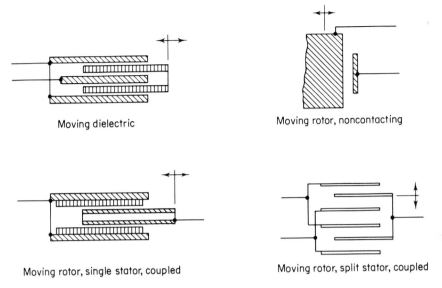

Moving dielectric

Moving rotor, noncontacting

Moving rotor, single stator, coupled

Moving rotor, split stator, coupled

Fig. 3-3. Capacitive displacement transduction.

ary. A sleeve is inserted between inner and outer electrode. The sleeve is made from an insulating material having a dielectric constant different from that of air. It slides between the electrodes with a minimum of friction. As the sleeve is pulled out of the electrode assembly, an increasing amount of electrode surface, on both cylinders, sees air as dielectric and a decreasing amount of surface sees the sleeve material as dielectric. This results in a change of capacitance proportional to sleeve motion (axial displacement). The outer cylinder can be an integral part of the transducer's case and can then be electrically grounded so as to provide shielding from external stray fields.

3.3.1.2 Coupled moving-rotor type. In its single-stator (unbalanced) version this type can consist of one cylinder within another, each constituting an electrode, with a dielectric material covering the inside surface of the outer cylinder. The inner cylinder slides in and out of the outer cylinder. The sensing shaft or the coupling must be electrically insulated from the moving electrode. The split-stator (balanced) version can have its sensing shaft linked to a number of ganged capacitor plates, each of which moves between two stationary (stator) plates. The upper stator plates are electrically interconnected to form one section of the split stator. The lower plates, interconnected, form the other section. As the rotor plates move, their capacitance to one section increases while their capacitance to the other section decreases.

3.3.1.3 Noncontacting moving-rotor types. This is the simplest form of capacitive displacement transducer. The transducer forms one plate; the measured object, the other plate of a capacitor. The object must be either metallic or, if insulating, must have a metallic backing material. The measured object should be grounded. The transducer plate is insulated from its mechanical support which, in turn, should also be grounded.

3.3.1.4 Capacitance-to-dc conversion.

Since it is usually impractical to use a transducer output which is purely a capacitance change, most capacitive displacement transducers are available with either integrally or separately packaged signal-conditioning circuitry. Such conversion circuitry typically operates from a battery or other d-c power supply and furnishes a d-c output voltage proportional to the displacement. If the transducer is of the split-stator type, it forms by itself the two capacitive arms of an a-c bridge. Single-stator types require a fixed capacitor as second bridge arm, either as part of the conversion circuitry or, preferably, within the transducer case. The value of this fixed capacitor should be manually adjustable so that the bridge network can be nulled as needed. The signal-conditioning package provides an oscillator, necessary for bridge operation, as well as conversion of the a-c bridge output to d-c.

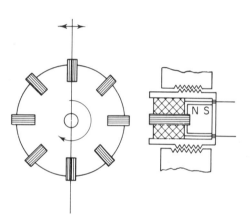

Fig. 3-4. Electromagnetic displacement transducer.

3.3.2 Electromagnetic displacement transducers

Since electromagnetic transducers produce an output only in response to a *change* in magnetic flux, their application to displacement measurement is limited. They are usable primarily where the measured object is in constant oscillatory or rotational motion (see Figure 3-4) or when a number of metallic objects move by the transducer at a fairly high speed. Although electromagnetic displacement transducers are frequently referred to as *proximity transducers*, they respond actually to a *change* in the proximity of a measured object to the transducer. Transducers of this type are sometimes used as incremental digital displacement transducers (see 3.3.8.1). A more detailed discussion of their characteristics as angular speed transducers is included in Chapter 12.

3.3.3 Inductive displacement transducers

Transducers converting displacement into a change of the self-inductance of a single coil can be grouped into coupled and noncontacting versions (see Figure 3-5). Both are used mainly for linear displacement measurements.

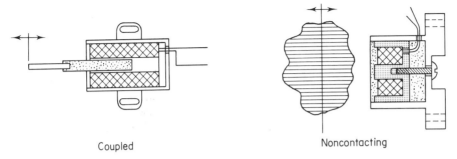

Coupled   Noncontacting

Fig. 3-5.   Inductive displacement transducer types.

3.3.3.1 Coupled inductive displacement transducers. The coupled designs consist essentially of a single coil (bobbin) into which a magnetically permeable core is inserted so that it can slide freely. The nonmagnetic sensing shaft is attached to the movable core. The self-inductance of the coil changes as the core is pushed into the coil. The coil is usually enclosed in a metallic housing (case). The two connecting leads are brought out of the case through an opening which is then filled with sealant. Alternately, an electrical receptacle, to which the leads are soldered or welded, can be made integral with the case.

3.3.3.2 Noncontacting inductive displacement transducers. These are similar to the coupled version except that their core is replaced by the measured object itself. Although measurements are feasible when these objects are of a highly conductive diamagnetic or paramagnetic material, they are considerably more successful when the objects are ferromagnetic; i.e., when they have a high permeability, regardless of their specific conductivity. As the ferromagnetic object, which can be an insulator (e.g., ferrite), semiconductor, or conductor is brought in closer proximity to the coil, the self-inductance of the coil changes. The selection of coil dimensions and characteristics require special considerations of the measured object's material, dimensions, configuration, and range of displacement.

3.3.3.3 Output conversion. Two types of circuits are used in conjunction with inductive displacement transducers. The more prevalent one is the a-c bridge, or impedance bridge, in which the inductance of the sensing coil (L_S) is compared with the inductance of a reference or balancing coil (L_R). Two resistors (R_1, R_2) complete the bridge configuration. The bridge network is excited from a source of alternating current. The a-c output voltage, which is proportional to bridge unbalance in its amplitude as well as its phase, can then be demodulated into a d-c voltage. The refer-

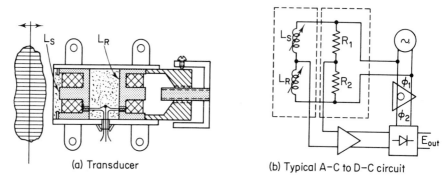

(a) Transducer   (b) Typical A–C to D–C circuit

Fig. 3-6. Inductive displacement transducer with integral bridge-completion coil.

ence coil (bridge-completion coil) should preferably be mounted within the transducer [see Figure 3-6(a)]. This reduces undesirable effects due to long connecting leads, causes both coils to see approximately the same temperature, and affords ready adjustability of the bridge-completion coil. In the design illustrated, the inductance of the coil L_R is manually adjusted by turning the end cap to which a threaded core (slug) is attached so that it moves to or from the coil as the cap is rotated. A set screw allows fixing the end-cap position at the desired coil adjustment. The bridge and signal-conversion circuit in which this transducer is typically used is shown in Figure 3-6(b). The demodulator shown is sensitive to the phase difference between excitation and bridge-output (phase demodulator). However, an amplitude demodulator can be used instead.

A design variation of this type of transducer is shown in Figure 3-7. The bridge-completion coil is not manually adjustable. Bridge-balance as well as gain and cable-length-compensation controls are provided on the panel of a separately packaged electronics unit. This unit also furnishes the 1MHz excitation to the transducer and performs the a-c to d-c demodulation. The coils are of ceramic-insulated magnet wire wound on ceramic forms. A cable-matching network module, which also contains the required two resistive bridge arms, is connected between the electronics unit and the transducer.

The inductive displacement transducer illustrated in Figure 3-8 is used in an entirely different circuit, an L-C oscillator in which the variable inductance (L) and a fixed capacitor (C) form the tank circuit. The small variable capacitor (C_A) provides adjustment of the oscillator's output frequency to the desired value. In an alternate circuit the two capacitors can be in the transducer. In this frequency-modulated-output configuration, the output frequency changes over a specified span as the sensing shaft responds to the displacement to be measured.

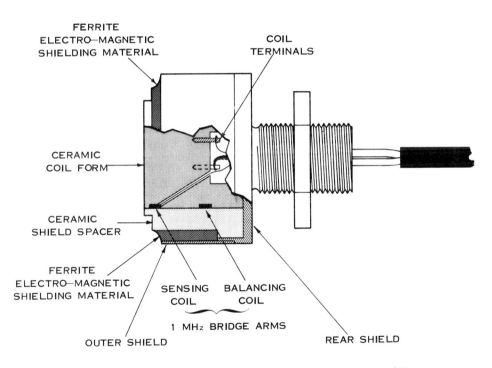

Fig. 3-7. Noncontacting inductive displacement transducer (courtesy of Kaman Nuclear).

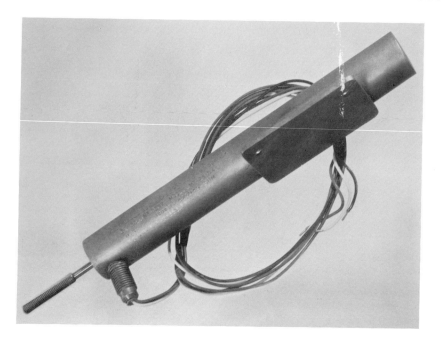

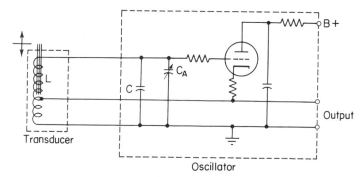

Fig. 3-8. Inductive displacement transducer in oscillator tank circuit (courtesy of G. L. Collins Corp.).

3.3.4 Photoconductive displacement transducers

Coupled as well as noncontacting transducers exist also in those designs utilizing the change in resistance of a semiconductor material due to a change, with displacement, in the amount of illumination incident upon the material or specific portions of the material.

3.3.4.1 Coupled photoconductive displacement transducers. An example of a photoconductive displacement transducer intended primarily for use with a sensing shaft is illustrated in Figure 3-9. The sensing shaft is attached to a moving shutter or mask within the transducer. This design acts as a frictionless potentiometer when the excitation voltage applied across the high-resistance strip is divided by light falling upon the photoconductive gaps between this strip and two low-resistance (high-conductance) film electrodes. Terminals on the latter act as the "wiper" terminal on a potentiometric device. The resistance of the high-resistance electrode is usually held at a specified value between 2 and 8 kilohms. The photoconductor material is cadmium selenide (CdSe). A light source is located within the transducer. The light intensity incident upon the photoconductor is held to between 400 and 800 ft-candles at a color temperature between 1800 and 2200°K. The transducer can respond to 99% of a step change in displacement in 15 ms. In an alternate version of this transducer the beam-focused light source itself is displaced and the nonmovable shutter is provided with a longer slit.

A related design contains a prefocused light bulb, a movable shutter, and two photoconductive sensors in an arrangement whereby light intensity incident upon one sensor decreases as that on the other sensor increases. The two sensors are connected in a bridge circuit so that the oppositely varying outputs of the two sensors become essentially additive.

Fig. 3-9. Photoconductive displacement transducer (courtesy of Conrac Corp.).

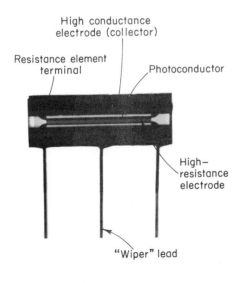

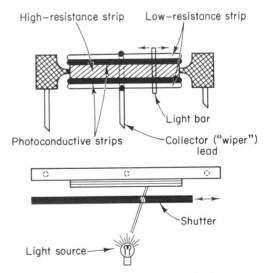

(a) Element (shutter and light source not shown)

(b) Typical arrangement for linear displacement transduction

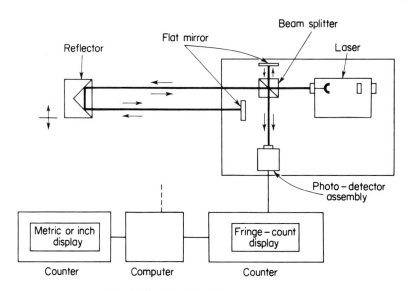

Fig. 3-10.   Laser interferometer system.

3.3.4.2 Noncontacting photoconductive displacement transducers. A number of devices have been developed for the measurement of linear as well as angular displacement by the change in intensity, phase, or position of an incident light beam on a photoconductive element. This light beam either emanates from the measured object and moves with it, or emanates from the transducer and is reflected by the measured object.

The light beam can be kept narrow by optical lens arrangements or by using fiber optics ("light-pipes"). Laser beams have proved very suitable for such applications because of their extremely small beamwidth as well as their narrow bandwidth. Light beams can be of constant intensity (d-c), or modulated (a-c), interrupted (chopped), or pulsed at their source. Polarized light is used for some of these instruments. Infrared light is utilized in the majority of noncontacting photoconductive displacement transducers.

If reflection from the measured object is necessary for sensing purposes, a highly polished reflecting surface or a specially shaped reflector is often attached to, or formed on, the object. Other photoelectric elements, such as photomultiplier tubes, are sometimes substituted for the simpler photoconductive elements.

One of the most valuable of these electro-optical devices is the *laser interferometer*, a digital-output photoelectric displacement transducing system (Figure 3-10). Fringe theory and optical interferometers are described in most optics textbooks and will not be explained in detail here.

As the reflector on the measured object is displaced, an *interference fringe* is seen by the photodetector for each quarter of the laser wave length, a distance of about 6.2 microinches. The output of the photodetector is fed to a high-speed bidirectional counter which displays the *fringe count*, i.e., one count per each 6.2 microinches of displacement. The fringe count can then also be fed to a computer which converts it into a digital inch or metric display. Additional sensing and computing subsystems can be added to correct this display for measured-object temperature as well as for ambient room temperature, humidity, and pressure.

3.3.5 Potentiometric displacement transducers

Displacement transducers with potentiometric transduction elements are almost invariably sensing-shaft coupled. They are relatively simple devices (see Figure 3-11) in which a sliding contact (*wiper*) moves over a resistance element (*potentiometric element*). The wiper is attached or mechanically linked to the sensing shaft but electrically insulated from it. The electrical connection from the wiper is made either by connecting a lead (wire) directly to the wiper, allowing sufficient play in the lead to accommodate wiper travel, or by a second wiper, connected to the first one, which slides over a bus bar to which the lead is then connected.

Potentiometric elements in displacement transducers are usually wirewound,

Fig. 3-11.   Examples of potentiometric displacement transducers.

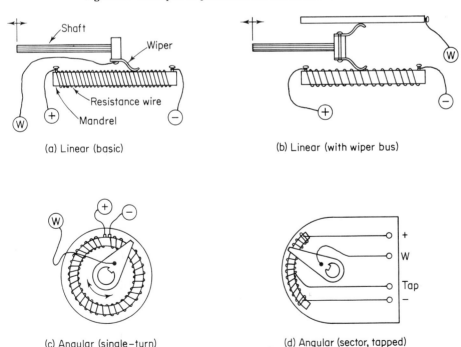

(a) Linear (basic)

(b) Linear (with wiper bus)

(c) Angular (single-turn)

(d) Angular (sector, tapped)

sometimes plastic-film, carbon-film, metal-film, or cermet (ceramic-metal mix). Resistance wire is made from various metal alloys, such as platinum or nickel alloy. The wire, when wound over the mandrel, is normally covered with a varnish which must subsequently be removed from the wiping area. Proper tensioning of the wire, as well as equal spacing, deserve special attention during the winding process. Wire diameter is usually on the order of 0.5 to 3 mils. Mandrels are made either from an insulating material, such as glass or plastic, or from insulation-coated metal, such as copper magnet wire or anodized aluminum.

Where mechanical travel is required beyond the specified displacement range (*overtravel*), the wiper slides over a bussed portion of the element (bus strip or wire turns soldered together) provided for this overtravel. The active portion of the element

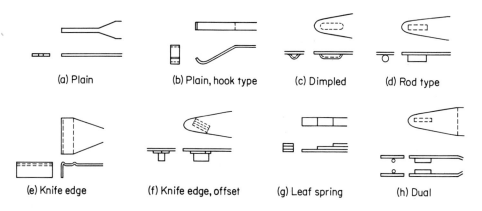

Fig. 3-12. Typical wipers.

can also contain one or more *taps* (additional electrical connections) at prescribed positions along the element.

Wipers exist in a number of different shapes and configurations (see Figure 3-12). Leaf-spring and dual wipers are used when high shock or vibration levels are expected in the transducer's application. Wipers are typically made from precious-metal alloys or from spring-tempered phosphor bronze or beryllium copper. Wiper contact force (against the potentiometric element) is generally between 4 and 15 gm. Conductive lubricants, such as niobium diselenide, have been used on wipers to reduce friction, even when in a vacuum environment.

Because of the internally exposed transduction element, sealing of the two openings in the transducer case (shaft and electrical connections) can be somewhat more critical than for some of the other types of displacement transducers. Various gland seals and double O-ring seals have been employed at the shaft opening to protect the internal portion from severely contaminating atmospheres. Sealed-off electrical

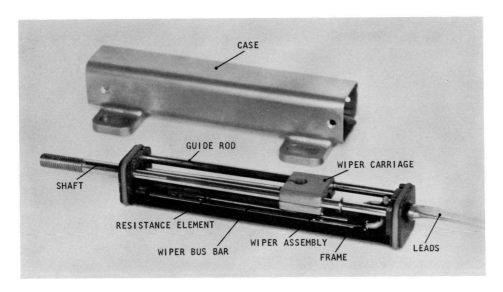

Fig. 3-13. Potentiometric linear-displacement transducer (courtesy of Bourns, Inc.).

connections can be made by using hermetically sealed header terminals or receptacles, or by epoxy potting of pigtail leads in the case-exit region.

3.3.5.1 Potentiometric linear-displacement transducers. A typical transducer is shown in Figure 3-13. In this example, the sensing shaft is threaded; electrical leads are potted at their case exit; both end plates are provided with gaskets which seal by contact with the inside surface of the case when the latter is slipped over the frame; and the wiper connection is made by means of a wiper bus bar.

Other designs incorporate two or more resistance elements and ganged wipers to furnish output signals to more than one measuring system, or one or more switches, in addition to the resistance element, to provide discrete signals at preset displacement values besides the normal analog output.

Large linear displacements, up to about 6 m, can be measured with single-turn or multiturn angular-displacement transducers (rotary potentiometers) whose shaft is provided with a pulley and, usually, a spring return. A cable is wrapped over the pulley. One full turn of the pulley corresponds to a known length of cable unreeled from it. The spring return serves to rewind the cable on the pulley. The pulley can be geared to drive a single-turn potentiometer over its full rotation for a given translational (linear) displacement of the cable end away from the transducer. Alternately, the pulley can be designed to drive a multiturn potentiometer over its full travel without use of gears. Another version uses a single-turn potentiometer without gears

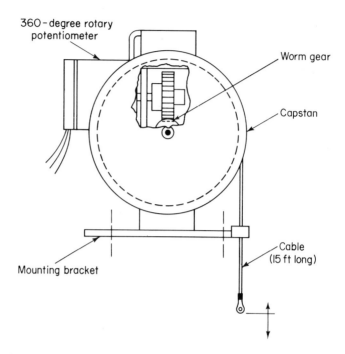

Fig. 3-14. Potentiometric transducer for large linear displacement; 15 ft of cable-end travel produce one revolution of potentiometer (developed by General Dynamics Corp., Convair Div.).

between its shaft and the pulley; as the cable unwinds, the potentiometer is driven over several full rotations. This type of transducer provides a "sawtooth" output, each tooth corresponding to one full turn of the potentiometer and a known translational displacement of the cable end. A fractional turn can be determined from the ramp of the last (incomplete) sawtooth. A string-operated potentiometric transducer for large displacements is shown in Figure 3-14.

3.3.5.2 Potentiometric angular-displacement transducers. This category includes instruments for the measurement of displacements of less than 10 degrees and over 2360 degrees. These devices are related to the general family of rotary potentiometers which include small "trimming" potentiometers as well as volume controls in radio receivers. It should be noted, however, that not all types within this family qualify as angular-displacement transducers. Only a limited number of rotary potentiometers can meet the various requirements and provide the performance characteristics expected from such transducers under a number of operating and environmental conditions.

Single-turn transducers can be designed for continuous rotation starting again with the zero-degree position after completing one turn. They can also be furnished with end stops. Since a certain amount of gap is necessary between the two excitation terminals [see Figure 3-11(c)], such as small tabs soldered or welded to the ends of the potentiometric element, a full 360-degree range is usually not feasible. Maximum spans are typically 355 to 358 degrees. Transducers with a much smaller range than this may incorporate an extensive bussed portion of the element if continuous rotation is required.

Angular displacement transducers for ranges less than 355 degrees, especially those less than 180 degrees, are sometimes referred to as "sector" potentiometers [Figure 3-11(d)]. For very low ranges, the sensing-shaft rotation can be mechanically amplified to achieve more wiper travel. An example of this is shown in the transducer of Figure 3-15 which has a range of ± 5 degrees.

Multiturn rotary potentiometers have been used when ranges exceed 360 degrees. These have their resistance element wound or applied over a helical mandrel, with the wiper sliding along the outside or inside of the helix. Three-turn and ten-turn potentiometers are standard multiturn configurations.

3.3.5.3 Autotransformer-type displacement transducers. Although not properly classifiable as potentiometric transducers within the strict definition used in measure-

Fig. 3-15. Potentiometric angular-displacement transducer for narrow-range displacements (courtesy of Bourns, Inc.).

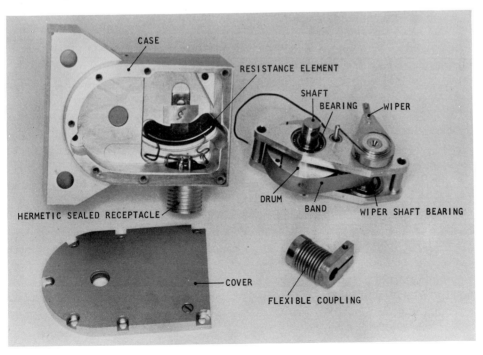

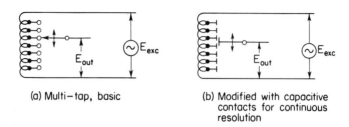

(a) Multi-tap, basic

(b) Modified with capacitive contacts for continuous resolution

Fig. 3-16. Autotransformer-type displacement transducers.

ment terminology, the a-c potentiometer has been used, in the form of an autotransformer, in some types of displacement transducers. This device shares with the resistive potentiometer the ability to provide an output varying from 0% to 100% of the excitation voltage applied across the transduction element depending on the position of a sliding contact [Figure 3-16(a)]. The resolution steps in such a tapped autotransformer may be less than acceptable for many measurement applications. A modified version [Figure 3-16(b)] provides continuous resolution by replacing the fixed tap contacts as well as the sensing-shaft-actuated sliding contact with capacitor plates. The plates can be shaped and arranged so as to provide essentially linear output with displacement.

3.3.6 Reluctive displacement transducers

This group includes all transducers which convert displacement into an a-c voltage change by a variation in the reluctance path between two or more coils or separated portions of one coil when a-c excitation is applied to the coil(s). The *differential transformer* (*LVDT*—"linear variable differential transformer") and the *inductance bridge* are used in linear-displacement transducers. The *synchro, the resolver*, the *induction potentiometer*, the *microsyn*, the *shorted-turn signal generator*, the *differential transformer*, and the *inductance bridge* are used for angular-displacement measurements. These different types are schematically illustrated in Figure 3-17.

3.3.6.1 Differential-transformer displacement transducers. A basic differential transformer consists of a primary winding and two secondary windings [Figure 3-17(a)]. The windings are arranged concentrically and next to each other. They are wound over a hollow mandrel (coil form, *bobbin*) which is usually of a nonmagnetic and insulating material. A ferromagnetic core ("armature") is attached to the transducer sensing shaft or acts as sensing shaft by itself. This core slides freely within the hollow portion of the bobbin (see Figure 3-18). The core is typically made from high-permeability ferromagnetic alloy and has the shape of a rod or cylinder. The winding assembly is potted within a cylindrical case, also typically of ferromagnetic metal.

In operation, a-c excitation is applied across the primary winding, and the movable core varies the coupling between it and the two secondary windings. When the core is in the center (*null*) position, the coupling to the two secondary coils is equal. As the core moves away from the null position, the coupling to one secondary,

Fig. 3-17. Reluctive displacement transducers.

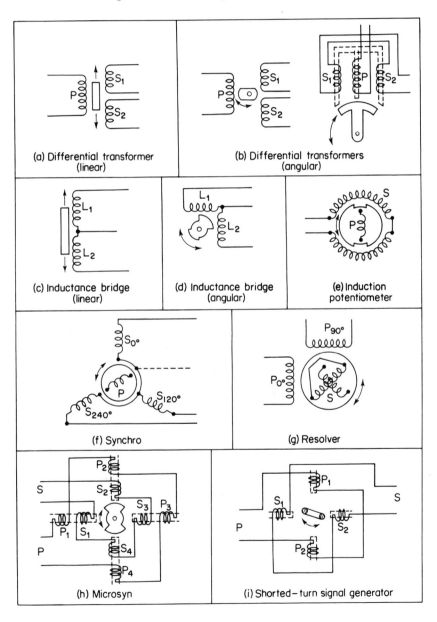

(a) Differential transformer
(linear)

(b) Differential transformers
(angular)

(c) Inductance bridge
(linear)

(d) Inductance bridge
(angular)

(e) Induction
potentiometer

(f) Synchro

(g) Resolver

(h) Microsyn

(i) Shorted—turn signal generator

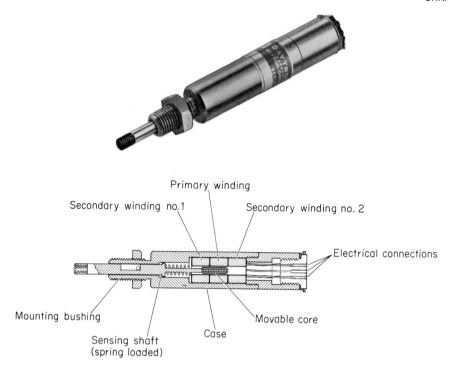

Fig. 3-18. Differential-transformer linear-displacement transducer (courtesy of Daytronic Corp.).

and hence its output voltage, increases while the coupling and the output voltage of the other secondary decreases. The phase of the secondary voltages also shifts with core motion, from 0 degrees at one end, through 90 degrees at the null position, to 180 degrees at the other end of travel.

Similar considerations apply to angular-displacement transducers using the differential transformer principle. A basic configuration and a typical "*E*-core" configuration (with secondaries connected in series) are illustrated in Figure 3-17b. Windings on an *E*-core are also used in some linear-displacement transducers.

In the simplest winding configuration, Type 1 of Figure 3-19, the two secondaries are connected in series opposition. The net output is an a-c voltage increasing in amplitude as the core is moved away from the null position (Figure 3-20). The phase change, however, is not detected by a (high-impedance) a-c voltmeter connected across the output. The amplitude is near zero at the core "center" position and reaches an equal maximum at the core "in" and "out" positions. Hence, this arrangement is useful only when the displacement range is one-half or less of the available core travel. The minimum voltage obtainable at the core "center" position is the *null voltage* (E_{null}), which is usually somewhat greater than zero.

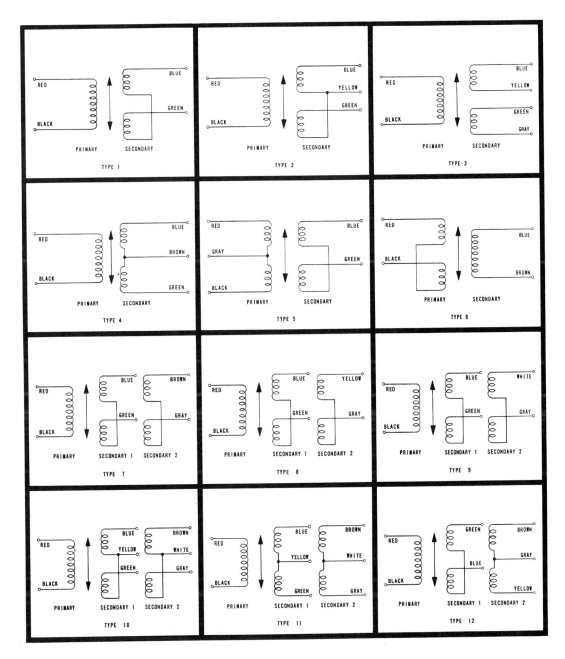

Fig. 3-19. Differential transformer winding configurations (courtesy of Columbia Research Laboratories, Inc.).

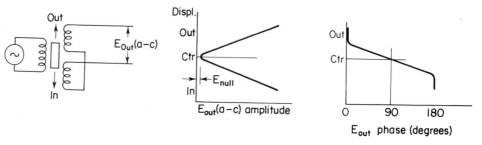

Fig. 3-20. Differential transformer output amplitude and phase.

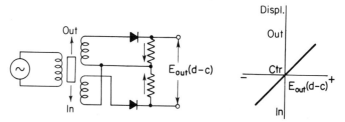

Fig. 3-21. D-C output from a differential transformer.

If the winding configuration shown as Type 2 in Figure 3-19 is used in conjunction with a simple demodulator circuit, a d-c output voltage is produced which is zero at the core "center" position, increasingly negative as the core moves "in" and increasingly positive as the core moves "out" (see Figure 3-21).

When operation of a reluctive transducer from a d-c power supply is required, a transistorized d-c to a-c converter is integrally or separately packaged and connected to the transducer. This version must usually provide a d-c output as well. The output-conversion circuitry is packaged with the excitation converter. Typical conversion and full-wave demodulation circuitry as integrally packaged with a differential-transformer linear-displacement transducer is shown in Figure 3-22. The winding configuration is per Type 3 of Figure 3-19.

The other winding configurations of Figure 3-19 are used for various applications including output-signal compensation and correction, for phase-shift control, and for null balance and potentiometer balance in servo control systems. Types 7, 8, and 9 differ in the turns ratio between Secondary 1 and Secondary 2, which is 1 : 1 for Type 7, 2 : 1 for Type 8, and 5 : 1 for Type 9. Primary windings in transducers for ranges larger than about one inch are often distributed over the full length of the bobbin, including those portions containing the secondary winding, and the core is somewhat shorter than the bobbin.

3.3.6.2 Inductance-bridge displacement transducers. The inductance bridge [Figure 3-17(c) and (d)] is used in linear- as well as angular-displacement transducers.

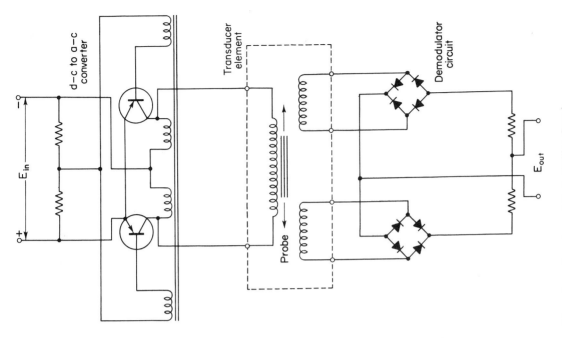

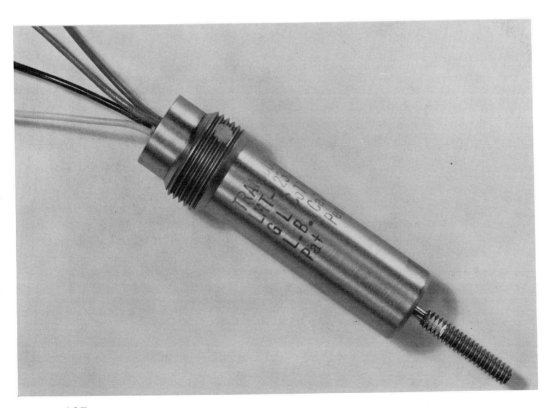

Fig. 3-22. D-C to d-c differential-transformer linear-displacement transducer (courtesy of G. L. Collins Corp.).

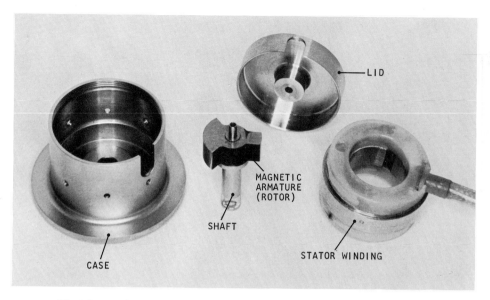

Fig. 3-23. Inductance-bridge angular-displacement transducer (courtesy of Bourns, Inc.).

Two coils and the moving core are so arranged that the inductance of one coil increases while the inductance of the other coil decreases with motion of the core. The matched set of two coils forms two arms of a four-arm a-c bridge. Hence, the output of this bridge is much greater than that of the inductive transducer circuit in which the inductance of the second (bridge-completion) coil does not vary.

Figure 3-23 shows a typical inductance-bridge angular-displacement transducer. Designed for a displacement range of ±45 degrees, the transducer consists of a fixed stator, containing the two coils, and a magnetic rotor attached to the sensing shaft. A case with a lid encloses the two subassemblies. The shielded cable is moisture-sealed to the stator. When 24 V, 400 Hz excitation is applied to a four-arm bridge—two of which are formed by the transducer, the other two by a matched pair of 500 ohm resistors—the full-scale bridge output is 6 V. The output can also be modified to d-c by use of a phase demodulator, integrally or separately packaged. Shaft rotation is continuous.

3.3.6.3 Induction-potentiometer displacement transducers. These devices are found more often in computing and control systems than in measurement systems, but are commercially available and usable as angular-displacement transducers. Their rotor, to which the sensing shaft is attached, contains the single primary winding [Figure 3-17(e)]. The stator is wound with the single secondary (which can be two secondary windings in series). The induction potentiometer is inherently capable of providing an output linear with angular displacement (shaft rotation) over a range of about ±35 degrees. Shaft rotation is normally continuous. A number of design refinements have been devised to overcome some of this type of transducer's limitations which include: a variable output impedance that can result in considerable loading error, the limited angular-displacement range, and errors due to case deformation, mechanical asymmetry, thermal effects, and aging.

3.3.6.4 Synchro-type displacement transducers. Synchros are widely used in servo control and other automation systems. The basic components of a synchro are a single-phase rotor in a three-phase stator. The stator windings are physically spaced at 120 degree intervals [Figure 3-17(f)].

When applied to angular-displacement measurements the synchro transducer is most frequently connected to another synchro which drives a readout device such as a dial. A pair of synchros so connected is sometimes called a "synchro chain." In such measurement systems, the transducer has been referred to as the "synchro transmitter" and the readout device as the "synchro receiver" (Figure 3-24). The synchro rotor windings are typically excited from a 26 V, 400 Hz supply. The stator windings are connected together and properly phased as shown. When the rotor of the transducer synchro is rotated through a given angle, the stator of the readout synchro will cause a torque to act on its rotor. This torque will be reduced to zero only when the readout rotor shaft has taken the same angular position as the transducer rotor shaft. A pointer can be linked to the readout rotor so that a visual display of angular displacement is presented on a graduated dial. Alternately, other display or recording devices can be linked to the readout rotor.

Fig. 3-24. Two-synchro angular displacement measurement.

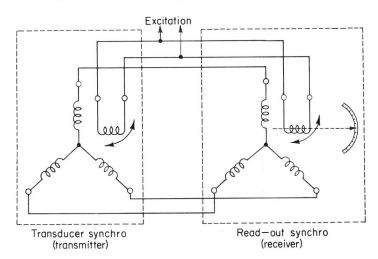

Transducer synchro
(transmitter)

Read—out synchro
(receiver)

An a-c voltage (variable amplitude) output can be obtained from a single synchro when a single-phase a-c voltage is used to excite the rotor winding. The output voltages from the stator (secondary) windings will then vary with angular displacement of the sensing shaft attached to the rotor. A constant-amplitude variable-phase output can be obtained from the rotor by exciting the stator from a three-phase supply.

3.3.6.5 Resolver-type displacement tranducers. Resolvers are similar to synchros in their design and their applications primarily in computing and control systems. They differ mainly in the number and spacing of their windings. The resolver [Figure 3-17(g)] has a two-phase stator (two stator windings spaced 90 degrees apart) and two rotor windings, also spaced 90 degrees apart. However, when used as a transducer, the resolver has one of its two rotor windings shorted. Use in measurement systems is similar to that of the synchro.

3.3.6.6 Microsyn-type displacement transducers. Microsyns are generally grouped with synchros and resolvers in the family of rotary transformer devices used mainly in computation and control systems but usable as measuring transducers. Of these, the microsyn and the related *shorted-turn signal generator* are somewhat more suitable as transducers because they require only one pair of excitation leads since excitation is single-phase a-c, and because they provide a single-phase output voltage whose amplitude is proportional to rotor displacement.

The microsyn and the shorted-turn signal generator both have a four-pole stator, usually made from laminated iron. Each of the *microsyn* poles is wound with two coils [Figure 3-17(h)]. The first is one of the four primary coils, all connected in series. The second coil is one of the four secondary windings so connected that the voltage induced in coils S_1 and S_3 opposes the voltage induced in coils S_2 and S_4. The rotor has no windings. It has a special "butterfly" configuration and is made from ferromagnetic material such as laminated iron. With the rotor in its neutral (null) position, the reluctance paths across all four rotor-stator pole gaps are equal, and the output from the secondary is as close to zero as mechanical and electrical construction will allow. Angular displacement of the rotor then causes a reluctance-path change to coils S_1 and S_3, opposite to that to coils S_2 and S_4. The resulting unbalance creates a net output voltage in the secondary. Microsyns are related to differential transformers and behave in a similar manner. They also exist with rotor and winding configurations slightly different from that shown.

The four poles of the *shorted-turn signal generator* [Figure 3-17(i)] are each wound with only one coil. One mutually opposite pole pair carries the primary, the other the secondary windings. The ring-shaped rotor is actually a one-turn shorted coil. Flux is induced in the shorted turn by the primary winding. An output is produced in the secondary winding when the rotor is displaced from a null position and flux linkages are produced. The shorted turn can be machined integrally with the sensing shaft or can be attached to it in some manner.

3.3.7 Strain-gage displacement transducers

The strain-gage transduction principle has been used in a limited number of angular- and linear-displacement transducers. Most of these incorporate a bending beam. Deflection of the beam results in strain which is transduced into a resistance change by either two or four strain gages bonded to the beam and connected in a Wheatstone-bridge circuit. This operating principle is illustrated in Figure 3-25. Mechanical overtravel stops are provided to prevent permanent deformation of the beam and damage to the gages. The beam-and-gage assembly is so constructed that the resistance changes linearly with displacement over a given range. The linkage between beam and sensing shaft can be by direct coupling (Figure 3-26) or by use of an eccentric shaft, a cam, or a lead screw. Metal-wire, metal-foil, and semiconductor gages have been used. Strain gages are further described in Chapter 13.

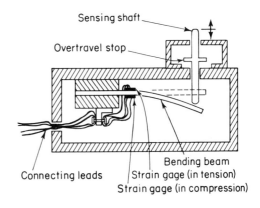

Fig. 3-25. Strain-gage displacement transducer (basic design).

Fig. 3-26. Strain-gage angular-displacement transducer (courtesy of BLH Electronics).

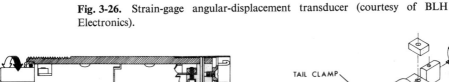

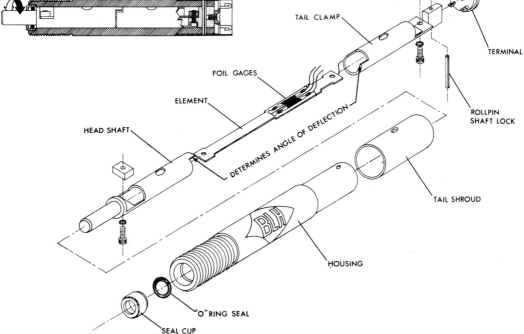

3.3.8 Digital-output displacement transducers

Displacement is one of the few measurands which can be transduced into a digital output by relatively simple means and without intermediate analog-to-digital conversion. Transducers whose output represents displacement by a number of discrete increments are true digital transducers. They exist in two forms:

1. *Incremental digital displacement transducers* which measure displacement with respect to the starting point of the given displacement only. The term "incremental encoder" has been used for this type, but a recent NEMA standard tends to discourage use of this term.
2. *Absolute digital displacement transducers* which measure displacement with respect to an internal fixed reference point. This transducer type represents displacement in coded discrete output increments. It is therefore referred to as *encoder*. Absolute digital linear-displacement transducers are called *linear encoders*. Absolute digital angular-displacement transducers are referred to as *shaft-angle encoders*, "shaft position encoders," or just "shaft encoders."

3.3.8.1 Incremental digital displacement transducers. These transducers exist in a variety of designs for angular-displacement measurements and to a lesser extent for linear-displacement measurement. The basic operating principle is illustrated in Figure 3-27. In the angular version, the sensing shaft is attached to a disc or other form

Fig. 3-27. Basic incremental digital displacement transducers.

(a) Angular   (b) Linear

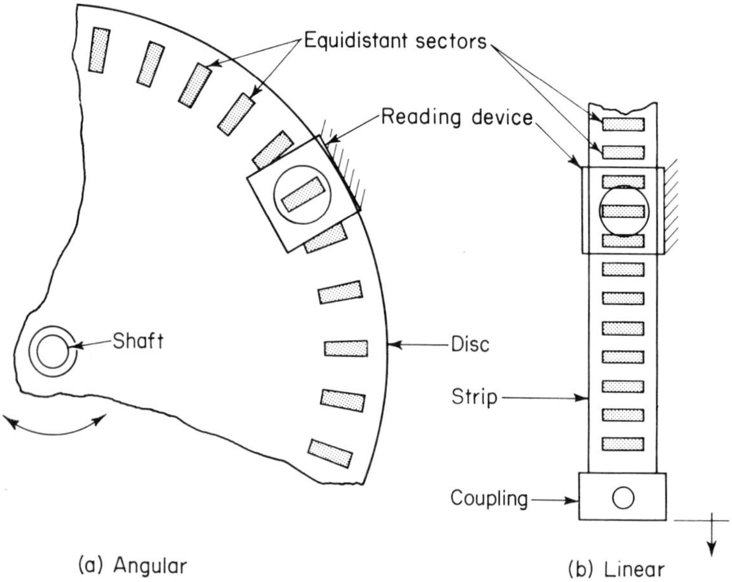

of rotor (the secondary sensing element) which is divided near its circumference into a number of equal sectors. The disc rotates past a reading device, fixed in position, which is capable of producing an electrical output for each sector passing it. The nature of the sectors and the reading device is given by the transduction principle utilized. The linear version operates similarly.

3.3.8.1.1 Transducers with toothed secondary sensing elements. When electromagnetic transduction is used (see 3.3.2), the rotor usually takes the shape of a gear with a large number of teeth. The material of the gear, or at least of the teeth, is ferromagnetic (e.g., soft iron). Each tooth is so shaped as to produce a sinusoidal output from the associated transduction coil. This output can then be fed to an electronic counter. With rotor motion starting from any position one count will be indicated for each tooth passing the coil. When motion stops, the accumulated count will indicate the total angular displacement.

When the direction of rotor motion can be either clockwise or counterclockwise, a direction-sensing feature must be incorporated in the transducer. Direction sensing can be achieved by shaping the teeth in some asymmetrical manner so that the output waveshape can indicate direction. More commonly, direction sensing is obtained by a second transduction coil so placed that its sinusoidal output will be 90 degrees out of phase (quadrature output) with respect to the first coil. The output of the second coil will lead or lag the output of the first coil. Phase-detection and associated logic circuitry can then be used to furnish a "clockwise" as well as a "counterclockwise" pulse train to a bidirectional ("up-down") counter. If desired, circuitry can be added to detect axis crossings of the two sine waves so that two counts, instead of one, per tooth are produced in each pulse train ("2-times multiplication") or four counts for both pulse trains ("4-times multiplication") when both are available.

Electromagnetic transduction used in conjunction with toothed rotors provides a usable output only when angular speed is above a certain minimum value (*minimum electrical speed*). Performance may also be affected by excessive angular speed. The requirement for a minimum electrical speed places a limitation on the use of these devices. This limitation can be obviated by use of other transduction principles, such as capacitive or reluctive transduction.

The reluctive principle is commonly employed in linear-displacement transducers of the incremental digital type. Their operation is quite similar to that of the angular version described above except that the gear disc is replaced by a toothed bar ("lug bar") of ferromagnetic material. Use of an *E*-core or similar transduction element offers good resolution within the area of one tooth and provides phase information suitable for direction sensing.

3.3.8.1.2 Transducers with coded secondary sensing elements. No minimum electrical speed requirements exist for incremental digital displacement transducers using a coded flat disc or strip secondary sensing element and different transduction principles. The code is established by sectors, spaced at equidistant intervals (Figure

3-27). The sectors can be electrical contacts, spaced around the disc or along the strip, insulated from each other at the surface but electrically connected together. A sliding contact (contact finger, wiper, *brush*, or pin) constitutes the reading device. As the disc or strip moves, the brush contacts alternately the conducting and nonconducting portions. Usually, the secondary sensing element is made of conductive material with insulating areas formed on its surface [Figure 3-28(a)]. With brush and contacts used as electrical switching device, a pulse train is generated, consisting of alternating "on" and "off" indications [Figure 3-28(c)]. The same type of pulse train is produced in photoelectric (electro-optical, "optical") transducers [Figure 3-28(b)], which use a disc or strip with alternating opaque and transparent sectors and a light source with associated light sensor as reading device. The light sensor can be photovoltaic or photoresistive. A third method, less common, uses alternate magnetic and nonmagnetic sectors with a saturable core or reluctive reading device.

Fig. 3-28.   Transduction of incremental-code elements.

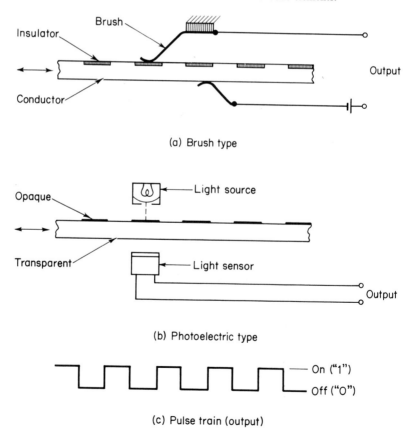

(a) Brush type

(b) Photoelectric type

(c) Pulse train (output)

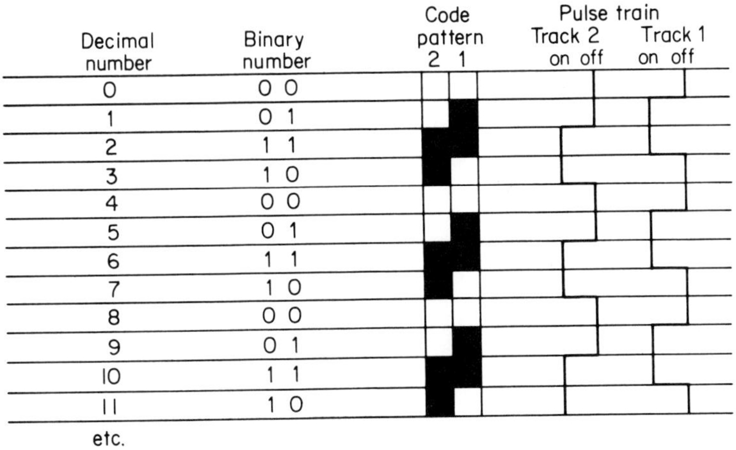

Decimal number	Binary number	Code pattern 2 1	Pulse train	
			Track 2 on off	Track 1 on off
0	0 0			
1	0 1			
2	1 1			
3	1 0			
4	0 0			
5	0 1			
6	1 1			
7	1 0			
8	0 0			
9	0 1			
10	1 1			
11	1 0			

etc.

Fig. 3-29.   Two-bit incremental code.

The output of a transducer with a coded disc or strip yielding only alternating ON and OFF indications can not show the direction of motion where motion can be bidirectional. The code in such transducers is produced by a single *track* of sectors and only one brush (or light source-sensor set) is used. The single pulse train [Figure 3-28(c)] shows this code to be a *one-bit* code in a *binary system*. The binary system is a number representation system with a radix of two. Each *binary digit* (*bit*) has the value of either "1" or "0," equivalent to "True" or "False" in logic and to ON or OFF in circuit continuity. In a binary representation, then, this pulse train provides one bit. Its sequence is 0-1-0-1-0-etc. Direction sensing can be obtained by adding a second track of sectors and using two brushes so that two pulse trains are produced, such as by the two-bit incremental coding illustrated in Figure 3-29. It should be noted in this figure that binary "1" (contact closed) is shown by a black area and that binary "0" (contact open) is shown by a white area in the code pattern, in accordance with a commonly used convention. With the two tracks in parallel, motion in one direction will produce a number sequence 00-01-11-10-00, whereas the sequence 00-10-11-01-00 will result from motion in the opposite direction. This information, when conditioned by appropriate logic circuitry ("up-down detector"), can be used to operate a counter in the "up-count" mode as well as the "down-count" mode.

Resolution in photoelectric coded incremental displacement transducers can be improved, where required, not only by multiplication logic operating on the basis of output axis crossings but also by use of special optical measuring devices and techniques. The most useful of these is the *interfering pattern* produced usually by a stationary reticle or other mask placed over the moving pattern. This allows a substantial interpolation between code sectors on the secondary sensing element. Two interfering patterns are illustrated in Figure 3-30.

The "$N + 1$" pattern [Figure 3-30(a)] is mainly used with coded disc elements.

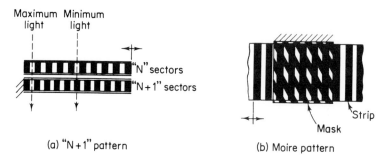

(a) "N+1" pattern   (b) Moire pattern

Fig. 3-30. Typical interfering patterns.

One of the two concentric discs is attached to the sensing shaft. It has N sectors. The other is stationary and has $N + 1$ sectors (one more sector than the rotating disc). In this configuration transmitted light intensity will be maximum at one point on the disc circumference and minimum at 180 degrees from that point. As the disc rotates, the sensor output due to light intensity is modulated in a quasi-sinusoidal manner. The number of periods per shaft revolution is equal to the number of code sectors on the rotating disc. Two light source-sensor sets, mounted 180 degrees apart, can provide for direction sensing and multiplication. This operating mode is illustrated in Figure 3-31; the additional set of light sources and sensors provide for further performance refinements.

The Moiré pattern [Figure 3-30(b)] has been used as a resolution-improving interfering pattern mainly in linear-displacement transducers, such as in the "Ferranti System." The stationary mask contains a light-bar pattern similar to that of the moving strip (secondary sensing element) but tilted at a small angle with respect to it. This produces a pattern characterized, with suitable illumination, by horizontal light bands made up of a series of parallelograms. As the strip moves laterally, the light bands move up or down depending on the direction of strip motion. The number of light bands as well as the distance between them is given by the angle between the mask pattern and the strip pattern. A number of vertically spaced photocells in front

Fig. 3-31. Incremental digital angular-displacement transducer using interfering $N + 1$ pattern (courtesy of Dynamics Research Corp.).

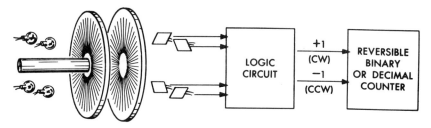

of the mask, corresponding to the number of light bands, can then be used to produce substantial output variations over the distance between strip sectors. The light source can be placed behind the moving strip if the strip contains alternating opaque and transparent sectors (as is always the case for the mask). Alternately, the light source can be placed in front of the mask if the moving strip has alternating reflecting and nonreflecting sectors. A collimating optics system is used in conjunction with the light source.

3.3.8.2 Absolute digital displacement transducers. These transducers are similar in operation to the incremental types discussed in 3.3.8.1.2, except that four or more tracks are used, and the sectors are so arranged that a binary code is produced which defines the displacement uniquely at each of a given number of positions with respect to the position at which the code starts. Generally, shaft-angle encoders are considerably more prevalent than linear encoders.

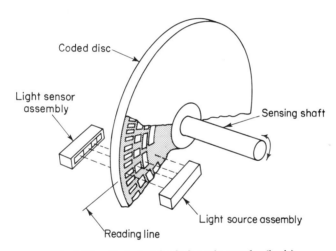

Coded disc

Light sensor assembly

Sensing shaft

Light source assembly

Reading line

Fig. 3-32.   Photoelectric shaft-angle encoder (basic).

In the contacting or brush-type encoder, one of the tracks, the *common* track, provides continuous contact to one brush throughout the transducer's range (360 degrees for angular encoders, various lengths for linear encoders). One brush is provided for each of the noncontinuous code-pattern tracks. Except for special code configurations, the brushes are usually in one radial line (*reading line*). The code pattern is formed by making certain portions of an otherwise conducting surface insulating, such as by selective oxidization.

In the photoelectric (brushless) encoder a light source and a light sensor are provided for each track. A "common" track is not required. The sectors are transparent ("slits") in an otherwise opaque strip or disc. The light source for each track can be an individual miniature light bulb or other light emitter such as a strobe lamp or a gallium-arsenide semiconductor emitter. More frequently, less light emitters are used than the number of tracks would demand, and fiber optics or other optical devices are used to distribute the light to each of the tracks. The light is in the visible or the infrared spectrum. Light sensors can be photovoltaic or photoconductive. Photo-diodes and phototransistors are frequently used as light sensors. The light-source and light-sensor assemblies are normally positioned on the reading line (Figure 3-32).

The relatively weak signals from the brushes or light sensors are either amplified or used to actuate a diode or semiconductor switching device to produce reliable inputs to computing circuitry.

Angular encoders (shaft-angle encoders) can utilize two code discs, one geared to the other with a gear ratio large enough to provide both a "coarse" and a "fine" output where the "fine" disc revolves many times for each revolution of the "coarse" disc. In a "hybrid" encoder one of the two discs is of the contact type whereas the other disc operates on the photoelectric principle. Gearing is also used to increase the total revolutions per count of a single disc. Some encoder designs have geared high- and low-speed discs linked by a scanning technique such as *V*-scan (see 3.3.8.2.2) which essentially eliminates gear train errors.

3.3.8.2.1 Code systems. A variety of codes are used in encoders. Most of them are equally usable in linear- and angular-displacement transducers. The choice of any one code depend on such factors as the needs of available computing circuitry, the nature of the required display, the total number of counts, the characteristics of the measur-and, and the necessary degree of reliability or certainty of output indications.

The *natural binary code*, or just *binary code*, is the most compact code. It requires fewer bits than other codes to represent a given range of numbers. It is the simplest code for use with arithmetic or comparison-computing circuits. It is easily understood by personnel working with digital electronic systems.

A four-bit binary code disc is illustrated in Figure 3-33. Shaded sectors signify ON ("1," "True," "yes") positions. Blank sectors signify OFF ("0," "False," "no") positions. In the binary system, a number representation system with a radix of 2, a given Arabic number is represented by a number of weighted "bits" (*binary digits*). The least significant bit, if it appears as "1" ("yes"), carries a weight of 2^0 ($= 1$). The next-to-least significant bit, if it appears as "1," carries a weight of 2^1 ($= 2$). The next following bit (third track, counting from the outside track), if it appears as "1," carries a weight of 2^2 ($= 4$). The fourth track bit, if it appears as "1," carries a weight of 2^3 ($= 8$). Any of the bits are zero when they appear as "0."

The number given by the brush positions in Figure 3-33 is 1011. This represents the Arabic number eleven because (counting from least to most significant bit, or right to left) the first digit has a weight of 1, the second a weight of 2, the third does not have a weight of 4 but is zero, the fourth has a weight of 8, and $1 + 2 + 8 = 11$. The four-bit binary code disc is usable only for numbers up through 15. A five-bit

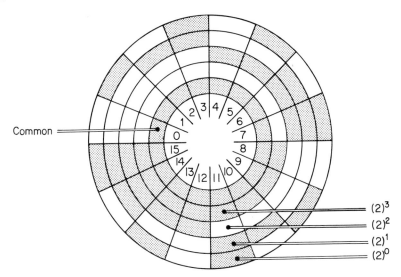

Fig. 3-33. Binary code disc (courtesy of Litton Industries).

disc extends the range of number to 31 ($2^4 - 1$), a six-bit disc to 63 ($2^5 - 1$), a seven-bit disc to 127, etc.

For an automatic detection of errors in indications from binary code discs by use of redundancy, an additional track for a parity bit can be added, so coded that at every position of the encoder an odd number of brushes are conducting (*odd parity*). Or, an even number of brushes can be made to be conducting at every position (*even parity*).

A disadvantage of binary code discs in shaft-angle encoders is that two or more bits can be required to change simultaneously during a single position change, as, for example, when changing from 7 (0111) to 8 (1000). Certain other codes were developed in which this disadvantage does not exist. Among these are the frequently used Gray and DATEX codes.

The *Gray code*, named after its inventor, Dr. Frank Gray of Bell Telephone Laboratories, is a "unit distance" or *monostrophic* code. The latter term signifies that only one bit changes in the transition between two adjacent numbers. This makes a reliable code pattern easier to produce. A basic Gray-code disc, providing 2^4 resolution, is shown in Figure 3-34.

The two basic codes, binary and Gray (binary), are further illustrated in Figure 3-35, together with a number of more advanced codes which are briefly described below.

The *binary coded decimal* code is a combination of the binary system and the Arabic decimal number system, since binary bits are used to form decimal numbers. The binary coded decimal (*BCD*) code most frequently used is the 8421 *BCD* system in which the Arabic numbers from 0 to 9 are represented by the first ten numbers of

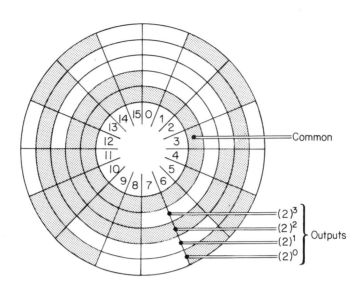

Fig. 3-34. Gray code disc (courtesy of Litton Industries).

the binary code. Each group of four bits (weighted 8, 4, 2, and 1), represents one decimal integer. For example, the number "835" is represented by "1000 0011 0101."

The *Excess 3 BCD* (binary coded decimal) code is formed by adding 3 to the value of the Arabic digit. This makes the binary numbers representing Arabic numbers 5, 6, 7, 8, and 9 mirror images of those representing 4, 3, 2, 1, and 0 (e.g., 0 = 0011, 9 = 1100). It allows subtraction by means of complementing and adding in arithmetic circuits. Also, by complementing all bits, the direction of rotation can be changed for an increasing count.

The *Gray coded Excess 3 BCD* (binary coded decimal) code is based on the cyclic decimal system used in the DATEX code, to which it is very similar. It is the Gray code version of the Excess 3 BCD code.

The *DATEX code* (patented by the Datex Division, Conrac Corporation) is a monostrophic (unit distance) code. In this code the ordinary (Arabic) decimal system is modified into an artificial one, the *cyclic decimal* system, in which an Arabic decimal digit that follows an odd Arabic digit is changed to the *9's complement*. This achieves a number sequence in which any two adjacent numbers differ in only one decimal digit: 0, 1, 2, 3, 4, 5, 6, 7, 8, *9*, *19*, 18, 17, 16, 15, 14, 13, 12, 11, *10*, *20*, 21, 22, 23, ...

In the *two-out-of-five code* each number is made up of two 1's and three 0's (even parity). Except for the (Arabic) zero, the bits are weighted 7, 4, 2, and 1. This code requires a parity track.

The *biquinary code* is a simplified (for an encoder) version of a direct decimal code. The *direct decimal code* requires a ten-bit number for each decimal digit, from 0000000001 for Arabic zero to 1000000000 for Arabic nine. The biquinary code accom-

Fig. 3-35. Single-readout code structures (courtesy of Datex Div., Conrac Corp.).

Arabic	Binary coded decimal Tens 8421	Units 8421	Gray coded excess 3 BCD Tens DCBA	Units DCBA
0	0000	0000	0010	0010
1	0000	0001	0010	0110
2	0000	0010	0010	0111
3	0000	0011	0010	0101
4	0000	0100	0010	0100
5	0000	0101	0010	1100
6	0000	0110	0010	1101
7	0000	0111	0010	1111
8	0000	1000	0010	1110
9	0000	1001	0010	1010
10	0001	0000	0110	0010
11	0001	0001	0110	0110
12	0001	0010	0110	0111
13	0001	0011	0110	0101
14	0001	0100	0110	0100
15	0001	0101	0110	1100
16	0001	0110	0110	1101
17	0001	0111	0110	1111
18	0001	1000	0110	1110
19	0001	1001	0110	1010
20	0010	0000	0111	0010

Cyclic decimal count	Datex cyclic code Tens ABCD	Units ABCD
0	1000	1000
1	1000	1100
2	1000	0100
3	1000	0110
4	1000	0010
5	1000	0011
6	1000	0111
7	1000	0101
8	1000	1101
9	1000	1001
19	1100	1001
18	1100	1101
17	1100	0101
16	1100	0111
15	1100	0011
14	1100	0010
13	1100	0110
12	1100	0100
11	1100	1100
10	1100	1000

Arabic	Binary $2^4 2^3 2^2 2^1 2^0$	Gray (Binary) $G_4 G_3 G_2 G_1 G_0$	Excess 3 BCD 8421	Two out of Five code 7421P (Parity)	Biquinary code 86420 odd
0	00000	00000	0011	11000	00001 0
1	00001	00001	0100	00011	00001 1
2	00010	00011	0101	00101	00010 0
3	00011	00010	0110	00110	00010 1
4	00100	00110	0111	01001	00100 0
5	00101	00111	1000	01010	00100 1
6	00110	00101	1001	01100	01000 0
7	00111	00100	1010	10001	01000 1
8	01000	01100	1011	10010	10000 0
9	01001	01101	1100	10100	10000 1

201

plishes the same purpose with only five bits plus one "odd-even" selector bit, a total of six bits. The five-bit number is the same for any two adjacent Arabic numbers, one of which is even, the other odd. The presence or absence of the "odd" sector in the odd-even selector track then indicates which of the two numbers is intended.

3.3.8.2.2 Anti-ambiguity provisions. In an ideal encoder all brushes are exactly on the reading line and make their transitions between sector-coded positions simultaneously. In an actual encoder, however, ambiguous output can be obtained in the transition between two adjacent decimal numbers (positions) when one or more segments are very slightly wider or narrower than their ideal prescribed width. Errors due to ambiguity can be eliminated by using certain inherently nonambiguous codes, such as the monostrophic ("unit distance") Gray and DATEX codes. Since only one bit changes between any two adjacent positions when a monostrophic code is used,

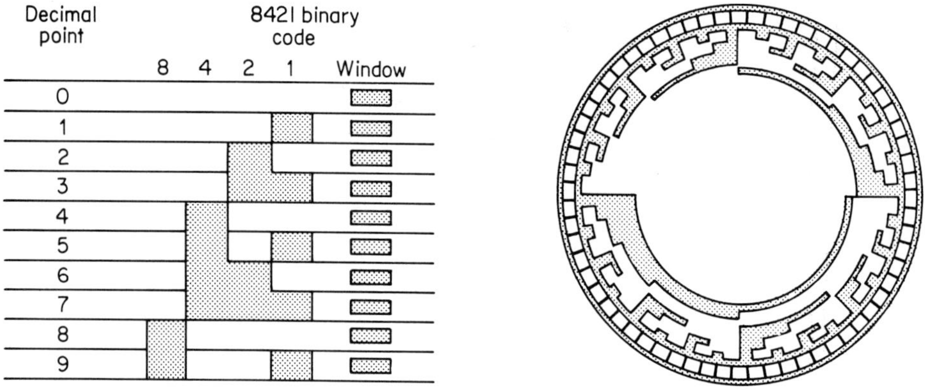

(a) Window track added to 8421 binary code

(b) Typical window code disc

Fig. 3-36. Anti-ambiguity window code (courtesy of Datex Div., Conrac Corp.).

an ambiguity error can be easily detected by logic circuitry when no bit or more than one bit changes.

Another means for preventing ambiguity is the *window code* (Figure 3-36), a form of electrical detecting where an outside track of special narrow segments is added and a readout is permitted only when reading "in the window," i.e., when the window sector is ON. Associated circuitry can also be so devised that it operates in the STORE mode (store the last signal) when the window is OFF, and in the FOLLOW mode when the window is ON.

A third method for eliminating ambiguity is the use of *V-scan, M-scan,* or *U-scan* reading devices which utilize more than one brush for each bit (except the 2^0 bit). The letters *V, M,* and *U* refer to an approximate appearance of the brush configuration. The *V*-scan is the most popular of these. It is frequently used with binary

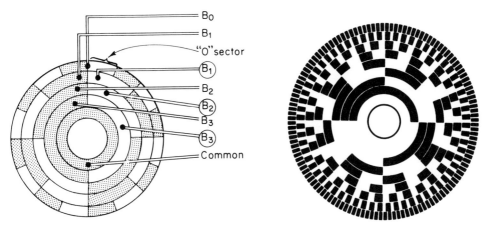

(a) Basic brush configuration

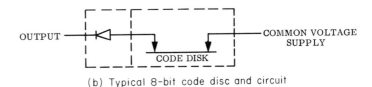

(b) Typical 8-bit code disc and circuit

Fig. 3-37. V-scan code (courtesy of Litton Industries).

codes. In the V-shaped configuration [Figure 3-37(a)] two brushes are used for each bit (except the least significant bit). External logic circuitry selects one or the other brush for readout on a given track on the basis of the readout just obtained from the next less-significant track. Of the two brushes on each track, one (circled letter B in the illustration) is called the "advance" or "leading" contact, the other the "retard" or "lagging" contact. The distance between the leading and the lagging contact, on each track, is one-half the width of the ON (or OFF) sector used on that track. The contact selected for readout, then, is always at least one-half the bit width of the least significant track from the transition point. Usually, the lagging contact will be selected for readout on the nth track if the $(n-1)$th track bit is ON, and the leading contact will be selected to read out the nth track if the $(n-1)$th bit is OFF. Figure 3-37(b) shows a typical V-scan code disc. Since this picture illustrates an actual disc rather than a code pattern (where the two colors are inverted), the ON sectors are shown in white, the OFF sectors in black (insulating or opaque).

3.3.8.2.3 Special codes. A number of code patterns have been developed to fill special display requirements. Among these are the codes used for indication of aircraft altitude in feet (*ICAO Code*, a combination of Gray and DATEX codes), of time in hours, minutes, and seconds, and of angle in degrees, minutes, and seconds of arc. Another pattern allows for bidirectional range (center zero).

3.4 Performance and Applications

3.4.1 Specification characteristics

Specifications for displacement transducers must state at the outset whether linear or angular displacement is to be measured. They should then show all applicable characteristics, appropriate values and tolerances, as outlined below, and additional limitations imposed if required for special transducer designs or special applications.

3.4.1.1 Mechanical design characteristics. An outline drawing should show the overall configuration and all its critical dimensions, details of sensing shaft and coupling (if any), mounting provisions and their dimensions, location of nameplate, electrical connections, and any external provisions for adjustments. Certain transducer case outlines have been standardized, and much of the above information can be omitted if the applicable standard is called out. This applies specifically to the "servo-mount" or "synchro-mount" configuration of many angular displacement transducers, a cylindrical case intended to be mounted by clamping it around a circumferential groove near the sensing shaft. The servo mount "size" gives the outside case diameter, (multiplied by ten and rounded off to the nearest one-tenth of an inch) as well as all mounting dimensions. The case of a "size 11 Synchro Mount," for example, has an outside diameter of 1.062 in. and a groove diameter of 1.000 in., etc. The case diameters for "size 18, 23, 27, and 35" are 1.750, 2.250, 2.750, and 3.500 in., respectively. The length must be specified separately.

Notes accompanying the drawing can then stipulate additional mechanical design requirements, including weight tolerances, mounting torque or force, case sealing, case and shaft materials, identification (nameplate contents), and characteristics of any external adjustment provisions.

The materials acceptable as measured-object sensing surface for noncontacting displacement transducers must be specified.

Overtravel of the sensing shaft beyond the transducer's range limits (see 3.4.1.3) is specified as the difference between these limits and the points at which overtravel mechanical stops are set, e.g., (Range: ± 5.0 cm) "Overtravel: -0.5, $+0.5$ cm, minimum" (if equal beyond both range limits), or "Overtravel: -1.0, $+3.0$ cm, minimum" (if greater for extended shaft than for retracted shaft); similarly, for angular ranges, "Overtravel: -5 degrees, $+5$ degrees, minimum" or (Range: 0 to 90 degrees) "Overtravel: -25 degrees, $+235$ degrees, minimum." Some specifications call out (total) "mechanical travel" instead of overtravel. This information is useful only when mechanical travel is referred to a reference point or if it is stated as "symmetrical beyond both range limits." Even when mechanical stops are not used (as in continuously rotary devices) overtravel can be specified to define the region where a specified over-travel output must be obtained (such as provided by busbars in potentiometric trans-ducers). Where overtravel stops are used, *overload* (in pounds or kilograms, pound-inches or kilogram-centimeters) can be used to specify the maximum force or torque which may be applied to the sensing shaft at the mechanical stop without causing damage or permanent deformation within the transducer.

Sensing shaft *concentricity*, *alignment* with respect to case, and *radial* as well as *axial play* should be considered for inclusion in a specification. The minimum range of acceptable radial or axial movement must be stated for "self-aligning" shafts. *Back-lash* should be covered in specifications for transducers containing gears. For spring-loaded shafts a requirement for *holding force* (at any point, after initial motion) should be included in addition to starting force.

Starting force is an important mechanical characteristic of coupled displacement transducers. This is the force required to initiate shaft motion, as indicated by a measurable change in transducer output (even after prolonged storage of the trans-ducer). It is specified as: "Starting force: __ oz (or lb, gm, kg), maximum." For angular transducers, *starting torque* is specified (in oz-in., gm-cm, etc.). The force needed to actuate the transducer after starting force or torque has been applied is the *running force* or *running torque*. It is always less than the starting force or torque but may also have to be specified if the application requires it. Additionally, for angular-displacement transducers, especially shaft-angle encoders, the maximum allowable moment of inertia (in gm-cm^2, etc.) of the entire rotating mass is often specified.

3.4.1.2 Electrical design characteristics. This listing begins with nominal and maximum excitation (voltage or current, d-c or a-c). If a-c excitation is used, its fre-quency range must be stated. For either excitation, the input impedance and current drain or power dissipation should be shown.

The external electrical connections (receptacle, terminals, or leads) should be described, with each individual connection identified as to function. The latter is sometimes combined with a simple internal wiring diagram (e.g., winding configura-tion of a differential transformer). Other transduction element specifications can de-scribe the number and location of taps on a potentiometric element or the cathode-tap location in an inductive element. Insulation resistance and breakdown voltage rating should not be neglected. Output characteristics include output impedance, allowable load impedance (including its reactive component, if any), output noise or ripple, and susceptibility to electromagnetic interference and output short-circuiting. For encoders, the maximum load voltage and current per digit, any internal drive circuits, and type of diodes and their purpose should be specified.

3.4.1.3 Static performance characteristics.

Range. The specification of range is relatively more complex for displacement transducers than for most other transducers because of the need for defining a reference point for displacement. The characteristic has, in the past, been specified as "electrical travel," "electrical stroke," "full-scale deflection," "linear range," "useful range," "input," "total stroke," "total range," "useful stroke," or "range (without reference point)." It is best specified as (unidirectional or bidirectional) range with respect to a reference position of the sensing shaft.

This reference position can be stated in one of two ways: 1. by a precise dimension between a point on the sensing shaft and a point on the transducer case; or 2. by a specific transducer output at the reference point. When method 1 is used, the referencing dimension is best shown on the outline drawing. There it can be a dimension between the end of the sensing shaft and either a mounting hole or the front end of the case for a linear-displacement transducer. For an angular-displacement transducer, a requirement can be shown for a scribe mark on the sensing shaft to be aligned with a matching scribe mark on the case. Method 2 usually defines the reference position much more closely and should be used, where feasible, for coupled transducer types. For these it is a perfectly acceptable method, because the sensing shaft is usually adjusted to an output-referenced position in its end-use installation. Frequently, method 2 is used with an additional approximate reference dimension shown to facilitate a preliminary coarse adjustment of the sensing shaft during installation.

The polarity of the range is always referred to direction of motion. Typical range specifications are shown by the following examples:

1. For a reluctive linear-displacement transducer: "Range: ± 3.0 cm; zero range at null-voltage position; $+$ range $=$ shaft extended" (unidirectional range callouts should be avoided).

2. For a reluctive angular-displacement transducer: "Range: ± 10 degrees; zero range at null-voltage position; $+$ range $=$ cw motion, looking at shaft end of case."

3. For a reluctive, d-c output, linear-displacement transducer: "Range: 0 to 2.0 in.; zero range at 0.000 V d-c output position; $+$ range $=$ shaft extended" (bidirectional range zero can be established at, e.g., the 2.500 V position, or for the null-voltage position if biasing is not used).

4. For a potentiometric linear-displacement transducer: "Range: 0 to 3.0 in.; zero range at position where first measurable output increase from 0.00% VR occurs; $+$ range $=$ shaft extended" (symmetrical bidirectional range can be based on the 50% VR position, unsymmetrical bidirectional range on another specified output).

5. For a noncontacting inductive linear-displacement transducer: "Range: 0.0 to 5.0 mm; zero range at 0.20 mm distance between transducer and measured surface; $+$ range $=$ away from transducer" (bidirectional range can be based on a different distance between transducer and measured surface).

6. For a contacting inductive linear-displacement transducer: "Range: 0 to 1.5 in.; zero range at 150 mH (approximately 2.5 in. between shaft and case end); $+$ range $=$ shaft retracted."

7. For a strain-gage angular-displacement transducer: "Range: ±30 degrees (zero range with zero force applied to sensing shaft); $+$ range $=$ cw motion, looking at shaft end of case."

Range specifications for certain transducer types differ somewhat from the above examples. The reference position for shaft-angle encoders is not separately stated since it is fixed within the transducer. However, either clockwise or counterclockwise rotation must be specified for "ascending count" (increasing displacement). For incremental digital angular-displacement transducers the range is sometimes specified only as the number of revolutions (turns) for "total count." The range of other multi-turn rotary devices is shown either in number of (full) turns or, preferably, in degrees (e.g., "0 to 3600 degrees"). The reference position for reluctive-transducer range is sometimes shown in terms of output phase rather than amplitude.

Output. Output is specified as full-scale output, as end points, or as sensitivity. Full-scale output is the algebraic difference between the end points which, in turn, are the outputs at the specified upper and lower limits of range. For bidirectional ranges the term *span* is applied to the algebraic difference between the upper and lower limits of range. Full-scale output is best defined over the transducers span and, when shown without polarity, denotes the output change over the full span. Unless it is only used to define theoretical end points, its specified value carries tolerances, e.g., "Full-Scale Output: $98 \pm 2\%$ VR," "Full-Scale Output: 5.00 ± 0.10 V d-c," or "Full-Scale Output: 200 ± 20 pF."

Alternately, only the end points and their tolerances are specified, e.g., "End Points: 1.0 ± 1.0 and $99 \pm 1\%$ VR," or "End Points: 600 ± 30 and 30 ± 20 mH" (if $+$ range $=$ away from a noncontacting inductive transducer). Specifying end points is useful when the range is bidirectional and the output changes polarity or phase after going through its null, e.g., "End Points: -2.50 ± 0.005 and $+2.50 \pm 0.05$ V d-c" or "End Points: 150 ± 10 (lagging phase) and 150 ± 10 (leading phase) mV a-c."

When the transducer's output is also a function of the excitation voltage, as in reluctive and strain-gage transducers, the excitation (including the frequency of a-c excitation) at which full-scale output and end-point specifications are applicable must be included, e.g., "Full Scale Output: 30 ± 2 mV at 10.00 V d-c excitation" or "End Points: 2.2 ± 0.1 (lagging phase) and 2.2 ± 0.1 (leading phase) V a-c at 6.00 V a-c, 2000 Hz excitation." Besides the nominal excitation, the maximum and preferably also the minimum excitation should always be specified (as electrical design characteristics).

Similar considerations apply when output is specified in terms of sensitivity, the ratio of the output change to the displacement change. Sensitivity is specified mostly for those types of transducers whose output is also a function of excitation, e.g., "Sensitivity: 1.800 ± 0.036 mV/V d-c/.001 in." (millivolts per volt of excitation per each .001 inch of displacement), "Sensitivity: 1.20 ± 0.005 mV/V a-c at 400 Hz/deg," or "Sensitivity: 3.00 ± 0.10 V/cm at 5.00 V a-c, 400 Hz excitation."

For some types of transducers the output is stated in terms of the phase angle. For inductive, capacitive, and some photoelectric transducers it is sometimes shown in terms of the output or the indication of associated signal-conditioning and display equipment in conjunction with which the transducer must always be used.

If performance characteristics are stated by an error-band specification, the specification for the error-band reference line or curve can, in some cases, replace separately specified full-scale output and end-point requirements.

Encoder full-scale output is specified as "total counts" or "total counts per turn" or, for multiturn shaft-angle encoders, as total number of revolutions and total counts per turn (revolution). Output in terms of number of counts can also be seen in specifications for certain photoelectric (electro-optical) transducers not classified as encoders, as well as for incremental digital displacement transducers.

Additional output specifications for encoders include the output code (binary, 8421 BCD, Gray, etc.), the antiambiguity logic required (window code, V-scan, etc.), the output notation (linear counts, hours-minutes-seconds, thermocouple curve, longitude, etc.), and output sign as well as requirements for either single or double zero count if the range is bidirectional.

When two or more equal or unequal outputs are provided by a transducer with a single sensing shaft, a requirement may exist to relate these outputs to each other (*output correlation*) at one or more points.

Static Errors. Linearity requirements are specified for all types of displacement transducers except encoders. The type of linearity (independent, terminal, end-point, etc.) should always be stated. When it is not stated, independent linearity is usually implied because it is easiest to achieve. Hysteresis and repeatability are relatively small in most displacement transducers but should, nevertheless, be specified. These characteristics are combined in some specifications which carry a single set of tolerances for "combined hysteresis and_____ linearity" or "combined hysteresis, repeatability, and_____ linearity."

Friction error and resolution are important characteristics to be specified primarily for potentiometric transducers. Resolution is sometimes specified for encoders although it is usually just the inverse of total count. The near-zero output at the reference position for bidirectional-range transducers is specified as a maximum tolerance for null voltage or "zero balance." For many reluctive transducers the specification of phase shift is essential. Maximum phase shift should be stated at the specified load impedance or range of load impedances and at a specified excitation (carrier) frequency or range of frequencies. Additionally, the frequency at which the phase shift (phase angle) is zero can be shown. In some reluctive transducers the amount of cross coupling between two physically separated coils is subject to specification tolerances.

Linearity, hysteresis, repeatability, friction error, resolution, and sometimes null voltage are expressed in percent of full-scale output ($\%$ FSO) for all analog-output displacement transducers except that percent voltage ratio ($\%$ VR) is used for output as well as errors of potentiometric transducers. Unless theoretical end points (without tolerances) or a theoretical slope are specified, "% FSO" always refers to the actual full-scale output of the transducer as determined by a static calibration. Null voltage is often expressed in millivolts per volt of excitation (at a specified frequency) or millivolts at a specified value of excitation.

A static error band can be specified in lieu of separate tolerances for linearity (all types), hysteresis, repeatability, and friction error. Certain error band specifications can also include null voltage or zero balance. The reference line for the static error bands (which also serves as reference line for any environmental error bands that may be specified) can be defined in a number of different ways. It can be the line between (actual) end points, the "best straight line" used with independent linearity, the least-squares line, the terminal line (between theoretical end points fixed at 0% and 100% FSO), the line between theoretical end points (other than 0% and 100% FSO), a theoretical slope, a theoretical slope passing through actual null voltage, or any definable curve that may or may not intercept a given point.

Mounting error should be stated as the maximum allowable performance changes to be observed after the transducer has been mounted using a specified mounting torque or force. Attitude error due to gravity in any transducer orientation when mounted is included in the specified acceleration error tolerances.

Overtravel output values are shown when the transducer must provide a fixed or varying output beyond the specified range limits. This is frequently required from potentiometric transducers when indications of 0% VR and 100% VR (usually with tolerances), respectively, are desired for specified travel beyond the lower and upper range limits.

3.4.1.4 Dynamic characteristics. When rapid motion, such as oscillatory motion, is applied to a displacement transducer's sensing shaft, a point will be reached with increasing speeds at which the output no longer follows the displacement correctly. Therefore, specifications often stipulate requirements for *maximum shaft speed* (in inches per second, centimeters per second, degrees per second, or revolutions per minute) at which the output can still follow the displacement correctly. In some specifications this is further defined as *sustained* and initial or *slewing* speed. The latter term refers to the initial application of motion to the sensing shaft of an angular-displacement transducer.

3.4.1.5 Environmental performance characteristics. Specifications for thermal effects are best expressed in terms of maximum temperature error or temperature error band over a given operating temperature range. The allowable behavior of the transducer during and after thermal shock or during a step change in temperature of the sensing shaft alone may be covered by additional specifications. When stating tolerances for vibration, shock, and acceleration error, the possible variation in magnitude of these errors due to shaft position should be considered. Effects of altitude or of immersion below sea level are shown as ambient-pressure error. The composition of the ambient atmosphere and its contaminants should be stated if they can have any effect on transducer performance. Usually, atmospheric environmental conditions (e.g., salt, humidity, smoke, soot, sand, dust, and ambient fluids other than air) primarily affect the method of case sealing.

3.4.1.6 Performance reliability characteristics. Storage life and stability requirements, though important, are often difficult to enforce after expiration of a transducer's warranty as furnished by the manufacturer. Sometimes they are stated as "design goal." Stability usually refers to repeatability over a certain period of time (e.g., 18 months), unless a different characteristic is designated (e.g., null-voltage stability).

Operating life is expressed as minimum number of full-range cycles. For linear-displacement transducers and for angular-displacement transducers with mechanical stops, each cycle is bidirectional. For continuous-rotation angular-displacement transducers, a total number of revolutions (in either direction) is usually specified.

3.4.2 Considerations for selection

Since a large variety of displacement transducers is commercially available, or can easily be modified, for virtually all applications, a selection on the basis of technical and economic requirements is usually quite simple. The primary criteria are given by the characteristics of the measuring system in which the transducer is to be utilized. These determine the type of transducer output needed and reduce the number of choices. The second most important criteria governing selection are range and overall system accuracy. Thirdly, the characteristics of the measurand must be considered, including the physical requirements for an optimum transducer installation. The latter will determine whether a coupled or noncontacting transducer should be used. A number of additional factors must usually be considered before one or more usable transducer designs can be defined. A final selection can then be made on the basis of cost, availability, and manufacturer's capabilities and reputation.

3.4.2.1 Use in d-c systems. Potentiometric transducers are most frequently used since they are simple devices capable of producing output levels up to 50 V, sometimes even higher. They are normally available for displacement spans between 0.5 and 24 in. or between 5 and 3600 degrees. Average resolution is approximately 0.5% FSO for the smallest spans and it becomes finer (smaller steps) with increasing spans because a greater number of (lower resistance) wire turns can then be used. Accuracy and friction error also improve with increasing spans.

Strain-gage transducers with metal-foil or metal-wire gages can be used where their relatively low-level output (normally 30 mV at 10 V excitation) is acceptable to the measuring system. When equipped with semiconductor gages their full-scale output is increased to between 2 and 4 V. Greater output can be obtained by use of an amplifier. Displacement spans are between 0.02 and 10 in. or 20 to 360 degrees. They have been applied primarily in those measuring systems in which strain-gage transducers and similar low-level-output devices are used for all measurands.

Reluctive transducers with integral a-c to d-c output conversion (and d-c to a-c excitation conversion if required) have become increasingly popular. Their usual displacement spans are between 0.01 and 120 in. and between 0.05 and 90 degrees. Full-scale output is frequently adjusted to 5 V since many telemetry systems require this level; however, lower and higher outputs can also be obtained.

Capacitive and inductive transducers with a-c to d-c output conversion and photoelectric transducers are sometimes used for measurement of relatively small displacements. They find their major usage as noncontacting transducers where displacement changes as low as one microinch must be resolved. Noncontacting capacitive and inductive transducers, whose excitation frequencies are sufficiently high, are also capable of following very high-frequency fluctuations in the displacement. Typical applications of noncontacting displacement transducers include measurement of rotating-shaft eccentricity and bearing film thickness, thickness and flatness of sheets in rolling mills, dimensions of machined parts, texture and finish of materials in process, and detection of proximity.

3.4.2.2 Use in a-c systems. Of all types of displacement transducers, the various reluctive designs are used almost to the exclusion of all others. Their displacement spans are the same as described for their d-c output versions. Output is not only proportional to excitation voltage, it also increases as the excitation frequency is increased (typically from 50 to 10,000 Hz). Some designs operate at even higher carrier frequencies.

Inductive, potentiometric, and strain-gage displacement transducers are also usable in a-c measuring systems.

3.4.2.3 Use in frequency-modulated systems. Inductive displacement transducers, used as the frequency-determining inductance in an oscillator circuit, have been used to provide a frequency-modulated output interchangeable with that of a typical subcarrier oscillator in an FM/FM telemetry system. The output of such transducer-oscillator combinations can also be displayed on a time-based counter (EPUT meter).

3.4.2.4 Use in digital systems. Ultimate displacement measuring accuracy is provided by incremental and absolute digital displacement transducers, and certain photoelectric devices such as interferometers, used in digital computation and display systems. Among these transducers, the encoders, especially shaft-angle encoders, are used in a majority of applications. They are capable of providing an overall system accuracy within less than 5 seconds of arc. However, their cost and the cost of the measuring system are considerably greater in most cases than analog-output transducers and analog measuring systems.

3.5 Calibration and Tests

3.5.1 Calibration

The determination of output-vs-displacement characteristics is considerably simpler for displacement transducers than for most other transducers, primarily because accurate dimensional measurements have been made for a long time and the state of the art of such measurements advanced along with machine-shop requirements. Accordingly, the most commonly used equipment for applying known displacements to transducers for their static calibration is derived from machine-shop control and inspection equipment: the micrometer (see Figure 3-38) for linear displacements, and the "dividing head" for angular displacements.

3.5.1.1 Calibration equipment. A typical setup for the static calibration of linear-displacement transducers is illustrated in Figure 3-39. The relative motion between transducer sensing shaft and case is provided by motor-driven rotation of the drive screw which moves the carriage to which the transducer case is clamped while the sensing-shaft end is fixed in position. An optical reader can be used to display carriage motion in very small increments. The equipment shown in this example is arranged for a comparison calibration of a potentiometric transducer. The ten-turn reference potentiometer is so geared (by interchangeable gears) to the drive screw that its output will change linearly from 0% to 100% VR with the transducer output. The two outputs are fed to a resistance-bridge comparison circuit. The output of this circuit is zero when the transducer output follows the reference output perfectly. When it does not, the output of the comparator is a function of the error in the test-specimen output.

Fig. 3-38. Manually operated micrometer arrangement for multiple calibration of linear-displacement transducers (courtesy of Bourns, Inc.).

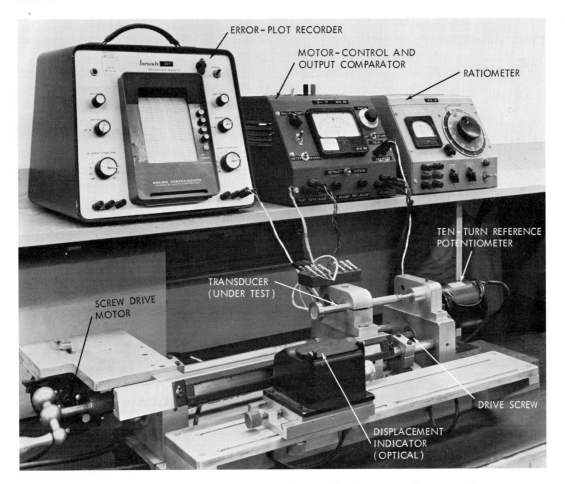

Fig. 3-39. Linear-displacement transducer calibration setup (courtesy of Servonic Instruments, Inc.).

The comparator output is fed to a strip chart recorder which provides a permanent record of the transducer error (*error plot*). The ratiometer is used instead of the comparator and error plotter when discrete transducer output readings must be recorded rather than the error in the output with respect to a reference output.

The rotating screw is used with a nonrotating nut (part of the carriage) whose materials and threads are designed to minimize backlash, wear, and sliding friction. The latter must also be minimized in the mutually contacting surfaces of the carriage and its guide rods. A *ball nut*, with recirculating balls, can be used to substitute the lower rolling friction for sliding friction. The ball-nut system is frequently used in automobile steering mechanisms. In the *hydrostatic nut*, friction and wear are reduced to an absolute minimum by use of a fluid film between screw and nut threads. The use of such a hydrostatic film requires a relatively complex gas or oil supply and return system capable of providing the required hydrostatic pressure between threads. *Rack-and-pinion* arrangements can be used instead of screw-nut systems where somewhat poorer accuracy is acceptable. Requirements for very close accuracy may dictate the use of optical devices such as interferometers or digital-output devices such as encoders.

Comparison calibrations are often used for angular-displacement transducers. For the calibration of most of these transducers, however, a mechanical dividing head is incorporated in the calibration fixture (see Figure 3-40). In this device the circumference of a drum is scribed with lines dividing the circle into 360 one-degree increments. A fixed scale of the same type is scribed on the top plate of the fixture so that one can be used as scale, the other as indicator. The drum is clamped to the transducer sensing shaft. A drive wheel is so geared to the drum—using a worm gear—that the wheel completes one full revolution for every one-degree motion of the angle-indicating drum. The drive-wheel drum is scribed with scale divisions, four divisions per minute of arc of the sensing shaft, so that each division equals fifteen seconds of arc. By reading between the divisions an overall reading accuracy within about five seconds of arc is provided. The backlash in this fixture is held to two seconds of arc.

The choice of output indicating equipment is dictated by the transduction principle employed.

3.5.1.2 Calibration methods. The static calibration consists of two or more calibration cycles. During each calibration cycle, known values of displacement are applied

Fig. 3-40. Angular-displacement transducer calibration fixture (courtesy of Servonic Instruments, Inc.).

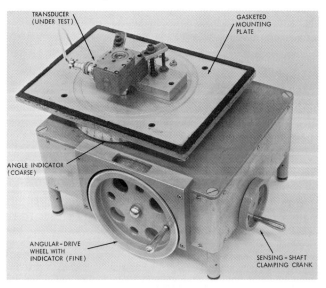

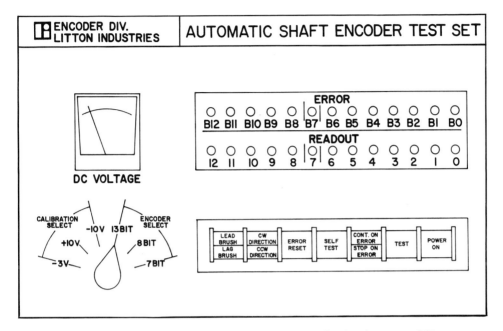

Fig. 3-41. Front panel of test set for encoder calibration (courtesy of Litton Industries, Encoder Div.).

to the transducer over its range in an ascending and descending direction and corresponding output (amplitude or phase, or both) values are recorded. During a comparison calibration the displacement of the transducer and of a reference transducer is varied continuously while the output of the transducer is compared to that of the reference transducer. During the much more common absolute calibration (referred to as just "calibration") the displacement is varied in predetermined increments, usually in increments of 10% or 20% of the transducer's span. Overtravel output can be determined during a calibration by exceeding the span by the specified amounts of travel.

The calibration cycle starts and ends at the point on which the transducer's range is based, usually the zero-percent-of-span or fifty-percent-of-span point. The latter is used for bidirectional ranges. For these, displacement is varied from 50% to 100%, to 50%, to 0%, and back to 50% of span. When the range is unidirectional, displacement is varied from 0% to 100% and back to 0% of span.

Close-accuracy, fine-resolution calibrations of digital-output transducers require special test equipment and digital techniques. Test sets, such as the type employing autocorrelation, are frequently automatic and can operate in a "self-test" mode (see Figure 3-41). Special test sets have also been developed for calibration and tests of synchros and related devices.

It is essential that the range reference point be determined very accurately before beginning a calibration cycle. When this reference point is stated in terms of output, it is good practice to "rock in" the reference point by sweeping past it in both directions with sweeps of decreasing amplitude, so as to minimize any hysteresis, friction, or backlash at this point. Only after the reference point has been obtained can the sensing shaft (or measured object for a noncontacting transducer) be clamped to the calibration fixture.

A subsequent examination of calibration data allows determination of compliance to static-performance-characteristic specifications.

3.5.2 Other static performance tests

Certain inspections and tests are often performed on each transducer prior to the static calibration. These include a visual inspection, overtravel determination, insulation-resistance and breakdown-voltage tests, input and output impedance determinations, continuous output plots (x-y plots) for potentiometric transducers, output-noise tests, and other checks for compliance to mechanical and electrical design-characteristic specifications.

Resolution tests are usually not performed on each potentiometric transducer but are performed on a sampling basis only. Other sampling tests include determination of weight and of mounting effects (mounting error).

3.5.3 Dynamic performance tests

"Frequency response" tests, intended to determine the ability of the transducer output to follow increasing cyclic or rotational sensing-shaft (or measured object) speed, are normally performed on a sampling basis only. Cyclic motion is applied as sinusoidal oscillatory motion. The wave shape and especially the amplitude of the motion must be carefully monitored during the test. The peak-to-peak amplitude should not exceed 90% of the transducer's span; preferably, it should be 10% or (for large spans) 1% of span.

3.5.4 Environmental tests

Because displacement calibration fixtures for coupled transducers are normally not intended for severe environments such fixtures are not used during most environmental tests. Instead, environments such as shock, vibration, acceleration, high sound-pressure levels, and nuclear radiation are applied to the transducer while the shaft is mechanically locked at a preselected position. This position is selected on the basis of the expected position while the maximum level of the environment is encountered during the transducer's intended use. When this cannot be determined, the position is either arbitrarily chosen near mid-span or selected, by analysis, as the position at which the environment may cause the largest effects on transducer output.

Temperature tests can be performed in the form of high-temperature and low-temperature calibrations if the driving shaft, coupled to the sensing shaft, is brought out through a small opening in the temperature chamber so that the calibration fixture is well outside the chamber and does not see the thermal environment to which the transducer is subjected. The fixture shown in Figure 3-39 can be used for temperature tests when a temperature chamber is inverted over the transducer so that the gasketed mounting plate forms its bottom lid. During ambient-pressure, immersion, and con-taminating- or explosive-atmosphere tests the calibration-fixture driving shaft can be coupled to the transducer sensing shaft through a bellows or similar motion-transmitting isolator, sealed to an opening in the test chamber.

Life-test methods are similar to those employed for dynamic performance tests except that shaft speed is kept constant and well below the rated maximum speed.

3.6 Specialized Measuring Devices

3.6.1 Transducers for small displacements

A number of transducing devices have been developed primarily for application to very small displacements. Since they are not too commonly used, they were not described in 3.3.

3.6.1.1 Capacitive-ionizing transducers.
This design, originally developed by Dr. K. S. Lion (Massachusetts Institute of Technology), uses a slim, gas-filled, sealed glass tube containing two parallel electrodes. When radio-frequency power is applied to two parallel external plates, between which the tube is located, a luminous discharge occurs in the tube and a d-c voltage appears across the two internal electrodes. This voltage is zero when the tube is centered between the plates but increases nega-tively or positively, respectively, when the tube is displaced toward one or the other external plates. This arrangement has been found useful for displacement ranges of less than 1 mm. Either the tube or the external plate pair (with the distance between them fixed) can be displaced. A modification of this design has been reported as usable for displacement ranges up to several centimeters.

3.6.1.2 Thermionic displacement transducers.
The variation of plate current with mechanical displacement of the anode in a space-charge-limited vacuum tube has been utilized in a triode developed by the Radio Corporation of America (Tube No. 5734). The cathode and grid are fixed in a near-vertical position within the metal-shell vacuum tube. The rod-shaped anode is brought out of the top of the tube through a flexible metal diaphragm. The metal shell is connected to the anode. At typical operating potentials and circuit constants, an output voltage of about ± 20 V is pro-duced by an angular displacement of ± 0.5 degrees of the protruding anode rod (sens-ing shaft), the maximum displacement range provided by this transducer.

3.6.1.3 Vibrating-wire displacement transducers. Displacement transducers using a vibrating wire ("vibrating string") are sometimes used for the measurement of small displacements. This type of transduction element has been used in pressure and acceleration transducers and is described in the chapters of this book dealing with these two measurands. Vibrating-wire displacement transducers have been applied primarily outside the U.S.A. The output of these transducers is an a-c voltage whose frequency is determined by the tension of a thin wire in a feed-back oscillator circuit. A change in displacement causes a corresponding change in wire tension and, hence, a change in the frequency at which it oscillates. The full-scale displacement ranges for this transducer type are usually less than 50 microns.

3.6.2 Displacement-altitude transducers

A group of relatively complex equipment exists, in the category of displacement measurement devices, whose purpose is to measure vertical displacement of a flying vehicle with respect to the terrain below the vehicle. Such *altimeters*, which must be differentiated from the commonly used pressure-altitude transducers, measure this vertical displacement continuously, while operated, in terms of reflection of energy (emitted from a point on the vehicle) by the terrain, sensed by a detector on the vehicle. The energy, usually emitted in pulsed form, can be in the form of radio-frequency (in the *radar altimeter*), X-ray, or nuclear-particle radiation.

3.6.3 Derived displacement measurements

As given by the interrelationships between displacement, velocity, and acceleration (see 3.1.2), displacement data can be obtained by integration of velocity-transducer output or by double integration of acceleration-transducer output.

3.6.4 Proximity switches

Although not normally considered as transducers, proximity switches are, in fact, noncontacting displacement transducers having a discrete-increment output at one point within their range. Electromagnetic, inductive, and reluctive transducers have been designed as proximity switches for applications ranging from "door closed" indications to counting the number of two-axled vehicles moving on a highway. Electromagnetic proximity switches provide an output when approached by a ferromagnetic-material object at a rate sufficient to cause an output-producing change in magnetic flux. Only when the output is stored in a "latching" mode (e.g., made to pull in a latching relay) can a continuous indication of proximity be obtained. Indication at any point within their range depends on the switching point at which an associated electronic gate circuit has been set to turn ON or OFF. The range of inductive transducers is usually limited to about one inch (away from the transducer). Larger ranges, reportedly up to five feet, can be obtained with reluctive transducer, e.g., those using an "*E*-core" transduction element.

BIBLIOGRAPHY

1. Bower, G. G., Smith, W. H., and Smith, R. O., "Performance Tests on Angular Position Pickup Model TTO-1A," *NBS Report 1B130*, Washington, D.C.: National Bureau of Standards, 1953.

2. Bartley, A. J., "The Use of Polarized Light for the Measurement of Small Angular Displacement," *Journal of Science Instruction*, Vol. 33, pp. 20–22, 1956.

3. Dummer, G. W. A., *Variable Resistors and Potentiometers*. London: Sir Isaac Pitman & Sons Ltd., 1956.

4. Lion, K. S., "Mechanic-Electric Transducer," *Review of Scientific Instruments*, Vol. 27, No. 4, pp. 222–225, April 1956.

5. Sink, R. L., "Angular Position Transducer," *U.S. Patent 2,775,755*, 1956.

6. Elliott, K. W. T. and Wilson, D. C., "An Optical Probe for Accurately Measuring Displacements of a Reflecting Surface," *Journal of Scientific Instruments*, Vol. 34, pp. 349–352, 1957.

7. Gibson, John E. and Tuteur, F. B., *Control System Components*. New York: McGraw-Hill Book Company 1958.

8. Ahrendt, W. R. and Savant, C. J., Jr., *Servomechanism Practice*. New York: McGraw-Hill Book Company, 1960.

9. Andrews, A., *ABC's of Synchros and Servos*. Indianapolis: H. W. Sams, 1962.

10. Brouillette, D. A., "A Shaft Multiplication Precision Angle Encoder," *Electrical Engineering*, Vol. 18, No. 2, February 1962.

11. Chass, J., "The Differential Transformer," *I.S.A. Journal*, Vol. 9, No. 5, pp. 48–50, and Vol. 9, No. 6, pp. 37–39, May and June 1962.

12. Marmorstone, R. J., "Digital Techniques in Precision Dimensional Measurement," *Automatic Control*, Vol. 17, No. 1, July 1962.

13. "Synchro and Resolver Evaluation and Test Equipment" (Editorial survey), *Electro-Technology*, Vol. 71, No. 5, pp. 201–210, May 1963.

14. Charnley, C. J. and Healey, M., "Position Transducers for Numerically Controlled Machine Tools," *CoA Note M & P No. 6*, Cranfield, England: The College of Aeronautics, December 1964.

15. Nelson, F. E., "Tests of a Motion Pickup of the Infra-Red Type," *Technical Note N-623*, Port Hueneme, California: U.S. Naval Civil Engineering Laboratory, October 1964.

16. Canfield, E. B., *Electromechanical Control Systems and Devices*. New York: John Wiley & Sons, Inc., 1965.

17. Spaulding, Carl P., "How to Use Shaft Encoders," Monrovia, California: DATEX Division of Conrac Corp., 1965.

18. Akhmedzhanov, A. A., "*Sistemy Peredachi Ugla—Povyshennoi Tochnosti*" (High-

Accuracy Systems for the Transmission of Angles), Moscow: ENERGIJA Publishing House, 1966.

19. Koppe, H., "Spirit Level Electronically Reads Angular Displacement," *Electronic Design News*, Vol. 11, No. 1, pp. 40–42, January 1966.

20. Whitten, L. G., "Position Monitoring Devices," *Report No. Y-1506*, Oak Ridge, Tenn.: Union Carbide Corp., Nuclear Division, 1966.

21. Baumann, E., "Probleme bei der Berührungslosen Wegmessung mit Inductiven Aufnehmern," (Problems in Noncontacting Displacement Measurement with Inductive Transducers), *Paper No. 21-DDR-230, Acta IMEKO 1964*, pp. 197–209; IMEKO, Budapest, Hungary, 1967.

22. Coleman, E. J., "Measuring Length Remotely and Accurately with a Laser Beam," *Instrumentation Technology*, Vol. 14, No. 1, pp. 45–48, January 1967.

23. Ross, E. A., "NEMA Standard Digital Position Rotary Incremental Transducers," *Control Engineering*, Vol. 14, No. 10, pp. 85–89, October 1967.

24. Starer, R. L., "Electro-Optical Tracking Techniques," *Instruments and Control Systems*, Vol. 40, No. 2, pp. 103–105, February 1967.

4 Flow

4.1 Basic Concepts

4.1.1 Basic definitions

Flow is the motion of a fluid, usually a confined fluid stream. *Note:* The term "flow" has often been applied to flow rate.

Flow rate is the time rate of motion of a fluid expressed as fluid quantity per unit time. *Notes:* (1) The measurand of flow transducers is almost invariably flow rate; (2) The fluid whose flow rate is measured by a transducer is usually contained in a pipe or duct.

Total flow is the flow rate integrated over a time interval.

Volumetric flow rate is flow rate expressed as fluid volume per unit time.

Mass flow rate is flow rate expressed as fluid mass per unit time.

Laminar flow (streamline flow) is motion of fluid particles along lines parallel to the local direction of flow; it can be represented as layers of fluid sliding steadily over one another. Laminar flow has also been referred to as **viscous flow**.

Turbulent flow is motion of fluid particles whose velocity at a point of observation fluctuates with time in a random manner.

Viscosity is a fluid's resistance to the tendency to flow.

Kinematic viscosity is the ratio of the viscosity to the density of a fluid.

Density is the ratio of the mass of homogeneous substance to its volume.

Specific gravity is the ratio of the density of a substance at a given temperature to the density of a substance considered as standard. Pure distilled water at 4°C as well as at 60°F has been used as standard for use with liquids and solids. Air at one atmosphere absolute pressure (14.7 psia) and at either 60, 68, or 70°F has been used as standard for gases.

The **Reynolds number** is a (dimensionless) number used to express the fluidity of a moving fluid (see 4.1.2). Low Reynolds numbers (below 2000) indicate laminar flow; high Reynolds numbers (generally above 4000) indicate turbulent flow. Flow associated with Reynolds numbers between 2000 and 4000 can be laminar or turbulent, depending on the configuration of the flow system.

The **Prandtl number** of a fluid is the ratio of its kinematic viscosity to its thermal conductivity.

A **flowmeter** is, generally, a flow-rate transducer.

4.1.2 Related laws

Reynolds number (of a fluid flowing through a pipe)

$$N_R = \frac{Vd\rho}{\mu}$$

where: V = average flow velocity
d = internal pipe diameter
ρ = density of fluid
μ = viscosity of fluid
N_R = Reynolds number

Bernoulli's equation (for horizontal flow)

$$p_s = p_0 + \tfrac{1}{2}\rho V_0^2$$

where: p_s = stagnation pressure (total pressure)
p_0 = static pressure
ρ = density of fluid
V_0 = flow velocity (upstream from stagnation point)

4.1.3 Units of measurement

Volumetric flow rate is most commonly expressed in **gallons per minute** (gal/min or *gpm*) for liquids and in **cubic feet per minute** or **per second** (ft³/min, ft³/s or *cfm, cfs*) for gases, or in metric units of volume per minute or per second. Very low rates are expressed in quantity **per hour**. Conversion of typical units is shown in Table 4-1.

Mass flow rate is usually expressed in **pounds (force) per minute** (*lbf/min*) or in equivalent metric units; mass-flow calibrations are performed on the basis of weight. Mass flow rate of gases is usually expressed as equivalent volume flow at standard conditions (20°C and 14.7 psia) in "standard" units, i.e., **standard cc per minute** (standard cm^3/min) or **standard cfm** (standard ft^3/min).

Total flow is expressed in the units of quantity used for the measurement of the flow rate from which the total flow is derived.

Viscosity (absolute viscosity) can be expressed in **pounds (force) per square foot per second** (*lb-s/ft²*). Other units for (the coefficient of) viscosity are the **poise** and **centipoise** (one poise = one dyne-s/cm²; one centipoise = 10^{-2} poise). The SI unit of viscosity is the newton-second per square meter (**N-s/m²**).

Table 4-1 CONVERSION OF TYPICAL VOLUMETRIC FLOW-RATE UNITS

1 cubic foot per hour	=	0.01667	ft³/min
	=	0.472	l/min
	=	28.317	l/hr
	=	0.1247	gal/min
	=	7.481	gal/hr
1 cubic foot per minute	=	28.317	l/min
	=	1700	l/hr
	=	28317	cm³/min
	=	7.481	gal/min
	=	449	gal/hr
1 liter per hour	=	0.0353	ft³/hr
	=	0.01667	l/min
	=	16.67	cm³/min
	=	1000	cm³/hr
	=	0.2642	gal/hr
1 liter per minute	=	0.0353	ft³/min
	=	2.118	ft³/hr
	=	1000	cm³/min
	=	0.2642	gal/min
	=	15.852	gal/hr
1 gallon per hour	=	0.1337	ft³ hr
	=	0.0631	l/min
	=	3.785	l/hr
	=	63.1	cm³/min
	=	3785	cm³/hr
1 gallon per minute	=	0.1337	ft³/min
	=	8.022	ft³/hr
	=	3.785	l/min
	=	227.1	l/hr
	=	3785	cm³/min

Note: Gallons are U.S. gallons

Kinematic viscosity is expressed in **stokes** or **centistokes** (one stoke $=$ one cm^2/s). The SI unit of kinematic viscosity is the square meter per second (**m^2/s**).

Density is expressed in units of mass per unit of volume.

Specific gravity is shown as "t_f/t_s specific gravity," expressed as a dimensionless number since it is a density ratio, where t_f is the temperature of the fluid under consideration and t_s is the temperature of the fluid considered as standard. The standard fluid for specific gravities of liquids is pure distilled water. Example: "the 60/60°F specific gravity of this liquid is 0.770." Specific gravities of a number of commonly used gases are shown in Table 4-2.

4.2 Sensing Elements

The sensing elements which respond directly to the flow rate of a fluid fall in three general categories:

1. A section of pipe or duct with a restriction of some sort which produces a differential pressure proportional to flow rate. The differential pressure can then be measured by a pressure transducer.

2. A freely moving or elastically restrained mechanical member which responds to the moving fluid either by rotating (e.g., a propeller), by deflecting (e.g., a cantilevered vane), or by a vertical displacement in a tapered tube (e.g., a float in a variable-area flowmeter).

3. The fluid itself, with one of its physical characteristics interacting with the transduction element (e.g., cooling of a heated wire by the fluid).

Table 4-2   68/68°F SPECIFIC GRAVITY VALUES
OF GASES (at 14.7 psia)

Acetylene	0.9073
Air	1.000
Ammonia	0.587
Argon	1.377
Carbon dioxide	1.516
Carbon monoxide	0.965
Chlorine	2.486
Ethane	1.0493
Helium	0.138
Hydrogen	0.0695
Methane	0.553
Nitrogen	0.966
Nitrous oxide	1.518
Oxygen	1.103

4.2.1 Differential-pressure flow-sensing elements

The flow-sensing elements in this group have been referred to as "head meters" or "variable-head meters" because the differential pressure across two points varies and pressure has been equated to "head," the height of the column in a differential manometer connected to the two points. They are characterized by maintaining a constant area of flow passage. The *Venturi tube, flow nozzle, orifice plate, pitot tube,* and *centrifugal section* are included in this group (see Figure 4-1). Since pressure transducers rather than flow-rate transducers are used to measure flow with these sensing elements their operation is described in 4.6.

Fig. 4-1. Differential-pressure flow sensing elements.

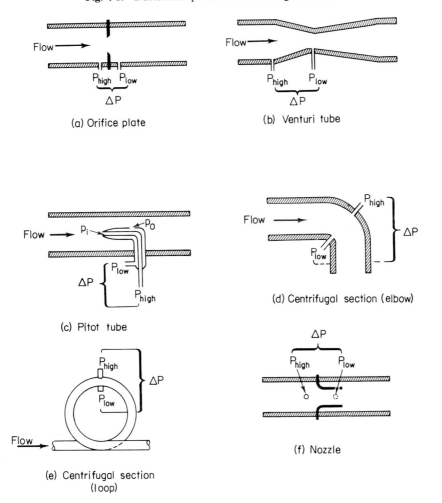

(a) Orifice plate

(b) Venturi tube

(c) Pitot tube

(d) Centrifugal section (elbow)

(e) Centrifugal section (loop)

(f) Nozzle

4.2.2 Mechanical flow-sensing elements

A number of different mechanical flow-sensing elements are used in flow-rate transducers. "Variable-area meters" use a *float* in a tapered section of tubing ("rotameter"), a spring-restrained *plug*, or a spring-restrained *hinged vane* (Figure 4-2). The displacement of these elements causes the area of flow passage to vary while the differential pressure or pressure drop ("head") remains constant. The displacement is transduced to provide an output proportional to flow rate.

The *drag body* is supported in a pipe section so that its displacement due to flow causes deflection of the supports or other mechanically linked members. The *cantilevered vane*, fixed at one end [Figure 4-2(d)], responds to the impact of flowing fluid by its own deflection.

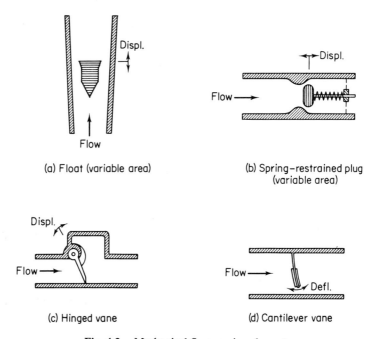

Fig. 4-2. Mechanical flow sensing elements.

Fig. 4-3. Rotating mechanical flow sensing elements.

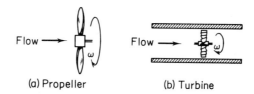

Rotating members (Figure 4-3) include the *turbine* installed in a pipe section and the *propeller* (as in an anemometer) installed in a free air stream. Both rotors turn at an angular speed proportional to flow rate.

4.2.3 Flow sensing by fluid characteristics

The moving fluid itself acts as flow-rate sensing element in conjunction with various transduction elements. A heated wire in a *hot-wire anemometer* transfers more of its heat as the flow velocity of its ambient fluid increases. The resultant cooling of the wire causes its resistance to decrease. A fluid containing a small amount of radioactivity causes an increase in *ionization current* (or photocurrent) in a nuclear radiation transducer, past which it flows, at increasing flow velocity. A mildly conductive fluid flowing through a transverse magnetic field has an increasing *electromotive force* induced in it at increasing flow velocity. When the boundary layer of a moving fluid is heated by a small heating element, the *convective heat transfer* to a temperature sensor located some distance downstream from the heater increases with increasing flow velocity.

4.2.4 Mass-flow-sensing elements

The mechanical sensing elements briefly described above are used for the measurement of volumetric flow rate. Mass flow rate can be calculated from this measurement and a simultaneous measurement of density. A computer can be fed density and volumetric-flow-rate inputs to provide a continuous output indicative of mass flow rate. Special mechanical sensing elements have been developed for use in flow-rate transducers whose output is directly proportional to mass flow rate. Examples of such transducers are described in 4.3.8.

4.3 Design and Operation

4.3.1 Electromagnetic flow-rate transducers

Electromagnetic transduction is utilized in two radically different types of flow-rate transducers, the fluid-conductor magnetic flowmeter and the turbine flowmeter.

4.3.1.1 The fluid-conductor magnetic flowmeter comprises a pipe through which the fluid flows, a set of electromagnetic coils, and a pair of electrodes in contact with the fluid. The electromagnetic coils produce a magnetic field perpendicular to the flow. The electrodes are located at right angles to both the flow and the magnetic field. The operating principle [Figure 4-4(a)] is based on Faraday's Law of Electromagnetic Induction which states that the electromotive force induced in a

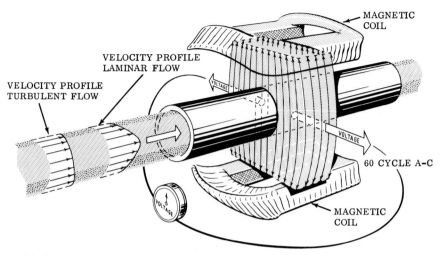

(a) Operating principle

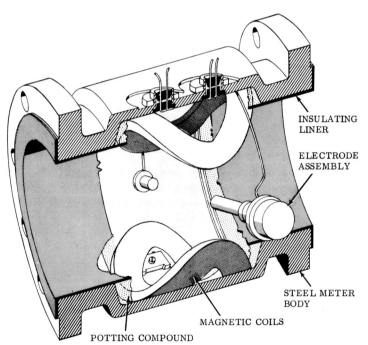

(b) Typical flowmeter with internal coils

Fig. 4-4. Fluid-conductor magnetic flow-rate transducer (courtesy of Fischer & Porter Co.).

conductor moving through a magnetic field is proportional to the time rate of change of the magnetic flux. Since the coils and electrodes are stationary and their geometry remains constant, the flux change is entirely due to the velocity of the conductor through the field. The fluid flowing in the pipe acts as the conductor.

The electromotive force proportional to flow velocity appears as a potential generated between the two electrodes. This potential $E_s = BdV$, where B is the magnetic flux density, d is the face-to-face spacing between the electrodes (mounted opposite each other in the pipe wall), and V is the average fluid velocity. With B expressed in gauss, d in cm, and V in cm/s, the voltage E_s is calculated in abvolts (1 abvolt $= 10^{-8}$ volt). This relationship applies equally to laminar, turbulent, or transitional flow. The two electromagnetic coils are usually connected in series and supplied with alternating current from a power transformer. A reference voltage, E_r, can be tapped off the series circuit, such as by a small reference transformer, so that E_r is always proportional to the magnetic flux density B. Substituting E_r for B, the signal voltage $E_s = E_r dV$, and the flow rate is proportional to the ratio E_s/E_r even in the presence of normally encountered fluctuations in excitation voltage and frequency. The magnitude of the output voltage (E_s) of a typical transducer is 3 mV a-c for a fluid velocity of 10 fps.

An industrial version of a fluid-conductor magnetic flowmeter was first reported in the U.S.A. by E. Mittelmann in 1950. Subsequent developments showed that the magnetic field need not be homogeneous and that the length of the coils (along the pipe) can be reduced when they are so shaped as to maximize the magnetic flux where the signal generating coefficient is lowest. Coils could also be placed on the inside of the pipe [Figure 4-4(b)]. Most designs have their coils outside the pipe. One version uses a C-shaped magnetic field pole piece for both coils to allow replacement of the pipe and electrode assembly without coil removal.

A limiting factor, in certain applications, is the requirement for the conductivity of the fluid to be greater than about 10^{-8} mho/cm^3. The presence of mixed-phase fluids (gas in liquids) can cause major measurement errors. D-C coil excitation is rarely used since it can cause electrolysis in a conducting fluid. D-C excitation has been used in special flow-rate transducers such as the arterial blood-flow meter which utilizes the (conducting) arterial wall as the "pipe." The linearity of output voltage with flow rate is a significant advantage of the fluid-conductor magnetic flow-rate transducer, as is the absence of any obstruction in the pipe.

4.3.1.2 The turbine flowmeter was originally developed for aerospace flow measurements but has since become popular in numerous other fields. A freely spinning bladed rotor (Figure 4-5) is used as sensing element. Its angular velocity is related to the volumetric flow rate by

$$\omega = \frac{V \tan \alpha}{R} = Q\frac{\tan \alpha}{(A_n - A_r)R}$$

where ω = angular velocity of rotor

$\quad\quad V$ = flow velocity

$\quad\quad \alpha$ = rotor blade angle relative to axis of flow

$\quad\quad R$ = average radius of rotor blade center of pressure

$\quad\quad Q$ = volumetric flow rate

$\quad\quad A_n$ = internal cross-sectional area of housing

$\quad\quad A_r$ = largest cross-sectional area of rotor

From this theoretical relationship it can be seen that $\omega = QK_d$ where K_d is a constant given entirely by fixed parameters of transducer design. Bearing friction, fluid and magnetic drag, swirl in the fluid stream, and velocity profile distortions can cause deviations from the theoretical behavior of the transducer. The error causes can be minimized by good mechanical design. Swirl in the stream, for example, is virtually taken out by flow straighteners such as the two triple-tube assemblies in the flowmeter of Figure 4-5 and the fixed blades on the shaft support of the flow-meter illustrated in the exploded view of Figure 4-6. Magnetic drag is usually neg-ligible except in small transducers. Thrust bearing friction has been eliminated in

Fig. 4-5.   Turbine flow-rate transducer (courtesy of Potter Aeronautical Corp.).

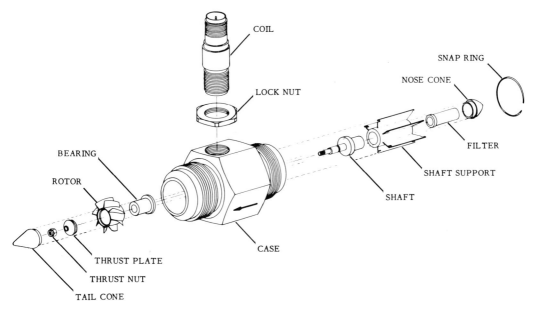

Fig. 4-6. Small turbine flow-rate transducer (courtesy of General Dynamics, Convair Div.).

designs using hydrodynamic forces to balance the axial loads on the rotor, thus obviating any need for a thrust bearing.

The transducer case is made of nonmagnetic metal and is usually threaded to receive the transduction coil (see Figure 4-6). Since the amplitude of the a-c output voltage from the coil is dependent on the gap between pole piece and rotor blade tips, the coil can be adjusted by turning it in its thread, then locking it in position, e.g., with a locknut, just before final calibration. Some designs use a permanent magnet encased in the rotor body, in conjunction with a transduction coil which has only a soft iron core. Most turbine flowmeters, however, use an electromagnetic transduction coil of the permanent-magnet type (illustrated in Figure 12-5). The rotor blades of such transducers are made of a ferroelectric material. As each blade passes the coil's pole piece, a voltage is induced in the coil. The transducer output taken across the two coil terminals is, therefore, an a-c voltage whose frequency is proportional to flow rate. The number of cycles per revolution is given by the number of blades on the rotor. The rotor blades can be so shaped as to produce a sinusoidal output voltage with very little harmonic distortion. Two or more separate outputs can be obtained by mounting two or more coil assemblies into the case. The transducer output frequency band for a given range of flow rates can often be made to coincide with an **IRIG**-standard subcarrier frequency band in an FM/FM telemetry system by selecting the appropriate number of rotor blades and effecting a design compromise between the pitch of the rotor blades and the number of blades.

Some designs are furnished with a sealable *spin port* in their case. This threaded opening allows the connection of a source of pressurized gas to the transducer in such a manner that the turbine will rotate at a speed typical for its intended use. Such a *spin check* is sometimes performed as a functional test on a turbine flowmeter in its as-installed condition. This test can indicate, e.g., whether the coil-to-rotor tip gap was adjusted to obtain proper output voltage amplitude.

A special design was developed to prevent damage to the rotor due to over-speeding, as during gas purge of a liquid system. This flowmeter is equipped with an electromagnetic brake which acts on the rotor with increasing force as its angular speed increases above a predetermined value.

The frequency output of turbine flowmeters is best measured with an EPUT (events-per-unit-time) meter or similar electronic counter. Overall system repeatabilities with 0.25% can be realized with such a system. Frequency-to-dc converters of various designs are available when a d-c analog representation of flow rate is required.

A substantial advantage offered by turbine flowmeters is their almost unique ability to act as a self-integrating transducer; i.e., each cycle of frequency output represents a discrete fraction of a volume unit.

4.3.2 Nucleonic flow-rate transducers

Several types of instruments have been developed to utilize radioisotopes and the detection of nuclear radiation for the measurement of flow rates. Such instruments have usually been custom-built for specific applications. One system uses a neutron source mounted to the outside of the pipe some distance upstream from a radiation detector (Figure 4-7). The relatively low-energy neutrons collide with atoms of the moving fluid and cause particle and electromagnetic radiation to be

Fig. 4-7. Schematic of nucleonic flow-rate measuring system (courtesy of Conrac Corp.).

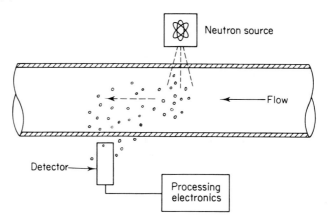

emitted. Although most of this radiation occurs in the fluid at the source location, some of it is still emitted when the fluid passes the detector. The number of counts produced by the detector are then indicative of the flow rate of the fluid. Another system obtains such an output by adding minute amounts of a radioactive trace element to the fluid. Nucleonic flowmeters, which offer no obstruction to the fluid and expose no portion of their transduction element to it, are useful for measurement of "difficult" fluids such as multiphase and variable composition fluids, slurries, and suspensions.

4.3.3 Reluctive flow-rate transducers

Reluctive transduction has been employed primarily in flowmeters of the variable-area type. The transducer shown in Figure 4-8 uses the displacement due to flow of a spring-restrained spool to change the inductances of the two coils embedded in the meter body. When the fluid moves from left to right, the spool moves to the right and increases the inductance of the right coil while decreasing the inductance of the left coil. The two coils can be connected as a two-active-arm inductance bridge. The bridge output voltage and phase, at constant a-c excitation, are then proportional to the flow rate. The a-c output can be converted into a varying d-c voltage or current by a phase-sensitive demodulator or similar rectifying circuit. The transducer responds equally to flow in either direction. Electrically conducting portions are isolated from the fluid. Reluctive transduction has also been used to obtain an output from the displacement of other mechanical flow-sensing elements

Fig. 4-8. Variable-area reluctive flow-rate transducer (courtesy of Lynch Corp., Cox Instruments Div.).

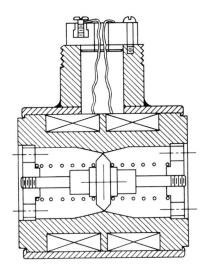

such as a rotameter float. Reluctance-bridge as well as differential-transformer configurations have been employed, sometimes with the armature mechanically linked to the sensing element through a bellows seal to isolate the coil assembly from the fluid.

4.3.4 Resistive flow-rate transducers

Transducers which convert flow rate into a change of resistance use the fluid itself as sensing element. The resistive transduction element responds to temperature changes in three transducer types: 1. the hot-wire anemometer, a specialized device used primarily for air-flow measurements and discussed in 4.6.2; 2. the thermal and boundary-layer flowmeters, most of whose commercial versions, however, employ thermoelectric rather than resistive transduction (see 4.3.6); and 3. the oscillating-fluid flowmeter, which is discussed below.

Under certain geometric and dynamic conditions flowing fluid will oscillate at a well-defined frequency that is proportional to a characteristic steady flow

Fig. 4-9. Oscillating-fluid resistive flow-rate transducer (courtesy of American Standard, Special Products Dept.).

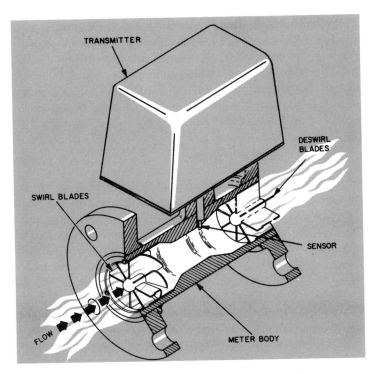

Fig. 4-10. Operating principle of oscillating-fluid flowmeter (courtesy of American Standard, Special Products Dept.).

velocity. Examples of this are the "aeolian tone" exhibited by "singing" telephone wires and the "edge tone" caused by jet flow over a sharp-edged surface. A related device is the "vortex whistle" in which a tone is generated near the exit of a reduced-diameter tube connected to a larger tube into which a fluid is admitted through a tangential inlet.

The flowmeter shown in Figure 4-9 utilizes the principle of the "vortex whistle" for flow-rate measurements of primarily gaseous fluids; however, this transducer type is equally usable for homogeneous liquids. The fluid entering the meter (Figure 4-10) is forced into a swirling motion by means of a fixed set of swirl blades at the inlet. The flow passage downstream from these blades is formed as a Venturi-like contracting and expanding cavity. In the region where the area is increasing the swirling flow oscillates (precesses) about the axis of the meter. The oscillations of the fluid create corresponding variations in fluid temperature seen by either a quartz-coated platinum-film sensor or a thermistor-type resistive temperature sensor. The deswirl blades serve to straighten the flow again as it leaves the flowmeter.

The temperature sensor has a very short time constant so that its usable frequency response extends up to 1000 Hz. The resistance changes are converted into voltage variations which are then amplified, gain-stabilized, filtered, and shaped into square wave pulses of an amplitude sufficiently high for display on a frequency counter. The output frequency (usually between 10 and 1000 Hz) is an analog of flow rate. The calibration curve of output frequency-vs-volumetric flow rate is reported to have an independent linearity within $\pm 1\%$ of full-scale output between flow-rate limits established by a minimum Reynolds number (typically 5000) and a maximum Mach number (typically 0.12).

4.3.5 Strain-gage flow-rate transducers

The deflections of cantilevered vanes and of supports of drag bodies due to the impact force of flow have been transduced by strain-gage bridges in a limited number of designs. The vane in such flowmeters is usually installed perpendicular to the direction of flow. A horizontally mounted wedge-shaped vane at the tip of a beam on which strain gages are installed has been used for air turbulence measurements.

4.3.6 Thermoelectric flow-rate transducers

The measurement of flow rate by heat-transfer methods is obtained by two types of transducers: the thermal flowmeter and the boundary-layer flowmeter. Both types use an electrical heater to increase the heat content of a portion of the moving fluid and two or more temperature sensors, sometimes resistive, usually thermoelectric, to measure the temperature of the fluid upstream and downstream from the heater. An indication of the heat transferred between heater and downstream sensor and, hence, of the flow rate, is obtained either by measuring the temperature differential between the two sensors at constant heat input or by measuring the change in heater power required to keep this temperature differential constant.

The first thermal flowmeter was described by C. C. Thomas in 1911 and later developments did not substantially alter the basic components of this design—a heating element immersed in the fluid and two temperature sensors, one upstream, the other downstream from the heater. The heat input (H) and the mass flow rate (Q_M) are related by $H = c\,\Delta t\,Q_M$, where c is the specific heat and Δt is the temperature difference between the upstream and downstream sensor locations. Since the entire core of the fluid was heated, the heater power required for some of the earlier designs was in the kilowatt region. The uneconomical power requirement was a severe limiting factor in the application of thermal flowmeters.

This disadvantage was overcome in the boundary-layer flowmeter by injecting the heat only into the very thin boundary layer of the fluid adjacent to the pipe wall. The temperature sensors are made flush with the inside pipe surface, and the heater is either mounted around the outside of the pipe or embedded in an internal groove flush with the inside of the wall and isolated from the fluid. The heat input varies with mass flow rate and temperature differential; however, the relationship is much

Fig. 4-11. Thermoelectric flow-rate transducer (courtesy of Hastings-Raydist, Inc.).

more complex than for the thermal flowmeter. Heater power requirements are usually below 50 W.

Figure 4-11 illustrates a thermal flowmeter design which uses a constant-ratio bypass section to sample the gas flow in the main duct for flow-rate ranges between 0 to 1.5 and 0 to 200 standard cfm. Ranges lower than these are measured without use of a bypass section. The full-scale output from multiple thermocouples is between 2 and 5 mV. Although this design is not a boundary-layer device, a number of design improvements effected a relatively low power consumption. Power requirements for the transducer and associated readout equipment are 15 W at 115 V a-c, of which only about 5 W are used by the transducer itself.

4.3.7 Ultrasonic flow-rate transducers

Piezoelectric ultrasonic flowmeters were developed for aerospace applications in the 1950's but have not reached wide commercial usage. One design uses two transducer pairs to establish an upstream and a downstream sonic path diagonally across the fluid. The difference in propagation velocity between the two paths is related to the volumetric flow rate.

4.3.8 Special mass-flow-rate transducers

A number of special transducer designs have been developed for the direct measurement of mass flow to avoid the separate density and volumetric-flow-rate measurements necessary to compute *inferential mass-flow rate*.

4.3.8.1 The twin-turbine mass flowmeter (Figure 4-12) uses two bladed rotors with different blade angles, coupled by a spring and capable of relative angular motion with respect to each other. As a result of the blade angle difference the two sets of blades tend to rotate at different speeds but are restrained from doing so by the spring coupling. They thus take an angular displacement with respect to each other, the magnitude of which is proportional to the flow momentum. The rotor assembly, considered as a unit, functions as a volumetric flowmeter, and the turbine

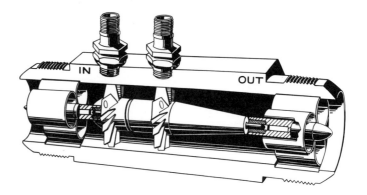

Fig. 4-12. Twin-turbine mass flow-rate transducer (courtesy of Potter Aeronautical Corp.).

angular speed is proportional to the average flow velocity. The angular displacement is observed as the phase angle between the outputs from the two transduction coils. The time interval in which the phase angle traverses a reference point is a direct measure of mass-flow rate. Digital time-interval measuring circuitry is usually employed to interpret the dual output.

4.3.8.2 Gyroscopic mass flowmeters. The operation of the gyroscopic mass flow-meter (Figure 4-13) is based on the gyroscopic vector relationship $\bar{T} = \bar{\Omega} \cdot \bar{H}$, where T is the applied torque, Ω is the processional velocity, and H is the angular momentum of the rotating mass. As shown in Figure 4-13(a), the liquid is caused to flow around a circular loop of pipe which is in a plane perpendicular to the input line. Following one pass around the loop, the fluid is returned to the input axis. During the rotation, the liquid develops angular momentum as is developed by the rotor of a gyroscope. The

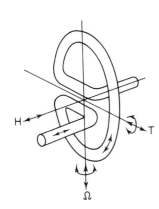

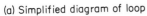

(a) Simplified diagram of loop

(b) Typical transducer design

Fig. 4-13. Gyroscopic mass-flow-rate transducer (courtesy of The Decker Corp.).

loop is vibrated through a small angle of constant amplitude about an axis in the plane of the loop. This vibration results in an alternating gyro-coupled torque about the orthogonal axis. The peak amplitude of this torque is directly proportional to mass flow rate. The torque acts against elastic restraints (torsion bars) to produce an alternating angular displacement about the torque axis. Appropriately mounted velocity sensors convert the alternating displacement to a proportional alternating electric signal. The peak amplitude signal is directly proportional to mass flow rate. An electronics unit conditions the signal and provides a d-c signal output linearly related to mass flow rate.

4.3.8.3 Angular-momentum mass flowmeters have been produced in several designs. One model incorporates two similar axial-flow rotors. The upstream rotor is driven at a constant low speed by a two-phase synchronous motor. This rotor (impeller) imparts constant angular momentum to the fluid. The downstream rotor (turbine) removes all angular momentum from the fluid. In so doing, a torque is exerted on the turbine in accordance with Newton's Second Law of Motion. This torque is linearly proportional to mass flow rate. The angular displacement of the turbine is proportional to the torque due to the spring restraint. A reluctive transduction element converts the angular displacement into a varying a-c output voltage.

In a related design the turbine is geared to a torque motor. The torque differential between motor and turbine drives a gear train. A potentiometer in the gear train develops an error-signal voltage which is amplified and used to balance the two torques. The torque-rebalancing signal is proportional to mass flow rate and appears also across the transducer output terminals.

4.4 Performance and Applications

4.4.1 Specification characteristics

Specifications can be prepared either for a flow-rate transducer alone or for a flow-rate transducer system. The latter can include signal-conditioning and display equipment. A brief statement as to what is being specified should precede the body of the specification. Unless the type of transducer is left open, the appropriate nomenclature should be included in this statement, e.g., "The specification covers a turbine flowmeter system with signal conditioning and total-flow display."

4.4.1.1 A listing of mechanical design characteristics should begin with the type of flow rate (mass or volumetric) to be measured by the transducer. The outline and the location of any mountings and all pipe and electrical connections and the nameplate should be shown on a drawing with all necessary dimensions. Pipe connections (fittings or flanges) and any electrical conduit connections should either be properly dimensioned or called out by reference to a standard. The materials of transducer case, pipe connections, and of all internal parts that can come in contact with the measured fluid should be stated.

Complete specifications of the measured fluid are of the utmost importance. The type of fluid should be stated together with its viscosity, density, vapor pressure, pH (acid or alkaline), nature and percent of solids in suspensions and their abrasive characteristics, and the ranges of operating pressures and temperatures of the fluid. Statements of the nominal (usually prevailing) operating pressure and temperature are useful. Some properties of the measured fluid can be omitted when it is an element (e.g., pure gaseous nitrogen) or when its properties can be defined by reference to a specification or standard. For mixed fluids the mixture ratio and properties of both fluids should be shown. When a flow-rate transducer can be used for several types of fluids they should be listed and accompanied by statements as to any limitations on their properties and as to any differences in transducer performance when any one of these fluids is the measured fluid. Attention should be called to any special flow characteristics such as pulsating or mixed-phase (partly liquid, partly gaseous) flow.

The pressure drop across the transducer should always be stated at the conditions for which its value applies.

Other mechanical characteristics to be included are transducer weight, external finish, type of sensing element, measured-fluid proof-pressure and burst-pressure ratings, and the nature of any special mechanical provisions (e.g., spin ports) for checking or servicing the transducer. Appropriate mechanical specifications should be added for signal-conditioning and display equipment when a complete transducer system is to be defined.

4.4.1.2 Electrical design characteristics will depend on the transduction principle employed. For example, input impedance is shown for strain-gage and reluctive transducers but not for electromagnetic transducers. The nominal and maximum excitation voltage and current or power drain are stated for all except self-generating (electromagnetic) flowmeters. Output impedance and nominal load impedance with its tolerances are usually specified at a nominal frequency for a-c output. If the output is derived from an inductive device, its ohmic resistance (measured across the output terminals) should be shown to facilitate functional checks of the transducer while not operating.

Output noise tolerances are specified especially for conditioned (e.g., demodulated and amplified) output. Waveshape and distortion are stated for a-c output, particularly frequency output. Insulation resistance and breakdown voltage rating should be specified unless one of the output terminals is grounded to the case. When the transducer is intended to operate only as a part of a specified system, the electrical characteristics can be specified as applicable just to the interface(s) with the user's power line and any other user-furnished equipment. All external electrical connections (connector pins, color-coded wires, terminals) should be identified as to their function.

4.4.1.3 Performance characteristics. The measuring range in appropriate units should be defined by both limits. Some types of flowmeters are not usable or do not perform acceptably below a certain value of flow rate. Hence the lower range limit is often greater than zero. The range limits denote the lowest and highest flow rate at which all specified performance characteristics still apply and at which output end points are specified. The maximum allowable flow rate which will not cause any performance deterioration or damage is specified as overload.

Output is usually specified by its end points or as sensitivity. If the output-vs-flow rate relationship is inherently linear, tolerances can be placed on (usually independent, end-point, or least-squares) linearity. The type of linearity should be specified. If the relationship is not linear, the curve should be described in general terms. Repeatability is specified for all types of flow-rate transducers and transducer systems. Tolerances should never be stated merely in "percent" but either in "percent of reading" or in "percent of full-scale output." When sensitivity is stated (always with tolerances), repeatability and linearity can also be shown in percent of sensitivity. Threshold specifications are often useful. Hysteresis is usually negligible and is rarely specified.

The performance of turbine flowmeters is conventionally stated as a calibration factor (K) expressed in cycles per gallon or liter or in cycles per cubic foot or cubic centimeter. A value for K should be accompanied by statements of essential fluid properties including the ranges of specific weight, viscosity, temperature, and pressure for which the calibration factor is considered applicable. A statement of line configuration, i.e., minimum length of straight pipe required upstream and downstream from the transducer and requirements for any flow straighteners other than those incorporated in the transducer, should also be included. Turbine flowmeter linearity and other performance characteristics can be expressed in percent of calibration factor.

The only essential dynamic characteristic is the time constant or the (95, 98, or 99%) response time stated at increasing or decreasing flow rate and for specific fluid properties.

Environmental performance characteristics include temperature error and allowable effects of operating as well as of nonoperating environmental conditions such as shock, vibration, acceleration (including mounting attitude effects), electromagnetic fields, nuclear radiation, high sound-pressure levels, ambient pressure variations, and contaminating, corrosive, or explosive ambient atmospheres.

Life characteristics can be shown as continuous or intermittent operating life and as long-term repeatability (stability).

4.4.2 Applications

4.4.2.1 Considerations for selection are based primarily on the properties of the fluid and the flow-rate range to be measured, on system accuracy requirements, and on any pipe configuration into which the flowmeter must fit. The pipe or duct is very often installed and in use before a flowmeter must be selected. Many transducer designs are intended for in-line installation, physically supported only by the pipe. Some of the more complex designs are too heavy for such installations and must be mounted to a supporting structure. Such limitations do not exist, of course, for free-gas-stream flowmeters (anemometers); however, other installation constraints may influence their allowable weight and size, especially in aerospace applications.

When mass flow rate of liquids must be measured, a determination should first be made as to whether the density or specific weight of the fluid will vary enough to prevent using a simple calculation, whether the density should be measured separately and inferential methods will be sufficiently accurate, or whether a mass flowmeter (usually more expensive) must be used. The mass flow rates of most gases can be measured with a system accuracy within 2 to 3% FSO by relatively inexpensive thermal flowmeters.

Pressure drop across the flowmeter must usually be minimized. In certain applications where this criterion takes precedence over most others, transducer selection may be limited to obstructionless types such as fluid-conductor magnetic flowmeters and boundary-layer flowmeters. Turbine flowmeters are among the most popular types primarily because they provide good system accuracy at a moderate cost. The differential-pressure types (see 4.6.1) are probably used in more applications than any true flow-rate transducer because of their long history of usage and the relative simplicity of their flow-sensing elements. However, the characteristics of the differential-pressure transducer which must be used with these have sometimes been insufficiently understood. Inadequate differential-pressure transducer specifications can cause poor data accuracy and unreliable measurement systems.

4.4.2.2 Installation. Most flowmeters require a certain length of straight pipe upstream as well as downstream primarily to avoid random fluid swirl at the sensing element. Flow straighteners (tubular, honeycomb, or vane type) in the pipe sections reduce the length of straight pipe required.

The introduction of mounting error, strain error, attitude error, and contamination of internal moving parts must be avoided in a flowmeter installation. Transducers which are installed in-line and which are supported only by the pipe are subject to case stresses introduced by bending, thermal deformation, and vibration of the pipe or by improper mounting. Such case stresses can affect the performance and operating life of the transducer. Because of earth-gravity (acceleration) effects, transducers with displacing or rotating mechanical sensing elements can exhibit performance changes (attitude error) when their as-installed attitude differs from that in which they were calibrated. To prevent contamination of the transducer by any small solid particles in the fluid a filter may have to be installed upstream.

Unidirectional flowmeters usually have an arrow marked on their case to indicate the intended flow direction. This indication must be observed for proper installation.

Unless an amplifier is integrally packaged with the flowmeter, the output amplitude is typically quite low. The interconnecting cable between transducer and signal-conditioning or display equipment should therefore be shielded, and the conductors should be twisted, if it is more than a few feet long, to avoid the pickup of noise caused by static or magnetic fields. The shield should be grounded only at the signal-conditioning end. Low-level frequency output can also be freed from most of the noise in it by use of a band-pass filter in the signal-conditioning unit. It is good practice to keep low-level signal wiring away from any power wiring or electrically-powered equipment. Cable connections should be moisture-proofed to avoid electrical leakage paths due to moisture condensation and to prevent eventual corrosive failure of the connections.

Overspeeding can damage flowmeters having rotating mechanical sensing elements. Secondary damage can be caused in downstream pumps, valves, and other equipment when portions of the rotor separate and travel downstream. The possibility of overspeeding should be considered during transducer selection. The presence of gas in a normally liquid-carrying line is the most common cause for overspeeding. Depletion of the measured liquid in a pressurized system, change of state of the liquid under certain conditions of temperature and pressure, and insufficient bleeding of gas prior to startup of a liquid system can introduce gas into the flowmeter.

4.5 Calibration and Tests

4.5.1 Calibration

As for most other transducers, two types of calibration methods are used for flow-rate transducers: absolute (primary) calibrations and comparison calibrations. Absolute calibrations use a precisely known mass (*gravimetric calibration*) or volume (*volumetric calibration*) of fluid passing through the flowmeter during a precisely known time interval. Since it is not too difficult to establish mass, volume, and time with very small uncertainties, absolute calibration methods are used more frequently on flow-rate transducers than on transducers for some other measurands. Comparison methods are applied mainly to repetitive-production calibrations.

The calibration fluid should be the measured fluid specified for the transducer. When use of this fluid presents severe calibration difficulties, a simpler fluid (e.g., air for gas flowmeters, water or naphtha for liquid flowmeters) can be substituted provided that conversion factors have first been determined by sufficient theoretical as well as experimental procedure to establish adequate confidence in them.

4.5.1.1 Absolute calibrations require that the factors which influence flowmeter performance at room conditions be known and recorded for each calibration. They comprise the density (or specific gravity), viscosity, temperature, vapor pressure, and absolute pressure of the fluid, upstream and (to a lesser extent) downstream disturbances, and the attitude (orientation) of the flowmeter being calibrated. When using a gravimetric method for calibrating volumetric flowmeters, the density must be measured with very close accuracy to allow a correct conversion of mass to volume.

The fluid system used for calibration, including the tank or cylinder used as source for the measured fluid, can be of the "open" type, i.e., vented to the atmosphere, or of the "closed" type, i.e., pressurized above the ambient atmospheric pressure.

The calibration methods are further classified as "static" and "dynamic." In the *static method*, the measurement of mass or volume occurs only while the liquid is not flowing into or from the measurement vessel. This method can provide very close accuracy under properly controlled conditions and is conveniently traceable to primary standards. In the *dynamic method* the measurement of mass or volume is performed while the liquid is flowing into or from the measuring vessel. Dynamic errors, difficult to detect by static checks referred to mass or volume standards, can be introduced in this method and additional verification tests on the dynamic calibration device are usually required.

Two operating procedures are used for flowmeter calibrations. The choice of either procedure is dictated by the type of end-use application which it duplicates most closely. The "running start-and-stop" procedure requires that a reasonably constant flow rate be maintained prior to, during, and immediately after a calibration run. The "standing start-and-stop" procedure requires that flow starts at the beginning of the calibration run and stops at its end and that at least 95% of the total flow through the transducer be at the specified flow rate.

Certain precautions apply to the design of a calibration system using liquid calibration fluids. The minimum permissible capacity of the measurement vessel is dependent upon the calibration procedure and its allowable uncertainty and upon the resolution of the transducer output indicator. The minimum capacity selected should be compatible with the desired accuracy of calibration. The piping between the transducer and the measurement vessel should be short, of small volume compared to the measured volume, and especially designed for the convenient elimination of all air, vapor, and of temperature gradients. It should be so constructed to assure that all of the liquid and only the liquid passing through the transducer is measured. Throttling or flow-control valves should be located downstream of the transducer to reduce the possibility of two-phase flow occurring within the transducer under test. When this is not practical, a back-pressure regulator or similar device should be installed downstream to maintain the required back pressure. Positive methods should be provided to assure that no leakage occurs during the calibration interval.

Flow of the liquid during a gravimetric calibration (see Figure 4-14) originates at a storage vessel and passes through a flow-control valve into a straight-pipe section equipped with a flow straightener located at a distance equal to about 8 times the pipe diameter upstream from the transducer. After flowing through the transducer,

Fig. 4-14. Simplified fluid diagram of gravimetric flow-rate calibration system.

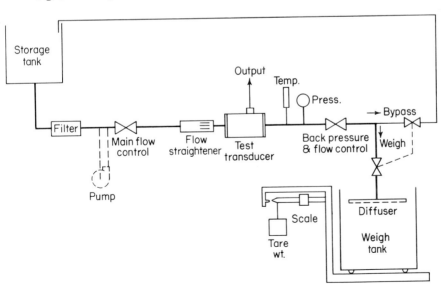

the liquid passes through any required control valves and back-pressure regulators and pours into the weigh tank through a flow disperser. The weigh tank is a vessel placed on a scale. The empty-tank weight is balanced by a tare weight. When the scale indicates that the tank has been filled to the required value, the calibration run is terminated. A timer which was started at the initiation of the calibration run is stopped at this point (usually by a switch attached to the scale), and the elapsed time is recorded. The flow rate which prevailed during the run can then be determined by dividing the total mass of the liquid gathered in the weigh tank by the elapsed time. The transducer output is recorded during the run.

When the transducer output can be conveniently totalized during the run without loss of accuracy, as is the case for frequency-output flow-rate transducers, the totalized output can be divided by the elapsed time to arrive at the average instantaneous output during the run. The totalized output can also be used to determine a calibration factor (e.g., cycles per gallon) directly, by mass to volume conversion on the basis of density.

The flow rate is varied by adjusting the main flow-control valve. Gravity flow is used wherever possible. When gravity flow is insufficient even with the control valve wide open, a boost pump can be installed between this valve and the storage tank.

Volumetric calibrations can be performed in the same manner as gravimetric calibrations except that the volume rather than the weight of the liquid in the measuring vessel is determined. The volume is measured by measuring the liquid level in a tank of known geometry. Liquid-level measurements can be performed with very close accuracy (see Chapter 8), and tank dimensions can be chosen for a simple accurately determinable geometry. Corrections to the measured value of the liquid volume may be necessary to allow for dimensional changes in the vessel due to changes in pressure and temperature and due to liquid-volume changes resulting from pressure and temperature differences between transducer and vessel.

Volumetric calibrations can also be performed by means of mechanical displacement devices such as a piston moving within a straight cylinder or an elastomer spheroid pushed through a straight or U-shaped cylinder.

4.5.1.2 Comparison calibrations require the use of a reference transducer which need not be of the same type as the test transducer. The reference transducer is installed in series with the test transducer with proper precautions observed to minimize flow pulsation and swirl at the inlet to both flowmeters and to assure equal temperature at both units. The flow rate through the test transducer is determined by observing the output or visual indication of the reference transducer. The comparison calibration should be performed under the same conditions which prevailed during the absolute calibration of the reference transducer. When this is not feasible, appropriate corrections must be made for differences in conditions which must, of course, be known.

4.5.1.3 Gas-flow calibrations require specialized equipment and methods. One technique consists of compressing air to a high pressure, storing it in a receiving vessel after cleaning and dehydration, then bleeding it down to about one standard atmosphere of pressure through a reference-standard-certified critical-flow nozzle which acts as flow reference. The transducer to be calibrated is installed downstream from the nozzle in a low-pressure test section which vents to the ambient atmosphere through a back-pressure hand valve. By measuring the absolute pressure upstream from the nozzle and maintaining constant temperature at the nozzle, the air flow rate can be determined by its known relation to upstream pressure.

Gas-flow calibrations, usually for ranges greater than 0.5 ft^3/min, are often performed by use of a *bell prover*. This device contains a bell which travels vertically along the inside wall of a vertical cylindrical tank, so that the bell forms the movable upper portion of the tank. A liquid sliding seal prevents leakage between tank wall and bell. As the bell moves downward in the tank, a discrete known volume of gas is delivered from a port on the tank to the flowmeter in the calibration system. A counterpoise mechanism augments a counterweight linked to the top of the bell by a cable to provide uniform gas pressure. The flow rate is established by using an elapsed-time indicator or a stopwatch.

4.5.2 Tests

4.5.2.1 Performance verification at room conditions is usually combined with a production calibration performed by a manufacturer or user. Mechanical characteristics are subject to visual inspection and, in some cases, to certification by the manufacturer. Electrical characteristics such as insulation resistance and power drain are verified by measurements whose results are recorded on a test-data form. Linearity and repeatability are determined by two identical sets of calibration runs, each set consisting of transducer output measurements at at least three different flow rates representing 0, 50, and 100% of the specified range. Pressure drop should be measured at the upper flow-rate range limit. A proof-pressure test should always be performed using a special test fixture when proof pressure is specified. A typical performance-verification test-data form intended specifically for turbine flowmeters is shown in Figure 4-15.

4.5.2.2 Environmental tests are normally performed only on a qualification or design-proofing basis since test setups for operating environmental tests are quite complex and costly. Fluid-temperature tests require stabilization of the test fluid at one or more temperature levels and maintaining the temperature as constant as possible during two or more calibration runs. The fluid system should be temperature-insulated for this purpose. Ambient-temperature tests call for calibrations while the test fluid is at one temperature and the flowmeter is inserted into a temperature chamber stabilized at a different temperature. During vibration tests the flowmeter

VENDOR'S MODEL NO.		PART NO.
PURCHASE ORDER NO.	**INDIVIDUAL ACCEPTANCE RECORD AND CALIBRATION FOR ROTATING VANE TYPE FLOW RATE TRANSDUCER**	SERIAL NO.
		NOMINAL RANGE

1. VISUAL: Mechanical ☐   Finish ☐   Nameplate ☐   Receptacle ☐

2. ELECTRICAL: Insulation Resistance _____ Megohms at _____ VDC

3. OUTPUT VOLTAGE: _____ Volts (RMS) at _____ cps

4. PRESSURE DROP: _____ psi at _____ gpm

5. PROOF PRESSURE: _____ psia for _____ Minutes

6. CALIBRATION: Calibration Fluid _____. Fluid Temperature _____ °F
Specified Output Frequency _____ cyc/gal

RUN NO.	FLOW RATE gpm	OUTPUT FREQ. cps	cyc/gal	DEVIATION FROM SPECIFIED cyc/gal
MEAN CYCLES/GAL.				

MAXIMUM DEVIATION _____ cyc/gal (—%)   ALLOWED: ± _____ cyc/gal (—%)

BY _____ DATE _____ APPROVED _____

Fig. 4-15. Example of test data form for performance verification of turbine flowmeters.

is connected to the fluid system with a flexible hose while mounted on a vibration exciter. The hose must be restrained from whipping during the test. Acceleration tests are difficult to perform as operating tests on a centrifuge and are usually limited to gravity-acceleration ("attitude" or "orientation" tests) on a tilting support. Some environmental tests are performed as nonoperating tests with the flowmeter capped at both ends and sometimes while it is filled with the test fluid. Nuclear-radiation and explosive-atmosphere tests should be performed on a filled nonoperating transducer. Filling is not required for humidity and corrosive-atmosphere tests.

4.6 Related Measuring Devices

This category covers the commonly known differential-pressure type of flow-sensing elements, which are connected to differential-pressure transducers whose output is then a measure of flow rate, and anemometers (literally "wind meters"), which are used almost entirely for the measurement of the flow velocity of free air streams. Visually-indicating flowmeters such as the vortex-velocity meter, the rotary displacement meter, the duplex-rotor meter, and the semi-rotary piston meter are not covered in this text.

4.6.1 Differential-pressure flow-rate measuring devices

Differential-pressure flow-measuring systems consist of a flow-sensing element which converts flow rate to differential pressure, a differential-pressure transducer, and (unless the transducer is packaged integrally with the sensing element) the necessary interconnecting tubing (Figure 4-16). The sensing element is one of the types shown in Figure 4-1, consisting of a constriction or a probe in a conduit. Constrictions used are those for which published coefficients are available.

The major groups of differential-pressure flow-sensing elements are the orifice plates, venturis, and nozzles. The *orifice* is a thin disc provided with a hole which is usually inserted between flanges in a pipe. The hole can be concentric, eccentric, or segmental (less that a full circle). The edge of the concentric hole is normally sharp, that of the eccentric hole is gently tapered, whereas the segmental orifice has a square-edged hole. A quadrant orifice has a nozzle-shaped hole, i.e., rounded on

Fig. 4-16.   Measurement of differential pressure due to flow rate.

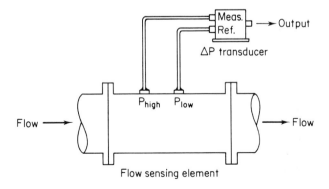

the upstream side. A *Venturi tube* does not always have an external configuration following its internal shape. It can also be an insert or a machined internal portion in a parallel-walled section. *Flow nozzles* are commonly arranged for clamping between flanges. *Centrifugal sections* are rarely used. Their main advantage is a minimal pressure drop. The *Dall tube* is a hybrid design combining the orifice and Venturi principles.

A characteristic common to all these devices is the proportionality of flow rate to the square root of the differential pressure. This relationship is, of course, reflected in the output of an inherently linear differential-pressure transducer. Some of these transducers, especially designed for operation with "ΔP" flow-sensing elements, perform the square-root extraction within their transduction element and thus provide an output which varies linearly with flow rate.

Differential-pressure transducers are covered in detail in Chapter 10. Three of their characteristics deserve special attention in flow-measurement applications. When the same range of flow rates must be measured at different absolute pressures of the measured fluid, the transducer's *reference-pressure error* must be within acceptable tolerances since the fluid pressure is the reference pressure of the transducer. When large flow surges can occur during normal operation, the *positive-proof pressure* of the transducer must be specified (and verified) accordingly. When reverse flow can occur even though it need not be measured, a negative differential pressure may be seen by the transducer, and an appropriate specification for the transducer's *negative-proof pressure* and its verification becomes important.

Further details of the operation of differential-pressure flow-sensing elements are not included here since the technical literature is replete with descriptions and explanations of these devices and their use (e.g., items 17, 19, 20, 22, 33, 37, and 45 of the bibliography). Useful reference material is also contained in two recommended-practice documents, ISA RP3.2 (Sharp-edged Orifice Plates) and ASME-PTC 19.5.4 (Instruments and Apparatus: Flow Measurement).

4.6.2 Anemometers

The measurement of free air-stream velocity as measured by anemometers is of particular importance to meteorology and aerodynamics.

The wind-speed *propeller-type anemometer* is a familiar sight in even the simplest weather stations. It is essentially a weather vane combined with a propeller whose angular speed is transduced by some tachometric device whose output is then proportional to wind speed. The angular position of the "weather vane" can be measured by an angular-displacement transducer whose output is a measure of wind direction.

The *hot-wire anemometer* is a resistive flow-velocity transducer which consists essentially of a thin heated wire supported at its ends so that it loses heat to the air stream which is being measured. This convective heat loss varies approximately with the square root of fluid velocity. Two operating modes are used for the hot-wire anemometer (see Figure 4-17). In both modes the wire is heated by the current

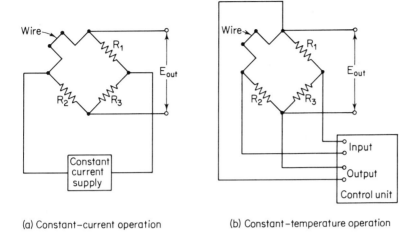

(a) Constant–current operation   (b) Constant–temperature operation

Fig. 4-17.   Hot-wire anemometer circuits.

flowing through it. When the wire is operated at constant current, its resistance increases with cooling and the resulting bridge unbalance produces an output voltage which can be related to fluid velocity. Faster response time is obtained by constant-temperature operation where a control unit containing a high-gain feedback amplifier compensates for any wire-resistance changes by a change in control-unit output, which is the bridge excitation, so that the resistance of the wire, and hence its temperature, is restored to its design value. The change in bridge-excitation voltage, which is also the output voltage, can be related to fluid velocity.

Hot-wire anemometers have been built in various freely-suspended and shaped-probe configurations. Thin (25 microns diameter) quartz rods plated with an even thinner platinum film have been used in some designs instead of the wire. Directional information can be obtained or transverse velocity fluctuations be determined by mounting two wires at right angles to each other.

BIBLIOGRAPHY

1. Reynolds, O., "An Experimental Investigation of the Circumstances Which Determine Whether the Motion of Water Shall Be Direct or Sinuous and of the Law of Resistance in Parallel Channels," *Philosophical Transactions of the Royal Society*, London, Vol. 174, Part III, p. 935, 1883.

2. Reynolds, O., "On the Dynamical Theory of Incompressible Viscous Fluids and the Determination of the Criterion," *Philosophical Transactions of the Royal Society*, London, A1, Vol. 186, p. 123, 1895.

3. King, L. V., "Precision Measurements of Air Velocity by Means of the Linear Hot-Wire Anemometer," *Philosophical Magazine*, Vol. 24, 1915.

4. Hazen, A., and Williams, G. S., *Hydraulic Tables*. 3rd Ed., New York: John Wiley & Sons, Inc., 1920.

5. Prandtl, L., "Ueber die Ausgebildete Turbulenz" (The Formation of Turbulence), *Proceedings of the Second International Congress on Applied Mechanics*, Zurich, p. 62, 1926.

6. Pigott, R. J. S., "The Flow of Fluids in Closed Conduits," *Mechanical Engineering*, Vol. 55, No. 8, p. 497, August 1934.

7. vonKarman, Th., "Turbulence and Skin Friction," *Journal of Aeronautical Sciences*, Vol. I, No. 1, p. 1, January 1934.

8. Prandtl, L. and Tietjens, O. G., *Applied Hydro and Aeromechanics*. New York: McGraw-Hill Book Company, 1934.

9. Bakhmeteff, B. A., *The Mechanics of Turbulent Flow*. Princeton, N.J.: Princeton University Press, 1936.

10. King, H. W., *Handbook of Hydraulics*. 3rd Ed., New York: McGraw-Hill Book Company, 1939.

11. Vennard, J. K., *Elementary Fluid Mechanics*. New York: John Wiley & Sons, Inc., 1940.

12. Thurlmann, B., "Methode zur Elektrischen Geschwindigkeitsmessung von Fluessigkeiten" (Method for the Electrical Measurement of Liquid Velocity), *Acta Physica Helvetica*, Vol. 14, pp. 373–419, October 1941.

13. Kolin, A., "An Alternating Field Induction Flowmeter of High Sensitivity," *Review of Scientific Instruments*, p. 109, May 1945.

14. Herning, F., *Grundlagen und Praxis der Mengenstrommessung* (*Fundamentals and Practice of Flow Measurement*). Düsseldorf, Germany: *Deutscher Ingenieur Verlag*, 1950.

15. Arnold, J., "Electromagnetic Flowmeter for Transient Flow Studies," *Review of Scientific Instruments*, Vol. 22, p. 43, January 1951.

16. Brand, D. and Ginsel, L. A., "The Mass Flow Meter," *Instruments*, Vol. 24, p. 331, March 1951.

17. Stearns, R. F., Jackson, R. M., Johnson, R. R., and Lawson, C. A., *Flow Measurement with Orifice Meters*. Princeton, N.J.: D. Van Nostrand Co., Inc., 1951.

18. Li, Y. T. and Lee, S. Y., "A Fast-Response True Mass Rate Flowmeter," *ASME Paper No. 52-A-170*, New York: American Society of Mechanical Engineers, December 1952.

19. Redding, T. H., *Bibliographical Survey of Flow through Orifices and Parallel-Throated Nozzles*. London: Chapman & Hall, 1952.

20. Grey, J. and Liu, F. F., "Methods of Flow Measurement," *American Rocket Society Journal*, pp. 133–140, May–June 1953.

21. Head, V. P., "Coefficients of Float-Type Variable-Area Flowmeters," *ASME Paper No. 53-A-208*, New York: American Society of Mechanical Engineers, 1953.

22. Galley, R. L., "Flow-Rate Measurement," *Instruments and Automation*, Vol. 27, Part II, Chapter V, No. 12, December 1954.

23. Christian, G. L., "Fuel Meter Shows Flow and Quantity," *Aviation Week*, July 4, 1955.

24. Grey, J., "Transient Response of the Turbine Flowmeter," *Jet Propulsion*, February 1956.

25. Shafer, M. R. and Ruegg, F. W., "Liquid Flowmeter Calibration Techniques," *Paper No. 57-A-70, Transactions of the American Society of Mechanical Engineers*, October 1958.

26. Hinze, J. O., *Turbulence*. New York: McGraw-Hill Book Company, 1959.

27. Miesse, C. C., "Study of Mass Flowmeters," *Final Report (ARF Project D173)*, Armour Research Foundation, July 1959.

28. Yard, J. S., "Characteristics and Uses of Turbine Flowmeters," *I.S.A. Journal*, Vol. 6, No. 5, pp. 54–59, May 1959.

29. Bowers, K., Galley, R. L., and Vincelett, P. S., "Flow Measurement in Rocketry," *Instruments and Control Systems*, April 1961.

30. Galley, R. L., "The State of the Art in Cryogenic Flowmetering," *Preprint No. 61.2.62*, Pittsburgh: Instrument Society of America, 1962.

31. Keller, G. D., "Precise Mass Flow Measurement Using Inferential Methods," *Preprint No. 46.3.63*, Pittsburgh: Instrument Society of America, 1963.

32. Kovasznay, L. S. G., Miller, L. T., and Vasudeva, B. R., "A Simple Hot-Wire Anemometer," *Project Squid Technical Report No. JHU-22-P*, Johns Hopkins University, July 1963.

33. Bollinger, L. E., "Transducers for Measurement," Part IV, "Fluid Flow," *I.S.A. Journal*, Vol. 11, No. 11, pp. 64–69, November 1964.

34. Crooks, W. M., Jr., "A Fixed Vane Sensor for Measuring Turbulent Air Flows," *Proceedings of the 10th National Aerospace Instrumentation Symposium*, Pittsburgh: Instrument Society of America, 1964.

35. Filban, T. J., Jr. and Shafer, M. R., "Flow Measurement Standardization," *Preprint No. 12.2-4-64*, Pittsburgh: Instrument Society of America, 1964.

36. Haffner, J. W., "Radioisotopes for On-Stream Analysis," *I.S.A. Journal*, Vol. 11, No. 5, p. 75, May 1964.

37. Hay, P., "Ein Neues Prinzip fuer die Durchflussmessung und -Regelung" (A New Principle for Flow Measurement and Control), *Acta IMEKO 1964*, pp. 77–89, IMEKO, Budapest, Hungary, 1964.

38. Laub, J. H., "The Boundary Layer-Type Flow Meter," *Acta IMEKO 1964*, pp. 179–194, IMEKO, Budapest, Hungary, 1964.

39. Potter, R. C., "The Hot Wire Anemometer," *Research Staff Technical Report WR 64-8*, Huntsville, Alabama: Wyle Laboratories, December 1964.

40. Schmoock, R. F. and Ham, D. L., "A Magnetic Flowmeter System with Digital Output," *Preprint No. 8.2-1-64*, Pittsburgh: Instrument Society of America, 1964.

41. Taylor, C. F. and Pearson, D. P., "A True-Mass Cryogenic Flowmeter," *Preprint No. 1.2-3-64*, Pittsburgh: Instrument Society of America, 1964.

42. Vasy, G. S., "Venturi Bodies for Flow Measurement," *Acta IMEKO 1964*, IMEKO, Budapest, Hungary, 1964.

43. "Meter Accurately Measures Flow of Low-Conductivity Fluids," *NASA Tech Brief No. 63-10280*, Pasadena, California: Technology Utilization Office, Jet Propulsion Lab., May 1964.

44. Brock, T. E. and Moon, C. J., Eds., "A Bibliography on Hot-Wire Anemometry," Cranfield, Bedford, England: The British Hydromechanics Research Association, October 1965.

45. Gambill, W. R., "Centrifugal Flowmeter Tests," *Report No. ORNL-TM-1101*, Oak Ridge, Tennessee: Oak Ridge National Laboratory, April 20, 1965.

46. Rodely, A. E., White, D. F., and Chanaud, R. C., "A Digital Flowmeter without Moving Parts," *ASME Paper No. 65-WA/FM-6*, New York: American Society of Mechanical Engineers, 1965.

47. Siddon, T. E., "A Turbulence Probe Utilizing Aerodynamic Lift," *Report No. UTIAS-TN-88*, Toronto: University of Toronto, June 1965.

48. "Liquid Hydrogen Mass Flowmeter Evaluation," *Final Report Contract NAS8-1526*, El Segundo, California: Wyle Laboratories, January 1965.

49. Minkin, H. L., Hobart, H. F., and Warshawsky, I., "Performance of Turbine-Type Flowmeters in Liquid Hydrogen," *Report NASA-TN-D-3770*, Cleveland: NASA Lewis Research Center, December 1966.

50. Brown, A. E., and Allen, G. W., "Ultrasonic Flow Measurement," *Instruments and Control Systems*, Vol. 40, No. 3, pp. 130–134, March 1967.

51. Mackenzie, D. D., "Mass and Volumetric Flow Measurement," *Instrumentation Technology*, Vol. 14, No. 3, pp. 23–29, March 1967.

52. Spink, L. K., *Principles and Practice of Flow Meter Engineering.* 9th Ed., Foxboro, Massachusetts: The Foxboro Co., 1967.

5

Force
and
Torque

5.1 Basic Concepts

5.1.1 Basic definitions

Force is the vector quantity necessary to cause a change in momentum. When an unbalanced force acts on a body, the body accelerates in the direction of that force. The acceleration is directly proportional to the unbalanced force and inversely proportional to the mass of the body.

Mass is the inertial property of a body. It is a measure of the quantity of matter in a body and of the resistance to change in motion of a body.

Weight is the gravitational force of attraction. On earth, it is the force with which a body is attracted toward the earth (mass times local acceleration due to gravity).

Load is the force applied to a body (or structure).

A **load cell** is a force transducer.

Torque is the moment of force. It is the product of the force and the perpendicular distance from the axis of rotation to the line of action of the force. *Note:* This distance is referred to as *lever arm*.

255

Torsion is the twisting of an object, such as a rod, bar, or tube, about its axis of symmetry.

5.1.2 Related laws

Force (in absolute system of units)

$$F = kma$$

where: F = force (acting on mass m)
   m = mass
   a = acceleration
   k = proportionality constant depending on units used

Force (in gravitational system of units)

$$F = \frac{Wa}{g}$$

where: F = force (producing acceleration a)
   W = weight (in same units as force)
   a = acceleration (acting on body of weight W)
   g = acceleration due to gravity (in same units as a)

Torque

$$T = Fl = I\alpha$$

where: T = torque
   F = force
   l = length of lever arm
   I = moment of inertia
   α = angular acceleration

Shear stress (due to torsion, in a solid cylindrical shaft)

$$S_s = \frac{16T}{\pi d^3}$$

where: S_s = maximum shear stress, psi
   T = torque, lb-in.
   d = shaft diameter, in.

Torsional deflection (of a solid cylindrical shaft)

$$\theta = \frac{32TL}{\pi d^4 E_s}$$

where: θ = helical angle of deflection
   T = torque, lb-in.
   d = shaft diameter, in.

L = length of shaft under torsion, in.

E_s = shear modulus of elasticity, psi.

Moment of inertia

$$I = \sum mr^2 = \frac{T}{\alpha}$$

where: I = moment of inertia

  m = mass (of one particle)

  r = crank length (between particle and axis of rotation)

Moment of inertia (of a solid cylindrical shaft, of mass M and radius r, about its own axis)

$$I = \tfrac{1}{2}Mr^2$$

5.1.3 Units of measurement

Units of force are defined in two systems. The **absolute system** of units employs *mass* as fundamental quantity. The **gravitational system** of units is based on *force* as the fundamental quantity.

Absolute systems. The now standard International System (*SI, Système International d'Unités*) is based on six fundamental units: the **meter** (m) as unit of length, defined as exactly 1,650,763.73 wavelengths in vacuum of the radiation corresponding to the transition between the energy levels $2p_{10}$ and $5d_5$ of the krypton-86 atom; the **kilogram** (*kg*) as unit of mass, defined as the mass of the international prototype kilogram kept in the custody of the Bureau International des Poids et Mesures (BIPM), Sèvres, France; the **second** (*s*) as unit of time interval, defined as the interval occupied by exactly 9,192,631,770 cycles of the radiation corresponding to the $(F = 4, M_F = 0)$ to $(F = 3, M_F = 0)$ transition of the cesium-133 atom when unperturbed by exterior fields; the **ampere** (*A*) as unit of electric current; the **degree Kelvin** (°K) as unit of temperature; and the **candela** (*cd*) as unit of luminance. All other SI units are derived from these six basic units and from two supplementary units, the **radian** (*rad*) as unit of plane angle and the **steradian** (*sr*) as unit of solid angle. The SI unit of *force* is the **newton** (*N*). Its interrelation with the basic units is: $1 \text{ N} = \text{kg} \cdot \text{m/s}^2$.

The metric absolute system is based on three units: the **centimeter** (*cm*) as unit of length, the **gram** (*gm*) as unit of mass, and the **second** (*s*) as unit of time. Because of the units used, this system has been referred to as the "CGS" system. A second metric absolute system has also been used, the "MKS" system, based on the meter, the kilogram, and the second. Both of these systems, as well as (eventually) all British systems, are intended to be superseded by the SI.

The unit of *force* in the CGS system is the **dyne**. Its interrelation with the fundamental units is: 1 dyne = gm $\cdot$ cm/sec². Its relation to the SI unit of force is: 1 dyne = 10^{-5}N.

The British absolute system is based on three units: the **foot** (*ft*) as unit of length, the **pound** (*lb*) as unit of mass, and the **second** (*s*) as unit of time. This system is sometimes called the "FPS" system.

The unit of force in the FPS system is the **poundal**. Its interrelation with the fundamental units is: 1 poundal = lb·ft/s².

Gravitational systems. These systems have been used widely in engineering work. They introduce the concept of *weight*, the force due to gravity exerted by the earth on a mass. The fundamental units of the *British gravitational system* are the **foot** (*ft*) as unit of length, the **pound** (*lb*) as unit of force, and the **second** (*s*) as unit of time. The unit of mass exists only as the gravitational unit of a 32.1740-pound mass, a unit called the **slug**. The fundamental units of the *metric gravitational system* are the same as used in the CGS or MKS systems except that the gram (or kilogram) is the unit of force, and that the "metric slug" is the unit for a 980.665 gm mass.

In the force and torque measurement field the gravitational systems (especially the British gravitational system) are still used more widely than the absolute systems in the U.S.A. and a number of other countries. However, because of the existing contradictions between the two systems in usage of force units, adoption of SI units has been urged more strongly in the area of force measurements than for most other measurands. Existing usage of units, then, can be summarized as below.

Force is expressed in **pounds force** (*lbf*, but often shown just as "lb"). It has sometimes been expressed in ounces force, also in grams force and kilograms force. The latter two units are slated for early obsolescence. The SI unit for force is the **newton** (*N*). A unit usable to express small forces is the **dyne** which equals 10^{-5} N.

One pound force is the force exerted by the earth on a one-pound mass at any location where the acceleration due to gravity is 9.80665 meters per second² (m/s²). Conversion: 1 lbf = 4.4482216152605 N.

Thrust is expressed in **pounds force** or **newtons**.

Mass is expressed in **pounds mass** (*lbm*, but usually shown just as "lb"). The SI unit for mass is the **kilogram** (*kg*). Mass has also been expressed in ounces (mass), drams (mass), or grains (mass). The British mass units are *avoirdupois* units. Conversion: 1 pound (mass) = 16 ounces (mass) = 256 drams (mass) = 7000 grains (mass). 1 lbm = 0.45359237 kg.

Weight is expressed in the units applicable to force (N, or lbf, usually shown as "lb"). A statement of weight should be accompanied by a statement of the gravitational acceleration (in m/sec²) at the location where the object was weighed or is assumed to be located. Unless otherwise stated, the gravitational acceleration is assumed to be one standard *g* (9.80665 m/sec²). Weight has also been expressed in metric units, such as the kilogram and gram.

Torque is expressed in **newton-meters** (*N-m*). Small torques can be expressed in **dyne-centimeters** (*dyne-cm*). It has also been expressed in pound (force)-inches, pound (force)-feet, ounce (force)-inches, kilogram (force)-meters, gram (force)-centimeters, and milligram (force)-millimeters. When so expressed, the term "force" has usually been omitted, as was done in the following conversions which are rounded off to three significant figures:

1 oz-in. = 0.0625 lb-in. = 0.0052 lb-ft   = 72 gm-cm

1 gm-cm = .0139 oz-in. = 10,000 mg-mm = 981 dyne-cm

Mass moment of inertia is expressed in kg-m^2 or slug-ft^2.

5.2 Sensing Elements

Most force and torque transducers use a separate sensing element which converts the measurand into a small mechanical displacement, usually a deformation of an elastic element. Two characteristics of elastic deformation are used to sense force: local strains and gross deflections. A maximum level of each occurs at some location in the sensing element but not necessarily at the same location. It is this maximum level of either strain or deflection (displacement) that is normally applied to the transduction element. Every sensing element is made of homogeneous material and is manufactured to very close design tolerances. Various types of steel are most commonly used as sensing-element materials.

Basic design parameters of force- and torque-sensing elements include relative size and shape, material density and modulus of elasticity, sensitivity in terms of local strain and gross deflection, dynamic response, and effects of loading by the transducer on the measured system. Mathematical models of force transducers have been established by various researchers on the basis of well-established laws of mechanics. The discussion of sensing elements in this chapter is limited to general descriptions intended to explain the components and operation of a number of existing transducer types. It should be noted that deflection characteristics are modified and measured-system loading is increased in most sensing elements when a mass, such as a coupling or a portion of a transduction element, is attached to them, usually at a point where force is applied.

5.2.1 Beams

Bending beams are the simplest force-sensing elements. Typical configurations are shown in Figure 5-1. The maximum deflection of a beam occurs at the point of force application; the maximum deflection of a cantilever beam always occurs at its free end.

The point of maximum strain in a simple cantilever beam (of constant section in both planes) is at the attached (fixed) end. Cantilever beams can also have a constant-strength configuration; i.e., they can have a triangular or parabolically tapered shape (narrowest at the point of force application) in the plane either perpendicular to or parallel to the direction of force and a constant section in the other

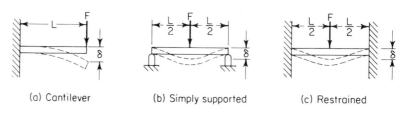

Fig. 5-1. Beam force-sensing elements.

plane. The advantage of a constant-strength beam is that the strain is constant along the top or bottom of the beam. In a simply supported beam the point of maximum local strain is at the point of force application. In a fixed-end ("built-in") beam maximum strain occurs at the fixed ends as well as at the point of force application; the center strain is opposite in direction to the end strains.

5.2.2 Diaphragms

Clamped circular plates (diaphragms) are used as force-sensing elements mainly for their favorable deflection characteristics. Maximum strain and deflection occur at the center of the diaphragm, where force is always applied. This element design has inherently good lateral stability.

5.2.3 Proving rings

The standard proving ring and the related flat proving rings or frames are examples of a frequently used type of force-sensing element (see Figure 5-2). Deflection is emphasized in the proving frame (flat proving ring) whereas local strain is

Fig. 5-2. Proving-ring force-sensing elements.

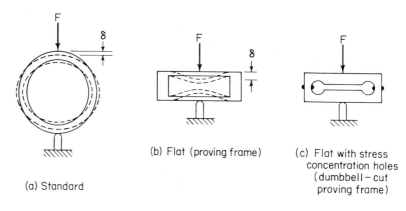

emphasized in the specially configured stress-concentration frame. The standard proving ring ("force ring") is usually of square or rectangular cross section. Maximum deflection as well as maximum strain occur at the point of force application; however, strain has an almost equally great magnitude at points 90 degrees of arc in either direction from the point of force application, and strain sensing is more convenient at these points.

5.2.4 Columns

Column force-sensing elements (Figure 5-3) normally have their point of maximum deflection at their vertical center (one-half of the column height) and their maximum strain at their lateral center. Their characteristics depend primarily on the

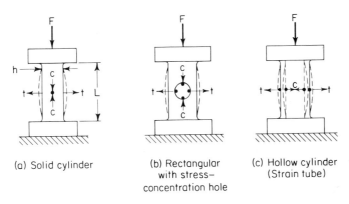

| (a) Solid cylinder | (b) Rectangular with stress— concentration hole | (c) Hollow cylinder (Strain tube) |

Fig. 5-3. Column force-sensing elements.

height-to-width (L/h) ratio and on wall thickness (for the hollow cylinder). Compression (c) and tension (t) in a column are normally sensed as strain (hoop strain in the hollow cylinder), sometimes as changes in magnetic characteristics or natural frequency, but rarely as displacement.

5.2.5 Torque-sensing elements

In-line torque-sensing elements for use with rotating shafts are special sensing shafts inserted between the mechanical power source and its load. The first of the typical examples shown in Figure 5-4 is a solid cylindrical shaft. If torque is applied to one of its ends while the other end is fixed, any line on the shaft surface originally parallel to the axis of rotation becomes a helix or portion of a helix. Similar twisting, but of much lower magnitude, occurs during normal use of the sensing shaft. It can be sensed as surface strain.

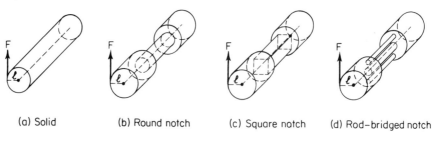

(a) Solid (b) Round notch (c) Square notch (d) Rod–bridged notch

Fig. 5-4. Torque-sensing shafts.

More strain can be produced by machining a section of reduced diameter ("notch") in the shaft and sensing the strain on the surface of this section only. This is achieved by the sensing shafts of Figure 5-4(b) and (c). The square notch offers the advantages of easier strain-gage mounting, placing of electrical connections in a low-stress area located near the corners of the square portion, higher resonant frequency, and less error due to side loads. The rod-linked notch of Figure 5-4(d) is used with special transduction elements. A design variation contains four beams of square cross section in place of the two cylindrical rods. This "squirrel-cage" sensing element offers certain performance improvements over many other types of elements when used with strain-gage transduction.

Reaction-torque-sensing elements are part of the case of a transducer rather than its rotating shaft. They sense the torque between two ends of the case when one end is attached to ground. Reaction torque has also been measured by a force transducer coupled between the end of a torque arm and ground.

5.3 Design and Operation

Several types of transduction elements have been successfully applied to force- and torque-sensing elements. Force-transducer designs, including those intended for reaction-torque measurements, are limited to those capable of utilizing either the local strain or the very small deflections of sensing elements. Torque transducers for in-line shaft torque measurements utilize surface strains (or stress patterns within the shaft) to a greater extent than deflection. All torque transducers are designed to measure torque regardless of direction of shaft rotation. Force transducers, however, can respond equally or unequally to applied tension and compression forces, and some designs are usable for one or the other direction of force application only.

A number of designs, additional to those described below and using other transduction methods, have been attempted at various times. Some of them have found limited applications in a few specialized areas. Most of these, however, have been temporarily or permanently shelved because they could not result in competitive hardware.

5.3.1 Capacitive force transducers

A relatively small number of transducers convert force into a change of capacitance. Such transducers require excitation and signal-conditioning circuitry somewhat more complex than strain-gage or even reluctive transducers. They are intended primarily for dynamic force measurements where a wide frequency response is required, as well as in specialized applications, e.g., measurements in low-ambient-pressure and high-temperature environments. Electrically grounded beams and diaphragms have been found usable as rotors in capacitive transducers.

The capacitance changes are frequently converted into corresponding changes of a d-c output voltage by signal-conditioning circuitry of various types. An a-c

Fig 5-5.   Capacitive force transducer (courtesy of Omega Dynamics Corp.).

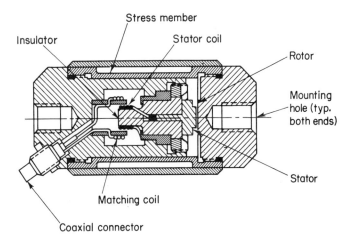

voltage is applied across the capacitive transduction element. Transducer operation tends to become more efficient at higher excitation frequencies since larger changes in capacitive reactance are achieved. The variable capacitor represented by the transducer can be one leg of an a-c bridge, a feedback element in an a-c amplifier, a shunting capacitor across a capacitance-operated ionization tube, or an element in a tuned high-frequency inductance-capacitance circuit.

In the transducer shown in Figure 5-5, the capacitive element is connected in parallel with a coil ("stator coil") to form a tuned circuit (tank circuit) at radio frequencies. One end of the coil, as well as the moving capacitor electrode ("rotor"), is grounded. The rotor forms one end cap of a modified hollow-cylinder force-sensing element ("stress member"). The transducer base forms the other end cap. Both end caps are provided with threaded mounting holes.

A radio-frequency voltage in the megahertz range is coupled to this tank circuit through the matching coil. Changes in capacitance are detected as variations in apparent impedance of the matching coil due to changes in tuning (and impedance) of the tank circuit. Additional signal conditioning after the oscillator and detector is used to provide an amplified d-c output voltage. The transducer is connected to the oscillator/amplifier package by means of a shielded coaxial cable.

5.3.2 Piezoelectric force transducers

Although some measure of quasi-static response can be obtained from piezoelectric force transducers when they are used with charge amplifiers, their application is almost entirely in dynamic force measurement. Force fluctuations occurring in less than 0.1 ms have been measured with such transducers.

Column-type sensing elements are normally used. In these the ceramic or quartz crystal transduction element is located at the center of a cylindrical column of sandwich construction. A typical design uses a pair of disc-shaped piezoelectric crystals separated by a thin electrode. This electrode is insulated from the case and is connected to the pin of the coaxial receptacle. The crystal assembly, in turn, is placed between two thick metal discs, each threaded at its center for mounting. A design variation is the washer-type transducer (force washer) illustrated in Figure 5-6 in which the sensing/tranduction element is an annular column (but not a hollow cylinder) of sandwich construction. The central hole is intended for a bolt or a stud. The force washer is typically used in threaded-fastener mountings of machinery and other equipment.

A piezoelectric transducer responds to compression forces only. However, it can be preloaded so that a compression force is continuously exerted on the transducer, constituting a static level. When a bidirectionally varying dynamic force is then applied, the compression force will be alternately relieved and intensified. The transducer is thus able to provide an output corresponding to bidirectional dynamic forces (tension and compression).

Piezoelectric force transducers do not require excitation but have a very high

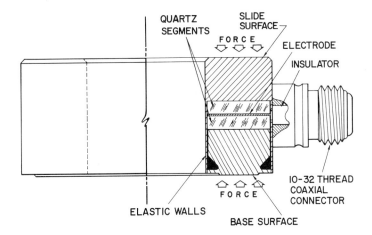

Fig. 5-6. Piezoelectric washer-type force transducer (courtesy of Kistler Instrument Corp.).

output impedance. Voltage amplifiers, emitter, cathode and source followers, and charge amplifiers are used in conjunction with these transducers. This signal-conditioning equipment is discussed in more detail in the sections of this book dealing with piezoelectric acceleration and pressure transducers.

5.3.3 Reluctive force transducers

Conventional reluctive force transducers using the deflection of a sensing element are widely used. Typical transducers using differential-transformer transduction elements are illustrated in Figure 5-7. Deflection of the proving ring or diaphragm, due to applied force, causes relative motion (vertical in the illustration) between a core and a concentric coil assembly. This results in changes in the amplitude and phase of a-c voltage in two secondary windings when a-c excitation is applied to the primary winding. The coil assembly is encapsulated and magnetically shielded. Excitation is usually between 2 and 10 V at between 50 and 10,000 Hz. Sensitivity increases with increasing excitation frequency.

When the secondary windings are connected in the most commonly used configuration, output voltage amplitude increases equally from a core center (null) position for sensing-element deflection in either direction (due to compression or tension), but output phase changes from 90 degrees towards zero degrees in one direction and towards 180 degrees in the other direction. A phase-sensitive detector in the output can eliminate the amplitude ambiguity.

The proving-ring type transducer of Figure 5-7(a) is used for tension and compression force (bidirectional range) measurement. The diaphragm-type transducer shown in the foreground of Figure 5-7(b) is also designed for tension and compres-

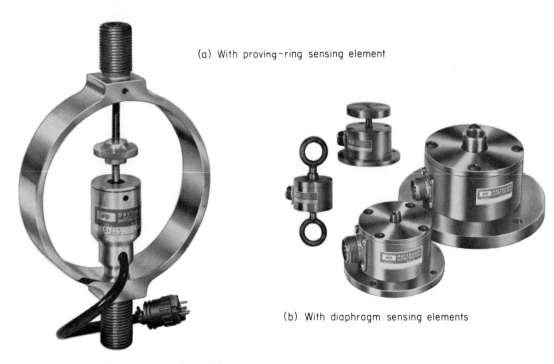

(a) With proving-ring sensing element

(b) With diaphragm sensing elements

Fig. 5-7. Reluctive (differential-transformer) force transducers (courtesy of Daytronic Corp.).

Fig. 5-8. Reluctive (inductance-bridge) force transducer (courtesy of Pace-Wiancko Div., Whittaker Corp.).

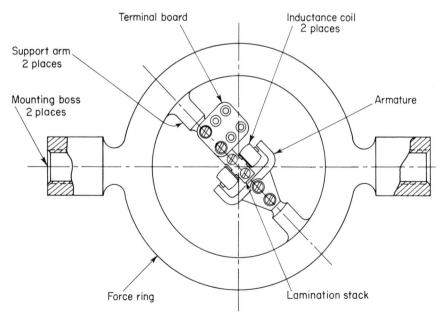

sion measurement; the base plate is provided with an internal mounting thread. The internal sensing element can deflect in either direction relative to the case. This transducer can be modified for tension measurement only by attachment of eyebolts and for compression measurement only by attachment of either a weight platform (weigh bridge) or a hard rounded-tip "load-button."

The proving-ring element has no lateral constraints. This transducer type should be loaded in such a manner that the resultant applied force vector is aligned as closely as possible with the vertical centerline. This limitation is minimized in transducers using a diaphragm which provides lateral stability.

A different design uses two thin-beam sensing elements of special configuration, one in each end cap of a bidirectional-range transducer. A core is supported in the center of a thin rod attached to and protruding beyond each of the two beams. This type of transducer can be used for ranges as low as ± 1 gm force.

The proving-ring sensing element is also employed in the reluctive transducer shown in Figure 5-8, but its deflection is utilized in a somewhat different manner for operation with the inductance-bridge transduction element. Relative motion between the two legs of the U-shaped ferromagnetic armature and the dual coil assembly is such that the proximity between one coil and its armature leg decreases while that of the other coil-armature-leg set increases when compression force is measured. The process is opposite for tension forces. The stack of soft-iron T-shaped laminations serves as core for both coils. The two coils are electrically connected as two arms of an a-c bridge. Matched resistors usually form the other two legs. The bridge is nulled with equal gaps between the armature and both coils and no force applied. Proving-ring deflection increases the inductance of one coil while decreasing the inductance of the other. The resulting bridge unbalance produces an a-c voltage proportional to deflection (also to bridge excitation) whose phase relationship can be used to indicate direction of deflection.

The operation of an entirely different type of reluctive force transducer is illustrated in Figure 5-9. It utilizes a rectangular column sensing element with a central stress-concentration section. The flat column is made from soft-iron laminations in various configurations depending on range, as shown in Figure 5-9(b). The operating principle of this "magnetoelastic" transducer is illustrated for the medium-range configuration. The stamped laminations are perforated in four places. After bonding into the finished column, two coils are wound crossing each other at right angles. The coils are used as the primary and secondary windings of a transformer. Variation in column permeability due to stress variations with applied force changes the coupling between the two windings. The coupling is zero with no force applied. Permeability decreases in the direction of an applied compression force, and flux increases in the transverse plane so that coupling between windings increases with applied force, and the output from the secondary winding increases proportionally as long as a constant excitation is applied to the primary winding. The opposite effect results from the application of a tension force. The output from the secondary winding also increases but with a phase relationship opposite to that obtained with compression.

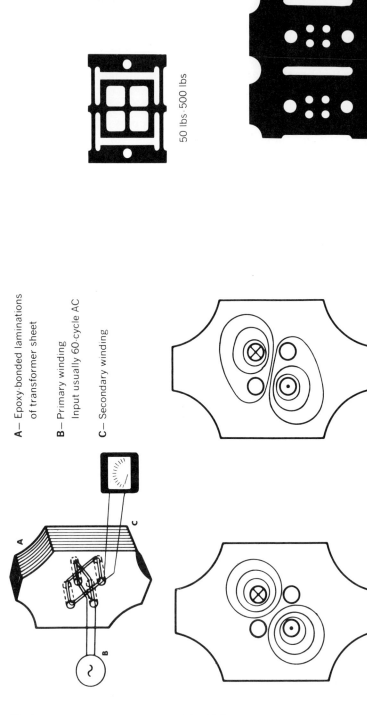

500 lbs-50 tons

50 lbs-500 lbs

50 tons 5000 tons

(b) Core designs for various force ranges

A— Epoxy-bonded laminations
of transformer sheet

B— Primary winding
Input usually 60-cycle AC

C— Secondary winding

Flux pattern in unstressed lamination
No voltage induced in secondary winding

When a load is applied, permeability in the
direction of stress decreases—rearranging
the flux pattern and coupling the secondary
winding. The induced voltage is a linear
function of the load.

(a) Principle of operation

Fig. 5-9. Reluctive (variable-permeability) transduction for force transducers
(courtesy of ASEA Electric, Inc.).

The output voltage amplitude varies with force in an essentially linear manner. Because of the relatively small number of turns in both windings the input and output impedances are quite low. A step-down transformer is normally used for excitation and a step-up transformer for the output. The low impedances allow long cables to be used between transducer and transformers (and ancillary equipment).

5.3.4 Reluctive torque transducers

Reluctive transducers utilizing the torsional deflection of a sensing shaft, e.g., to displace an armature between two coils in an inductance bridge, are relatively rare. Obtainable deflections tend to be too small to permit a simple transducer design. Lengthening of the sensing shaft can be used to increase deflection. Slip rings and brushes are necessary for electrical connections to the transduction element on the sensing shaft.

Torque transducers utilizing changes in shaft permeability due to stress promise to become more successful. Two design versions of this type of transducer have been manufactured, both using differential-transformer transduction wherein the coupling between an a-c excited primary and the two secondary windings is dependent on permeability changes within the relatively short rotating sensing shaft while the coil assembly is stationary. The permeability in the shaft to which torque is applied increases in a direction 45 degrees with respect to the axis of rotation due to tensile stress in that direction and decreases in the opposite 45-degree direction due to compression.

The first design version (see Figure 5-10) uses an X-shaped coil assembly mounted close to the shaft and parallel to the axis of rotation. The primary winding is located at the center of the X. One pair of secondary coils are placed at extreme ends of one leg of the X, and the other pair of secondary coils is placed at extreme ends of the other leg. The two legs are at 90 degrees to each other, and each leg is at 45 degrees to the axis of rotation. When torque is applied to the shaft, coupling from the primary winding increases to one secondary-winding pair and decreases to the other secondary-winding pair.

In the second design version (see Figure 5-11) three ring-shaped coil assemblies form a stator around the rotating sensing shaft. Each multipole ring holds a number of coils. The centrally located ring contains the primary-winding coils, with poles indicated as N and S in the illustration. The two outside rings contain the coils of the two secondary windings (poles indicated as A and B). With no torque applied to the shaft internal stresses are zero and the magnetic fields between the primary S and N poles are symmetrical so that the zero-equipotential lines are situated symmetrically under the secondary poles A and B. Secondary flux and, hence, secondary voltages are zero. When torque is applied, the principal stresses ($+\sigma$ and $-\sigma$ in the illustration) are obtained. The permeability in the direction of tension (poles B and S and poles A and N) is increased, whereas the permeability in the direction of

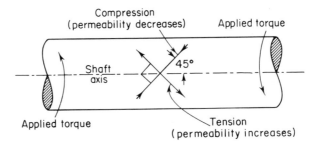

(a) Tension—compression stresses in surface of a circular shaft

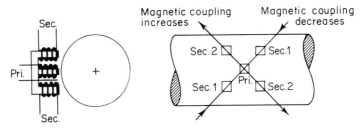

(b) Magnetic coupling through shaft

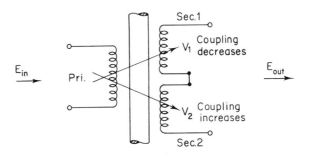

(c) Change in coupling due to stresses

Fig. 5-10. Operating principle of reluctive torque transducer employing variations in shaft permeability due to stresses (courtesy of Bergen Laboratories, Inc.).

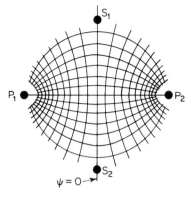
Without torque

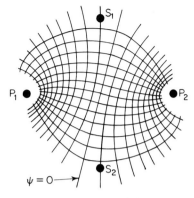
With torque applied

(a) Magnetic field pattern in rotating shaft

Pole ring cross section. Number
of poles is a multiple of four

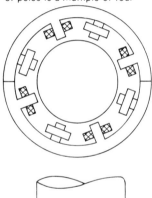

Evolution of shaft surface showing pole
positions. When no torque is applied, the
magnetic fields between primary N and S
poles are symmetric—with no coupling to
adjacent secondaries

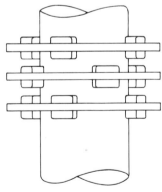

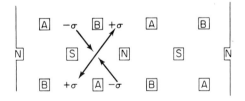

When torque is applied, permeability changes
occur at 45° shear planes—decreasing in the
plane of compression, increasing in the plane
of tension. The redistributed flux induces a
voltage in the adjacent secondary windings.
The voltage is a linear function of the torque

Primary poles on center ring
Secondary poles on outer rings

(b) Operating principle of non-contacting transduction element

Fig. 5-11. Reluctive (variable-permeability) torque transduction with ring-
shaped coil assemblies (courtesy of **ASEA Electric, Inc.**).

compression (poles *A* and *S* and poles *B* and *N*) is decreased. Corresponding voltages are then induced in the secondary windings which are connected in series.

5.3.5 Strain-gage force transducers

Transducers which convert force into a change of resistance due to strain in two or four arms of a Wheatstone bridge have been widely used in numerous applications for force ranges between 10 and 250,000 lbf and in some applications to over 8,000,000 lbf. They constitute the most popular type of force transducer. Unbonded metal-wire and bonded semiconductor strain gages are used in some designs; however, bonded metal-wire or metal-foil gages are employed most frequently. Transducers with a four-active-element strain-gage bridge outnumber those with a two-active-element bridge. Most designs incorporate fixed resistors for the purposes of end-point adjustment and temperature compensation. Sensing elements include

Fig. 5-12.   Strain-gage compression force transducer (courtesy of Revere Corp. of America).

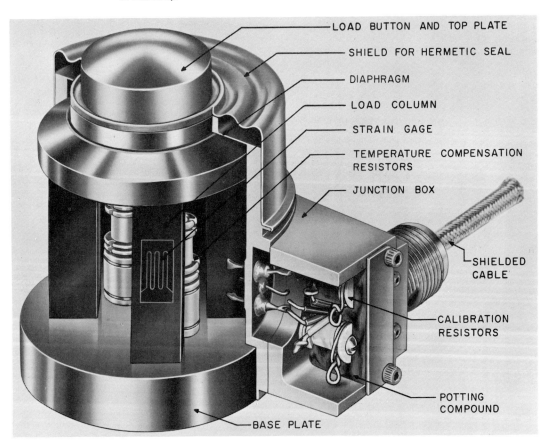

LOAD BUTTON AND TOP PLATE

SHIELD FOR HERMETIC SEAL

DIAPHRAGM

LOAD COLUMN

STRAIN GAGE

TEMPERATURE COMPENSATION RESISTORS

JUNCTION BOX

SHIELDED CABLE

CALIBRATION RESISTORS

POTTING COMPOUND

BASE PLATE

primarily proving rings and columns. Some models are designed to respond to compression forces only, but virtually all of them can be optionally or alternatively built as "universal" (tension and compression) force transducers. Low-range (under 500 lbf) transducers often incorporate mechanical overload stops.

A typical strain-gage compression force transducer, usable in applications such as weighing systems, is illustrated in Figure 5-12. A multiple-column sensing element is provided with a strain gage bonded to each column. A diaphragm provides lateral stability. The corrugated end cap ("shield") forms a hermetic seal between the case and the top plate. Temperature compensation resistors are mounted on the base plate in close proximity to the sensing element to minimize temperature-gradient errors. Calibration resistors are potted within the junction box attached to the case. Interconnections for temperature compensation and bridge adjustment resistors are schematically illustrated in Figure 10-40 for a typical transducer strain-gage bridge.

Figure 5-13 illustrates a different design, a strain-gage compression force transducer which operates as a universal force transducer when the locking pin and threaded load button are removed. Either a proving ring or a square column is used as sensing element. Bonded strain gages provide transduction. Both ends of the sensing element are internally threaded to enable use of the transducer for tension measurement. Two separate diaphragms stabilize the upper portion of the sensing element laterally. Deflection of either type of sensing element is less than 0.012 in. at the upper limit of any force range. Columns and proving-ring elements are normally made from high-strength steel. After completion of assembly the transducer is hermetically sealed. Immediately prior to sealing the air and entrapped moisture is often evacuated from the case and replaced with a dry inert gas such as nitrogen.

A few transducer designs utilize semiconductor strain gages rather than metal-wire or metal-foil gages. The deflection of their sensing elements for full-scale output can be smaller than that of conventional designs because of the large gage factor inherent in semiconductor gages. Full-scale outputs of about 50 mV per volt of excitation have been obtained with total deflections of between 0.0007 and 0.0001 in. However, temperature compensation is practical for only a relatively narrow temperature range, and strain-gage resistance is frequently larger than that of metal gages.

Strain-gage force transducers have been built in a number of special configurations. Annular ("hollow") transducers of various heights have outside diameters from about 20 in. to less than 0.5 in. ("force washers"). Most of these are usable for compression measurements only. Unusual designs have been developed for pedal-pressure and seat-belt force measurement in automotive testing. X-Y force transducers either contain two sensing elements at 90 degrees to each other, isolated by an integral flexure system, or incorporate a four-bar parallelogram structure. They provide a separate output for force vectors in each of two orthogonal planes at a single point of measurement. Triaxial (X, Y, Z) transducers are similarly furnished with three separate outputs.

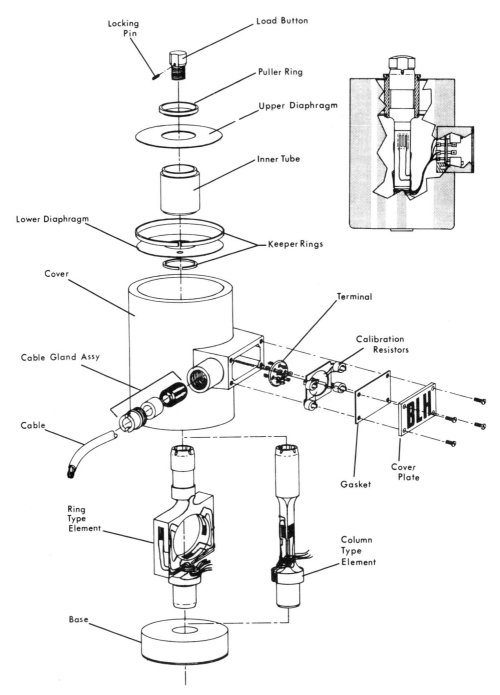

Fig. 5-13. Strain-gage universal force transducer (courtesy of BLH Electronics).

5.3.6 Strain-gage torque transducers

Torque transducers employing strain-gage transduction are more frequently used than any other type of torque transducer. The gages are usually bonded metal-wire or metal-foil gages mounted on the sensing shaft at 45-degree angles to the axis of rotation (shaft centerline). Semiconductor gages or combinations of these with metal-foil gages are used in some designs. Adjacent gages are placed at an angle of 90 degrees from each other. Four gages are mounted around the circumference of the shaft. When torque is applied to the shaft, one pair of gages (mounted parallel to each other) will increase in resistance due to tension strain while the other pair will decrease in resistance due to compression strain.

It is common practice to machine the shaft to either a reduced diameter or to an undercut portion of square cross-section, primarily for the purpose of increasing transduction-element sensitivity. Square-cut shafts have one gage mounted to each

Fig. 5-14.   Strain-gage torque transducer (courtesy of Lebow Associates).

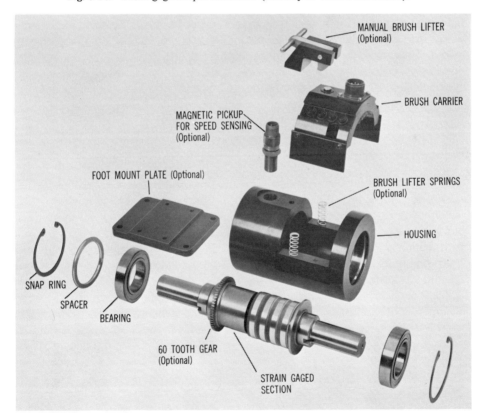

side of the square. A cylindrical protective cover is sealed over the strain-gage section of the shaft.

The sensing shafts of most strain-gage torque transducers are equipped with *slip rings* (rotating electrical contacts). The leads from the electrical network in the sensing shaft, which often contains adjustment and temperature-compensation resistors besides the four-arm strain-gage bridge, are attached to the slip rings from the inside. One or more sets of brushes are mounted to the inside of a portion of the transducer case so that they are forced against the slip rings. Connecting wires from the brushes are either brought out of the case as a multiconductor cable or connected to the pins of an electrical receptacle.

Fig. 5-15. Strain-gage torque transducer using rotary transformer instead of slip rings (courtesy of S. Himmelstein & Co.).

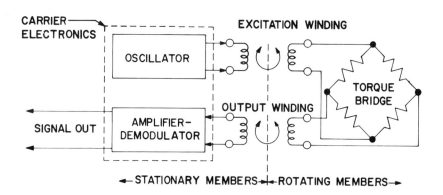

A typical strain-gage torque transducer is illustrated in Figure 5-14. This design incorporates several optional features such as an electromagnetic angular-speed transducer (see Chapter 12) and a device which permits lifting the brushes off the slip rings when no measurements are being made. Use of the brush lifter extends the operating life of both sets of contacts.

Brushes and slip-rings are subject to wear and dirt accumulation and require periodic maintenance. Ease of brush replacement is a major design factor for brush assemblies. Slip rings can introduce noise in the inherently low-level transducer output due to brush bounce, rubbing-surface contamination, and contact pressure variation. Such noise is minimized when a balanced four-arm bridge is used on the rotating shaft. Silver graphite is commonly used as brush material. Favorable results were reported for a compressed random arrangement of short molybdenum wires. Copper mesh and metal "wool" have also been used in brushes. Slip rings are usually made from sterling silver or coin silver and sometimes from monel or other alloys. The force exerted by the brushes on the slip rings must be kept low to avoid introducing significant measurement errors in the torque measurement.

A strain-gage torque transducer which uses a (noncontacting) rotary transformer instead of slip rings is shown in Figure 5-15. This unit contains two primary and two secondary windings. The excitation primary is on the stator, and the excitation secondary is on the rotor; whereas the output primary is on the rotor, and the output secondary is on the stator. Coupling between excitation and output windings can cause cross talk and must be minimized. Shaft and winding surfaces must be as concentric as possible to avoid amplitude modulation during operation. A typical design uses 120-ohm bonded metal-foil gages, a carrier frequency of 20 kHz, an excitation secondary voltage of 5 V rms, an oscillator source-impedance of 6 ohms, and an amplifier input impedance of 1000 ohms. A measurement bandwidth of 0 to 3000 Hz is provided.

Reaction-torque transducers with strain-gage transduction elements are more closely related to force transducers. Since they are typically used to measure the torque required to restrain the housing or stator of a rotary device from turning while its rotor or shaft is subjected to a torque, they do not contain any rotating sensing shafts. Instead, they are either of coaxial design (see Figure 5-16) or are furnished with a short coupling shaft. Coaxial transducers are mounted between the device under test and its normal mount; the device's drive shaft extends through the transducer. The rotating shaft does not contact the transducer's inside surface.

The case of a coupling-shaft type of reaction-torque transducer is bolted to a test frame or bench top while the coupling shaft is attached to the housing of a rotary device. Torque applied to the coupling—which is equal and opposite to the torque of the device's rotating shaft—causes strain in torque-sensitive flexures in the transducer. Transduction is achieved by strain gages on the flexures. Angular displacement of the coupling shaft is less than 0.25 degrees at full-range torque. Coaxial transducers contain torque-sensing elements between the two mounting flanges. The strain in these elements is transduced to provide the required output.

Fig. 5.16. Strain-gage reaction-torque transducer (courtesy of Lebow Associates, Inc.).

5.3.7 Encoder-type torque transducers

A number of torque transducers have been designed to utilize discs of the type used in incremental digital angular-displacement transducers (incremental encoders—see 3.3.8.1). Two discs are mechanically attached to the sensing shaft so that shaft deflection due to torque causes a relative angular displacement between the two discs. When toothed ferromagnetic discs are used, the phase difference between the outputs of the two electromagnetic or reluctive reading devices (one for each disc) is proportional to shaft deflection. The same technique can be used with optically-coded discs (discs with alternating transparent and opaque sectors) associated with photoelectric reading devices.

A different photoelectric technique employs a single light beam through both discs. The two discs are first so aligned that the combination of the opaque and transparent sectors on both discs, through which the light beam passes, produce a certain light intensity at the light detector (null output). As torque is applied to the sensing shaft, the second disc will pass more or less light whenever the sector in the first disc is fully transparent, depending on the amount of light selected to produce the null output. The output of the light detector will, therefore, be proportional to torque.

5.3.8 Vibrating-wire force transducers

When a wire between a pair of permanent magnets is stretched between two points of a deflecting force-sensing element and the wire is caused to vibrate at its resonant frequency in a feed-back oscillator, the oscillator frequency will change with wire tension. This principle has been applied in force transducers with frequency-modulated output, designed primarily in countries other than the U.S.A. When a second wire, not affected by sensing-element deflection, is made to oscillate at the same frequency as the sensing-element wire when no force is applied, the beat frequency (frequency difference) between the two frequencies of oscillation will be proportional to tension in the sensing wire only. By this scheme a convenient frequency output is provided by the transducer, which can be displayed by a frequency counter. A system employing the beat-frequency technique has been designed as commercial weighing equipment.

Use of vibrating-wire transduction in torque transducers has also been reported.

5.4 Performance and Applications

5.4.1 Specification characteristics

After deciding whether a universal (tension/compression), tension, or compression force transducer, or a shaft-torque or reaction-torque transducer is to be defined, and after selecting the required transduction principle, the following characteristics should be considered for inclusion in the specification.

5.4.1.1 Mechanical design characteristics. A drawing should show the outline of the transducer together with its essential case and mounting dimensions and their tolerances. Torque- and force-connecting provisions should be shown in detail. This applies to all forms of shaft ends for torque transducers. Threaded force connections should be specified by nominal size, number of threads per inch, thread depth (if female) or length (if male), thread series, and class of thread. Slot width, length of clevis, and hole diameter should be shown for clevis-type force connections; spherical or fixed bearing and bore should be indicated for bearing-type connections; and hardness and curvature should be shown for compression buttons (load buttons). Locations and types of electrical connections and of any optional devices (e.g., brush lifters and shaft-speed transducing coils on torque transducers or lifting handles on force transducers) should also be indicated on the drawing.

The information to be shown on the nameplate should include nomenclature (e.g., "Piezoelectric Force Transducer"), manufacturer's name, location, part number and serial number, range, nominal and maximum excitation, and identification as to function of all external electrical connections. Nomenclature can include additional terms such as "bonded-" (strain-gage) or "universal-," "compression-," or "tension-" (force).

Sensing-element material and certain dimensional characteristics, such as con-centricity with force or torque connections, method of sealing, and details of hermetic-seal headers are often specified, as are brush and slip-ring materials for shaft-torque transducers. Case material and finish should be specified when the transducer may be used in a corrosive or contaminating atmosphere. Weight (with tolerances), allowable mounting force or mounting torque, and the maximum allow-able angular misalignment of axial force are normally specified.

5.4.1.2 Electrical design characteristics. Electrical connections can be shown in a simplified schematic which also indicates certain transduction-element details such as the use of two or four active arms for strain-gage bridges, the use of one, two, or more separate bridges, internal adjustment and compensation resistors, or winding configurations of reluctive transducers. Nominal and maximum excitation current or voltage (and frequency of a-c excitation), input and output impedance and allowable load impedance, and insulation resistance or breakdown-voltage rating (or both) should always be stated. Characteristics and connections of any shunt-calibration provisions should be included when applicable.

5.4.1.3 Static performance characteristics are applicable at room conditions. *Range* is shown as unidirectional for torque transducers (since it is assumed to be equal for both directions of rotation) and for force transducers responding to either compression or tension force only, e.g., "0 to 500 kgf, tension." The lower range limit is sometimes shown as greater than zero for torque transducers. The range of universal force transducers is shown as bidirectional (e.g., "$\pm 10,000$ lbf"), except when the transducer responds unequally in the two directions where it is separately specified for tension and for compression. *Safe-overload rating* (without subsequent change in performance characteristics) and *ultimate-overload rating* (without ensuing structural failure) are shown in the same units chosen for range.

Output is specified, with tolerances, at a given value of excitation, either by its end points, as full-scale output and zero-force output ("zero balance" or "residual unbalance"), or as sensitivity. It can also be stated as sensitivity or as full-scale output and zero-force output per unit (usually per volt) of excitation. Output phase characteristics should be considered for reluctive transducers.

Accuracy characteristics include terminal, end-point, independent, or other types of linearity, hysteresis, and repeatability. The first two are sometimes combined and covered by a single set of tolerances. A static error band, referred to whatever line would be chosen to define linearity, combines the three individual characteristics. A limiting tolerance, together with a time of application, is often stated for creep, a small gradual change in output from that first obtained when a given force or torque is applied to the transducer. Warm-up time is sometimes included in specifications, as are zero and sensitivity shifts over relatively short periods of time. Long-term stability (repeatability over long periods of time) can be specified but is considered difficult to verify and is more meaningful as a design target.

The *deflection factor* (maximum sensing element displacement at the upper range limit) is frequently specified for force transducers as are tolerances for nonaxial loading effects at one or more angles. A low deflection factor is desirable when fast response is required from a force transducer. Nonaxial loading can cause significant measurement errors in most force transducers.

Maximum and minimum *shaft speed* must be stated for torque transducers. Moment of inertia, torsional stiffness, shaft runout, and axial end float are also frequently specified for these transducers.

Life requirements can be shown in terms of cycling life for force transducers and as continuous or intermittent operating life for torque transducers.

5.4.1.4 Dynamic performance characteristics of force transducers can be specified as frequency response, as resonant frequency, ringing period (when the transducer represents a system with two or more degrees of freedom), and damping ratio or, when overshoot is small, as time constant and overshoot. For torque transducers they are usually determined from mechanical characteristics rather than separately specified.

5.4.1.5 Environmental characteristics include the temperature error band or the temperature error over a stated operating temperature range—sometimes shown as separate tolerances for thermal zero and sensitivity shifts—and acceleration, ambient-pressure, and vibration errors. Limitations on effects of shock, high sound-pressure levels, atmospheric contaminants, or ambient fluids other than air, electromagnetic interference, magnetic fields, and nuclear radiation may also have to be included when one or more of such environmental conditions can be expected. Magnetic fields are of particular significance to torque transducers since they can induce undesirable signals and noise in electrical conductors rotating near them.

5.4.2 Selection and applications

Primary selection criteria are generally range, accuracy characteristics, and case dimensions for force transducers, and range, maximum shaft speed, and accuracy characteristics for torque transducers. Electrical characteristics of the associated measuring system and environmental conditions, especially temperature and, for torque transducers, magnetic fields are additional important criteria. Overload capability and dynamic characteristics also deserve careful attention. The choice of whether or not to use slip rings in torque transducers is mainly dictated by service and maintainability considerations. Associated signal amplifiers should provide sufficient filtering, preferably at their input, to minimize the transmission of noise and to prevent saturation which can be caused by output due to high-frequency measurand pulsations.

5.4.2.1 Force transducer applications include a large variety of thrust, stress, and weight measurements. Truck scales with remote readout are a well-known example. Automatic weighing devices, e.g., batching-hopper weighing systems and weight checking equipment under conveyor belts, are used in many industries. The weighing of tanks is used for measurement of their contents (see Liquid Level, 8.2.11). Three or more appropriately located force transducers can be used for center-of-gravity determinations. In cable attachments, they are used to measure load on the cable and its stress—important applications in well drilling and logging, antenna mast supports, nets used by fishing vessels, crane and winch operation, and numerous other activities. Simple dynamometers use a tension force transducer at the end of an arm radially clamped to a shaft of an electric motor. Jet and rocket engine thrust in one or more directions is measured by force transducers. Flexural pivots (*flexures*) can be used on force connections to minimize errors due to side and torsional loads.

5.4.2.2 Torque transducer applications are found in testing, evaluating, and monitoring of rotating components. In-line shaft-torque transducers are installed, for example, in main-drive shafts on ships for continuous monitoring of transmitted horsepower. They are used for testing motors, pumps, blowers and fans, clutches and brakes, gear boxes, speedometer cables, rotating joints and couplings, aircraft propellers, and machine tools of all types. Used as dynamometers in conjunction with a mechanical load, they are widely employed in testing of engines, power plants, and entire self-propelled vehicles.

Flexible couplings are used on both ends of the sensing shaft to minimize misalignment effects. Single flexible couplings are used with "floating shaft" transducers whose cases are not mounted to any structure but only restrained from rotation by a solid strap or bracket. Footmounting of the transducer case reduces the net length of an unsupported shaft and reduces the possibility of low resonant frequencies. Double flexible couplings are used with this mounting configuration to minimize the effects of any parallel or angular misalignment. Use of flange-mounted transducers with internal spline drive requires careful alignment during mounting and shaft coupling.

Vibration of the shaft due to its bending is maximum at the critical shaft speed, which can be approximated (expressed in rpm) as 4.6×10^6 times the ratio of shaft diameter over the square of the shaft length, with both dimensions expressed in the same units (inches or centimeters). Operation of a torque transducer near its critical shaft speed should be avoided. The factor 4.6×10^6 applies to shafts which are simply supported at each end. When flexible couplings are used, the critical speed of a "floating shaft" transducer will be somewhat lower, whereas that of a foot-mounted transducer will be considerably higher than that of a simply supported

Fig. 5-17. Reaction-torque transducer applications (courtesy of Lebow Associates, Inc.).

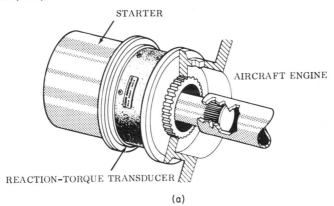

(a)

(b) Hydraulic pump testing

type. Torsional vibration is maximum at the resonant frequency, which is a function of the torsional stiffness of the shaft and the inertia of the masses on both shaft ends.

Reaction-torque transducers have no critical speed and can be dead-weight calibrated without disassembly. However, errors are introduced when the shaft is accelerating or decelerating during a measurement. Such transducers are used for a variety of measurements including bearing friction, stepping-switch torque, and viscosity—the latter by rotating a drum in a container of the liquid under test, with the container supported by a reaction torque transducer so that increasing viscosity results in increasing torque. They are also used as dynamometers and for testing rotating components such as automotive brakes, alternators, aircraft starters, and hydraulic pumps (see Figure 5-17) when connecting power cables or hydraulic lines do not interfere with reaction torque. In some cases errors introduced by such cables or tubing can be minimized by use of special calibration provisions and methods.

5.5 Calibration and Tests

5.5.1 Calibration

Force transducers are calibrated by applying known forces to them in discrete increments and recording their output. Torque transducers are calibrated by applying known forces at a known moment arm. Two types of calibration techniques are used. Primary techniques involve loading the transducer with a number of different weights. Comparison calibrations require the measurement of the force applied to the transducer by a force-measuring device of very close accuracy.

The dead-weight calibrator provides the best accuracy for primary force calibrations. Machines with capacities up to 250,000 lb have been installed in various facilities. Several machines, ranging in capacity up to 12,000,000 lb, are installed at the National Bureau of Standards facility in Gaithersburg, Maryland. The 1,000,000-lb calibrator extends over three floors. Its disc-shaped weights are stacked on the first floor and below it, in a 26-ft pit. Each weight (Figure 5-18) has a diameter of 10 ft and weighs 50,000 lb. A hydraulic jack on the third-floor level is remotely controlled from the second-floor control level (Figure 5-19) to lift the required number of weights. The machine can apply tension or compression force to the device under test depending on whether the upper or lower frame supports the weights. An uncertainty of less than 0.002% of reading is reported for this calibrator.

The principle of the lever arm is applied in another primary force calibration method using a *beam-balance calibrator*. This calibrator (Figure 5-20) applies force to a transducer being calibrated by means of weights suspended at the end of a multiplication arm. The arm extends from its fulcrum in both directions so that application of weights at one end will produce a compression force on one side of the fulcrum and a tension force on the other side of the fulcrum. Cross flexures are used at all pivot points, and additional plate flexures are used to stabilize the loading heads so that only axial force is transmitted from the multiplication beam to the transducer being calibrated.

Comparison calibrations can be performed by using either a proving ring or a reference transducer having close accuracy and good stability. Either reference device is first calibrated in a dead-weight machine. During the comparison calibration either is then mounted back-to-back with the transducer being calibrated.

Fig. 5-18. Weight stack of 1,000,000 lbf dead-weight machine (courtesy of National Bureau of Standards).

Fig. 5-19. Control level of 1,000,000 lbf dead-weight machine with proving ring in position for calibration (courtesy of National Bureau of Standards).

Fig. 5-20. Beam-balance force calibrator (courtesy of Ormond, Inc., and Transducers, Inc.).

Proving rings used for calibration permit measurement of their vertical deflection under load by a diametrically mounted reading device. A micrometer screw, protruding upward from the bottom boss, and a vibrating reed, protruding downward from the upper boss, are used in the proving ring originally developed by H. L. Whittemore and S. N. Petrenko shortly after the First World War. The reed is set into vibration by manual or electrical means. It is used to indicate, by a characteristic buzzing sound or by a reduction in vibration amplitude, the point at which the spindle of the micrometer screw contacts the reed end. This is the point at which the micrometer dial should be read. The inherent accuracy of proving rings is about one-tenth the accuracy of pure dead-weight calibrators.

During all types of calibrations it is essential that the direction of force application be precisely aligned with the transducer's sensing axis so that the apparent force seen by the transducer equals the actual force applied to it. If the transducer is slightly tilted, the fraction of the actual load seen as apparent load varies as the cosine of the misalignment angle.

Various types of carefully selected transducers have been used as reference-calibration devices, primarily in portable systems used for the calibration of transducers in fixed installations.

Piezoelectric force transducers can be calibrated dynamically by mounting them between a vibration exciter and a mass load. The acceleration of the mass in response to the force applied to it through the transducer is measured and used to determine the dynamic force applied to the transducer by the vibration exciter.

Selection of output measuring equipment depends on the transduction principle employed in the transducer being calibrated.

5.5.2 Tests

Since typical applications of force and, especially, torque transducers are in environments at or near room conditions, relatively few tests are normally performed besides a calibration.

The calibration, comprising at least two full-range calibration cycles, generally serves to establish the static performance characteristics: the static error band, end points and full-scale output, linearity, hysteresis, and repeatability, and zero balance, zero shift, and sensitivity shift if these were specified. Calibration cycles are performed at the beginning and end of the time during which the shift specifications apply. Creep can be determined at one or more points during the calibration if repeated output readings are taken for some time after a given level of force or torque was applied. Safe overload rating is verified by performing at least a three-point calibration cycle to verify end points and mid-range output after application of the specified overload.

A mechanical inspection is usually performed on each transducer to verify dimensions, smoothness, flatness, and parallelism of mounting surfaces (except on shaft-torque transducers), angularity, concentricity, and dimensions of force-connection threads and torque sensing-shaft connections, mounting holes, and dimensions, finish of case and of visible portions of the sensing element, nameplate information, slip rings, and brushes on shaft-torque transducers, and general evidence of good workmanship. Weight is verified on a sampling basis only.

An electrical inspection of each transducer should include determination of input and output impedance, insulation resistance or break-down-voltage rating or both, and output noise when slip rings or internal a-c to d-c conversion are employed.

Environmental tests, normally performed on a qualification or sampling basis, are often limited to temperature tests. The equipment used to apply force or torque to the transducer in the temperature chamber must be well isolated from the temperature in the chamber by using appropriate flexible seals in the walls of the chamber.

Tests to determine performance in a contaminating, conductive, or corrosive ambient fluid should be performed when the transducer is not hermetically sealed. When it is sealed, a simple immersion test can be used to check the seal and to prove that such ambient fluids, as well as variations in ambient pressure, will not affect transducer performance. Another method of verifying a hermetic seal is usable on force transducers with ranges up to $\pm 10,000$ lbf. The transducer is placed in a bell jar fully connected for readout. The bell jar is evacuated. A hermetically sealed transducer will immediately produce an output, caused by "bulging," as if a tension load were applied, since its ambient pressure is now -14.7 psig. A leaky transducer will show no change in its zero-force output.

Dynamic tests can be performed by the application of a step change in force or torque to determine time constant and overshoot. Sinusoidally varying forces can be applied, if necessary, to transducers with very little damping. In this test the transducer is sandwiched between the exciter and a mass load. Special techniques are required for sinusoidal torque variation.

5.6 Special Measuring Devices

In addition to the multicomponent (X-Y and X-Y-Z) force transducers described in 5.3.5, combination devices have been designed for simultaneous measurement of force and torque by a single transducer and for the convenient performance testing of small motors and pumps.

Reaction-torque as well as shaft-torque transducers are available with an integral axial-force transducer. Their typical application is in machine tools. The reaction-torque force transducer, for example, is used as a drilling dynamometer. Bolted to the table of a drill press, it can measure the thrust (force) of the drill as well as the torque produced in the work piece. Such a device is useful in establishing drilling forces and studying tool configurations and lubricant characteristics.

Special test fixtures have been developed for torque testing of rotary devices on a production basis. An example of such a fixture is the Torque Table®, which consists of a plate supported from a fixed mounting base by means of torque-sensitive flexure straps to which strain gages are bonded in a four-arm bridge configuration. The flexure straps are mounted at 45-degree angles. The intersection of lines drawn through the straps describe the center of rotation of the table. A reaction torque is sensed by the table when a rotating device mounted to the table has its input or output shaft aligned exactly at the center of rotation. Torque can be measured without speed limitations or friction losses. A d-c output angular-speed transducer can be installed on the same fixture (Figure 5-21) to provide an *x-y* plot of torque vs speed.

Developments in the solid-state field include new approaches toward force transduction, such as a silicon planar transistor to whose emitter-base junction an applied force can be coupled directly by a mechanical link. A device of this type is reported to produce significant output changes in response to force variations as low as 5 gm.

Fig. 5-21. Test fixture for torque-vs-speed testing of motors (courtesy of Lebow Associates, Inc.).

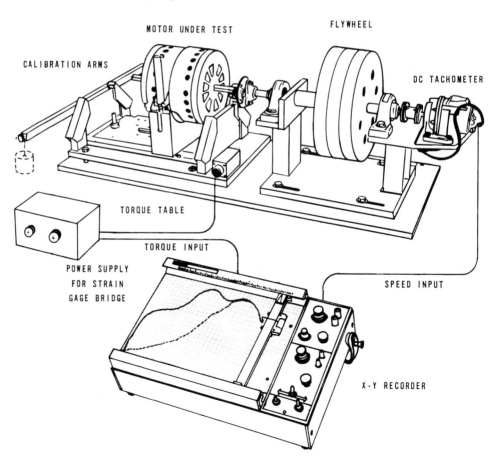

BIBLIOGRAPHY

1. Wilson, B. L., Tate, D. R., and Borkowski, G., "Dead-Weight Machines of 111,000- and 10,100-Pound Capacities," *N.B.S. Circular C446*, Washington, D.C.: National Bureau of Standards, 1943.

2. Wilson, B. L., Tate, D. R., and Borkowski, G., "Proving Rings for Calibrating Testing Machines," *N.B.S. Circular C454*, Washington, D.C.: National Bureau of Standards, 1946.

3. Gibson, R., *et al.*, "Instruments for Measuring Small Forces," *Journal of Scientific Instruments*, Vol. 30, pp. 159–162, May 1953.

4. Lashof, R. W. and Macurdy, G. B., "Precision Laboratory Standards of Mass and Laboratory Weights," *N.B.S. Circular 574*, Section 1, Washington, D.C.: National Bureau of Standards, 1954.

5. Roark, R. J., *Formulas for Stress and Strain*. New York: McGraw-Hill Book Company, 1954.

6. Bell, R. E., "Electrical Transducers for Automatic Weighing," *Electrical Engineering*, Vol. 74, No. 10, pp. 926–928, October 1955.

7. Scalise, D. T., "Measure Heavy Forces to Avoid Excessive Bolt Loading," *Iron Age*, Vol. 176, pp. 85–87, October 27, 1955.

8. Clouston, J. G., "Transducer for the Sequential Measurement of Two Levels of Thrust," *Journal of Scientific Instruments*, Vol. 33, p. 321, August 1956.

9. Judson, L. V., "Units and Systems of Weights and Measures," *N.B.S. Circular 570*, Washington, D.C.: National Bureau of Standards, 1956.

10. Ruge, A. C., "Precision Measurements of Force," *Metal Progress*, Vol. 70, pp. 92–93, July 1956.

11. Lauer, J. L. and Fiel, P. J., "Device for Measuring Friction," *Review of Scientific Instruments*, Vol. 28, pp. 294–295, April 1957.

12. Mason, W. P. and Thurston, R. N., "Use of Piezoresistive Materials in the Measurement of Displacement, Force, and Torque," *Journal of the Acoustical Society of America*, Vol. 29, pp. 1096–1101, 1957.

13. Green, M., "Using the Load Cell in High-Accuracy Weighing Systems," *Control Engineering*, Vol. 6, No. 7, pp. 121–124, October 1959.

14. Harris, C. J., Kaegi, E. M., and Warren, W. R., "Pressure and Force Transducers for Shock Tunnels," *I.S.A. Journal*, Vol. 7, No. 8, August 1960.

15. Carleton, R. J., Jr., "High-Capacity Force Standards," *I.S.A. Journal*, Vol. 8, No. 6, pp. 38–42, June 1961.

16. "Standard Load Cell Terminology and Definitions," Chicago: Scientific Apparatus Makers Association, 1962.

17. Emmerling, A. A., "A Torque Measurement Transducer System," *Electrical Engineering*, pp. 621–625, October 1963.

18. Lebow, M. S., "Some Principles of Transducer Design," *I.S.A. Transactions*, Vol. 2, pp. 85–92, 1963.

19. Harting, D. S., "Compensating a Semiconductor Load Cell," *Strain Gage Readings*, Vol. VI, No. 5, December–January 1963–64.

20. Bryzzhev, L. D., "Certain Problems in Measuring Forces by Means of Piezoelectricity," *Measurement Techniques* (U.S.S.R), No. 8, pp. 31–33, 1964.

21. Hockersmith, T. W. and Ku, H. H., "Uncertainties Associated with Proving Ring Calibration," *Preprint No. 12.3-2-64*, Pittsburgh: Instrument Society of America, 1964.

22. Karrer, H. E., Reed, D. J., and Virgin, G. L., "Dynamic Force Measurement," *Preprint No. 16.4-3-64*, Pittsburgh: Instrument Society of America, 1964.

23. Karrer, H. E. and Reed, D. J., "Measurement of Thrust Vector Control Forces on Solid Rocket Motors," *Paper 6AS64, Proceedings of the 10th National Aerospace Instrumentation Symposium*, Pittsburgh: Instrument Society of America, 1964.

24. Ormond, A. N., "Unique Linearization of Strain Gage Transducers," *Preprint 16.8-2-64*, Pittsburgh: Instrument Society of America, 1964.

25. Raines, E. G., "An Application of Semiconductor Strain Gauges and Low Noise Level Slip Rings for the Measurement of Torsional Stresses in a Rotating Shaft," *Tech. Memo ARL/ME.265*, Melbourne: Department of Supply, Australian Defence Scientific Service, Aeronautical Research Labs., May 1965.

26. Chapin, W. E. and Mitchell, R. K., "A Summary of Force-Transducer Technology: Measurement Techniques, Calibration, Element Design, Availability, and Research Activity," *TIC Summary Report No. 2*, Columbus, Ohio: Transducer Information Center, Battelle Memorial Institute, 1966.

27. Eide, R. H., "Use and Abuse of Load Cells," *Test Engineering and Management*, Vol. XVI, No. 1, pp. 14–17, July 1966.

28. Glenn, D. C., "Force Measurement in a Vacuum Environment below 10^{-6} Torr," *NASA TM X-1209*, Washington, D.C.: National Aeronautics and Space Administration, 1966.

29. Himmelstein, S., "The MCRT, a Practical Electrical Coupling for Measuring Strain in Rotating Members," *Preprint No. 16.17-2-66*, Pittsburgh: Instrument Society of America, 1966.

30. Stone, L. E., "Five-Million Pound Load Cell Design and Calibration," *Preprint No. 16.9-2-66*, Pittsburgh: Instrument Society of America, 1966.

31. Zimmerman, W. D., "Triaxial Force Transducer for Determining Thrust Angle Deviations of Canted Nozzle Rocket Motors," *Preprint No. 1.1-5-66*, Pittsburgh: Instrument Society of America, 1966.

32. "Load Cells, Bonded Strain Gage, General Specification for," *Specification NOTS-PD-101*; China Lake, California: U.S. Naval Weapons Center, 1966.

6 Humidity and Moisture

6.1 Basic Concepts

6.1.1 Basic definitions

Humidity is a measure of the water vapor present in a gas. It is usually measured as absolute humidity, relative humidity, or dew-point temperature.

Absolute humidity is the mass of water vapor present in a unit volume.

Specific humidity is the ratio of the mass of water vapor contained in a sample (of moist gas) to the mass of the entire sample.

The **humidity mixing ratio** is the mass of water vapor per unit mass of the dry constituents.

Relative humidity is the ratio of the water-vapor pressure actually present to water-vapor pressure required for saturation at a given temperature; this ratio is expressed in percent. Relative humidity is always temperature-dependent.

Moisture refers to the amount of liquid adsorbed or absorbed by a solid. It has also been used to refer to the water chemically bound, adsorbed, or absorbed in a liquid.

The **dew point** is the temperature at which the saturation water-vapor pressure is equal to the partial pressure of the water vapor in the atmosphere. Any cooling of the atmosphere below the dew point would produce water condensation. The relative humidity at the dew point is 100%. The dew point has also been defined as the temperature at which the actual quantity of water vapor in the atmosphere is sufficient to saturate this atmosphere with water vapor.

6.1.2 Related laws

Partial pressure (by Dalton's Law)

$$p = p_a + p_w$$

where: p = total atmospheric pressure
p_a = partial pressure due to dry air
p_w = partial pressure due to water vapor

Relative humidity (based on vapor pressure)

$$\% \text{ RH} = \frac{p_w}{p_s} \times 100$$

where: $\%$ RH = relative humidity (expressed in percent)
p_w = partial pressure of the water vapor actually present in the atmosphere
p_s = partial pressure which would be exerted by the water vapor if the atmosphere were saturated at the existing temperature

Percent saturation

$$\% \text{ Saturation} = \frac{W_e}{W_s} \times 100$$

where: W_e = existing weight of water vapor
W_s = saturation weight of water vapor

Absolute humidity (based on vapor concentration)

$$d_v = \frac{m_v}{V}$$

where: d_v = density of water vapor in moist air mixture = absolute humidity
m_v = mass of water vapor in a given sample
V = volume occupied by sample

Specific humidity   *Humidity mixing ratio*

$$q_s = \frac{m_v}{m_{ma}} = \frac{m_v}{m_v + m_a} \qquad R_q = \frac{m_v}{m_a}$$

where: R_q = humidity mixing ratio

q_s = specific humidity of moist air

m_v = mass of water vapor in sample

m_{ma} = mass of moist air in sample

m_a = mass of dry air in sample

6.1.3 Units of measurement

Relative humidity is expressed in **percent.**

Absolute humidity is typically measured in **grams per cubic meter.**

Humidity mixing ratio is usually expressed in **pounds per pound**, **grams per pound,** or **grams per kilogram.**

Moisture is expressed in **percent by weight** (either with respect to the total weight or to the dry weight) or in **percent by volume.**

Dew point is shown in **degrees** Fahrenheit (°F) or Celsius (°C).

6.1.4 Hygrometry, psychrometry, and dew-point measurement

The *hygrometer* is an instrument which measures humidity directly. It can be calibrated in terms of absolute humidity. More frequently, its output is used to indicate relative humidity.

The *psychrometer* is a humidity-measuring instrument which utilizes one "wet-bulb" thermometer and one "dry-bulb" thermometer. The "dry-bulb" thermometer measures ambient temperature. The "wet-bulb" thermometer measures temperature reduction due to evaporative cooling. A wick, a porous ceramic sleeve, or a similar device saturated with water is in close physical contact with the "bulb" (sensing portion) of the "wet-bulb" thermometer to keep it moist. Relative humidity is determined from the two temperature readings and a reading of the barometric

Table 6-1 SIMPLIFIED TYPICAL PSYCHROMETRIC TABLE†

DIFFERENCE BETWEEN DRY-BULB AND WET-BULB READINGS, °F

DRY-BULB READING, °F	1	3	5	10	15	20	25	30	35	40
	RELATIVE HUMIDITY, PERCENT									
40	92	76	60	—	—					
50	93	80	68	38	12	—				
60	94	83	73	48	26	6	—			
70	95	86	77	55	37	20	3	—		
80	96	87	79	61	44	29	16	4	—	
90	96	89	81	65	50	36	24	13	3	—
100	96	89	83	68	54	42	31	21	12	4
110	97	90	84	70	57	46	36	27	18	11

†Relative-humidity values apply at a barometric pressure of 29.24 in. Hg (990 millibars).

pressure, usually by means of a *psychrometric table*. At any given ambient temperature ("dry-bulb" reading), the relative humidity decreases as the difference between "dry-bulb" and "wet-bulb" readings increases (see Table 6-1). This temperature difference is referred to as the wet-bulb *depression*.

Dew-point measurement is measurement of the temperature of a surface at the instant when moisture is first precipitated on it as the surface is artificially cooled. *Note:* The *hydrometer* is an instrument used to measure the density of a liquid. The prefix *hydro-* (from Greek *hydor*, water) refers to water in liquid form, whereas the prefix *hygro-* (from Greek, *hygros*, moist, wet) refers to water in vapor form. The prefix *psychro-* is derived from the Greek word *psychros*, cold, and refers to the cooling of the wet-bulb thermometer.

6.2 Sensing Elements

Many of the commonly used humidity-sensing elements are of the types that also perform the transduction. A typical example of this is the resistive element. Such elements are here referred to as sensing elements although they are actually combination sensing/transduction elements.

The sensing elements of humidity and moisture transducers can be classified into three major categories on the basis of the directness of their measuring ability. Hygrometers measure humidity directly. Among these, the resistive hygrometer element is most widely used. The use of displacement-producing elements in transducers has been decreasing sharply since reliable resistive elements became available. Other types, such as the oscillating-crystal element, appear promising for certain applications.

Psychrometers measure humidity indirectly. Resistive temperature transducers are most frequently used as "wet bulb" and "dry bulb." Dew-point elements yield a direct indication of dew point from which humidity can also be derived. In this last major category the cold-mirror devices appear to be most common.

6.2.1 Hygrometer elements

6.2.1.1 Resistive hygrometer sensing elements. Among the most popular sensing elements are those in which a variation of ambient relative humidity produces a variation in their resistance. This resistance change occurs in certain materials such as hygroscopic salts and carbon powder. These materials are usually applied as a film over an insulating substrate and are terminated by solid metal electrodes (see Figure 6-1). Hygroscopic-salt (aqueous-electrolytic) films are ionizable materials whose resistance varies as a function of the vapor pressure of water in the ambient air. The best known hygroscopic-salt element is the *lithium-chloride element*. This element uses a film consisting of an aqueous solution of (usually less than 5%) lithium chloride (LiCl) in a plastic binder.

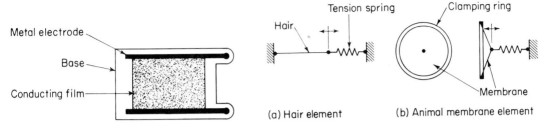

Fig. 6-1. Resistive humidity sensing element.

(a) Hair element   (b) Animal membrane element

Fig. 6-2. Mechanical humidity sensing elements.

Among other variable-resistance film materials, barium fluoride, potassium dihydrogen phosphate, cerium titanate, and carbon have been found usable for humidity-sensing elements.

6.2.1.2 Dielectric-film hygrometer elements. The hygroscopic film has been used in certain capacitive humidity-sensing elements where it acts as part of the dielectric. Changes in the dielectric cause a change in capacitance with humidity. An example of such elements is one using a heterogeneous mixture of a hygroscopic liquid and water on the surface of an inert granular solid as the dielectric of a capacitor. The capacitance of the element changes with the partial pressure of water vapor in a gas sample.

6.2.1.3 Mechanical hygrometer elements. Among the various materials which change their dimension with adsorbtion and desorption of water from the air, two organic materials have been used for humidity sensing elements considerably more than others. The first and most popular material is hair, particularly human hair. The second is animal membrane. Other materials, such as paper, wood, bone, textiles, and plastics have been found less satisfactory.

Human hair [Figure 6-2(a)], as well as animal membrane [Figure 6-2(b)], expands with increasing humidity. This dimensional change has been utilized for the operation of transduction elements in a number of different designs. The change in the length of human hair over the range 0% to 100% RH is quite small, about one part in fifty, and nonlinear. Sensitivity decreases from approximately 0.0004 in. per in. of hair length per % RH at 15% RH to about 0.00005 in./in. per % RH at 85% RH at one gram tensile force on the hair.

6.2.1.4 Oscillating-crystal hygrometer elements. Promising results have been obtained in the measurement of humidity by means of a quartz crystal with a hygroscopic coating (Figure 6-3) operating in an oscillator circuit. The mass of the crystal changes with the amount of water sorption on the coating. This changes

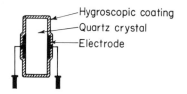

Fig. 6-3. Oscillating-crystal humidity sensing element.

the oscillator frequency so that it becomes a measure of humidity ambient to the crystal. Hygroscopic polymers appear to be the most suitable coating materials.

6.2.1.5 Aluminum-oxide hygrometer elements. The electrical properties of anodized aluminum are used for humidity-measurement purposes in small strip, needle, or rod types of elements. The element consists of aluminum whose surface is anodized so that a thin layer of aluminum oxide is formed. The structure of such a film has been determined to consist of a multitude of fibrous pores [Figure 6-4(a)]. A very thin metal coating—sometimes aluminum, usually gold—is vacuum-deposited on the outside surface of the aluminum-oxide layer. It acts as one electrode, and the aluminum base acts as the other electrode. Water vapor is transported through the gold layer and equilibrates on the pore walls in a manner functionally related to the vapor pressure of water in the atmosphere ambient to the sensing element. The number of water molecules absorbed on the oxide structure determines the change in impedance of the element as shown in the equivalent circuit of Figure 6-4(b). The transduction in the element is both capacitive and resistive. The output is an impedance (or admittance) change which can be measured with an a-c bridge or similar circuitry.

6.2.1.6 Spectroscopic hygrometer elements. A number of successful hygrometers have been developed in which the determination of humidity is based on analysis

Fig. 6-4. Aluminum-oxide humidity sensing element.

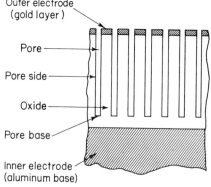

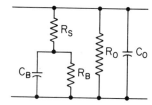

C_O = capacitance across oxide layer
R_O = resistance of solid portion of oxide layer
R_S = resistance of pore side
R_B = resistance between pore base and inner electrode
C_B = capacitance between pore base and inner electrode

(a) Section through element

(b) Equivalent circuit of single pore

of absorption bands of water vapor in a gas sample. Such instruments employ a *sensing path* as sensing element, such as a cylinder of predetermined length, through which a beam of radiation passes (usually infrared, sometimes microwave, ultraviolet, or visible). The absorption characteristics are determined in the gas sample which extends over this sensing path, e.g., that contained in the cylinder.

Spectroscopic humidity-measuring apparatus is more appropriately categorized as analysis equipment and will not be covered in further detail in this chapter.

6.2.2 Psychrometer elements

The sensing elements of transducers which measure humidity by the "wet- and dry-bulb" method are temperature-sensing elements. These are described in more detail in Chapter 14. Psychrometric sensing elements are always dual elements with normally separate outputs (Figure 6-5). One element ("dry bulb") measures the temperature of its ambient fluid. The other element ("wet bulb") is enclosed by a wick which is saturated with distilled water. When the measurement is made, the ambient fluid (e.g., air) ventilates over the wick and cools the sensing element below the ambient temperature by causing evaporation of water from the wick. The evaporation of moisture from the wick depends on and thus is a measure of the vapor pressure or moisture content of the ambient fluid.

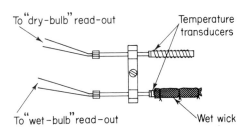

Fig. 6-5. Psychrometric sensing element.

The elements are usually resistive, but are sometimes thermoelectric, and are primarily platinum- or nickel-wire windings or thermistors. Two different types of sensing elements may make up the required dual element of a psychrometric humidity transducer. The wick is made from cotton or other textile materials. It can also be a porous ceramic sleeve, fitted over the resistive element.

6.2.3 Dew-point-sensing elements

The *dew point* is a discrete temperature. Specifically, it is that temperature at which liquid water and water vapor, or ice and water vapor, are in equilibrium. At this temperature only one value of saturation vapor pressure (of water vapor) exists. Hence, the absolute humidity can be determined from this temperature and knowledge of the pressure. The temperature at which the vapor and solid phases of water (vapor and ice) are in equilibrium is often referred to as the *frost point*.

To determine the dew point at any given air (or other ambient gas) temperature, the temperature of a surface is artificially lowered until dew (or frost) first condenses on it. As soon as this point is reached, the temperature of the surface is measured. A dew-point-sensing element must therefore perform the function of temperature sensing as well as sensing the change from vapor to liquid (or solid) phase. It is essential to the accuracy of humidity data produced that the temperature of the surface be measured at the exact instant where condensation first occurs since the characteristics (e.g., appearance) of the condensate will not change appreciably if the surface continues to be cooled below the dew point.

Dew-point-sensing elements sense relative humidity only indirectly, as do psychrometric sensing elements. In both cases, tables must be consulted from which relative humidity values can be read off or calculated on the basis of the measurement obtained and knowledge of the ambient temperature (or temperature of the gas sample). Relative humidity is determined from dew-point data by use of *saturation vapor-pressure tables*. More precisely, it is determined from a knowledge of dew-point and gas-sample temperatures by using a table showing saturation vapor pressures over water and from a knowledge of frost-point and gas-sample temperatures by using a table showing saturation vapor pressure over ice. When frost crystals do not form at temperatures below the ice point (°C) because of the absence of a crystal nucleus on the surface, and dew condenses instead of frost, a third table can be used which gives saturation vapor pressure over supercooled water. The Smithsonian Meteorological Tables and the International Critical Tables contain these saturation vapor-pressure tables.

A simplified saturation vapor-pressure table is represented by Table 6-2. Using this type of table, percent relative humidity (% RH) can be determined as in the following example:

Example: At an air temperature of 95°F the dew point was measured as 45°F. From Table 6-2, the saturation vapor pressure is 1.660 in. Hg at 95°F and 0.300 in. Hg at 45°F. Thus

$$\%RH = \frac{.300}{1.660} \times 100 = .181 \times 100 = 18.1\%$$

Table 6-2 SATURATION VAPOR PRESSURE (e_w) OVER WATER; SIMPLIFIED TABLE FOR RELATIVE HUMIDITY DETERMINATIONS BASED ON DEW-POINT DATA.

Temperature (°F)	e_w (in. Hg)	Temperature (°F)	e_w (in. Hg)	Temperature (°F)	e_w (in. Hg)
35	0.203	60	0.522	85	1.214
40	0.248	65	0.622	90	1.422
45	0.300	70	0.739	95	1.660
50	0.362	75	0.875	100	1.933
55	0.436	80	1.032	105	2.244

6.2.3.1 Instant-of-condensation-sensing function. This function is usually provided by a thin disc or plate with a smooth surface closely coupled thermally to a cooling element and a condensation detector. Thermoelectric (Peltier-effect) coolers have

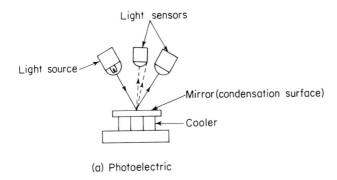

(a) Photoelectric

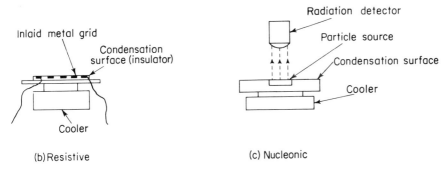

(b) Resistive   (c) Nucleonic

Fig. 6-6. Dew-point condensation detection elements.

proved very satisfactory. Condensation .detectors include: a photoelectric type
[Figure 6-6(a)] which uses a mirror as condensation surface, a light source to illumi-
nate the surface with a light beam, and one or more light sensors which respond
to light reflection from the surface; a resistive type [Figure 6-6(b)] with a metal grid
inlaid in the smooth condensation surface, in which a change in surface resistance
(conductivity) occurs when condensation forms; and a nucleonic (ionizing) type
[Figure 6-6(c)] in which an alpha- or beta-particle radiation source is located flush
with the condensation surface and a radiation detector senses the drop in particle
flux when condensation forms over the radiation source. Specific versions of these
condensation detectors are further described in 6.3.6.

6.2.3.2 Temperature-sensing function. The temperature at which condensation
first occurs upon cooling of the surface is sensed by resistive or thermoelectric sensing
elements. These are described in more detail in Chapter 14. Platinum-wire resistive
temperature sensors are sometimes preferred over thermistors or thermocouples,
but all three types have been used successfully. It is essential that the temperature
sensor be mounted as close to the surface as possible (see Figure 6-7). Wherever

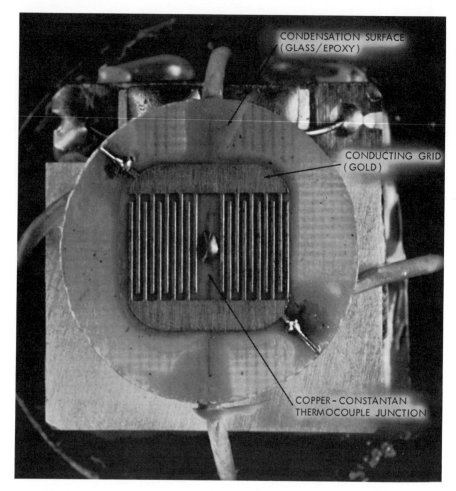

Fig. 6-7. Thermoelectric temperature sensing element in resistive condensation detector of dew-point transducer (courtesy of Vāp-Air Div., Vapor Corp.).

feasible, it should be flush with the surface and kept very small in size since its intended use is to measure the temperature of the condensate itself.

6.2.3.3 Energy-balance sensing elements. An entirely different sensing principle is used in certain dew-point elements which use, in part, the hygrometric operation of a saturated lithium-chloride solution. This solution is used to impregnate a wick around a cylindrical mandrel. A bifilar winding is wound around the mandrel to serve as electrodes. Current flow between the electrodes is a function of moisture sorption. A power supply is connected across the two electrodes so that the wire-lithium-chloride element acts as self-regulating heater. The element heats itself until the vapor pressure of the lithium-chloride solution is in equilibrium with the vapor pressure of the ambient atmosphere. A resistive or thermoelectric temperature-sensing element is contained within the hollow mandrel. The output of this element is a function of dew point since the vapor pressures for saturated lithium chloride are well-established.

6.3 Design and Operation

6.3.1 Resistive humidity transducers

The development of small humidity transducers with a resistive sensing/transduction element received its impetus in the late 1930's when more accurate and reliable humidity transducers were needed for *radiosondes*, the small telemetry packages carried aloft by weather balloons and used to obtain meteorological data. The telemetry circuits were such that a resistance change in the transducer from less than 6 megohms at 15% RH to less than 5 kilohms at 100% RH was desirable. Another target requirement was that the time constant of the transducer be less than 25 s at an air flow of 200 m/min over the range 40% to 90% RH.

Successful resistive humidity transducers were developed starting in 1938 using an aqueous solution of a hygroscopic salt in a plastic binder coated onto an insulating substrate and provided with metallic electrodes. Development of the carbon-powder element for radiosonde use was initiated in 1942.

6.3.1.1 Hygroscopic-salt humidity transducers. The two most popular forms of this transducer, the cylinder type and the wafer, type, are both derived from earlier developments. F. W. Dunmore developed a resistive hygrometer using a thin-walled glass tube coated with a hygroscopic-salt film in 1938 at the National Bureau of Standards. The film was a 2% to 5% aqueous solution of lithium chloride (LiCl). In 1939 the design was refined. A thin-walled polystyrene-coated aluminum tube, wound with a bifilar winding of platinum or palladium as electrodes, was coated with a lithium-chloride solution in a polyvinyl-alcohol (partially hydrolized polyvinyl acetate) binder. The precious-metal electrodes were substituted for the previously used copper wire which tended to corrode rapidly. Named after its originator, the lithium-chloride element is sometimes referred to as the "Dunmore element" or the "Dunmore hygrometer."

The Dunmore element was modified into the wafer-type or flat-strip element in 1942 by J. Onderdonk at the Friez Instrument Division of the Bendix Corporation, a major contractor to the U. S. Government for radiosonde equipment. The insulating substrate was a polystyrene strip onto whose long edges a strip of tin was hot-sprayed while the center of the strip was masked. The two tin strips formed the required

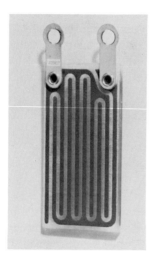

(a) Standard version   (b) Miniature version

Fig. 6-8. Wafer-type resistive humidity transducer (courtesy of Phys-Chemical Research Corp.).

electrodes for the hygroscopic film, a solution of 4% lithium chloride and 3% polyvinyl alcohol. The film was applied by dipping the tinned-edge strip into a tank containing the coating liquid.

In humidity transducers subsequently developed by several companies, the constituent materials were further modified, special production processes were established to improve transducer performance and reliability, and electrodes were shaped so as to be more effective than tinned wafer edges. It was found, for example, that making the film thinner or the film surface wider resulted in less *polarization*, the drift of an element's resistance-vs-humidity characteristics caused by repeated on-off cycling with d-c excitation. Design compromises were then needed to improve the time constant of the transducers since increased film area also resulted in slower response. Storage of the finished element together with silica-gel desiccant in hermetically sealed containers was found to greatly increase the storage life of the transducers.

A wafer-type humidity transducer in its standard and miniaturized versions is illustrated in Figure 6-8. The wafer is made from a chemically treated styrene copolymer. Electrodes are printed on both faces of the resistive surface in the pattern shown and are brought to solder terminals. The surface resistance varies with adsorption and desorption of moisture resulting from changes in relative humidity, from 4 megohms at 10% RH to 1 kilohm at 100% RH in the standard version. In the miniaturized version the resistance change is from 25 megohms at 10% RH to 50 kilohms at 100% RH. The time constant is shorter for increasing relative humidity than for decreasing relative humidity, because adsorption is more rapid than desorption. For this tranducer it is about 30 s in still air and less in moving air.

Other designs use electrodes of various patterns printed or deposited on a plastic wafer and coated with the humidity-sensitive resistive film.

An example of a cylindrical resistive humidity transducer is shown in Figure 6-9. It consists of a polystyrene or Kel-F® coil form wound with a bifilar winding of palladium wire. The humidity-sensitive resistive coating is applied over the outside of the coil form and over the wires which serve as electrodes and which terminate in plug-type pins. The element is enclosed in a metal jacket provided with holes to afford good ventilation over the sensing surface. The entire sensing/transduction subassembly can be installed in various enclosures or can be assembled into a transducer with integral electrical receptacle and perforated protecting sheath such as the flange-mounted version shown.

In order to prevent calibration shifts due to polarization, hygroscopic-salt humidity transducers are usually excited with a-c voltages having a frequency of 20 Hz or higher.

6.3.1.2 Carbon-element humidity transducers. The development of resistive humidity transducers using a carbon-film sensing/transduction element has been fostered primarily by the Electronics Research and Development Laboratories of

Fig. 6-9. Cylinder-type resistive humidity transducer, unassembled and assembled in flange-mount case (courtesy of Hygrodynamics, Inc.).

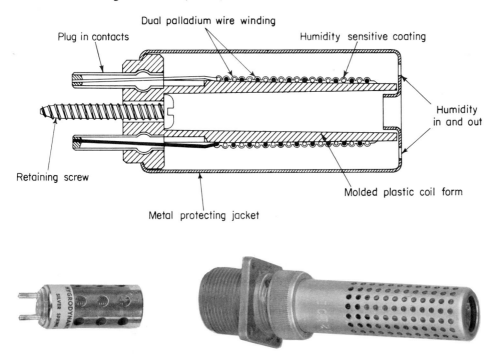

the U. S. Army for radiosonde applications. This transducer has been produced in both the wafer and the cylinder types. A typical wafer sensor is 2.5 in. long, 0.68 in. wide and 0.040 in. thick, made from acrylic plastic with metallized long edges to serve as electrodes, and is coated with a carbon-powder suspension in a gelatinous cellulose carrier. Cylindrical sensors have plastic rod diameters from 0.12 to 0.75 in. and lengths from 0.25 to 4 in. with rod-end or full-length (spiral) metallized electrodes, also coated with the carbon powder in gelatinous cellulose film.

The resistance of the carbon-film humidity transducer at +25°C changes from about 15 kilohms at 10% RH to about 3 megohms at 100% RH. The transducer appears to be free from polarization effects over typical use periods and can therefore be excited with a-c or d-c; however, a-c excitation is normally used for extended-time use and extensive testing. It has a time constant of less than 0.5 s at 25°C and at an air flow of 200 m per min.

6.3.1.3 Other single-element resistive humidity transducers.

Certain types of humidity transducers with single resistive elements have been developed but were found to require further work to make them commercially practicable. Among these are: the "Humistor" (Registered Trademark, Philco Corporation), a polyelectrolyte resistive element in which the conductivity of ion-exchange resins varies with moisture contained in the resin; the cerium titanate and other humidity-sensitive-ceramic elements; and elements consisting of a glass substrate onto which a thin film of primarily lead iodide is vacuum-deposited. In several instances further development was discontinued because of intrinsic difficulties and because of the increasing availability of more promising transducer types.

6.3.2 Resistive-capacitive humidity transducers

This type of transducer uses an aluminum-oxide element whose impedance, equivalent to that of a capacitance-resistance network, changes with humidity. The element was originally developed by F. Ansbacher, A. C. Jason, and others in England and Scotland between 1952 and 1960, although a related device was patented in the U.S.A. by L. R. Koller in 1941. In the early 1960's the element was further developed in the U.S.A. for radiosonde purposes by C. M. Stover. A multipurpose commercial version of this transducer is illustrated in Figure 6-10. The small rod-shaped element is mounted by its sheathed connecting leads which in turn are supported over most of their length by a Teflon® cylinder. The latter is further stiffened at the probe mount (flanged or threaded bulkhead mount), which is equipped with a miniature bayonet-type electrical connector for external connections to the element. The transducer probe assembly is surrounded by a thin stiff mechanical-protection frame.

The resistive-capacitive transducer is reportedly unaffected by ambient pressure and by a great number of gases other than air. It is operable over a wide temperature range. Its time constant is less than five seconds in near-still air.

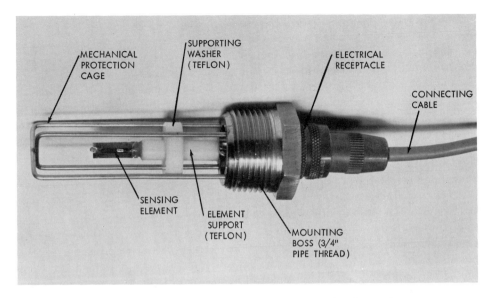

Fig. 6-10. Humidity transducer with aluminum-oxide sensing element (courtesy of Panametrics, Inc.).

The commercial model is normally furnished with associated readout equipment calibrated in terms of dew point and frost point and correlatable to a calibration in terms of moisture. The portable readout box also contains the excitation and signal-conditioning equipment for the transducer.

6.3.3 Oscillating-crystal humidity transducers

This transducer utilizes essentially a common radio-frequency oscillator crystal coated with a hygroscopic material. Changes in water sorption on the coating cause corresponding changes in the frequency at which the crystal oscillates. The crystal is normally mounted together with its oscillator circuit. The oscillator output frequency can be read out directly on a frequency meter or counter. Alternately, a sealed reference crystal can be mounted with the transducer and so connected that the output of the assembly is a beat frequency (difference in the frequency of oscillation of the two crystals) in the audio-frequency range. The reference crystal also provides temperature compensation.

The transducer can be constructed as a self-contained humidity telemeter if provided with a radiator (antenna) and if the oscillator output frequency and its variation are chosen so as to fall within an accepted telemetry band. The crystal case can be made part of a flow-through plumbing arrangement for humidity and moisture measurement in process gas samples. The crystal type humidity transducer was developed primarily by the Esso Research and Engineering Company in the U.S.A.

6.3.4 Electromechanical humidity transducers

These transducers depend on the displacement obtained from the expansion and contraction of a natural material to operate a transduction element. Materials used in such hygrometers are essentially limited to two, human hair and "goldbeater's skin" (an animal membrane), although nylon thread and other plastic materials have been used at times. Transduction elements used include reluctive and potentiometric elements, the latter sometimes connected as a (two-terminal) variable resistor. The hair or membrane is maintained under a small amount of tension. The displacing link of the transduction element, often requiring mechanical amplification, is attached to the sensing material or to the point on it to which the tensioning spring is attached.

Hair hygrometers were first developed by H. B. deSaussure in the late 18th century. Modern versions use either a single hair, two or three hairs in a balanced mounting configuration, or a bundle of hairs. Tension on the hair is kept slightly below one gram. The surface of the hair must be kept free from atmospheric contaminants. Prior to its installation in the transducer all oil and dirt are removed from the hair by suitable solvents. Considerable improvements in performance—greater sensitivity and shorter response time—can be obtained by rolling the hair into an elliptical (1:4 ratio of axes) cross section, a process originated by E. Frankenberger (German Weather Bureau) around 1944. Tension on rolled hair should be kept

Fig. 6-11. Animal-membrane humidity transducer (courtesy of Bacharach Industrial Instrument Co.).

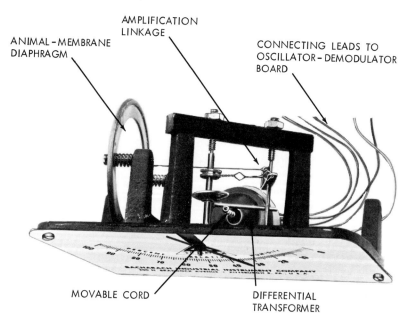

ANIMAL-MEMBRANE
DIAPHRAGM

AMPLIFICATION
LINKAGE

CONNECTING LEADS TO
OSCILLATOR-DEMODULATOR
BOARD

MOVABLE CORD

DIFFERENTIAL
TRANSFORMER

below 0.3 grams since rolled hair is not as strong as natural hair. Treatment of natural hair with certain chemicals can result in improvements similar to those gained by rolling.

A typical animal-membrane hygrometer is illustrated in Figure 6-11. The animal-membrane ("goldbeater's skin") diaphragm is clamped in an edge-supporting ring whose lateral position can be adjusted by two spring-loaded bolts. A thin rod is attached between the diaphragm center and an amplification linkage. The mechanical system is kept under slight tension, pulling the diaphragm into a shallow cone. The linkage actuates the movable core of the differential-transformer reluctive transduction element. A circuit board installed in the same case with the transducer contains a temperature-stabilized 4 kHz oscillator as excitation supply and a demodulator for a-c to d-c output conversion. The linkage simultaneously drives a pointer over an indicator scale for an additional indication at the point of measurement.

6.3.5 Psychrometric humidity transducers

The cooling produced by the evaporation of water, a phenomenon well understood in our oldest civilizations, has been applied to the measurement of humidity since the early 19th century. In the late 19th century two types of psychrometers came into use: the sling psychrometer, which is ventilated by swinging it while holding it by an attached hand sling, and the whirled psychrometer, ventilated by spinning (whirling) around a shaft in its nonsensing end. The psychrometer itself consists of two mercury-in-glass thermometers clamped to a board. The bulb of one thermometer is covered with a wet cotton or linen wick. This "wet-bulb" thermometer often protrudes slightly further into the air stream than the "dry-bulb" thermometer so that the evaporating water cannot contact the "dry bulb." A minimum ventilation rate of 300 m/min is used to get a meaningful reading of the wet-bulb temperature.

Sling psychrometers and liquid-filled humidity indicating and recording systems are still in common use, as are the terms "wet-bulb" and "dry-bulb" even when applied to electronic psychrometric transducers which do not use liquid-filled sensing elements. Ventilation of these transducers is achieved by forced-air systems, such as blowers or fans which are often packaged integrally with the transducers. Psychrometers so ventilated have been referred to as "aspiration psychrometers."

Although thermoelectric psychrometric humidity transducers have been developed and found useful where ventilation is very low or where relative humidities to be measured are very low, most psychrometric humidity transducers are resistive. One hybrid design uses both principles, a resistive element to measure dry-bulb temperature and a differential thermopile for measuring wet-bulb depression. One set of thermopile junctions is covered by a wick, and the other set is placed close to the resistive element.

Resistive transducer types use platinum-wire or nickel-wire windings, usually on cylindrical mandrels. Because of the narrow temperature ranges often encountered,

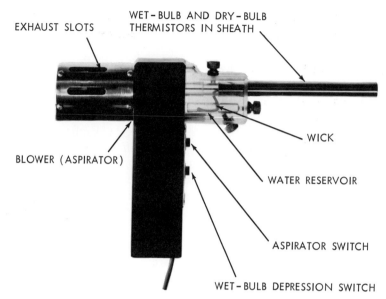

EXHAUST SLOTS

WET-BULB AND DRY-BULB
THERMISTORS IN SHEATH

BLOWER (ASPIRATOR)

WICK

WATER RESERVOIR

ASPIRATOR SWITCH

WET-BULB DEPRESSION SWITCH

Fig. 6-12. Psychrometric humidity transducer with thermistor elements (courtesy of Atkins Technical, Inc.).

thermistors have been found very useful since they are capable of providing a large resistance change over a relatively narrow range of temperatures. Thermistor sensing elements are used in the psychrometric humidity transducer shown in Figure 6-12. This instrument, intended for hand-held use, operates in conjunction with a battery-operated portable readout unit which incorporates a resistance bridge whose relative unbalance, as caused by thermistor resistance changes, is read off a meter. The meter indicates dry-bulb temperature until the wet-bulb depression switch is pushed. It then indicates wet-bulb temperature. A small electric aspirator fan (blower) is mounted in the transducer. When activated (by pushing the aspirator switch), it draws air over the two thermistors mounted within the probe sheath. The wet-bulb thermistor is covered by a thin cotton wick which dips in the small water reservoir also integral with the transducer.

A different design uses equalization of the two temperatures, attained by means of a resistance-type heating element wound between the wet-bulb element and its wick, to provide an indication of wet-bulb temperature in terms of the current required to heat the wet-bulb element to the same temperature as the dry-bulb element.

6.3.6 Dew-point humidity transducers

The design and operation of modern dew-point transducers are based on concepts developed during the middle and late 19th century involving polished silver

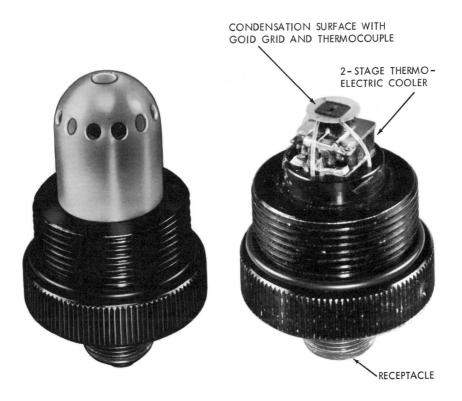

CONDENSATION SURFACE WITH
GOID GRID AND THERMOCOUPLE

2-STAGE THERMO-
ELECTRIC COOLER

RECEPTACLE

(a) Assembled transducer   (b) Transducer with cover removed

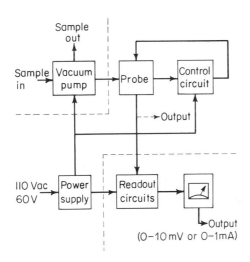

(c) Block diagram of transducer
("probe") with associated circuitry

Fig. 6-13. Thermoelectric dew-point transducer with resistive condensation
detection (courtesy of Vāp-Air Div., Vapor Corp.).

condensation surfaces, visual observation of the formation of condensate, and temperature determinations with mercury-in-glass thermometers. The electrical methods currently used originated essentially in the 1930's.

The sensing elements described in 6.2.3 have been used in a number of successful transducer designs. In some of these, closed-loop control between condensation sensor and cooler is used to stabilize the condensation surface at the dew point and to continuously track the dew point from then on. The output of the temperature sensor is then always indicative of the dew point. Most designs also incorporate a heater for the condensation surface so that the condensate is removed and the surface can subsequently be re-cooled to produce new dew-point indications.

Closed-circuit cooler control is used in the transducer shown in Figure 6-13. The change in surface conductivity (resistance of the condensation surface) is used to regulate a two-stage thermoelectric cooler. A unique change in the slope of the graph of surface resistance vs temperature-above-dew-point occurs at the dew-point temperature. Cooler regulation is so controlled that over successive cooling cycles the condensation-surface temperature is sensed by the thermocouple whose junction is embedded at the condensation surface. The output from the thermocouple can be read out directly or conditioned by readout circuits to deflect a pointer on a meter as well as provide an input for other ancillary indicating, recording, or control equipment. A vacuum pump, with associated plumbing, can be added to draw a gas sample over the sensor surface so that the dew point of this gas can be measured.

The transducer illustrated in Figure 6-14 also uses closed-circuit cooler control. The internal sensing subassembly [Figure 6-14(b)] contains the lamp which illuminates the rhodium-plated-mirror condensation surface and, behind a cover shield, the two photosensors whose connecting leads and terminals are visible in the illustration. A thermoelectric (Peltier-effect) cooler is mounted under the mirror surface, closely coupled thermally to the surface. The mirror is provided with the resistive (thermistor or platinum-wire) temperature sensing element as well as a resistive (nichrome) heating element. The two photosensors have one common terminal since they are used in a comparison-bridge network. Circuitry for amplification of the signal from this bridge and for cooler control is usually packaged in a separate control unit. With the photosensors in closed-loop proportional control of cooler temperature, the mirror temperature is stabilized at the dew point, and variations in dew point are continuously tracked and indicated by the output of the temperature sensor.

Among other promising dew-point transducer developments is a design which incorporates a nucleonic condensation detector and a platinum-wire resistive temperature sensing element. A radioactive foil intended to emit alpha particles (e.g., polonium foil) or beta particles is the condensation surface or is integral and flush with a condensation surface. When condensate forms it will partially absorb the particles emitted. This condition is sensed by an internal semiconductor-type radiation detector which is in a closed-loop control circuit for the two-stage Peltier-effect (thermoelectric) cooler to which the condensation surface is mounted. The result of

(a) Assembled transducer   (b) Internal sensing subassembly

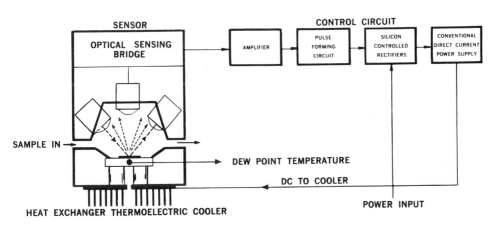

(c) System block diagram

Fig. 6-14. Resistive dew-point transducer with photoelectric condensation detector (courtesy of Cambridge Systems, Inc.).

such cooler control is maintenance of the surface at the dew point, as in the two types of dew-point transducers described above. An internal heater is also provided. The radioactive source is so chosen and mounted as not to present a radiation hazard. The resistive sensing element can be connected into a bridge-circuit so that a d-c output voltage indicative of dew point is produced.

A dew-point transducer using an energy-balance (hygrometric/dew-point) sensing element is illustrated in Figure 6-15. The heater winding consists of two gold wires bifilarly wound over lithium-chloride-impregnated glass-cloth tape. As in a hygrometer, moisture sorption on the lithium chloride varies the resistance between the wire electrodes. However, the resistance change between the electrodes is not used as output indicative of relative humidity. Instead, it is used to regulate the current between the electrodes which in turn are connected to a power supply. It is this

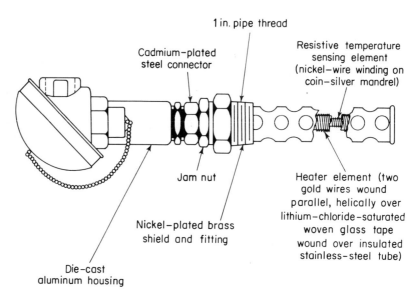

Fig. 6-15. Dew-point transducer with energy-balance sensing element (courtesy of the Foxboro Co.).

current that causes self-heating in the element. The heating progresses until the lithium chloride is in vapor-pressure equilibrium with the measured fluid (e.g., ambient air, gas sample, etc.). Since saturation vapor pressures for the lithium chloride are known, a calibration chart can be prepared for indications of absolute humidity.

6.4 Performance and Applications

6.4.1 Specification characteristics

The nature of the characteristics to be specified for humidity transducers will depend, to a limited extent, on the type of transducer required (hygrometer, psychrometer, or dew-point sensor). It will usually be necessary to state whether the transducer output should vary with relative humidity, with dry-bulb and wet-bulb temperatures, or with ambient and dew-point temperatures.

6.4.1.1 Mechanical design characteristics. These include dimensions, electrical and plumbing connections, mounting provisions, weight, constraints on construction and materials, and identification. Much of this information is best specified on a drawing which shows the desired configuration with all critical dimensions, the type of electrical connections and of any required plumbing connections, the type of mounting required and its dimensions, and the provisions for appropriate ventilation over the sensing element(s). Other pertinent information, including minimum name-plate content, and weight tolerances can be shown in the form of notes on the drawing. Additional configurational information, in lieu of mounting specifications, is given for hand-held transducers. If associated circuitry is separately packaged, a drawing of the circuitry unit (amplifier, control unit, etc.) is additionally required.

For measured fluids other than air, the type of fluids and their chemical com-position should be stated. When measured air is expected to contain substantial amounts of contaminants, such as salt or combustion products, this information must be included in the specification. Allowances may have to be made for limitations on the choice of transducer design and materials on the basis of either the type of measured fluids or the expected ambient atmosphere or both.

For "immersion probes," transducers intended to measure humidity, dew point, or moisture in pressurized fluid systems, the proof-pressure and burst-pressure ratings must be specified, preferably with a maximum allowable leakage rate for one or both of these fluid-pressure levels.

6.4.1.2 Electrical design characteristics.

Specifications for excitation voltage and power consumption and for output characteristics of humidity and moisture transducers are primarily influenced by the type of associated circuitry required for their operation and by the source (supplier or user) of this circuitry. Most manu-facturers of humidity transducers also offer a variety of power-supply, signal-con-ditioning, and display equipment to be used with their transducers. A few transducer types, in fact, are not normally available without such equipment. When a power supply is furnished, the a-c line voltage, frequency, and power consumption are specified.

Electrical design characteristics specified for the transducer itself include type (d-c or a-c, or either) and range of excitation voltage or current, a-c excitation fre-quency, the nominal resistance(s) at one point in the range and at one ambient tem-perature for transducers using resistive elements, type of thermocouple materials (or standard letter designation for them) for thermoelectric elements, operating principle and essential characteristics of condensation-temperature-sensing elements of dew-point transducers, and equivalent characteristics of other transducer types. Sufficient information must be stated to enable the transducer user to select appro-priate excitation and signal-conditioning circuit constants for use of the transducer in his measuring system.

Additional requirements are shown, when applicable, for insulation resistance and breakdown-voltage rating, allowable output noise or ripple, warm-up time, short-circuit and polarity-reversal protection, and electromagnetic interference limits including susceptibility to conducted and radiated interference.

6.4.1.3 Static performance characteristics.

Range. The range of hygrometric transducers is shown in percent relative humidity (% RH) and that of psychrometric and dew-point transducers in terms of temper-

ature (°F or °C). For psychrometric transducers the dry-bulb and wet-bulb temperature ranges should be shown or the wet-bulb-depression range should be shown
for a specified operating temperature range (see below). The dew-point-temperature
range is shown for dew-point transducers. Range can also be shown in units of
moisture, e.g., in parts per million of volume per unit volume of air, in grains per
pound of dry air, etc.

Operating temperature range (in °F or °C) is a performance characteristic rather
than an environmental one for humidity transducers, because their performance is
almost invariably related to measured-fluid (e.g., ambient) temperature.

Output can be specified as end points with tolerances, e.g., (Range: 20% to 90% RH)
"Output: 600 ± 100 to 2 ± 1 kilohms." It can also be shown as a ratio of the outputs at the range limits (and intermediate levels) to the output at a specific level
within the range, and at a specific temperature. However, this reference output
(with tolerances) must also be stated, e.g., (Range: 10% to 95% RH) "Output:
70 ± 5 to 5940 ± 200 percent of resistance at 33% RH, 25°C; Resistance at 33% RH,
25°C: 20 ± 5 kilohms." It should be noted that the tolerances on the reference resistance increase the actual resistance tolerances at the end point above those shown.

Similarly, the end points of capacitive transducers can be shown in microfarads,
those of oscillating transducers in megahertz, of capacitive-resistive transducers
in kilohms or megohms of impedance at a given carrier frequency, and of thermoelectric transducers in millivolts.

It should be noted that a relatively small change in dew point represents a
considerable change in water content. For example, a dew-point range of $+20$ to
$-110°C$ represents a change in moisture content from 20,000 to 0.001 μgm H_2O/l.

Because of the usually nonlinear calibration curves, however, output is best
specified in the form of a family of curves or group of tables, each showing several
output values for corresponding values of relative humidity, dew point, or moisture
over the specified range, at a specific temperature (of the measured gas sample or
the ambient atmosphere.)

Output tolerances (over a specified range) also establish the *calibration interchangeability* of all transducers of the same design (same part number) defined by
the same specification. For psychrometric transducers resistance-vs-temperature (or
voltage-vs-temperature) characteristics are tabulated for the dry-bulb element as
well as for the wet-bulb element unless the two elements are of the same type and
are so closely matched to each other that a single tabulation can serve for both elements. When a transducer output is to be fed solely into one type of signal-
conditioning, indicating, or recording equipment and when this transducer is
always required to be calibrated and tested with a standardized version ("bogey")
of such equipment, the output can also be stated in terms of signal-conditioner
output, indicator scale divisions, or recorder scale divisions, respectively.

Accuracy, as such, is meaningful only when used to describe the maximum error
in the indication of a measuring system, as referred to the "true" humidity value such

as can be certified by a government standards agency. It should not be stated for transducers. Furthermore, the output-vs-measurand characteristics of humidity transducers are usually not linear. Hence, accuracy characteristics are best stated as (1) repeatability, over consecutive calibration cycles, as referred to any one calibration curve or table furnished with the transducers, and (2) hysteresis. These characteristics should be expressed in percent of full-scale output but are usually expressed in units of measurand.

Resolution is shown only for the rare mechanical-element hygrometers with a potentiometric transduction element. Threshold (sometimes erroneously called "sensitivity") is frequently specified, however.

Error-band specifications can be applied to humidity transducers when close calibration interchangeability is achievable. The static error band includes calibration interchangeability, repeatability, and hysteresis. It is applicable at a specific temperature (e.g., 25°C). The temperature error band includes all ambient-temperature effects on these performance characteristics. However, care must be taken to assure that the tolerances shown are the largest tolerances applicable at any point in the operating temperature range or, when so specified, over specific portions of that range. A nominal (theoretical) output-vs-measurand curve, applicable at a stated temperature, serves as the error-band reference curve. This can be illustrated by the following example of typical vertical-column headings (no further specifications of output and accuracy characteristics are required): "Relative Humidity; Theoretical Output at $-20°C$ (ohms); Theoretical Output at 0°C (ohms); Theoretical Output at $+10°C$ (ohms); Theoretical Output at $+25°C$ (ohms); Theoretical Output at 40°C (ohms); Temperature Error Band, -20 to 0°C ($\pm$ ohms); Temperature Error Band, $+1$ to $+40°C$ ($\pm$ ohms)." Horizontal columns in this example would then list all the theoretical outputs and tolerances for the specified range (e.g., 10% to 90% RH) shown in increments of 10% RH.

6.4.1.4 Dynamic performance characteristics.

The time constant and, frequently, response time or rise time are specified at one or more ventilation rates (sometimes also for fluids other than air), at one or more fluid temperatures, and for one or more partial-range step changes in relative humidity, dew point, or moisture, and at a given ambient atmospheric (or gas sample) pressure. The percentage of the total output step must always be given for response time and rise time, e.g., "90% Response Time" or "5% to 95% Rise Time."

6.4.1.5 Environmental performance characteristics.

The operating temperature range and any limitations of effects of these temperatures on transducer performance are usually shown as performance characteristics. An exception to this exists in the case of "immersion probes" used in pressurized fluid systems, where an additional requirement is shown for the temperature environment ambient to the "head" of the probe (as differentiated from the "stem" which is sealed in the pressurized duct or vessel). Tolerances may be specified, e.g., for errors due to temperature differences between sensing elements and probe "head" (conduction error). The same exception can exist for an ambient-humidity probe whose "head" is maintained in a controlled environment. Allowable heat-sink temperatures should be considered for coolers in dew-point transducers.

The ambient-pressure range (or pressure range of another measured fluid) is frequently specified together with any tolerances for ambient-pressure error and sometimes for effects of rapid changes in ambient pressure. Ambient-pressure specifications (sometimes shown in terms of altitude) are of particular significance for meteorological transducers carried aloft on balloons, aircraft, or sounding rockets as part of a radiosonde.

Other environmental requirements can include shock, vibration, acceleration, high sound-pressure levels, and nuclear radiation.

6.4.1.6 Performance reliability characteristics. Stability tolerances (repeatability after a given storage period under specified environmental storage conditions) can be specified. Storage conditions are usually connected with a requirement for keeping the transducer in a sealed container, together with a desiccant such as silica gel and a color-change type of humidity indicator within the container. Operating life can be stated in terms of intermittent or continuous operation over the specified range or portions thereof, or in terms of a cycling life. Operational as well as post-storage stability can be improved by preshipment aging of the transducer.

6.4.2 Considerations for selection

Before selecting a humidity transducer for a given application with a given measuring system a determination should be made whether relative humidity or dew point is to be measured. Although relative humidity can be calculated from dew point (and ambient temperature), dew-point transducers are best suited for dew-point measurements, and hygrometric transducers are usually selected for relative humidity measurements. Psychrometric transducers are also used for relative-humidity measurements when indirect measurements and associated simple manipulations are acceptable, as is the case when the user is accustomed to thinking of humidity in terms of wet-bulb and dry-bulb readings. An advantage of the psychrometric over the hygrometric transducer is that ambient (dry-bulb) temperature can never be neglected.

In a considerable number of applications measurements of either relative humidity or dew point yield equally useful information. In others, one or the other is preferred. Relative humidity is measured in most meteorological work, in materials storage, drug processing, air conditioning, printing, fabric dying, food processing, and many other industrial and scientific applications. Dew point is usually measured in specific processes, such as concrete-block curing, drying of chemicals, lumber, and other materials, heat treating and certain other metal processing operations, and detection of incipient freeze-up. Both are measured in most test laboratories, in aerospace testing, in textile manufacturing, in gas-utilities monitoring, in paper manufacturing, and in a number of other research and industrial applications.

Selection is limited to only a few designs of relatively high cost when very small quantities of moisture must be measured with close accuracy.

6.4.2.1 Psychrometric humidity transducers are available in relatively few designs. They are capable of providing very close data accuracy when certain precautions are observed in their use and when the two sensing elements are extremely repeatable and their output is read out with a minimum of error.

Whenever possible the sensing elements should be closely matched to each other. The largest errors in humidity data occur when the error in dry-bulb measurement has a polarity opposite to that of the error in wet-bulb measurement. For example, at a true dry-bulb temperature of 50°F and a wet-bulb depression of 1.0°F, the error in relative humidity would be 12.7% if the dry-bulb reading is 1°F above its true value, and the wet-bulb reading is 1°F below its true value. When one reading is correct but the other is 1°F above or below its true value, the relative-humidity error in the above example would be reduced to about 6.5%. However, if both readings were exactly 1°F below their true values, the relative-humidity error in this example would only be 0.11%. When common signal-conditioning equipment and a single indicating device is used for both elements their close mutual matching is mandatory.

Other conditions which can cause errors in psychrometric transducers include insufficient ventilation rate, wick contamination, improper length or fit of the wick, radiant heating effects, and failure of the operator to allow for ambient pressure or to obtain the correct wet-bulb depression.

When closely matched, thermistor elements usually offer a shorter time constant but less accuracy than the platinum-wire elements. The latter can be used in reference psychrometers for production calibration of hygrometric transducers.

A variety of tables and slide rules are available for the conversion of psychrometric data to relative humidity at different atmospheric pressures. Most of them were developed by governmental meteorological agencies such as the U.S. Weather Bureau.

6.4.2.2 Hygrometric humidity transducers. Transducers using mechanical sensing elements are not used as much as in the past, before the development of reliable transducers with electrical sensing/transduction elements. Hair elements are rarely found in transducers, although they are still used in low-cost, e.g., residential humidity dial indicators. Their time constants at room temperature are about 200 s with adequate ventilation but longer at temperatures below 60°F. They exhibit considerable hysteresis, especially when a near-full-range humidity increase is followed by a humidity decrease. Considerable improvements in performance can be obtained from rolled-hair elements including reduction of time constants to less than 30 s and less hysteresis. Animal-membrane elements, when properly selected and treated, offer reasonably good performance with combined hysteresis and repeatability within $\pm 3\%$ RH between 15% and 90% RH at temperatures between 32°F and 130°F and a time constant approximately the same as that of rolled hair. The animal-membrane is stronger than hair and can stand higher tensional loading.

Resistive hygrometric transducers are small and rugged. Their ranges extend up to 95% RH. Humidity above this level can cause excess water absorption in the film and dissociation of materials of lithium-chloride elements, but normal operation can usually be restored by air drying. Combined repeatability and hysteresis is within $\pm 1.5\%$ to $\pm 3\%$ RH at temperatures between $+40°F$ and $+120°F$ if the element is kept free from contamination. Some chemical contaminants, e.g., acids and sulfides, cause permanent output shifts. Others, such as alcohol and ammonia, cause a temporary shift while they are present. Dusty or dirty atmospheres will also contaminate the element. Many elements, however, can be cleansed to remove dirt and almost completely restore their original characteristics. Time constants of the best aqueous-electrolytic film-element transducers are around 3 s (with adequate ventilation) but can be as long as 30 s for some element designs; those of carbon-element transducers are less than 1 s, all at room temperature. Time constants increase at lower temperatures but again to a lesser extent for carbon-element transducers. Combined hysteresis and repeatability for the latter are within $\pm 3\%$ RH between 5% and 95% RH at temperatures near room temperature. Excitation and signal conditioning are very simple for resistive humidity transducers. They are typically connected as one arm of a Wheatstone bridge excited by a-c through a step-down transformer. The bridge output is demodulated by a full-wave rectifier bridge. Their operating characteristics at temperatures above $+200°F$ and below $-50°F$ and at various ambient pressures (including those at high altitudes) have been sufficiently established.

Aluminum-oxide humidity transducers offer performance characteristics as good or better than most resistive transducer types. They are rarely subject to contamination and have better stability over a longer storage life as well as a greater cycling life, especially when aged for two to three months before calibration and shipment. Variations of temperature over considerable portions of the operating temperature range as well as ambient-pressure changes have relatively little effect on the transducer output compared to the output at room temperature and at sea level. A disadvantage compared to resistive transducers is the complex impedance change of the output instead of a simple resistance change. This output characteristic requires excitation at audio frequencies and an impedance-bridge circuit for signal conditioning.

Piezoelectric humidity transducers offer very low threshold because of their high-frequency oscillation. Only a small change in humidity or moisture is needed to produce an easily measurable change in the frequency of their output. They are also capable of responding to very low moisture levels. Their accuracy characteristics can be expected to improve with further developmental work. Because of their relatively complex excitation, signal-conditioning, and shielding requirements, they are normally used only in conjunction with manufacturer-supplied power-supply and display equipment.

6.4.2.3 Dew-point transducers are used primarily for continuous dew-point measurements in industrial gaseous-material systems to which they are connected by plumbing. System pressures up to about 80 psig are acceptable to most transducer designs, and some designs will accept up to 200 psig. When used for meteorological, research, and laboratory-test purposes their sensing elements are directly exposed to the measured fluid. When intended for relative-humidity determinations they can be furnished with an additional fluid-temperature ("dry-bulb" temperature) sensing element with separate output.

Most transducers are equipped with a heater as well as a cooler, (sometimes with a single Peltier-effect heater/cooler) so that dew-point readings can be checked and so that increasing as well as decreasing dew-point temperatures can be followed.

Dynamic response characteristics are given primarily by the maximum cooling rate of the condensation surface and are usually expressed as the maximum rate of change of dew point without exceeding specified error limits. This rate of change ("tracking rate") is typically 4°F per second.

Hysteresis plays no important role in dew-point transducers since the dew point is always approached from the same direction. When dew points increase, the condensation surface is always heated above the new dew point so that the condensate evaporates completely and then cooled down until condensation occurs again. Repeatability of dew-point transducers is usually within ±0.5°F to ±1.0°F (as referred to their supplier-furnished calibration curve) between 32°F and temperatures above room temperature and within ±1.0°F to ±2.0°F between −20°F and 32°F (frost-point range). The best repeatability is obtained from high-purity platinum-wire resisting elements. Thermistor and thermocouple elements yield slightly poorer repeatability. Transducers are available with dew-point ranges as large as −70°F to +170°F.

Inherent errors of platinum-wire-element dew-point transducers above 32°F include a condensation error of about −0.2°F due to the thermal gradient existing across the dew or frost deposit, temperature-sensing error of about ±0.1°F due to thermal gradients across the thickness of the condensation surface, a self-heating error of about +0.1°F and a hysteresis of ±0.1°F, both within the platinum-wire element. It is therefore difficult to obtain repeatabilities better than +0.3°F, −0.4°F from even the best available dew-point transducers. Additional errors can, of course, be contributed by signal-conditioning and readout equipment.

6.5 Calibration and Tests

6.5.1 Calibration

Because the state of the art of commercially produced humidity transducers with electrical output is relatively new and requirements for close transducer accuracies are of fairly recent origin, calibration equipment and calibration methods have not yet reached the degree of standardization and simplicity existing for a majority of other measurands, e.g., pressure, temperature, displacement, or force. Calibration equipment is generally quite complex and not widely available. Calibration methods also require high operator skill.

There are two basic approaches to humidity-transducer calibration. The first relies upon the creation of a gas whose humidity remains constant for a sufficiently long period of time to allow the determination of the absolute humidity of the gas, using an instrument of known close accuracy, as well as of the transducer output at that humidity level. Reference humidity-measuring instruments include the gravimetric "train," the pneumatic bridge, and special laboratory types of dew-point measuring apparatus.

The second approach involves the creation of a gas whose humidity is known and held constant by controlling the basic gas parameters which define the humidity. Atmospheres of known humidity are produced by such apparatus as the two-pressure generator, two-temperature-bath equipment, and aqueous salt solutions.

A number of individual calibration cycles are usually performed on each transducer, each cycle at a different temperature. Selected methods and equipment are briefly described below.

6.5.1.1 Salt solutions.

Saturated as well as unsaturated aqueous solutions of certain salts, maintained at known constant temperatures, and contained in sealed enclosures, are commonly used to maintain known levels of relative humidity. Solutions of acids and some other substances have also been used for this purpose. Saturated solutions containing an excess of the undissolved solid crystals can provide a test atmosphere of sufficient volume for the calibration of many humidity transducers having a known equilibrium relative humidity. Such a solution has a definite water vapor pressure at a given solution temperature.

A typical salt solution is prepared by adding distilled water gradually to the crystals of the salt chosen, while constantly stirring, until about half the crystals are dissolved. The salt must be chemically pure. The solution should be kept in a chamber so designed that the surface of the solution is kept large but the vapor space is minimized. The use of hygroscopic chamber materials should be avoided. The air should be adequately circulated over the surface. Solutions should be permitted to cool after preparation and before use until the temperature rise of the water due to heat of chemical reaction has subsided. Solution and vapor should be at the same temperature.

Table 6-3 Equilibrium Relative Humidities for
Saturated Salt Solutions†

Saturated salt solution	Formula	Percent relative humidity at stated temperatures		
		68°F(20°C)	77°F(25°C)	86°F(30°C)
Lithium chloride	$LiCl \cdot H_2O$	12.4	12.0	11.8
Potassium acetate	$KC_2H_3O_2$	23.3	22.7	22.0
Magnesium chloride	$MgCl_2 \cdot 6H_2O$	33.6	33.2	32.8
Potassium carbonate	$K_2CO_3 \cdot 2H_2O$	44.0	43.8	43.5
Potassium nitrite	KNO_2	49.0	48.1	47.2
Magnesium nitrate	$Mg(NO_3)_2 \cdot 6H_2O$	54.9	53.4	52.0
Sodium nitrite	$NaNO_2$	65.3	64.3	63.3
Sodium chloride	$NaCl$	75.5	75.8	75.6
Ammonium sulfate	$(NH_4)_2SO_4$	80.6	80.3	80.0
Potassium nitrate	KNO_3	93.2	92.0	90.7
Potassium sulfate	K_2SO_4	97.2	96.9	96.6

†(Courtesy of Hygrodynamics, Inc.)

The relative humidity in the vapor space of the chamber is determined by the water-vapor pressure and the saturation water-vapor pressure at the temperature of the vapor. Typical values of equilibrium relative humidity for a number of saturated salt solutions are shown in Table 6-3. These values were selected from various published tables and were adjusted by interpolation to the temperatures listed. They are shown as examples and are not intended to be used for actual transducer calibration. Instead, the latest available tables of these characteristics should be consulted. The effect of the transducer and of the possible reference transducer on the vapor should not be neglected. Either may add or subtract a significant amount of moisture to or from the vapor or may introduce a temperature unbalance. The latter can be significant for relative-humidity calibrations. Sufficient time must be allowed for vapor equilibration.

When all precautions are observed, as described above, relative humidities within ±1% RH of their true value can usually be produced by the best salt solutions and by using the latest published reference tables. The record of a calibration performed on this basis should be annotated showing the salt and reference table used. For accuracies closer than ±1% RH a reference-quality humidity-measuring instrument must be used together with the transducer being calibrated.

6.5.1.2 Two-pressure generators of various designs are used to produce an atmosphere of known humidity in a chamber in which the transducers are mounted for their calibration. A typical two-pressure humidity generator consists of a one-stage or multistage saturator into which air at atmospheric pressure is introduced and saturated with water vapor. The saturated air is then expanded to a lower pressure. This lower pressure is the unsaturated water-vapor pressure in the calibration chamber. Varying this pressure, such as by a needle valve, regulates the humidity in the chamber which is exhausted to a vacuum pump through a second regulator valve.

The pressure of the saturated air (saturated water-vapor pressure) is measured by a manometer or close-accuracy pressure transducer at the outlet of the saturator. The pressure in the calibration chamber is similarly measured, either as an absolute pressure or as pressure difference between chamber and saturator. The percent relative humidity is then 100 times the ratio of the chamber pressure to the saturator outlet pressure. It should be noted that this simple relationship applies only to an ideal gas. When it is used for nonideal gases, such as real air-vapor mixtures, a correction factor must be applied. Investigations of corrections have been carried on by various researchers. Any errors in pressure measurement will, of course, cause much more significant errors in the relative humidity values so determined.

Certain refinements are incorporated in most two-pressure generators. A heater is usually provided for a portion of the pipe section between saturator outlet and chamber-pressure regulating device to prevent condensation after cooling due to expansion. A heat exchanger is then connected between regulator (expansion valve) and chamber to reduce the gas temperature to the desired value. The chamber temperature is monitored by a temperature transducer (thermoelectric or resistive). An air filter is installed in the inlet to the saturator. Instead of using ambient atmospheric pressure and a vacuum pump, a compressor can be used to raise the pressure at the saturator inlet sufficiently to allow test chamber venting to the ambient atmosphere at even the largest pressure ratios. Air-circulation provisions can be incorporated in the chamber. Other design refinements include improvements in accuracy and validity of pressure measurements and in the length of the time required for a full-range humidity calibration.

6.5.1.3 Constant-humidity atmosphere producers are used for humidity-transducer calibration in conjunction with a reference humidity-measuring instrument which determines the absolute humidity (moisture content) of the atmosphere being produced. In such systems, different amounts of dry and water-vapor-saturated air are mixed at constant pressure and temperature to produce the desired atmosphere in a test chamber. The mixing ratio is controlled very accurately and is known sufficiently well to establish a coarse value of absolute humidity in the mixture.

High-quality atmosphere producers, in conjunction with a dew-point measuring device equipped with a microscope for visual condensation observation, are capable of providing constant dew-point temperatures known to be within $\pm 0.2°F$ between $-20°F$ and $+100°F$. This error increases to $\pm 2°F$ for frost point temperatures between $-20°F$ and $-100°F$.

The gravimetric "train" or "gravimetric hygrometer" is considered the best available reference humidity-measuring instrument since it provides moisture-content measurements in gases, in terms of mixing ratio, on an absolute basis. In this equipment the water vapor from a given mixture is absorbed by a solid desiccant, and the mass of the water vapor is determined by close-accuracy weighing of the desiccant before and after a test run. The pressure and temperature of a known volume of the dry gas are determined to allow calculation of gas density and the mass of the dry gas. The term "train" is applied to a number of cascaded drying tubes through which the gas sample must flow.

6.5.2 Other static performance tests

Prior to calibration a number of mechanical and electrical characteristics of a humidity transducer are usually verified. Mechanical inspections include a visual inspection, dimensional, material, and nameplate checks, verification of correctness of electrical and any pneumatic connections, and inspection of the transducer's sealed container, if any, and its desiccant. Electrical inspections comprise tests of insulation resistance, breakdown-voltage rating, current or power drain, and maximum excitation. On certain hygrometric transducers, polarization tests are performed to determine any performance changes which may be caused by d-c excitation applied continuously or intermittently for a certain length of time. The polarization test is normally performed on a sampling basis only.

6.5.3 Dynamic performance tests

These involve the determination of the transducer's time constant and, sometimes, its 90, 95, 97, 98, or 99% response time under one or more sets of carefully controlled conditions as stated in the applicable specification. Such test conditions as the two limits of the step change in humidity, and ambient temperature, pressure, and flow rate (ventilation rate) of the air or other specified gas must be monitored and recorded.

6.5.4 Environmental tests

Among environmental qualifications tests on humidity transducers the most important, when specified, are ambient-pressure and vibration tests. Ambient-pressure levels are sometimes combined with different ambient temperatures during "temperature-altitude" tests.

Attention must be paid to limitations of test equipment and reference transducer, if any, in providing and monitoring humidity levels at reduced ambient pressures. Vibration tests are usually performed only to detect structural weakness. When they are performed to determine performance changes during vibration, such as by coupling vibration to a transducer in a humidity chamber through a soft sealing membrane, the effect of transducer (probe) vibration on the immediately ambient humidity must be taken into consideration. Other environmental tests such as shock, electromagnetic-interference, acoustic-noise, and nuclear-radiation tests are performed while the transducer is not measuring known levels of humidity. Usually, however, the transducer is connected to normally used excitation and readout equipment during most environmental tests. If desired, the ambient atmospheric humidity can be used as a roughly known measurand level.

6.5.5 Reliability tests

Stability tests are often performed on a sampling basis. Such tests can determine differences in performance characteristics after long-term storage in specified containers (storage-life tests) or after a specified number of calibration cycles (cycling-life tests). Stability tests can also be performed to determine performance changes while the transducer is exposed to one known level of humidity for a specified period of time (continuous-operating-life tests).

6.6 Specialized Measuring Devices

6.6.1 Measurement of moisture content in liquids and solids

In the description of transducers covered in this chapter, primary emphasis was placed on measurement of humidity and moisture in air and other gases, their most frequent application. The determination of moisture content in nonaqueous liquids and in solids often requires special analysis techniques using gas chromatographs and spectrometers or spectroscopes. A number of moisture spectroscopes have been successfully developed and applied. Their operation is usually based on the partial and selective absorption, or changes in particle velocity, due to moisture content, of ultraviolet, infrared, visible-light, neutron, gamma, or other radiation. Moisture content of some solids, e.g., textiles and paper, can sometimes be determined by pressing against their surface a humidity transducer of special design using a shallow cavity between surface and sensing element for creating a gas sample whose relative or absolute humidity can then be measured.

In some applications the water activity or moisture condition of a liquid or solid is considered more important than the moisture content or total water quantity. Required measurements of absorbed water can be expressed as percent equilibrium relative humidity which is 100 times the ratio of the water-vapor pressure of the liquid or solid to the saturation water-vapor pressure in air at the temperature of the liquid or solid.

6.6.2 Ice detectors

The timely detection of ice formation is important to the proper function of aircraft, helicopters, radio and radar antennas, and to the usability of many ground

Fig. 6-16. Ice detector for aeronautical applications (courtesy of Rosemount Engineering Co.).

structures, including highways. Nonaeronautical applications of ice detectors are considerably more recent than aeronautical applications because the need for ice detection in the air, e.g., while flying through supercooled clouds, is not seasonal and has been recognized for a much longer time.

A typical airborne ice detector is shown in Figure 6-16. The thin probe protruding from the streamlined portion of the case above the flange is the sensing element of the detector. It is set into axial vibration at its resonant frequency of about 40 kHz. As ice builds on the probe the added mass decreases the resonant frequency. This frequency change is detected by appropriate electronic circuitry and the error signal is used to indicate icing as well as to energize a heating element in the probe.

This heater deices the probe rapidly and readies the detector for its next cycle of operation. The cycles repeat as long as icing occurs.

Because the magnitude of the frequency change is related to the rate of icing, it can be used to produce an indication of icing intensity or severity. The electronic circuitry associated and integrally packaged with the detector provides an output signal when the ice thickness reaches a preselected level. This level can be adjusted between about 0.020 and 0.200 in. of ice thickness. The output can be in one of a number of optional forms. The switching level adjustment can be at the detector or can be remotely located.

A different design concept is used in a device which detects ice, sleet, or snow by melting it and then sensing the conductivity of the resulting water by two silver electrodes ("moisture contacts"). The ambient temperature at which the heater and moisture contacts start operating is sensed by a thermostatic switch whose set point can be adjusted as required for a particular application.

Another approach toward the detection of ice and snow is employed in a sensing unit which, in conjunction with a control box, indicates incipient snow or ice conditions on the basis of the combination of low temperature and precipitation.

BIBLIOGRAPHY

1. Dunmore, F. W., "An Electric Hygrometer and its Application to Radio Meteorography," *Journal of Research of the National Bureau of Standards*, Vol. 20, pp. 723–744, 1938.

2. Dunmore, F. W., "An Improved Electric Hygrometer," *Journal of Research of the National Bureau of Standards*, Vol. 23, p. 701, 1939.

3. Dunmore, F. W., "Humidity Variable Resistance," U.S. Patent 2,285,421, 1942.

4. Dember, A. B., "Humidity Responsive Resistor," U.S. Patent 2,481,728, 1945.

5. Zimmerman, O. T. and Lavine, I., "Psychrometric Tables and Charts," Dover, N.H.: Industrial Research Service, 1945.

6. Wexler, A., "Divided Flow, Low-Temperature Humidity Test Apparatus," *Research Paper RP1894*, Vol. 40, Washington, D.C.: National Bureau of Standards, 1948.

7. Wexler, A., "Recirculating Apparatus for Testing Hygrometers," *Journal of Research of the National Bureau of Standards*, Vol. 45, No. 5, 1950.

8. Wexler, A. and Brombacher, W. G., "Methods of Measuring Humidity and Testing Hygrometers," *National Bureau of Standards Circular 512*, Washington, D.C.: National Bureau of Standards, 1951.

9. List, R. J., "Smithsonian Meteorological Tables," 6th Rev. Ed., Washington, D.C.: Smithsonian Institution, 1951.

10. Wexler, A. and Daniels, R. D., Jr., "Pressure-Humidity Apparatus," *Journal of Research of the National Bureau of Standards*, Vol. 48, No. 4, April 1952.

11. Harmantas, C. and Mathews, D., "Sensing Element for the Electric Hygrometer," U.S. Patent 2,710,324, 1953.

12. "Humidity Element, Resistance, ML-379/AM," *Military Specification MIL-H-17654,* 1953.

13. Smith, W. J. and Hofflich, N. J., "The Carbon-Film Electric Hygrometer Element," *American Meteorological Society Bulletin,* Vol. 35, No. 2, pp. 60–62, 1954.

14. Wexler, A. and Hasegawa, S., "Relative Humidity-Temperature Relationships of Some Saturated Salt Solutions in the Temperature Range 0° to 50°C," *Journal of Research of the National Bureau of Standards,* Vol. 53, No. 1, July 1954.

15. Cutting, C. L., Jason, A. C., and Wood, J. L., "Capacitance-Resistance Hygrometer," *British Journal of Applied Physics,* Vol. 32, pp. 425–431, 1955.

16. Forsythe, W. E., "Smithsonian Physical Tables," 9th Rev. Ed., Washington, D. C.: Smithsonian Institution, 1956.

17. Wexler, A., "Electric Hygrometers," *Circular 586,* Washington, D.C.: National Bureau of Standards, 1957.

18. Hollander, L. E., Jr., Mills, D. S., and Perls, T. A., "Evaluation of Hygrometers for Telemetering," *I.S.A. Journal,* Vol. 7, No. 7, pp. 50–54, 1960.

19. Jones, F. E. and Wexler, A., "A Barium Fluoride Film Hygrometer Element," *Journal of Geophysical Research,* Vol. 65, pp. 2087–2095, 1960.

20. Kobayashi, J., "Investigations on Hygrometry," Tokyo: Meteorological Research Institute, 1960.

21. "Radiosonde Observation Computation Tables," Washington, D.C.: Weather Bureau, U. S. Department of Commerce, Revised 1960.

22. "Radiosonde Set AN/AMT-11 ( ) X," *Military Specification MIL-R-18990A,* 1960.

23. Grote, H. H. and Marchgraber, R. M., "The Dynamic Behavior of the Carbon Humidity Element ML-476," *USAERDL Technical Report 2379,* Fort Monmouth, N.J.: U.S. Army Electronics Research & Development Labs., 1963.

24. Quinn, F. C., "Equilibrium Hygrometry," *Instruments and Control Systems,* Vol. 36, No. 7, pp. 113–114, July 1963.

25. Stover, C. M., "Aluminum Oxide Humidity Elements for Radiosonde Weather Measuring Use," *Review of Scientific Instruments,* Vol. 34, No. 6, p. 632, 1963.

26. Bisberg, A., "A Miniature Solid-State Dew Point Sensor," *I.S.A. Preprint No. 11.2-1-64,* Pittsburgh: Instrument Society of America, 1964.

27. Quinn, F. C., "Measuring Moisture in Gases," *Instruments and Control Systems,* September 1964.

28. "Design and Testing of an Aerospace Dew Point Hygrometer," *NASA Report CR-63620,* Newton, Massachusetts: Cambridge Systems, 1964.

29. Crowley, J. E. and Rohrbrough, S. F., "Results of Test Program on Aircraft Dew Point Hygrometer," *Report No. AFCRL-65-873,* Springfield, Virginia: Clearinghouse for Federal Scientific and Technical Information, 1965.

30. Quinn, F. C., "Calculations of Moisture under Vacuum Condition," *Journal of Environmental Sciences,* December 1965.

31. Wexler, A., Ed., *Humidity and Moisture Measurement and Control in Science and Industry.* Vols. 1–3, New York: Reinhold Publishing Corp., 1965.

32. Bridgman, R. C., "Conductivity-Type Hygrometer Has Fast, Continuous Response," *Instruments and Control Systems,* Vol. 39, No. 1, pp. 95–98, January 1966.

33. Charlson, R. J., Buettner, K. J. K., and Maykut, G. S., "Liquid Film Hygrometry," *Final Report AFCRL-6-244*, Seattle, Washington: University of Washington, Dept. of Atmospheric Sciences, March 1966.

34. Hedlin, C. P., "A Device for Calibrating Electrical Humidity Sensors," *Materials Research and Standards*, Vol. 6, No. 1, pp. 25–29, January 1966.

35. Reim, T. E., "Advanced Nuclear Gage Controls Process Moisture," *Instrumentation Technology*, Vol. 14, No. 7, pp. 47–51, July 1967.

7   Light

7.1 Basic Concepts

7.1.1 Basic definitions

Light is a form of radiant energy. By common definition it is an electromagnetic radiation whose wavelength is between approximately 10^{-2} and 10^{-6} cm (see Figure 7-1). By strict definition only visible radiation can be considered as "light," and infrared light as well as ultraviolet light must then be referred to as "radiation."

Visible light is radiant energy which can be detected by the human eye; its wavelengths are between about 0.4 and 0.76 microns.

Infrared light is radiant energy in the band of wavelengths between about 0.76 and 100 microns. The portion of the band between 0.76 and 3 microns is sometimes referred to as *near infrared light*, and that portion between 3 and 100 microns has been called *far infrared light*. Most of the infrared light band overlaps the *heat radiation* band of electromagnetic radiation.

Ultraviolet light is radiant energy in the band of wavelengths between about 0.4 and 0.01 microns. *Note:* X rays occupy the adjacent portion of the electromagnetic radiation spectrum where wavelengths are shorter than 0.01 microns.

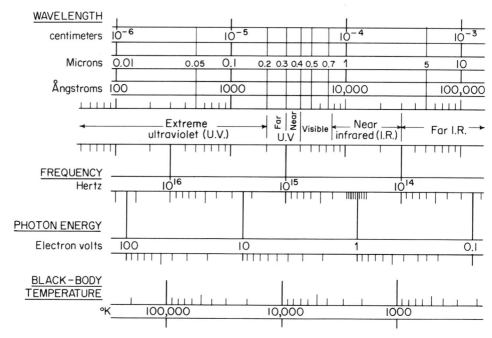

Fig. 7-1. The light spectrum.

Luminous flux is the time rate of flow of light. *Note:* Luminous flux and other luminance-related terms are intended to pertain primarily to visible light. These terms are not commonly used in relation to infrared and ultraviolet radiation.

Luminous intensity, in any direction, is the ratio of the luminous flux emitted by a point source in an infinitesimal solid angle, containing this direction, to the solid angle.

Luminance is the luminous intensity of a surface in a given direction per unit of projected area of the surface as viewed from that direction. It is the photometric quantity equivalent to "brightness."

Illumination (illuminance) is the luminous flux density on a surface, expressed as luminous flux per unit area (of a uniformly illuminated surface).

Luminosity is the ratio of the luminous flux to the corresponding radiant flux.

Responsivity is the ratio of the output amplitude to the product of the effective radiation flux density (at a given wavelength) and the detector area (active surface area).

Quantum efficiency is the number of photoelectrons (or similar electronic charges) produced in the light-sensitive portion of a light sensor by each incident photon.

Noise equivalent power (NEP) at a given wavelength is the ratio of the product of the effective radiation flux density (J_λ) at this wavelength and the detector area (A) to the signal-to-noise ratio or $\text{NEP} = (J_\lambda A)(S/N)^{-1}$.

Dark current is the current flowing in a photoconductive or photoemissive light sensor in total darkness, i.e., in the absence of any incident optical photons or other radiation.

7.1.2 Related laws

Distance from point source (Inverse Square Law)

$$E = \frac{I}{r^2}$$

where: E = illumination (1m/m²) at distance r
I = luminous flux (1m)
r = distance from point source (m)

Photon energy

$$\mathscr{E}_p = hf$$

where: $\mathscr{E}_p$ = photon energy (ergs)
h = Planck's constant, 6.63×10^{-27} erg-s
f = frequency (Hz)

Note: 1 electron volt = 1.6×10^{-2} ergs

Blackbody temperature (Wien's Displacement Law)

$$\lambda_m T = 0.2897 \text{ cm-deg (a constant)}$$

where: λ_m = wave length at which radiant energy density is maximum (cm)
T = absolute temperature of blackbody radiator (°K)

7.1.3 Units of measurement

The unit of luminance is the **candela** (*cd*), one of the six basic units of the International System of Units (SI). One candela is one-sixtieth of the luminous intensity of one square centimeter of a blackbody radiator which is at the temperature of solidifying platinum.

The unit of luminous flux is the **lumen** (*lm*). One lumen is the flux emitted within a solid angle of 1 steradian by a point source having a uniform intensity of one candela.

The unit of illumination is the **lux** (*lx*). One lux is an illumination of one lumen per square meter.

The unit of luminous intensity, prior to 1948 when the candela became the standard unit, was the **candle power** (*cp*), equivalent to the luminous flux of one **candle** when viewed in a horizontal plane. One candle was the unit of luminous flux from a spermaceti candle burning at the rate of 120 grains per hour. The candle power is still used extensively, as are the related foot-candle and foot-lambert.

The **foot-candle** is the unit of illumination of one candle per square foot as well as of the illumination at a spherical distance of one foot from a one-candle source.

The **foot-lambert** is the unit of luminance equal to $1/(4\pi)$ candle per square foot as well as of luminance of a surface emitting or reflecting light at the rate of one lumen per square foot.

Luminosity is expressed in **lumens per watt.**

The unit of wavelength of light is the **Ångstrom** (Å or A—the diacritical mark over the A is often omitted). One Angstrom unit equals 10^{-10} meter. Another frequently used unit of wavelength is the **micron** (μ or μm), which equals 10^{-6} meter (10^{-4} centimeter).

Photon energy is expressed in **ergs** or **electron volts.**

Noise equivalent power is expressed in **watts** and responsivity in **volts per watt,** when the surface area is given in cm² and the radiation flux in watts/cm².

7.1.4 Related measurements

The quantities of primary concern in the area of light measurement are emission, reflection, refraction, diffusion, scattering, absorption, and polarization of light. Related to light measurements are the determinations of color (*colorimetry*), opacity, turbidity (*turbidimetry*), and smoke and haze density.

Color is a characteristic of visible light associated primarily with its wavelength (see Figure 7-2) and also with luminance and purity. *Monochromatic* light is light in a narrow band of wavelengths. *Hue* is a subjective attribute of color perception which determines whether the color is red, orange, green, bluish green, reddish purple, etc. The names of the many different color combinations are usually listed in national (e.g., U.S.A. Federal) standards in connection with numbered color chips. This makes it possible to duplicate any desired color combination at any time and location by specifying a color-chip number selected from such a standard.

The *opacity* of a medium is the reciprocal of its *transmission* (the ratio of transmitted to incident light flux). *Turbidity* is the cloudiness in a liquid caused by the presence of suspended powdered or granular solids.

Fig. 7-2. The color spectrum.

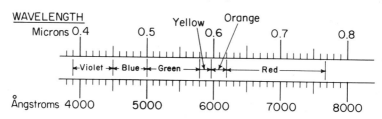

7.2 Transduction Methods

Since light transducers (light detectors, photosensors, light sensors) convert an electromagnetic radiation (light) into a usable electrical output, their transduction element usually acts also as a sensing element. Several basic methods are used for the transduction of light.

7.2.1 Photovoltaic transduction

Photovoltaic light sensors [Figure 7-3(a)] are self-generating. Their output voltage is a function of the illumination of a junction between two dissimilar materials. The materials are nonmetallic semiconductor elements or semiconductive metal compounds. Photons incident on the junction after passing through a very thin conductive layer cause an electron flow across the junction area so that the conductive layer becomes the negative terminal of the device.

7.2.2 Photoconductive transduction

Photoconductive (photoresistive) light sensors [Figure 7-3(b)] are semiconductive materials which change their resistance in response to incident light. The material, usually a metal salt, is contained between two conductive electrodes to which connecting wires are attached. The change in conductance results from a change in the number of charge carriers created by the absorption of photons. Polycrystalline films (e.g., lead salts and indium antimonide), as well as bulk single crystal materials (e.g., doped germanium), are used as photoconductors.

7.2.3 Photoconductive-junction transduction

In junction-type photoconductors the resistance across the junction in a semiconductor device changes as a function of incident light. The junction photocurrent increases with incident photon flux. This form of transduction of light is provided by photodiodes as well as by phototransistors [Figure 7-3(c)]. N-P-N phototransistors exist in addition to the P-N-P types, two versions of which are shown.

7.2.4 Photoemissive transduction

In photoemissive light sensors such as the phototube [Figure 7-3(d)], electrons are emitted by a cathode when photons impinge on it. An anode close to the cathode collects some of these electrons when it is at a positive potential with respect to the cathode. This causes current flow which can be used to produce an output voltage across a load resistor in series with the anode. The envelope around cathode and anode, typically a lime-glass or fused-silica bulb, can be either evacuated or gas-

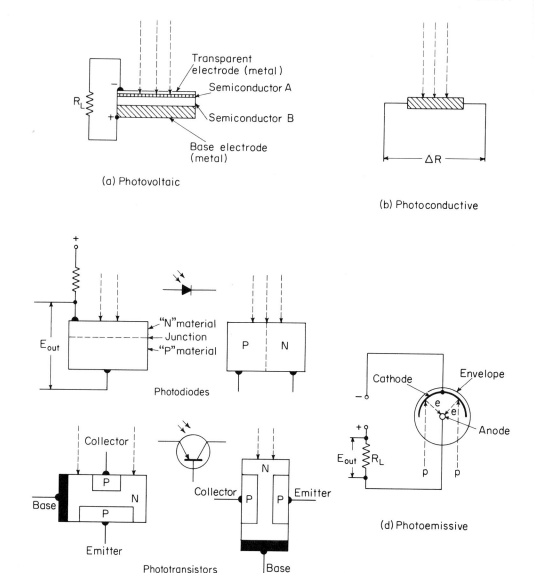

Fig. 7-3. Basic methods of light transduction.

filled. The output of a gas-filled phototube is greater than that of a vacuum phototube since additional electrons, freed from gas molecules by electron collisions, are drawn toward the anode. In the *photomultiplier* tube additional electrodes are used between cathode and anode to amplify the electron current by use of secondary emission.

7.2.5 Photoelectromagnetic transduction

This specialized transduction method is effected in a semiconductor placed in a magnetic field. When photons are absorbed near the front surface of the semiconductor, the resulting excess of carriers at that surface and their absence at the rear surface cause a diffusion of carriers toward the rear. The force due to the application of a transverse magnetic field then deflects the holes toward one end of the semiconductor and the electrons toward the other end. This action results in an emf developed between the two end terminals. The advantage of this transduction method is a reduction of internally generated thermal noise.

7.3 Design and Operation

7.3.1 Photovoltaic light transducers

7.3.1.1 The selenium photovoltaic cell is probably the best known of these devices. A typical selenium photosensor (Figure 7-4) consists of an iron base plate coated with a selenium layer having a thickness of about 70 microns. A cadmium film whose bottom surface is oxidized contacts the upper selenium surface. The cadmium and cadmium oxide layers are of molecular thickness so that they are quasi-transparent. A deposited-silver-alloy collector ring forms a usable electrical termination for the cadmium electrode. Incident light is converted into an electron flow from the p-type selenium to the n-type cadmium oxide at the region of contact between these two materials. Because the cadmium oxide acts as barrier to any reverse flow of electrons, this type of sensor has been referred to as a "barrier-layer photocell."

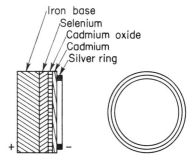

Fig. 7-4. Selenium photovoltaic sensor.

The output current depends on the load resistance as well as on the illumination. The current-vs-illumination relationship is approximately linear with a load resistance under 100 ohms. It becomes increasingly nonlinear as saturation occurs with higher load impedances. The open-circuit output voltage saturates quite rapidly, and its curve becomes asymptotic to about 650 mV. The time constant varies between 0.1 and over 5 ms depending on load impedance and illumination. Frequency response can be optimized by using a relatively small surface area (to reduce internal capacitance) and a low load-impedance value. The spectral response peaks at about 5700 Å. The temperature effect on current output at typical illumination and load values is a deviation of about $\pm 0.25\%/°C$ from the room temperature current over an operating temperature range of 5°C to 45°C.

7.3.1.2 Silicon photovoltaic sensors ("silicon solar cells") are better known as solar power converters (e.g., power sources on satellites) than as light transducers in

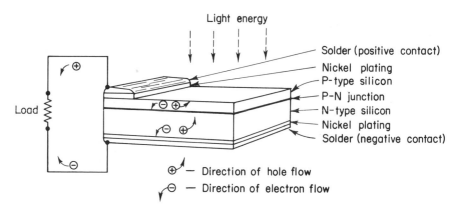

Fig. 7-5. Typical silicon photosensor structure (courtesy of Solar Systems Div., Tyco).

measurement systems. However, their characteristics are quite suitable for photometric purposes. The basic structure of a typical silicon photocell is illustrated in Figure 7-5. A slice of silicon approximately 0.5 mm thick is doped with a small amount of arsenic to form n-type silicon, which contains an abundance of electrons (negative charges). Boron is then diffused into the upper surface of the wafer to create a very thin p-type silicon layer, which has an excess of "holes" (positive charges). The resulting p-n junction between the two types of silicon acts as a permanent electric field. When the thin p-type layer is illuminated, the incident photons cause a flow of holes and electrons. The electric field constituted by the p-n junction directs the flow of holes into the p-type material and the flow of electrons into the n-type material. The resulting unbalance of charge carriers within the sensor causes an emf to be developed between the two surfaces, which are nickel plated to provide adhesion for solder contacts. When a load (resistance) is connected across the two terminals, the hole and electron carriers flow through the circuit until a balance condition is achieved. This current flow is then proportional to illumination.

N-P junctions operate in the same manner as p-n junctions. Their construction differs in that a molecular layer of phosphorus is diffused into the surface of boron-doped p-type silicon. The n-type surface has an abundance of electron charges, and the p-type material has an excess of holes, as in the p-n junction; however, the photons strike the n-type layer.

The spectral response of silicon photosensors peaks near 8000 Å, in the near-infrared region. The open-circuit output voltage varies logarithmically with illumination. Its amplitude at any illumination level is dependent on the load resistance but is independent of sensing-surface area. The output current is a function of surface area, load resistance, and illumination. Two or more individual sensors can be connected in parallel to increase output current and in series to increase output voltage. The time constant depends on the load resistance and is typically around 20 μs.

Temperature effects are relatively small on the short-circuit and maximum-power currents. The open-circuit and maximum-power output voltages, however, decrease almost linearly with increasing temperature at a slope of about 2mV/°C. Temperature compensation has been accomplished in some sensor packages by connecting a resistor with a positive temperature coefficient to the sensor, e.g., across one of two series-connected sensors.

Silicon photocells are produced in a variety of configurations. Bare cells are circular, rectangular, or circle segments. Enclosures are square, disc-shaped, or tubular and can be hermetically sealed. Typical sensors are illustrated in Figure 7-6. The rectangular bare cell can be used in "shingle" assemblies where a number of such cells are connected in series. The segmented enclosed cell and the miniature (0.215 in. diameter × 0.202 in. high) sealed cell are intended for photometric applications.

Special silicon photocell arrays (Figure 7-7) have been built for reading heads in punched-card and punched-tape systems. Such assemblies are intended for on-off operation. Another discrete-output photovoltaic device is the binary point-source-location detector element shown in Figure 7-8. When optically coupled to a cylindrical lens, it sees a distant point source of light energy as a line extending across all binary element lines. The sensitive areas of a single binary line are connected in parallel. By monitoring the output of each binary line (bit), the location of the point source can be determined on one axis to one part in n where n is the number of elements in the last binary line. The resolution of the 10-bit detector is one part in 512. The size of the detector is 1×1 in. Using a second detector of this type with an associated lens or split-image optics for location of the same point source in a transverse axis, a resolution of one part in n^2 within the viewed field is accomplished. Analog rather than digital location of light-emitting targets (e.g., in optical tracking) is achieved by means

Fig. 7-6. Silicon photosensor configurations (courtesy of Solar Systems Div., Tyco).

Fig. 7-7. Solar-cell array for punched-tape reader; center-to-center spacing of cells is 0.1 in. (courtesy of Solar Systems Div., Tyco).

Fig. 7-8. Silicon P-N junction photovoltaic binary detector element (courtesy of Electro-Nuclear Laboratories, Inc.).

of a four-quadrant null detector. The four elements of this circular sensor each subtend a 90-degree arc of the circle and are separated from each other by a thin gap. Sequential scanning produces four pulses of equal amplitude when a small circular illuminated area is at the center of the disc. Off-center locations of the light spot result in a different amplitude for each pulse. The pulse height decreases when a smaller portion of the illumination is received by the quadrant. This effect is even more pronounced when the quadrants are photodiodes (see 7.3.3).

Silicon photosensors can operate as photoconductive sensors when they are reverse biased.

7.3.1.3 Germanium photovoltaic sensors are similar to silicon types in construction and operation. Their spectral response, however, peaks near 15,500 Å (1.55 microns), and their operating-temperature range is much narrower.

7.3.1.4 Indium-arsenide (InAs) and indium-antimonide (InSb) photovoltaic sensors are intended for infrared-light transduction. InAs detectors are single-crystal p-n junction semiconductors designed to cover the near-infrared range. Their response peaks at about 3.2 microns and drops off sharply at longer wavelengths. Their time constant is in the vicinity of 1 μs. InSb detectors operate primarily in the far-infrared region with a peak spectral response at 6.8 microns. Their time constants vary between 0.3 and 20 μs. depending on their junction impedance and capacitance at the zero-bias point.

Both types are frequently used while artificially cooled to increase their sensitivity. Cooling can be attained by placing the detector in good thermal contact with a Peltier-effect (reverse thermoelectric) cooler. Typical coolers of this type require a 30 A, 0.2 V power supply for sensor operation at −40°C; however, more complex coolers with low current (2-3 A) drain have also been designed. Another method of cooling is complete immersion of the sensor in either a Dewar flask filled with liquid nitrogen or in a cryostat cooled by liquid nitrogen or hydrogen, or both. Operation at this temperature can increase the sensitivity of an InAs or InSb detector to ten times that at room temperature. It also tends to lower the wavelength of peak spectral response. Such cooling is also used for some other types of semiconductor infrared light sensors.

7.3.2 Photoconductive light transducers

This type of light sensor is probably best known for its widespread usage in photographic exposure meters and for automatic exposure controls in cameras. The transduction element is essentially a light-sensitive resistor which decreases its resistance with increasing illumination. The absolute value of resistance of a photoconductive sensor depends on the illumination level, the photoconductive material, the thickness, surface area, and geometry of the sensitive material, the geometry of its electrodes, and the spectral composition of the incident light.

Fig. 7-9. CdS and CdSe photoconductive light sensors (courtesy of Clairex Corp.).

The bulk single crystals of cadmium sulfide (CdS) and cadmium selenide (CdSe) are the most popular photoconductive materials because their spectral sensitivity is close to the visible-light region and because of their relatively high sensitivity to illumination changes. The spectral response peaks are near 6000 Å for CdS and near 7200 Å for CdSe. The pass band of spectral response is quite narrow for CdSe and only slightly wider for CdS. The peaks are sharp enough to make both materials useful for differentiation between colors. Compared to CdS photocells, CdSe detectors generally have a shorter time constant (approximately 10 ms as compared to about 100), a lower resistance, more sensitivity, a greater temperature coefficient of resistance, more creep ("light history effect"), and better linearity in the position of the range below 1 ft candle. Neither type exhibits a linear calibration curve over a wider illumination range. Typical CdS and CdSe photosensors are illustrated in Figure 7-9. The electrodes are deposited in special patterns by evaporative techniques to obtain a large active-surface area and the close electrode gaps needed for low element resistance. The maximum power-dissipation rating (between 50 and 500 mW depending on detector type) can usually be doubled when a heat sink is used in the detector mounting. Hermetically sealed cases are most frequently used.

Lead sulfide (PbS) and to a lesser extent lead selenide (PbSe) are used in infrared photoconductive sensors. The spectral response of both materials peaks between 2 and 2.5 microns; however, the roll-off of the spectral response curve of PbSe at longer wavelengths is very shallow up to about 4 microns so that the response is fairly flat between 1.8 and 3.6 microns. The time constants are between 100 and 700 μs for PbS and close to 10 μs for PbSe.

Among other photoconductive sensor materials are mercury-doped germanium (HgGe) for far-infrared measurements and CdS-CdSe mixed crystals for a visible-light response intermediate to the individual crystal responses.

7.3.3 Photodiodes and phototransistors

The electrical behavior of semiconductor diodes and transistors is normally affected by any light striking their junction. The junction of these devices is therefore well protected from any incident light. Most transistors are enclosed in a sealed metal can partly for this purpose. If this can is removed and the junction is at (or very close to) a surface on which light impinges, the output of the transistor will show the effects of this light. Photodiodes and phototransistors are so constructed that this otherwise undesirable effect is enhanced by purposely exposing their junctions to light in a controlled manner.

Silicon is used in most of these devices. Germanium is employed as sensing material primarily when better spectral response in the infrared region is required. The spectral response peaks near 1.0 microns for silicon sensors, and near 1.6 microns for germanium photoconductive-junction sensors. Silicon devices have a much wider operating-temperature range than germanium designs. Time constants are typically less than 1 μs and values in the 10-50 ns range are not uncommon. Transistors provide amplification of the light-induced electrical signal. The general characteristics of germanium and silicon diodes and transistors are covered in a number of texts dealing with such semiconductor devices.

Fig. 7-10. Silicon photodiode (courtesy of Electro-Nuclear Laboratories).

Diode as well as transistor types are most frequently hermetically sealed in metal cans, whose size varies in accordance with "TO-" transistor case sizes, and come equipped with a lens (see Figure 7-10) or window. Disc-shaped configurations with sealed windows have also been used for diodes. A silicon surface-barrier photodiode of special design has been developed for the 600 Å to 2500 Å (ultraviolet) spectral range.

7.3.4 Photoemissive light sensors

This group of transducers comprises diode phototubes and photomultiplier tubes. The phototube is well known from its utilization in sound-movie projectors as well as from numerous photometric applications. In the latter, however, it has been

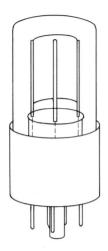

Fig. 7-11. Diode phototube.

largely replaced by other types of light transducers. Photomultiplier tubes, on the other hand, are widely used where low levels of illumination must be measured. They are also used in conjunction with scintillators for nuclear-radiation sensing (see Chapter 9).

The phototube (Figure 7-11) consists of a curved photosensitive cathode (photocathode) and an anode slim enough not to cast a significant shadow on the cathode. The photocathode is made from a thin metal sheet, e.g., a silvered steel sheet, coated on its concave surface with a photoemissive material. The operating characteristics of the phototube are dictated to a large extent by the type of photoemissive material selected. Cesium oxide coatings respond in the visible and near-infrared region. Sodium coatings provide a response in the blue and ultraviolet portion of the spectrum. The sensitivity of gas-filled phototubes is considerably larger than that of vacuum types, but the photocurrent-vs-luminous flux characteristics of gas-filled phototubes are less linear.

Photomultiplier tubes are multi-electrode phototubes. A number of secondary-emission *dynodes* are placed between cathode and anode. An additional dynode is located behind (or around) the anode. By use of a voltage-divider network, successively larger voltages, in steps of about 100 V, are applied to the dynodes so that the dynode nearest the anode has a potential close to the high anode-voltage itself (about 2000 V). A linear arrangement of grid-type "venetian-blind" dynodes (Figure 7-12) is frequently used. Other internal configurations include

Fig. 7-12. Schematic internal arrangement of electrodes in typical photomultiplier tube (courtesy of Electro-Mechanical Research, Inc.).

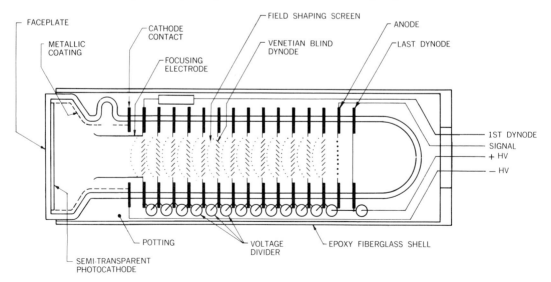

curved plate-type dynodes facing each other, with or without intermediate grids, on a common mounting base or along a portion of the tube length as in the "box and grid" and "focusing" systems. When photons strike the cathode, free electrons are liberated and drawn to the first (lowest-voltage) dynode because its potential is positive with respect to the cathode. At dynode "1" several electrons are liberated by each electron emitted from the cathode. These secondary-emission electrons are drawn to dynode "2" whose potential is positive with respect to dynode "1." This process is multiplied by each successive dynode until all electrons are finally collected by the anode. The amplification of the cathode photocurrent (the *current gain*) is typically between 10^5 and 10^7 when between 9 and 14 dynodes are used. Typical dynode materials are silver-magnesium and beryllium-copper.

The spectral response, luminous sensitivity, and other essential characteristics of photomultiplier tubes are dictated by the materials of the photocathode and the envelope (e.g., glass bulb) of the cathode end of the tube. Typical photoemissive cathode materials include cesium-antimony (Cs-Sb), bismuth oxide (BiO-Ag-Cs), cesium oxide (AgO-CsO-Cs), "bi-alkali" (Sb-Na-K), and "multi-alkali" (Sb-K-Na-Cs). Transparent photocathodes of cesium telluride (CsTe) and opaque rubidium telluride (RbTe), cesium-telluride (CsTe), and cesium-iodide (CeI) photocathodes are used in special ultraviolet-radiation photomultiplier tubes. The cathode envelope (window) is usually made of lime glass or fused silica; lithium fluoride (LiF) and sapphire windows are used for special (primarily ultraviolet) spectral-response requirements. The combinations of photocathode and window materials are frequently referred to by an "S" (spectral response) designation number, as shown in Table 7-1. This table lists typical window and cathode materials which have been used in photomultiplier tubes (some also in phototubes) and typical spectral-response-peak values, based on various literature. Where two response-peak wavelengths are obtained, both are shown (the less significant one in parentheses). Tubes with "solar-blind" cathodes are used where only ultraviolet radiation is to be measured.

7.3.5 Photoelectromagnetic light sensors

A few detectors of this type have been developed for measurements in the far-infrared spectrum. This device can measure radiation between 2 and 6.5 microns without artificial cooling. A permanent magnet is used to establish the required magnetic field for the indium antimonide photovoltaic sensing material. The time constant is less than 1 μs.

7.4 Performance and Applications

7.4.1 Specification characteristics

7.4.1.1 Design characteristics. Light sensors are usually mounted by the user in a housing assembly of his own design together with other optical and electronic compo-

Table 7-1 PHOTOCATHODE MATERIALS AND SPECTRAL-RESPONSE PEAKS

Designation No.	Photocathode material	Window material	Response peak (Å)
S-1	AgO-Cs	Lime glass	(3700), 8000
S-3	AgO-Rb	Lime glass	4200
S-4	Cs-Sb	Lime glass	4000
S-5	Cs-Sb	"9741" glass (special glass)	3300
S-8	Cs-Bi	Lime glass	3650
S-9	Cs-Sb (Semi-opaque)	Lime glass	4800
S-10	BiO-Ag-Cs (Semi-opaque)	Lime glass	4500
S-11	Cs-Sb (Semi-opaque)	Lime glass	4400
S-13	Cs-Sb (Semi-opaque)	Fused silica	4400
(S-13)	Cs-Sb (Semi-opaque)	Sapphire	4400
S-17	Cs-Sb	Lime glass (refl. substrate)	4900
S-20	Sb-K-Na-Cs (Semi-opaque)	Lime glass	4200
(S-20)	Sb-K-Na-Cs (Semi-opaque)	Fused silica	2600, (4200)
S-21	Cs-Sb (Semi-opaque)	Special "9741" glass	4400
—	Sb-Na-K	Lime glass	3700
—	CeTe (Transparent)	Fused silica	1900
—	RbTe (Opaque)	LiF	1900
—	CsTe	Sapphire	2600
—	CsTe	LiF	2600
—	Cs-Sb	Fused silica	1800, 4200

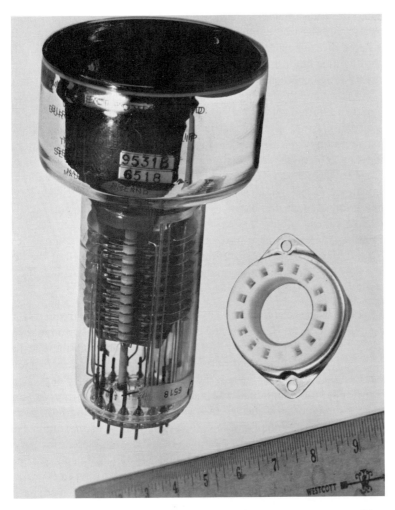

Fig. 7-13. Photomultiplier tube and mating socket (courtesy of Gencom Div., Whittaker Corp.).

nents. Small sensors are often mounted by their leads, as is typical for electronic components such as resistors and capacitors, and no special mountings are provided. The specification of mechanical design is therefore limited to such basic dimensions as case length and diameter (or width), the size and location of any window or lens, the type and location of connecting leads or pins, and the degree of sealing. Case material and finish are sometimes shown. Many small sensors are built in "TO-" series semiconductor configurations which are standardized (e.g., "TO-18"). The connector-pin bases of photomultiplier tubes (Figure 7-13) are commonly shown by EIA (Electronic

Industries Association) standard designation numbers (e.g., "B14-38"). The active (light-sensitive) area is stated numerically. Identification is often limited to manufacturer's name or trademark and model number. Polarity of leads is marked on the case of photovoltaic and photodiode devices.

Electrical design characteristics include the nominal load resistance (or allowable range of load resistances), element resistance and capacitance for photovoltaic detectors, maximum excitation voltage and power for photoconductive sensors, and nominal current, voltage, biasing, and impedance characteristics for photodiodes and phototransistors, preferably in graphical or tabular form. Photoemissive sensors require the specification of nominal and maximum supply voltage and peak cathode current. Photomultipliers also need values assigned to the allowable range of cathode-to-first dynode and anode-to-last dynode voltages and to essential inter-electrode capacitances (e.g., anode-to-last dynode and anode to all other electrodes). The number of dynodes must always be stated.

The sensing material and transduction principle are always stated for photoconductive and photovoltaic sensors in the nomenclature employed at the heading of the specification, e.g., "CdSe Photoconductive" or "Silicon Photovoltaic."

7.4.1.2 Performance characteristics

7.4.1.2 Performance characteristics at room temperature are frequently shown in graphical or tabular form. Such specifications usually state only nominal characteristics. Performance tolerances are rarely shown, except for "minimum" callouts and occasional "minimum-nominal-maximum" values for sensitivity, detectivity, or equivalent conversion characteristics. The output at zero illumination ("dark current") is often shown for nonphotoconductive devices. Spectral response is always specified, with tolerances sometimes assigned to the peak-response wavelength. Time constants are normally shown as nominal, and sometimes (preferably) as maximum, for one or more load-impedance values. Where the time constants are different for dark-to-light and light-to-dark conditions, both values should be shown. Response time or rise time, when properly specified (e.g., "98% response time," "10% to 90% rise time") can be substituted for time constant. The field of view should be specified, in steradians or by a polar plot, particularly when the sensor has a lens.

Conversion characteristics of photovoltaic detectors are shown either as a plot of open-circuit output voltage and short-circuit output current-vs-illumination or as minimum output values at a fixed illumination wavelength and level. Similarly, the effects of load resistance on output can be shown as minimum (sometimes also maximum) output power, current, and voltage at a fixed illumination wavelength and level and an optimum load resistance value (with tolerances) or as a family of output curves at various load resistances.

Some specifications show only the detectivity, the noise-equivalent-power, and sometimes the responsivity, at the peak-response wavelength, by minimum, nominal, and maximum values. A statement of detectivity should be accompanied by its definition and the test conditions under which it was determined.

Photoconductive-sensor performance is stated in terms of nominal sensor resistance (preferably with tolerances) at a fixed illumination level and wavelength. An additional graph then shows the variation of sensor resistance with illumination changes.

Photodiode and phototransistor characteristics are best shown as graphical representations of voltage and current characteristics at various illumination levels, different bias voltages, and external circuit constants.

Phototube characteristics are usually shown as output-current variation with changes in illumination at different anode-to-cathode potentials for one or more values of load impedance.

Photomultiplier performance specifications include the current gain (amplification), the cathode luminous sensitivity (in μA/lm), and the anode sensitivity. Photocathode characteristics are stated either by "S" number (see Table 7-1) or by cathode and window materials. Some specifications also show the cathode radiant sensitivity (in A/W or μA/μW) and the quantum efficiency (in %).

7.4.1.3 Environmental performance characteristics are specified mainly in the area of temperature effects. An operating temperature range is usually stated and thermal effects on performance are often shown by curves. Other environments, e.g., shock and vibration, are usually considered nonoperating conditions.

7.4.2 Selection and applications

Primary selection criteria for light transducers are sensitivity and spectral response (see Figure 7-14). Next in importance are associated-circuitry characteristics, operating-temperature range, and cost. Photovoltaic sensors have the advantage of being self-generating; i.e., no external power supply is needed. Photoconductive and photoconductive-junction sensors do require a simple power supply but are considerably more sensitive than photovoltaic types. Photodiodes are somewhat more versatile than bulk-effect photoconductive sensors because of the variety of output characteristics obtainable with different biasing arrangements, and usually have shorter time constants. Phototransistors offer the advantages of photodiodes and provide amplification of the light-induced signals as an additional advantage. Phototubes, no longer in common use for light-measuring applications, provide characteristics similar to those of photodiodes. Photomultiplier tubes, which require a high-voltage power supply, are used wherever very weak illumination levels must be measured. Some photomultipliers are sufficiently sensitive to provide a measurable output for each individual photon incident at their photocathode and can therefore be used for photon counting.

The spectral response characteristics of any light sensor can be modified by placing it behind a window having the required transmittance characteristics in the portion of the wavelength spectrum where measurements are to be made. Windows are passive components and are also used as filters which reduce the illumination on the sensing

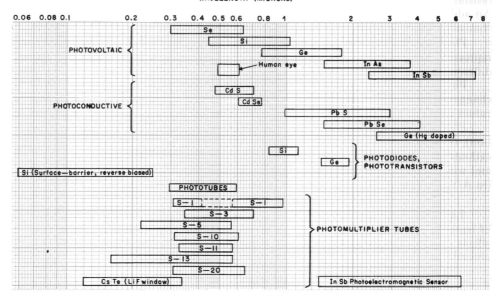

Fig. 7-14. Spectral response (within −10 db of peak) for typical light transducers.

surface at undesired wavelengths. Commonly used window materials and their spectral responses are: quartz crystal (0.2 to 1.4μ), fused silica (0.2 to 1.4μ), borosilicate glass (0.4 to 1.2μ), cultured sapphire (0.15 to 1.6μ), lithium fluoride (0.11 to 1.8μ), calcium fluoride (0.12 to 11μ), barium fluoride (0.15 to 15μ), sodium chloride (0.2 to 23μ), thallium bromide (0.5 to 40μ), and cesium iodide (0.25 to 70μ). The latter has also been used as photocathode material in photomultipliers for special applications.

A "chopper" (oscillating or rotating shutter) is often used to convert the d-c illumination into a-c light detected by the sensor, primarily to facilitate the amplification of the sensor output.

Applications of light sensors range from photographic exposure meters to sophisticated spectroscopes for astronomy and planetary research. Ruggedized ultraviolet-radiation sensors have become useful tools in space research for observations from vehicles located beyond earth's atmosphere, which is virtually opaque to radiation in the ultraviolet spectrum below 3000 Å. Light detectors are used even more frequently in conjunction with light sources for nonphotometric applications where only an on-off signal is required, e.g., in light-sensitive alarms, door controls, counting and sorting systems, and card and tape readers. When associated with a fixed light source they are also used in a number of transducer types for other measurands, such as attitude (horizon scanners, sun seekers, star seekers), acceleration, displacement (for analog and digital output), liquid level, and pressure, and are covered in the chapters of this book dealing with these measurands. Applications of infrared light sensors overlap those of radiant-heat-flux transducers (see 14.6).

7.5 Calibration and Tests

7.5.1 Calibration

The light source used for the calibration of light sensors was originally a candle of special construction (spermaceti candle) which burned at an approximately known rate (120 grains per hour). This "standard candle" was later replaced by a tungsten lamp which is normally operated at a color temperature of 2854°K. By use of special liquid filters (Davis-Gibson filters) this color temperature can be converted to 4800°K and to 6500°K. Since the introduction of the candela as unit of luminance, the second-ary-standard tungsten lamps are calibrated against the luminous intensity of one square centimeter of a blackbody radiator which is at the temperature of solidifying platinum. One-sixtieth of this luminous intensity is one candela, or "new candle." Calibrations in terms of candle power can be performed on the basis of calling a candela a candle. They can also be performed in terms of lumens instead of candles and in terms of lux instead of foot-candles.

The output of a certified tungsten lamp is usually stated in horizontal candle power (HCP). With the lamp mounted vertically, the sensor to be calibrated is mounted in a plane perpendicular to the axis of the lamp and with the sensing surface facing the lamp filament. Different illumination levels can then be produced by varying the distance between lamp and sensor and using the inverse square law to determine the illumination at the sensing surface. When illumination is to be expressed in foot-candles, the HCP of the lamp is divided by the square of the distance (in feet). Since the lamp is intended to act as a point source, the distance should never be less than six times the longest source dimension, as a "rule of thumb." Background illumination should be minimized by performing the calibration in a dark chamber or dark room.

Infrared-light sensors can be calibrated using a blackbody radiation source. Special sources are used for ultraviolet-light detectors. Variations in spectral response can be obtained by varying the temperature of the source and by the use of filters. D-C as well as a-c calibrations are performed, the latter by use of choppers or pulsed or modulated radiation sources. Self-luminous light sources, containing radioactive isotopes with phosphor crystals, are sometimes employed for calibration purposes.

7.5.2 Tests

Frequency-response tests can be performed by using a chopper of special design to obtain the desired waveshape and by varying the rotational speed of a disc chopper or the frequency of a vibrating-reed chopper. Time constants can be measured by using a quick acting shutter. The chopper or shutter are placed between light source and light detector. Instead of such mechanical devices, semiconductor or gas-filled light sources can be employed, excited from a sine-wave modulator or a pulse generator. Among semiconductor sources the gallium-phosphide (GaP) and gallium-arsenide

(GaAs) sources are most frequently used. The former emits light in the visible portion of the spectrum, and the latter emits light in the near-infrared portion. The 10% to 90% rise time of such devices is less than 10 ns.

Temperature tests are easiest to perform with the sensor mounted on a heated or cooled block while the temperature of the sensor is monitored by a thermocouple. When sensors are tested in a temperature chamber, with the source outside the chamber, the transmittance and spectral characteristics of the required window in the chamber must be known and considered. A chamber of this type or a bell jar must, however, be used for ambient-pressure tests.

During other environmental tests the light transducer should be installed in its end-use housing or other assembly.

7.6 Related Measuring Devices

Measuring devices related to light transducers include primarily those intended for the measurement of light characteristics other than its intensity and those using light-measuring systems, including a light source, for analysis of the transmittance of the light path through a medium which is to be tested.

Spectrometers and colorimeters belong in the first category. The incident (for spectrometers) or reflected (for colorimeters) light is passed through continuously changing gratings or filters whose scanning frequency is synchronized with readout equipment or a telemetry signal conditioner. The light sensor, whose spectral response is precisely known, receives light at wavelengths or in narrow wavelength bands which change continuously or discretely at a known rate. The output of the light detector, when synchronized, shows the illumination received in each of the sampled portions of the spectrum. Scanning can be omitted when several light sensors, each with a different filter or a different spectral response peak (or both), are arranged to receive equal amounts of the incident light through suitable optics.

The second category includes turbidimeters, opacity meters, and smoke and haze detectors. With a fixed amount of luminous flux produced by a light source and by use of special optical devices, the output of the detector varies with changes in the transmittance of a liquid or gaseous medium.

Certain fire detectors are also related to light transducers. Infrared sensors can be used to detect a sudden increase in heat flux in a specific wavelength band produced by a fire. An ultraviolet sensor with a narrow band-pass output filter has been used for flame detection. The band-pass filter is tuned to the characteristic flicker frequency of certain flames.

BIBLIOGRAPHY

1. Bendz, W. I., *Electronics in Industry.* New York: John Wiley & Sons, Inc., 1947.

2. Simpson, O. and Sutherland, G.B.B.M., "Photoconductive Cells for Detection of Infrared Radiation," *Science*, Vol. 115, pp. 1–4, January 4, 1952.

3. *Lighting Handbook.* 2nd Ed., New York: Illuminating Engineering Society, 1952.

4. Robertson, J. K., *Introduction to Optics.* New York: John Wiley & Sons, Inc., 1954.

5. Heavens, O. S., *Optical Properties of Thin Solid Films.* London: Butterworths Scientific Publications, Ltd., 1955.

6. Jenkins, F. A. and White, H. E., *Fundamentals of Optics.* 3rd Ed., New York: McGraw-Hill Book Company, 1957.

7. Strong, J., *Concepts of Classical Optics.* San Francisco: W. H. Freeman & Co., Publishers, 1958.

8. Engstrom, R. W., "Absolute Spectral Response Characteristics of Photosensitive Devices," *R.C.A. Review*, Vol. XXI, No. 2, pp. 184–190, June 1960.

9. Hackforth, H. L., *Infrared Radiation.* New York: McGraw-Hill Book Company, 1960.

10. Gould, H. J., "Photoconductivity of a Mixed Crystal," *Technical Note*, Duarte, California: Conrac Corp., March–April 1961.

11. Riesz, R. P., "High-Speed Semiconductor Photodiodes," *Review of Scientific Instruments*, Vol. 33, p. 994, September 1962.

12. Williams, R. L., "Fast High-Sensitivity Silicon Photodiodes," *Journal of the Optical Society of America*, Vol. 52, p. 1237, November 1962.

13. Görlich, P., Krohs, A., and Pohl, H.-J., "Problems in the Application of New Photo-Electronic Elements to Detection and Demodulation of High-Frequency Modulated Laser Beams" (in German), *Acta IMEKO 1964*, pp. 317–327, IMEKO, Budapest, Hungary, 1964.

14. Naugle, A. B., Merriam, J. D., and Eisenman, W. L., "Properties of Photodetectors," *NOLC Report 602*, Corona, California: Naval Ordnance Laboratory, June 1, 1964.

15. Lucovsky, G. and Emmons, R. B., "High Frequency Photodiodes," *Applied Optics*, Vol. 4, p. 697, June 1965.

16. Merriam, J. D., Eisenman, W. L., and Naugle, A. B., "Properties of Photodetectors," *NOLC Report 621*, Corona, California: Naval Ordnance Laboratory, April 1, 1965.

17. Udalov, N. P., *Semiconductor Transducers* (in Russian), Moscow: ENERGIYA Publishing House, 1965.

18. Eisenmann, W. L., "Properties of Photodetectors," *Photodetector Series*, *NOLC Report 637*, Corona, California: Naval Ordnance Laboratory, February 15, 1966.

19. Mandalakas, J. N., "A Temperature-Compensated Light Sensor," *I.S.A. Journal*, Vol. 13, No. 9, pp. 37–39, 1966.

20. Rome, M., "New Developments in Multiplier Phototubes for Space Research," *Applied Optics*, Vol. 5, No. 5, pp. 855–861, May 1966.

21. Selgin, P. J., "The Split-Window Light-Beam Comparator," *I.S.A. Journal*, Vol. 13, No. 6, pp. 53–55, 1966.

22. Wolf, W. L., "Handbook of Military Infrared Technology," Washington, D.C.: Office of Naval Research, 1966.

23. "Photoconductive Cell Manual," New York: Clairex Corp., 1966.

24. Chow, K. T. and Bates, C. E., "Capabilities of Solid-State Detectors," *Laser Focus*, pp. 33–37, May 1967.

8 Liquid Level

8.1 Basic Concepts

This chapter covers transducers for continuous and discrete ("point") measurements of the *level* of liquids and quasi-liquids (e.g., slurries and powdered or granular solids) in tanks and other vessels. From level measurements the *volume* of the liquid can be determined if the tank geometry and dimensions are known and the *mass* of the liquid can be established if its density is additionally known. Discrete-level transducers can also be used to detect the presence or absence of a liquid in a pipe or duct. The *interface* whose location is determined by a level measurement is usually between a liquid and a gas; however, it can also be between two different liquids.

Level measurements have an obvious economic significance. The knowledge of how much liquid is in a tank has been important to mankind for many centuries. Devices with a long history, such as the "dipstick" and the sight glass, are still in use today. The increasing utilization of electronic measuring systems fostered the development of a multitude of transducer designs representing a number of radically different approaches. Various sensing methods are described in 8.2 together with basic concepts and related laws.

Units of level measurement are either in terms of tank height (in meters, centi-

354

meters, feet, inches, etc.), in terms of volume of liquid (in liters, gallons, etc.) represented by the level, or in terms of mass (kilograms, pounds, etc.)

8.2 Sensing Methods

Sensing methods of various degrees of complexity are employed in measurements
of continuous and discrete liquid levels. Not all methods are equally usable for continuous and point sensing. Discrete-level transducers ("point sensors") are generally
more numerous than continuous-level transducers. Two or more point sensors are
often used in the same vessel to determine different discrete levels. Some methods
are applied to the direct or indirect measurement of the total mass of liquid in a vessel.

8.2.1 Buoyancy sensing

Archimedes' principle—a body submerged wholly or partially in a fluid is buoyed
up by a force equal to the weight of the fluid displaced—is put to use in transducers
whose sensing element is a float, either hollow or made of a material lighter than the
liquid whose level is to be measured (Figure 8-1). The float tends to follow the
liquid/gas interface. The transducer case, containing a sealed-off transduction element,
is mounted to the top or side wall of the tank. The motion of the float relative to the
case causes an appropriate output from the transduction element. Floats are used in
continuous-level as well as discrete-level transducers. Permanent magnets are often
embedded in floats to actuate magnetic contacting mechanisms in a sealed transducer
case.

Fig. 8-1.   Level-sensing float.

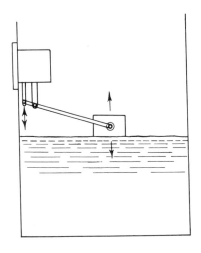
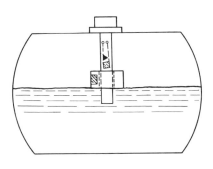

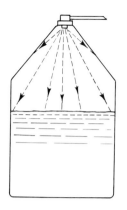

Fig. 8-2. Cavity-resonance liquid sensing.

8.2.2 Cavity-resonance sensing

Several related sensing methods are applied to the determination of liquid volume or level in a tank whose characteristics as a resonant cavity change with a varying amount of liquid in it. Electromagnetic oscillations at infrasonic, ultrasonic, or radio frequencies are excited within the cavity from a coupling element placed at the top of the tank (Figure 8-2). As the liquid rises in the tank the gas-filled cavity formed by liquid surface, tank walls, and tank top shrinks and its resonant frequency changes accordingly. When the resonant frequency of the empty tank is known and a scaling factor is applied, the level or volume of the liquid in the tank can be determined. The radio-frequency method is used with dielectric liquids. Variable-frequency (manually-controlled or sweep-frequency) oscillators can be used for the resonant-frequency search.

8.2.3 Conductivity sensing

The level of electrically conductive liquids can be sensed by immersing two electrodes into the liquid and monitoring the change in resistance between them (Figure 8-3). The electrical conductivity of the liquid is used in continuous- as well as discrete-level transducers. Some transducers contain only a single electrode placed close to the metallic tank wall which serves as the other electrode.

8.2.4 Dielectric sensing

Since the dielectric constant of a liquid in a tank is usually different from that of the gas, this characteristic can be used in capacitive liquid-level transducers. The variable-dielectric capacitive principle is widely used in continuous- and discrete-

Fig. 8-3. Conductivity level sensing.

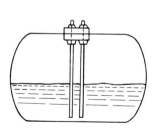
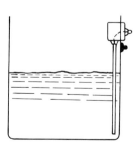
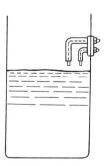

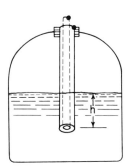 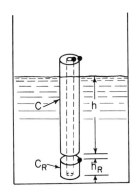 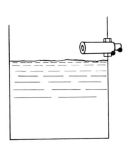

Fig, 8-4. Level sensing by dielectric variation.

level sensing devices (Figure 8-4). Single-electrode transducer designs using the tank wall as the other electrode have been used in narrow tanks where a relatively coarse measurement accuracy is acceptable. The sensing element is most frequently configured as two (sometimes four) coaxial tubes. Alternate tubes are connected together ("ganged") when more than one pair of tubes is used. A four-arm a-c bridge network converts the capacitance change due to rising or falling level into an a-c voltage. The level-sensing element forms one arm of the bridge. A fixed (reference) capacitor is used as its opposite bridge arm. To compensate for changes in characteristics of the liquid during measurement the reference capacitor (C_R) can be mounted in the tank together with the sensing capacitor (C) so that C_R is always fully submerged. Capacitor C_R, located near the bottom of the tank, has a fixed height (h_R) which is a small fraction of the height of capacitor C. The measured level (h) is measured to the top of capacitor C_R. The ratio of the capacitance change then equals the ratio of the measured level to the height of the reference capacitor, or $\Delta C/\Delta C_R = h/h_R$.

8.2.5 Heat-transfer sensing

The rate of heat transfer from a heated element is larger in liquid than in gas. This principle is applied to the operation of various discrete-level transducers (Figure 8-5). Resistive transducers use a wirewound or thermistor element whose resistance decreases due to cooling when contacted by liquid. Thermoelectric transducers sense the cooling of a wirewound heater element by a thermoelectric sensing junction attached to the element.

8.2.6 Nuclear-radiation sensing

Radiation emanating at a constant rate from a radioactive source in one side of a tank will reach a radiation detector in the opposite side of the tank to a lesser degree

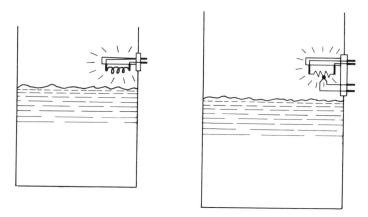

Fig. 8-5. Level sensing by heat transfer rate.

when the path is through liquid than when it is through gas (Figure 8-6). Gamma radiation from a source such as Cs^{137}, Co^{60}, or Ra^{226} is most commonly used. Its attenuation by liquid is caused mainly by absorption. Continuous-level sensing can be achieved with a vertical radiation path by monitoring the gradual change in detector output counts with rising or falling level. Single or multiple discrete levels can be determined by use of one radiation source and one or more detectors in hori-

Fig. 8-6. Radiation sensing of liquid level.

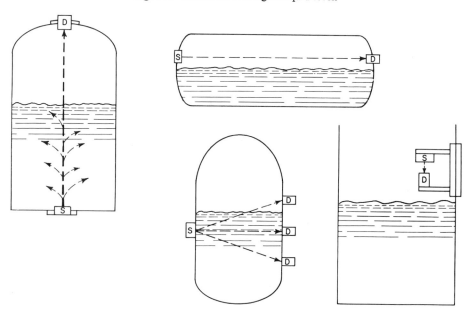

zontally opposite tank walls. When tank dimensions are small enough to allow a relatively short path length, the source as well as the detectors can be mounted against the outside of the tank.

8.2.7 Optical sensing

Discrete-level transducers use the interaction of a liquid with a light beam in two ways (Figure 8-7). Light incident upon a glass/fluid interface is transmitted through the interface when the fluid is a liquid but will be reflected by the interface when the fluid is a gas. When the interface is so shaped that a reflected light beam falls on a suitably located light detector, the output of that detector will indicate the presence or absence of liquid at the interface. Since the light beam is either reflected or not reflected, the transducer acts as "on-off" device.

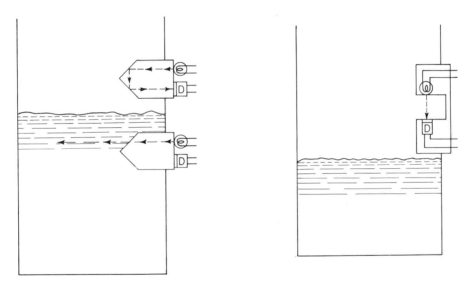

Fig. 8-7. Level sensing by optical means.

A related method uses the reduction in light seen by a light detector when liquid in the path between it and the light source causes partial scattering or absorption of light.

8.2.8 Pressure sensing

Light level is often measured in terms of *head*, the height of a liquid column at the base of which a pressure, p_L, is developed (Figure 8-8). When the specific weight

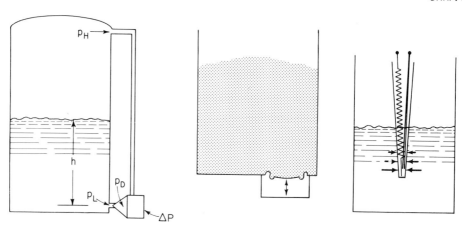

Fig. 8-8. Pressure sensing of level.

of the liquid, w, is known, the level, h, above the point at which p_L is measured can be determined from the difference between the pressure on the surface, p_H, and the pressure p_L in a stationary tank by

$$h = \frac{p_L - p_H}{w}$$

The pressure difference ($p_L - p_H$) is customarily measured by a differential-pressure transducer pneumatically connected between the top and bottom of the tank. For a liquid of any given specific weight the output of this transducer is then directly proportional to level.

Head measurement has been used for level monitoring of powdered and granular solids in tanks vented to the ambient atmosphere by making a sensing diaphragm integral with the bottom of the tank and transducing the diaphragm deflection.

A special pressure-sensing method is applied in a proprietary device which relies on the compression of an immersed flexible sleeve to press a portion of a continuous electrical contact against a continuous strip-type electrical resistance element so that the resistance between strip and contact terminals decreases with rising level.

8.2.9 Sonic-path sensing

Level sensing by beamed sound energy is accomplished by two different methods (Figure 8-9). Continuous level changes can be determined by an echo-ranging technique (see 11.6) using either a sound projector in combination with a sound receiver or a single-element sound transceiver operating alternately in the "transmit" and "receive" mode. The "target" in this system is the liquid/gas interface, whose distance from the transceiver (or projector and receiver combination) can be determined by

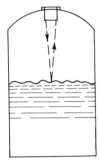
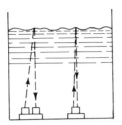

Fig. 8-9. Sonic sensing of liquid level.

measuring the time required for a transmitted pulse to be reflected by the interface and return to the receiver and by knowing the velocity of sound in the fluid through which the sound travels.

Discrete levels are sensed by a sound projector and a sound receiver placed opposite each other. When the sonic path between the two devices is through gas, the receiver output represents normal reception of sound. When the level rises so that the sonic path is through liquid, the output of the receiver either changes radically or drops to zero, depending on circuit adjustments. The frequency of sound employed in these methods is usually in the ultrasonic range (20 to 100 kHz).

8.2.10 Damped-oscillation sensing

The change in damping caused by a change from gas to liquid of the fluid ambient to an oscillating element (Figure 8-10) is used for level indications by two types of point sensors. One type uses a vibrating paddle whose oscillation amplitude is reduced, due to increased viscous damping, when the paddle is submerged in liquid. The amplitude changes are applied to a (typically reluctive) transduction element. The other type, which exists in more design variations, utilizes a piezoelectric or magnetostrictive element which oscillates in a gaseous medium but stops oscillating, due to acoustic damping, when the medium changes to liquid. Oscillation frequencies are chosen in the ultrasonic range for the latter type, and in the low audio-frequency range for the former type.

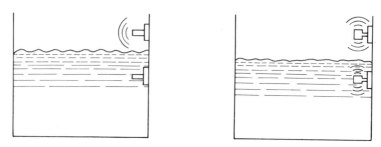

Fig. 8-10.   Level sensing by oscillation damping.

8.2.11 Weight sensing

The level, volume, or mass of a liquid in a tank can be determined by weighing a tank of known geometry continuously or at discrete intervals and by subtracting the weight of the empty tank (tare weight) from the value obtained. Alternately, the empty tank is balanced by a mass, and the force sensed by a force transducer in the weighing system is a measure of the tank contents. These methods are illustrated in Figure 8-11.

8.3 Design and Operation

8.3.1 Capacitive liquid-level transducers

Continuous- and discrete-level transducers whose capacitance changes with variations in the dielectric between one or more pairs of electrodes exist in a variety of configurations. Capacitance increases with rising level of the liquid, whose dielectric constant is greater than that of the gas it replaces.

Fig. 8-11.   Level determination by weighing.

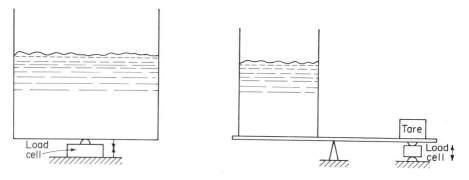

8.3.1.1 Capacitive continuous-level transducers use either a single probe, two or four concentric cylinders, or special electrode shapes. Single-probe transducers rely on the tank wall to act as the other (grounded) electrode. When intended for use in electrically conductive liquids they are coated with a thin film of plastic or ceramic material. The conductive liquid itself then acts as the grounded electrode.

Spacing between concentric cylinders is maintained by rings of a hard insulating material such as Teflon®. Some designs use discs of the same material in the inner cylinder to stiffen it and to help maintain its diameter. When probe-type transducers are installed through the tank wall, a glass or ceramic seal is provided in the mounting portion of the probe. Four concentric cylinders, with alternate electrodes connected together (Figure 8-12), can be used where a larger capacitance change is required.

The probe length is always chosen slightly longer than necessary to reach the lowest level to be sensed. Capacitive level probes have been built in lengths exceeding 10 m. When severe sloshing of the liquid is expected, a cylindrical *stillwell* can be installed firmly in the tank. The probe is then inserted into the stillwell which is slotted or perforated or, when a specific amount of damping is required, provided with an opening at its bottom end only. The capacitance of one pair of concentric cylinders is typically 120 pF per m of length. Capacitive elements can be contoured to match tank geometry.

Fig. 8-12. Capacitive continuous level transducer (developed by Convair Div., General Dynamics Corp.).

8.3.1.2 Capacitive discrete-level transducers have been produced in a multitude of shapes, such as four coplanar concentric rings, two, three, or four circular or rectangular discs, two-cylinder probes, single-rod or dagger-shaped probes, and several ganged rods forming a squirrel-cage around a central cylindrical electrode. An on/off signal is provided by associated circuitry when the capacitance changes due to presence or absence of liquid. Liquid-after-gas detection is generally more reliable than gas-after-liquid (falling level) detection since some of the liquid will tend to adhere to the probe. This problem can be minimized by appropriate probe design, material, and finish, chosen to prevent any significant adherence of liquid.

8.3.1.3 Signal conditioners generally fall into two categories, open-loop and closed-loop. Discrete-level transducers use open-loop signal conditioners which convert and amplify the transducer output to provide a definite on-off signal, such as

Fig. 8-13. Capacitive discrete level sensor with integral control unit in junction box (courtesy of Controlotron Corp.).

by use of a relay or electronic switch operated by the amplified output. The necessary circuitry can be packaged integrally, as in the point-sensor shown in Figure 8-13, or separately. When the signal conditioner is separately packaged, the transducer is best connected as a three-terminal device, using a separate shielded cable for each of the two electrodes. The shields are grounded. Since the fixed capacitance between the conductor and shield of each cable is much greater than varying stray capacitance would be between an unshielded conductor and structure ground, the system is essentially insensitive to cable length and random capacitance variations.

Typical frequencies of the a-c excitation to the transducer are 400 Hz (the frequency of airborne generators and inverters) or 10 kHz (where maximum system sensitivity is obtained in most level-sensing applications); however, excitation frequencies up to 200 kHz have been successfully used.

Continuous-level transducers use either open-loop or closed-loop signal conditioners. The latter function as self-balancing electronic bridges in which the capacitive transducer forms one leg of the bridge. An a-c voltage controlled by capacitance changes acts to rebalance the bridge when it is fed back to it. Signal conditioners for continuous-level transducers are usually packaged separately. They can provide either an analog or digital output.

8.3.2 Magnetostrictive liquid-level transducers

Magnetostrictive transducers are used for continuous- and discrete-level measurements using the sonic-path method, and for discrete-level sensing, only, when the damped-oscillation method is applied. The latter is the method most frequently employed; hence, most magnetostrictive transducers are point sensors.

The coil assembly of a damped-oscillation point sensor (see Figure 8-14) consists of a drive coil and a feedback coil, both wound on a ferromagnetic rod. Either a solid or hollow rod can be used. When an iron rod is subjected to a longitudinal magnetic field, it increases slightly in length (*Joule effect*). A nickel rod decreases in length under these conditions. Due to the inverse Joule effect (*Villari effect*), a change of magnetic induction occurs within the rod under longitudinal stress. The combination of these effects is applied in a *magnetostrictive resonator* in which the rod is maintained in longitudinal elastic vibration. The rod is retained against the probe tip. Drive and feedback coils are connected in a current-driven feedback oscillator. When in oscillation, the rod and probe tip vibrate at the frequency controlled by the characteristics of the rod, typically 40 kHz. Signal levels are so adjusted as to maintain oscillation in the rod only when the probe tip is exposed to a compressible fluid such as air, froth, or foam. As soon as the probe tip encounters a noncompressible fluid—a liquid—the vibration is damped so that oscillation stops. When the level of the damping fluid falls below the probe, oscillation begins again. The probe sheath is made of either aluminum or stainless steel and is hermetically sealed.

The oscillator and output-conditioning circuitry required to produce an on/off signal is usually packaged in a separate control unit connected to the transducer by coaxial shielded cables, one cable for each of the two coils. The capacitance which the cables may present to the oscillator circuit may be quite high; hence, the length of the cables is limited. Possible installation problems due to this limitation are eliminated when the control unit, which usually incorporates a relay, is integrally packaged with the transducer (see Figure 8-15).

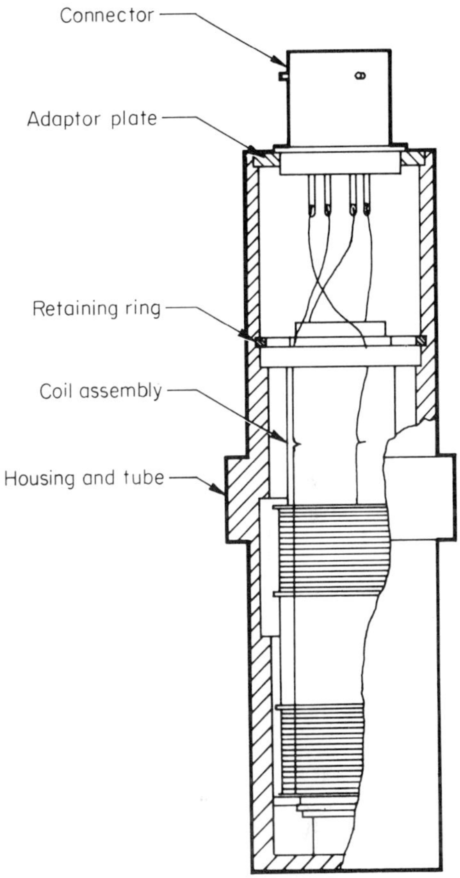

Sonic-path type discrete-level transducers use the principle of the vibrating magnetostrictive resonator in the sound projector but rely on the Villari effect in the sound receiver. Both devices must have the same resonant frequency within very close tolerances. Continuous sonic-path ("Sonar") level sensors use pulsed sound energy.

Fig. 8-14. Basic internal construction of magnetostrictive discrete-level transducer (courtesy of Powertron Div., Conrac Corp.).

8.3.3 Nucleonic liquid-level transducers

Transducer systems using one or more radioisotope sources and one or more radiation detectors have been applied primarily to discrete-level sensing, although quasi-continuous level measurements have been made by a strip containing a large

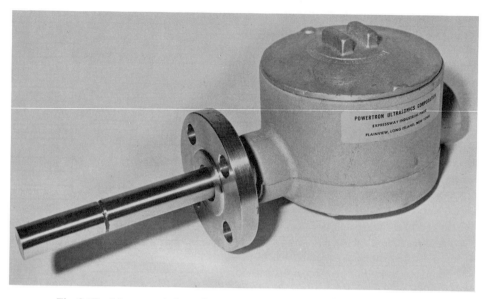

Fig. 8-15. Magnetostrictive point sensor with integral control unit (courtesy of Powertron Div., Conrac Corp.).

number of detectors installed vertically on the outside of a tank. Few commercial liquid-level transducers employ the nuclear-radiation method of sensing since the existence of a multitude of other usable devices makes competition difficult. In specialized applications, such as level measurements of caustic slurries, molten metals, scrap metals, and molten glass, a system containing one source and one Geiger-Mueller detector, mounted to opposite outside surfaces of the tank wall, has been reported usable.

Gamma-emitting radioactive isotopes of cobalt, cesium, and radium have been used as sources. The source holder can be provided with a remotely actuated shield to avoid radiation exposure of maintenance personnel. The detector output, which increases when the level of the radiation-absorbing material is below the radiation path, can be used to close a valve or actuate an alarm, a pump, or other control device.

8.3.4 Photoelectric liquid-level transducers

The optical level-sensing method is applied in the design and operation of primarily those transducers typified by the sensor shown in Figure 8-16. A light source and light detector (photoelectric cell) are mounted side by side but are separated by an opaque partition in the transducer housing from which a glass prism extends. When no liquid is present at the dihedral end surfaces of the prism, the beam is reflected toward the photo electric cell and a relatively large output signal is produced.

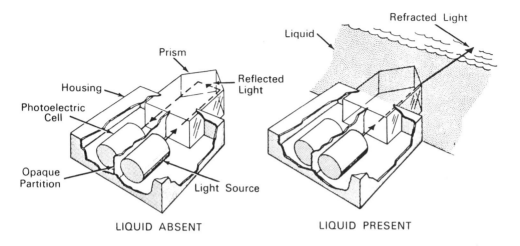

(a) Design principle

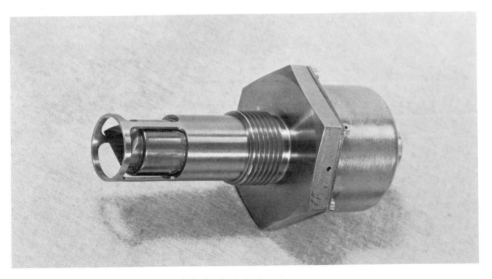

(b) Typical design

Fig. 8-16. Photoelectric discrete-level transducer (courtesy of Bendix Corp.).

When liquid is present at the prism end, the light beam is refracted at the incident face of the prism and passes into the liquid instead of being internally reflected to the photoelectric cell. As a result, a very small output signal will be produced.

Discrete-level sensors of this type have been used for liquid level sensing at liquid temperatures from $-430°F$ to $+160°F$, where the index of refraction of the liquid is

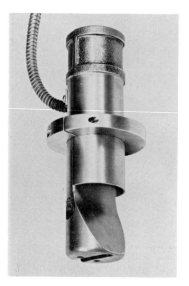

(a) Filling-nozzle gas liquid sensor

(b) Filling-nozzle sensor
installed in tank top

(c) Gas liquid interface sensor for pipelines

Fig. 8-17. Sonic-path liquid-level transducers using piezoelectric transmitter
and receiver (courtesy of National Sonics Corp.).

greater than 1.09. The housing is usually hermetically sealed, and a metal sheath within a mechanical protecting cage covers all but the tip of the prism. The shape of the probe tip is intended to prevent adhesion of liquid with falling level. A silicon "solar cell" is commonly used as light detector. Its output in the absence of liquid at the prism is between 15 and 20 times that produced when liquid is present. This permits the use of a simplified direct-coupled switching circuit as output amplifier. No amplifier is needed when only a d-c voltage indication is required.

8.3.5 Piezoelectric liquid-level transducers

Piezoelectric transducers are used for level measurement in the same manner as the magnetostrictive transducer described in 8.3.2 from which they differ only in that they use a piezoelectric crystal as oscillating element (at ultrasonic frequencies) instead of a magnetostrictive rod. In damped-oscillation sensors the crystal is normally mounted at the tip of a hermetically sealed probe sheath; however, exposed crystals in free-flooding perforated sheaths have been designed for special applications. Oscillation is damped out by the presence of liquid on the probe tip or on the exposed crystal.

Various designs exist using the damped-oscillation sensing method as well as the sonic-path method. The latter is applied in the special versions of discrete-level sensors illustrated in Figure 8-17. The filling-nozzle sensor shown in Figure 8-17 (a) and (b) has its liquid outlet above the slanted portion (deflector) which forces the liquid away from the sensor to avoid any false indications due to spray or splash. During tank filling, gas escapes through the vertical gap (air vent) which appears at the probe tip. This gap continues through the length of the probe, beyond the mounting flange. A piezoelectric transmitter and receiver face each other across the gap. The sensor characteristics are so selected that the sonic path exists only through liquid, not through air. Hence, the receiver will provide an output only when the tank is so full that liquid starts to flow through the air vent. The distance between the mounting flange and the two piezoelectric crystals determines the level in the tank which produces an output from the transducer. A separately packaged control unit, connected to the sensor by a dual coaxial cable, usually furnishes a command signal to a valve to shut off liquid flow.

The transducer system shown in Figure 8-17 (c) provides an output from the receiver when the gas/liquid interface is high enough in the pipe to provide a sonic path through liquid. When an attenuation mode is possible, the output signal can then also be used to determine which of several known liquids is flowing through the pipeline.

8.3.6 Potentiometric and related float-actuated liquid-level transducers

The buoyancy of liquid has been utilized in float-actuated continuous- and discrete-level sensors for a long time. The float ("displacer") is cylindrical or annular and can be a hollow brass, aluminum, stainless-steel, ceramic, or plastic shape, or a solid shape made of some material generally lighter than the liquid to be measured. Displacers heavier than the liquid are also used, restrained by a spring which pulls up the tall displacer when a rising level reduces its weight.

Continuous-level transducers have reluctive or, more frequently, potentiometric transduction elements. The float is connected or linked to an actuating rod whose motion is transmitted to the reluctive core or potentiometric wiper either through a sealed bellows or membrane or by noncontacting magnetic action. Figure 8-18 shows the actuating rod in a sealed tube tipped with a ferromagnetic "attraction ball." A follower magnet in the transducer case moves with this ball along the outside of the tube, due to magnetic attraction through the tube wall. The permanent magnet is mechanically linked to the potentiometer wiper arm. The potentiometer can be connected as a rheostat to provide an output in terms of resistance rather than voltage-ratio.

Discrete-level sensors normally have the permanent magnet embedded or mounted directly in an annular float. A typical design is shown in Figure 8-19. A magnetic reed switch contains a reed contact attracted toward the magnet when the float shifts it into the actuating position so that the contact closes. The contact rating of the switch is sufficient for all indications and most control requirements. A special design variation of this type of sensor is a potentiometric device which has a number

Fig. 8-18. Potentiometric liquid-level transducer (connected as rheostat) (courtesy of Thomas A. Edison Industries).

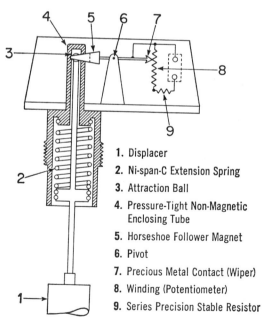

1. Displacer
2. Ni-span-C Extension Spring
3. Attraction Ball
4. Pressure-Tight Non-Magnetic Enclosing Tube
5. Horseshoe Follower Magnet
6. Pivot
7. Precious Metal Contact (Wiper)
8. Winding (Potentiometer)
9. Series Precision Stable Resistor

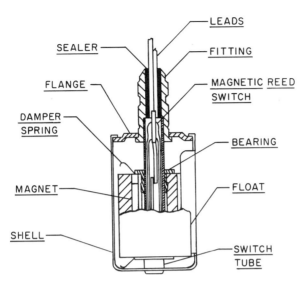

Fig. 8-19.   Float-actuated liquid-level switch (courtesy of Revere Corp. of America).

of taps on its element. Each tap is connected to a magnetic reed switch. Sequential closing of the switches as the float rises with increasing level in a tank provides an output change in discrete steps.

8.3.7 Resistive liquid level transducers

Discrete liquid-level sensing by the heat-transfer method is usually accomplished with resistive transducers, although thermoelectric transduction elements have been used in some designs (see 8.3.8). Wire elements, thermistors, and carbon resistors are used as sensing/transduction elements. Their resistance undergoes a step change when their ambient fluid is changed from gas to liquid since heat is transferred more rapidly from the element to the fluid when the latter is a liquid. Resistive devices are also used for discrete-level sensing by the conductivity method. Applications of either method to continuous-level measurements are rare.

8.3.7.1 Carbon resistors have been employed only in special sensors for cryogenic liquid-level detection. Since the temperature coefficient of resistance of carbon is negative, a step increase in resistance is observed when the level of the liquid reaches the resistor. Satisfactory results are reported for resistors having a nominal room-temperature value of between 500 and 700 ohms when used in gaseous and liquid helium. Performance is improved when the mass of the resistor is reduced by cutting its leads as short as possible, by removing the phenolic insulation around the carbon element, and by mounting the resistor in a stagnation shield.

8.3.7.2 Thermistors have a large negative coefficient of resistance and undergo a step increase in resistance when immersed in liquid. Small bead or disc thermistors are used as one arm of a Wheatstone bridge. Some configurations use a second thermistor in the opposite arm of the bridge. This reference thermistor is mounted close to the sensing thermistor so that both see the same ambient temperature; however, it is sealed in an air- or inert-gas-filled enclosure to minimize transfer of heat from it to the liquid. A number of commercial thermistor-tipped probes have been manufactured with mountings and seals similar to those of probe-type resistive temperature transducers.

8.3.7.3 Resistance-wire elements are used primarily for cryogenic liquid-level sensing. Platinum wire, 0.001 in. or less in diameter, is commonly used in "hot-wire probes" for liquefied gases to provide a short time constant, good sensitivity, and resistance change linear with temperature. Platinum wire has a positive temperature coefficient of resistance. The magnitude of the coefficient is considerably smaller than that of a thermistor. Metals other than platinum, but of similar behavior, have also been used. The operating temperature of the wire, for a given power, is a function of the heat transfer coefficient between the wire and the ambient fluid. The wire is heated by passing sufficient current through it.

When a wire immersed in liquid is heated, convection currents will first circulate the superheated liquid and evaporation will occur only at the surface of the liquid. When heating is increased, bubbles of vapor will form at minute spots on the wire surface (*nucleate boiling*). They will tend to rise in the liquid but will condense before reaching the surface. With further heating more and larger bubbles will form and reach the surface. Heat transfer will have increased from the initial phase through the two nucleate boiling phases. A peak is reached beyond which heat transfer drops due to an unstable film of vapor forming at the wire surface. With a further increase in heat, the stable *film boiling* phase is reached. Heat transfer is greatly reduced in this phase since a film of vapor insulates the hot wire thermally from the liquid. Hot-wire sensors are therefore designed to operate at a sufficiently low temperature to assure that the nucleate boiling phase will not be exceeded.

Two types of electrical circuits are used with wire-element liquid-level transducers. The element can form one arm of a Wheatstone bridge. When so connected, a second element, maintained at the same ambient temperature as the first but thermally isolated to avoid heat transfer from it to the liquid, is sometimes connected as the opposite arm of the bridge. The element can also be excited from a constant-current supply so that the transducer output originates from the voltage change due to the change of element resistance upon immersion in liquid.

To improve response time, as much wire surface as practical is exposed to the ambient fluid. Figure 8-20 illustrates a special sensing element used in a discrete-level sensing probe intended for cryogenic liquids. The probe sheath is about seven inches long and is perforated near its tip, where the sensing element is installed. The 0.001-in. diameter platinum wire is loosely supported from wire hooks to avoid wire strain. The active element is contained entirely in a free-flooding area. The wire can be

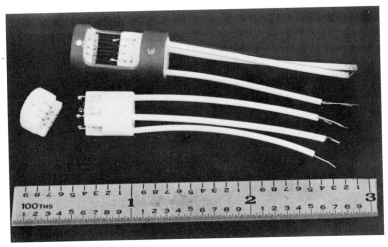

Fig. 8-20. Resistance-wire sensing element of discrete-level transducer (developed by Convair Div., General Dynamics Corp.).

gold plated to avoid its acting as a catalyst with certain fluids. A similar design, but a different transducer configuration, is shown in Figure 8-21. The 0.005-in. diameter gold-plated platinum wire is wound as a grid over small insulating studs to form a "planar" winding. Proper orientation during installation assures that the grid is parallel to the surface of the liquid. Since the grid is very thin, an accuracy within ± 0.015 in. of level is claimed by the manufacturer of this device.

The associated excitation, signal-conditioning, and output circuitry of resistive liquid-level transducers is separately packaged in a compact control unit. Shielding of interconnecting wires is normally not required. The room-temperature resistance of wire elements is typically between 10 and 50 ohms.

8.3.7.4 Liquid conductivity is used to measure level by resistive transducers in which the change in resistance between two electrodes, or between one electrode and the grounded tank wall, is caused by the conductive liquid itself. When the liquid comes in contact with the electrodes, the resistance between them is drastically reduced. A portion of the metallic electrode is usually insulated with a glass or plastic coating. Figure 8-22 illustrates a transducer system which can be used for sensing two discrete levels of a conductive liquid in a metal tank. Immersion length can be adjusted at the mounting which also provides a seal. The control unit supplies a sensing current sufficiently small to avoid causing electrolysis or a potential explosion hazard.

8.3.8 Thermoelectric liquid-level transducers

Liquid-level transducers with thermoelectric transduction elements are not commonly used. Thermocouples have been used to detect discrete levels of hot

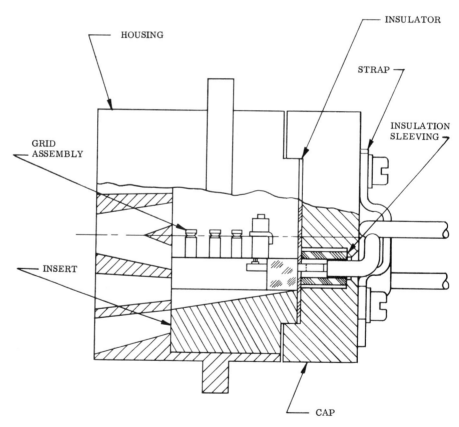

Fig. 8-21. Wire-element resistive discrete-level transducer (courtesy of Acoustica Associates, Inc.).

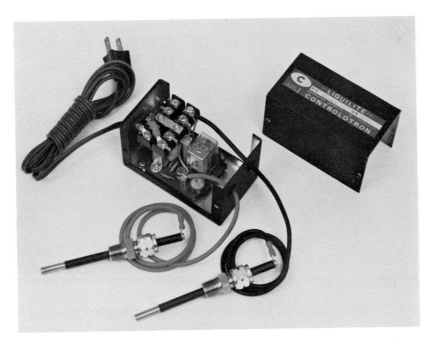

Fig. 8-22. Single-electrode conductivity-type discrete-level transducers with control unit (courtesy of Controlotron Corp.).

liquids, including liquid metals; however, when operating in this mode, the thermocouples are actually used as temperature transducers.

One point-sensor design has a thermoelectric sensing junction welded to the center of a steel wire which is heated by current from a constant-current supply. The wire is mounted near the tip of a perforated probe sheath. The thermoelectric reference junction is also contained within the probe but is located a short distance away from the wire. When the probe is immersed in liquid, the sensing and reference junctions are at the same temperature. When it is in gas, the temperature of the wire rises due to the reduced heat transfer to its ambient fluid, the sensing junction attains a temperature higher than that of the reference junction, and a thermoelectric electromotive force is generated. A magnetic amplifier and a rectifier are used to condition the small signal so that the transducer system has no output when the probe is in liquid and an output as large as 28 V d-c when it is in gas.

8.3.9 Vibrating-element liquid-level transducers

Unrelated to the magnetostrictive (8.3.2) and piezoelectric (8.3.5) transducers, whose oscillation at ultrasonic frequencies stops when damped in liquid, is one type of transducer whose mechanical sensing element is caused to vibrate at a certain ampli-

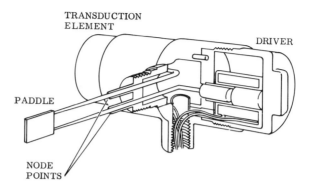

Fig. 8-23. Vibrating-paddle discrete-level transducer.

tude when in gas and at a reduced amplitude when in liquid. A simplified example of such a design is illustrated in Figure 8-23. The vibrating element is a paddle excited into mechanical oscillation at the line-voltage frequency, typically either 60 or 400 Hz. The paddle structure is designed to have a natural frequency close to the excitation frequency. The displacement of a portion of the paddle structure is transduced by a reluctive element. When the paddle is immersed in liquid, the amplitude of its vibration is damped. This causes a substantial reduction in the a-c output of the transducer.

8.4 Performance and Applications

8.4.1 Specification characteristics

Characteristics to be considered for inclusion in specifications will depend, to some extent, on the sensing method and transduction principle employed. Most characteristics, however, are applicable to all types of liquid-level transducers.

8.4.1.1 Mechanical design characteristics include all outline dimensions, mounting dimensions, and details of mounting provisions, relative locating dimension of the sensing element, of electrical connections, of any tubing connections, and of the nameplate, and weight. Most of these should also be specified for any separately packaged control or signal-conditioning unit. Materials and construction of the transducer, type of sealing, acceptable measured and ambient fluids, measured fluid operating pressure, proof-pressure and burst-pressure ratings, and type of end-use installation should be stated. The flow rate of the fluid should be considered. The electrical connector and its mating connector and the required interconnecting cable should be described. Any limitations on cable length should be stated. Nameplate information should be listed in detail.

8.4.1.2 Electrical design characteristics to be specified for the transducer comprise excitation voltage or current, frequency of a-c excitation, power required or maximum current drain, insulation resistance and breakdown-voltage rating, output impedance, input impedance when excitation conditioning is not incorporated, and the electrical characteristics of any coaxial or shielded interconnecting cable. For the signal-conditioning or control unit, when separate, the excitation requirement and all pertinent characteristics of the output should be described. The load impedance which can be seen by the control-unit output terminals should always be shown with tolerances.

8.4.1.3 Performance characteristics can be shown for the transducer alone. When a separate signal-conditioning or control unit is required, end-to-end performance is usually specified for the entire transducer system including any special or length-limited interconnecting cables.

Specifications for continuous-level transducers should show the measuring range and either end points or sensitivity as output characteristics. Linearity, hysteresis, and repeatability or an error band about a stated reference curve comprise the accuracy characteristics. Tolerances for additional errors should be considered for certain sensing methods or transduction principles. Examples of such errors are friction in potentiometric elements and effects of changes in the physical or chemical characteristics of the measured fluid.

Performance characteristics of discrete-level transducers are generally limited to repeatability and a description of output "in gas" and "in liquid." When repeatability is shown, any additional "accuracy" tolerances should be explained.

Dynamic performance characteristics are shown as response time for discrete-level transducers and as time constant and, sometimes, a 95, 98, or 99% response time for continuous-level transducers or transducer systems. It is essential that additional statements explain whether the dynamic characteristics apply for rising or falling level or for the liquid-to-gas or gas-to-liquid transition (sometimes a tolerance is shown for each). It is also essential to state the flow velocity or absence of flow or turbulence of both fluids as well as the type and temperature of both fluids when the response is affected by these characteristics. An example of a correct callout for a point sensor would be "95% response time: 50 ms, maximum, when immersed from still air into still automotive lubricating oil at any temperature between $+40$ and $+120°F$."

8.4.1.4 Environmental characteristics. An operating temperature range should be shown for the measured fluid and for the ambient atmosphere around the probe head (portion external to the tank or duct) as well as around any separate signal-conditioning or control unit. The allowable thermal effects on accuracy characteristics of both should be specified. The same considerations apply to measured-fluid pressure and ambient atmospheric pressure. Acceleration, vibration, shock, acoustic noise, magnetic fields, electromagnetic interference, and nuclear-radiation levels, and the allowable effects of such environmental conditions should be stated whenever they can be expected during operation of the transducer. Certain additional environmental conditions, e.g., exposure to ambient humidity and a corrosive or contaminating atmosphere, are specified for the associated package and the probe head only.

8.4.1.5 Reliability characteristics are shown in terms of operating life or cycling life, with the nature of a cycle described in detail for the latter. Long-term stability, when shown, refers to repeatability over a specified long period of time.

8.4.2 Applications and selection

8.4.2.1 Applications. Liquid-level transducers are used in virtually all industries for measurement and automatic-control applications such as tank metering, high- and low-level warning, liquid-leak detection, flow monitoring, tanking and voiding operations, and for fuel management and propellant utilization in vehicles ranging from automobiles to manned-space-vehicle boosters (Figure 8-24). Examples of liquids whose level is measured are: beverages, dairy products, distillery products, fruit juices, vegetable juices, soups, sweet and salt water, petroleum and coal-tar products, pharmaceuticals and chemicals, jet and rocket fuels, liquified gases, and liquid metals. Many types of sensors are also used in solid suspensions, slurries, and powdered and granular solids.

Discrete-level sensors are used more frequently than continuous-level types. Some installations use a number of point sensors spaced at equal intervals along the height of a tank for a quasi-continuous determination of level. The same purpose is served by a single probe assembly of a number of equally spaced sensing elements (e.g., annular electrodes of a capacitive transducer whose second electrode is a long metal cylinder), each providing a separate output. Signal-conditioning logic circuitry for digital display of multi-point sensors has been developed, in some instances with computation networks for interpolation between points.

8.4.2.2 Considerations for selection after continuous or discrete sensing has been chosen are based primarily on the characteristics of the measured liquid, including its temperature, viscosity, and conductivity. The next most important factors are dynamic and accuracy characteristics and cost. The following considerations are generally applicable to the more frequently used transducer types.

Capacitive, damped-oscillation magnetostrictive and piezoelectric, nucleonic, buoyancy, and vibrating-element transducers can be used for all liquids over a wide range of liquid temperature. Photoelectric transducers can be used in most liquids, but the index of refraction of the liquid is a limiting factor. High temperatures can affect the light source and light detector adversely. Resistive transducers are used where no explosion hazard is caused by their use. Those employing thermistors are

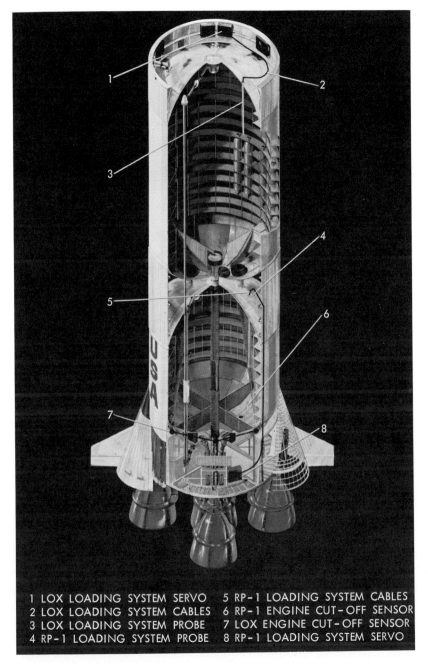

Fig. 8-24. Liquid-level sensors and control systems installed in Saturn S-1C booster (courtesy of Bendix Corp.).

1 LOX LOADING SYSTEM SERVO   5 RP-1 LOADING SYSTEM CABLES
2 LOX LOADING SYSTEM CABLES   6 RP-1 ENGINE CUT-OFF SENSOR
3 LOX LOADING SYSTEM PROBE   7 LOX ENGINE CUT-OFF SENSOR
4 RP-1 LOADING SYSTEM PROBE   8 RP-1 LOADING SYSTEM SERVO

used in the medium range of liquid temperatures, between $-80°F$ and $+200°F$. Hot-wire sensors are intended for cryogenic use. Conductivity sensors can be used only with conductive liquids but are not limited by temperature.

The cost per transducer system of resistive, photoelectric, vibrating-element, and float-type sensors is relatively lower than that of capacitive and nucleonic designs. Ultrasonic transducers span a wide range of cost depending on complexity, especially of the associated circuitry. The time constants of liquid level transducers are dictated by their design; however, most resistive transducers, with the exception of the conductivity types, are somewhat slower in response than other types.

Acceleration effects are most pronounced in buoyancy-type transducers. Nucleonic transducer systems whose detectors require high-voltage connections need protection against corona discharge at low ambient pressures. They can also be affected by external radiation at very high altitudes. Photoelectric transducers can give false indications due to incidence of light from an external source. Damped-oscillation sensors can be adversely affected by high or fluctuating pressures in the tank. Magnetostrictive, piezoelectric, and capacitive transducers require special attention as to impedance, length, and grounding of their coaxial interconnecting cables.

All transducers must be designed to minimize the possibility of erroneous indications due to slosh, spray, splash, or adherence of liquid during operation in gas but close to the surface of the liquid. Their time constant is often considerably longer for the liquid/gas transition than for the gas/liquid transition.

8.5 Calibration and Tests

Calibrations and tests are normally performed on an end-to-end basis; i.e., the entire transducer system is tested whenever a specific separately-packaged signal-conditioning or control unit and associated interconnecting cabling of a specific type and length must be used with the transducer. Certain tests can be performed on the associated system by substituting an electrical quantity (e.g., a fixed capacitance or resistance) for the transducer.

8.5.1 Calibration

The calibration of liquid-level transducers is comparatively simple since the reference quantity—the height of the surface of a liquid—is a linear dimension whose measurement with a certainty of ±0.005 in. (about ±0.1 mm) is quite simple and adequate even for laboratory calibrations. The allowable uncertainty of many production calibrations can be considerably larger than this amount. The level at which a discrete or analog output is obtained from the transducer can be determined visually by using either a sight glass or an observation port and measuring the height on a vernier scale. Most transducers can be calibrated either by keeping the level fixed and mechanically immersing and retracting the test specimen or by raising and lowering the level while the specimen remains in a fixed position. Immersion must be referenced to a specific dimension or alignment mark on the transducer since the sensing element has a finite vertical thickness when installed in a tank or duct.

8.5.2 Mechanical and electrical inspection

Certain inspections are performed on each transducer or transducer system before calibration. Nameplate information is verified. Critical dimensions, connections, and attachment provisions are checked for conformance to the applicable drawing. Care must be taken not to damage sensing elements within perforated sheaths with sharp inspection tools. Proof pressure is applied while the transducer is mounted in a fixture designed to simulate the intended installation, and absence of leakage is verified. An inert gas or filtered air should be used for pressurizing. Insulation-resistance or voltage-breakdown tests and any required impedance and continuity checks are made between specified connector pins or other cable terminations. Excitation and output characteristics should be verified during calibration. Caution should be exercised not to exceed the current rating of any switch contacts during electrical checks.

8.5.3 Performance verification

Accuracy characteristics are determined from the calibration record, which should provide sufficient data for this purpose. Time constant and response time are verified during qualification and sampling tests. Test conditions should be fully described on the test-data sheet. They should be in accordance with any condition, e.g., fluid temperatures, which may be contained in the specification. When no such conditions are specified, the actual conditions expected to prevail during the transducer's intended use should be simulated as closely as possible. It is good practice to determine dynamic characteristics for rising as well as falling level even if only one of these is specified.

8.5.4 Environmental tests

When specific levels of various environments are applied to liquid-level transducers, the environmental effects may be more severe in gas or in liquid or at different depths of immersion, depending on transducer design and materials. When feasible, and when justified by the intended application, the fluid characteristics should be so chosen as to maximize environmental effects. If a transducer performs adequately under these conditions, it can be relied upon to perform at least as well when fluid characteristics are such that they tend to lessen the environmental effects.

Fluid-temperature tests can be performed with either gas or liquid (or both) at first one, then the other, limit of the operating temperature range. Ambient-temperature tests on the transducer head and on the associated package and cables can require that the fluid at the sensing element be either at the same, or at a different, temperature. The portion of the transducer system subjected to the test should be well stabilized at the test temperature before an output reading is taken. Additional temperature-gradient tests may be necessary when, e.g., the liquid is much colder or warmer than the gas in the end-use of the transducer, and rapid level changes are expected.

Fluid-pressure tests are performed with the transducer mounted into a pressurized tank. During ambient-pressure tests on the probe head or on the associated subsystem, the transducer sensing element may or may not be subjected to a different pressure, depending on specification requirements or end-use conditions. Similar considerations apply to humidity and contaminating- or corrosive-atmosphere tests.

During high-sound-pressure, nuclear-radiation, magnetic-field, and electromagnetic-interference tests the transducer is usually exposed to the ambient atmosphere only; however, if necessary, the sensing element can be partly or wholly immersed in liquid. Vibration, acceleration, and shock tests are almost never performed in liquid.

8.5.5 Life tests

Compliance with specified operating- and cycling-life requirements are verified only during a qualification test. Simple automatic equipment can be devised for mechanically immersing and retracting a transducer and for leaving it immersed and retracted for specific periods of time.

8.6 Special Measuring Systems

8.6.1 Level sensing with differential-pressure transducers

Although the application of the pressure-sensing method to liquid-level measurements employs pressure transducers (described in Chapter 10) rather than liquid-level transducers, this technique warrants further elaboration because it is so frequently used.

A number of different techniques of varying complexity are shown in Figure 8-25. All are based on the relationship $hw = p_L - p_H$, explained in 8.1.2. In the simplest configuration, Figure 8-25 (a), differential pressure between the liquid at a measuring point near the bottom of the tank and the ambient atmosphere is measured on a tank completely vented to the same atmosphere. This allows the use of a gage-pressure transducer flush-mounted to the tank so that its sensing element is normally covered by the liquid. The head of air between the transducer gage vent and the surface of the liquid is assumed to have a negligible effect on the measurement. The pressure (P_G) sensed in this manner is directly equal to hw. In the following examples the head of any gas in either line is assumed to have a negligible effect on measurement accuracy. The specific weight of the liquid is assumed to be much larger than

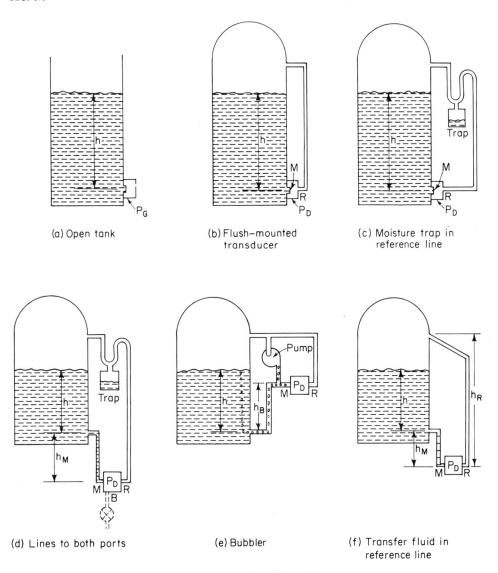

(a) Open tank

(b) Flush–mounted
transducer

(c) Moisture trap in
reference line

(d) Lines to both ports

(e) Bubbler

(f) Transfer fluid in
reference line

Fig. 8-25.   Level-sensing techniques using differential-pressure transducers.

that of the gas. When this assumption cannot be made, the gas head must be included in the calculations. The configuration shown in Figure 8-25 (b) can be used for a closed tank when there is no possibility of moisture condensation in the line connecting the gas in the top of the tank (*ullage*) to the reference port (*R*) of the differential-pressure transducer. The measurand port (*M*) is flush-mounted to the tank. The differential pressure (P_D) equals hw regardless of the absolute value of the ullage pressure. The transducer, however, must either be free of any reference-pressure effects, or the errors introduced by different values of ullage pressure must be known. This applies to all installations in closed tanks where a differential-pressure transducer is used for liquid-level measurements. When vapor in the ullage gas may enter and condense in the reference line, a trap can be inserted in this line as shown in Figure 8-25 (c). The trap prevents a liquid head of unknown height from building up against the transducer's reference port.

When the transducer cannot or should not be flush-mounted to the tank—as is the case in a majority of installations due to factors involving serviceability, high or low tank temperature, etc.—the system shown in Figure 8-25 (d) can be used. The transducer must be physically located below the level of the measuring point (*tap*) in the tank. To assure that only liquid fills the measurand line and the cavity enclosed by the sensing element of the transducer, the latter can be equipped with a "bleed port" (*B*) in this cavity. Before starting system operation all gas entrapped in the line is bled out of the transducer. When only liquid is vented the bleed port is sealed or closed by a valve. The differential pressure sensed by the transducer equals $(h + h_M)w$ and a correction for h_M must be made in the determination of liquid level.

When the transducer must be mounted above the level of the measuring tap, a *bubbler* system can be used to prevent a liquid head in the measurand line. A pump [Figure 8-25 (e)] can be used to blow compressed ullage gas through the line, or a separate gas supply can be used for this purpose. The gas vents into the tank at the measurand tap; hence, the pressure of the gas at the measurand port of the transducer equals the pressure in the liquid at this tap and the differential pressure equals hw, if the gas head in the line can be neglected. If it can not be neglected, the differential pressure equals $hw - h_B w_B$ where w_B is the specific weight of the gas.

An entirely different operating mode applies to the system shown in Figure 8-25 (f), where the reference line is filled with a transfer fluid. This may be necessary, e.g., when the ullage gas is corrosive to the reference side of the transducer. The fluid, usually a liquid at least equal in specific weight to the measured liquid, presents a head, h_R, to the reference port of the transducer; therefore, the reference pressure will always be higher than the measured pressure and a negative differential pressure is sensed which is zero when the level is at the ullage tap and maximum when it is at the measuring tap. The head of measured liquid in the measurand line, h_M, must also be considered.

Transducers of all types have been used to measure the differential pressure. Some manometric transducers were designed primarily for liquid-level measurements. Among these is the capacitive manometer, in which a mercury column, as one electrode, rises within a tube of dielectric material externally coated or surrounded by metal forming the other electrode. A related device contains the mercury column in a glass tube graduated internally with metal rings or bands, each used as electrical contact with the mercury acting as a common (shorting) contact.

8.6.2 Time-domain reflectometry for level sensing

Development of continuous-level sensing systems employing a coaxial probe and time-domain reflectometry has been pursued by research organizations such as Battelle Memorial Institute. Pulsed radio-frequency energy is applied to a coaxial transmission line with a solid extension in the form of a probe. The probe is installed vertically in a tank. The r-f pulse is square and has a quasi-instantaneous rise time and relatively long width. A portion of the pulsed energy is reflected by any impedance mismatch in the transmission line, which includes the probe. An oscilloscope is used to monitor the pulse as well as any reflection, which appears as a step in the pulse. The distance of the reflection-causing phenomenon can be determined from the time of its return. The liquid/gas interface is such a phenomenon. Fixed impedance mismatches can be placed in the coaxial probe at known locations to provide calibration reference marks in the oscilloscope display. The simple construction of the probe—a solid center conductor held coaxially in a conducting sleeve, both of stainless steel—permits level measurement under adverse environmental conditions such as liquid temperatures up to 1000°C.

8.6.3 Level sensing with variable-coupling transformers

Devices using the decrease in mutual inductance between two windings of a transformer due to the relative presence of liquid have been developed primarily for liquid-metal applications. An example of a continuous-level sensor is a vertical-immersion ceramic mandrel wound bifilarly with primary and secondary windings of nickel-alloy wire. The wound mandrel is inserted in a metallic protective sheath similar to a thermowell. The alternating current creates a fluctuating magnetic field around the primary winding. This field induces eddy currents in the secondary winding as well as in the thermowell and in any liquid (metal) in contact with the thermowell. As the level of the liquid rises, the increasing amounts of eddy currents induced in it cause a corresponding decrease of current induced in the secondary winding. The transducer (secondary winding) output, therefore, varies inversely with liquid level. This type of device has been operated in various liquid metals having melting points below 700°C.

BIBLIOGRAPHY

1. Vigoureux, R., *Ultrasonics*. London: Chapman & Hall, Ltd., 1950.

2. Friedman, G. J. and DeBottari, L., "Which System for Propellant Level Sensing?," *Space/Aeronautics*, August 1960.

3. Blanchard, R. L. and Sherburne, A. E., "A Digital Capacitance System for Mass,

Volume, and Level Measurements of Liquid Propellants," *Paper No. 2639-62*, New York: American Rocket Society, 1962.

4. Palevich, L. G., "Principles of Construction of Resonant-Cavity Level Gages," *Priborostroyeniye* (U.S.S.R.), No. 11, pp. 6–8, 1962.

5. Burgeson, D. A., Pestalozzi, W. G., and Richards, R. J., "The Performance of Point Level Sensors in Liquid Hydrogen," Boulder, Colorado: Cryogenic Data Center, National Bureau of Standards, 1963.

6. Olsen, W. A., "A Survey of Mass and Level Gauging Techniques for Liquid Hydrogen," *Advances in Cryogenic Engineering*, Vol. 8, pp. 342–359, 1963.

7. Wright, D. E., "Nucleonic Systems for Space Vehicle Propellant Gaging and Utilization," *Giannini Technical Note No. 3*, Duarte, California: Conrac Corp., 1964.

8. Frederick, J. R., *Ultrasonic Engineering*. New York: John Wiley & Sons, Inc., 1965.

9. Herbster, E. J. and Roth, J. H., "How to Gage by Capacitance," *I.S.A. Journal*, Vol. 12, No. 6, pp. 81–83, June 1965.

10. Welsh, F. M., "Digital Level Transducer," *Report No. EDC-1-65-33*, Cleveland, Ohio: Engineering Design Center, Case Institute of Technology, 1965.

11. Becker, W. and Sorel, F., "Les Erreurs dans la Détermination du Niveau d'un Liquide par Différence de Pression" (Errors in Liquid Level Determinations by Pressure Difference), *Report EUR 2958.f* (English summary), Ispra, Italy: EURATOM, Nuclear Research Establishment, 1966.

12. Canter, K. F. and Roellig, L. O., "Low Temperature Liquid Helium Level Indicator," *Review of Scientific Instruments*, Vol. 37, No. 9, pp. 1165–1167, September 1966.

13. Jackson, R. C., "Liquid Metal Level Measurements," *Preprint No. 1.1-4-66*, Pittsburgh: Instrument Society of America, 1966.

14. Johnson, T. R., Teats, F. G., and Pierce, R. D., "An Induction Probe for Measuring Liquid Levels in Liquid Metals," *Report ANL-7153*, Argonne National Laboratory, Illinois: 1966.

15. Perkins, C. K., "Capacitance Sensing for Weight Engineering," *Technical Paper No. 571*, Los Angeles: Society of Aeronautical Weight Engineers, 1966.

16. Conison, J., "How to Apply Transmitters for Measuring Liquid Level," *Instrumentation Technology*, Vol. 14, No. 7, pp. 63–65, July 1967.

17. Dozer, B. E., "Liquid Level Measurement for Hostile Environments," *Instrumentation Technology*, Vol. 14, No. 2, pp. 55–58, February 1967.

18. Fontana, A. A., "True Volume Measurement," *Instrumentation Technology*, Vol. 14, No. 10, pp. 74–75, October 1967.

19. Kaminski, R. K., "Sonics Measure Level," *Instrumentation Technology*, Vol. 14, No. 12, pp. 55–61, December 1967.

20. Kulwiec, R. A. and White, R. J., "Liquid-Methane Level Pegged Ultrasonically," *Chemical Processing*, pp. 34–35, September 1967.

21. Levins, D. B., "Fuel Quantity and C/G System," *Instruments and Control Systems*, Vol. 40, No. 3, pp. 87–88, March 1967.

9

Nuclear
Radiation

9.1 Basic Concepts

9.1.1 Basic definitions

Nuclear radiation is the emission of charged and uncharged particles and of electro-magnetic radiation from atomic nuclei. *Charged particles* include alpha and beta particles and protons; the latter are not emitted during radioactive decay. *Uncharged particles* are typified by the neutron. Gamma rays and X rays are the most common forms of (nuclear) *electromagnetic radiation*.

Alpha particles (α-particles) are nuclei of helium atoms. An alpha particle consists of two protons and two neutrons and has a double positive charge.

Beta particles (β-particles) are negative electrons or positive electrons (positrons). They are emitted when *beta decay* occurs in a nucleus, a radioactive transformation by which the atomic number is changed by $+1$ or -1 while the mass number remains unchanged.

Gamma rays (γ-rays) are electromagnetic radiation quanta resulting from quantum transitions between two energy levels of a nucleus.

Neutrons are uncharged elementary particles of mass number 1. The energy ranges are up to 10^2 eV for *slow neutrons*, 10^2 to 10^5 eV for *intermediate neutrons* and above 10^5 eV for *fast neutrons*. The term "thermal neutrons" has often applied to slow neutrons, although proper classification limits the energy range of thermal neutrons to about 0.03 eV.

Protons are positively charged elementary particles of mass number 1.

X rays are quanta of electromagnetic radiation originating in the extra-nuclear part of the atom. Their wavelength spectrum has been designated as the region between 100 Ångstroms (0.01 microns) and about 0.1 Ångstroms. *Soft X rays* have a relatively longer wavelength, and *hard X rays* have a relatively shorter wavelength. Hard X rays have a greater penetrating power than soft X rays.

Bremsstrahlung is a continuous spectrum of X-radiation produced by the acceleration or deceleration applied to a high-velocity charged particle when it is deflected by another charged particle. The literal translation of this German word is "braking radiation," from *Bremse* (brake).

Ionization is the process of formation of subatomic particles with a positive or negative charge (*ions*).

Scintillation is the emission of light energy (photons) by a photoluminescent material (*phosphor*) due to the incidence of ionizing radiation upon the material.

Cerenkov radiation is the radiation emitted by a high-energy charged particle when it passes through a medium in which its speed is greater than the speed of light. The index of refraction of such a medium must be greater than 1.0.

Compton scattering is the elastic scattering of photons by electrons. A *Compton electron* is one set in motion by interaction with a photon.

Neutron density is the number of neutrons per unit volume.

Neutron flux is the product of neutron density and speed (for neutrons of a given energy).

Half-life is the time required for the disintegration of half the atoms of a radioactive substance.

9.1.2 Related laws

Special theory of relativity

$$E = mc^2 \qquad m = \frac{m_o}{\sqrt{1 - \dfrac{v^2}{c^2}}}$$

where: E = energy
m = mass (in motion)
m_o = mass (at rest)

$$c = \text{velocity of light}$$
$$v = \text{velocity of mass}$$

Radioactive decay law

$$N = N_o\, e^{-\lambda t}$$

where: $N =$ number of atoms present at time t
  $N_o =$ number of atoms present at $t = 0$
  $t =$ elapsed time
  $\lambda =$ decay constant
  $e =$ base of natural logarithms ($= 2.71828$)

Planck's law

$$E = hf$$

where: $E =$ energy (in ergs)
  $f =$ frequency (in hertz)
  $h =$ Planck constant ($= 6.6256 \times 10^{-27}$ erg-seconds)

Note: frequency is often represented by the symbol ν

9.1.3 Units of measurement

The **roentgen** (r) is a unit of X-ray or gamma-ray radiation (*exposure dose*); one r equals the quantity of radiation whose associated secondary ionizing particles produce ions, in air, carrying one electrostatic unit charge (of either sign) per 0.001293 gm of air. The milliroentgen (*mr*) equals 10^{-3}r.

The **rad** is a unit of *absorbed radiation dose;* it equals 100 ergs per gram.

The **rem** (*r*oentgen *e*quivalent, *m*an) is the unit of *RBE* (*relative biological effectiveness*) dose; it equals the absorbed dose in rads, times an agreed conventional value of the RBE. It was originally defined as the absorbed dose that will produce the same biological effect in human tissue as that produced by one roentgen. Typical RBE values are 1 for X rays, 10 for protons, and 20 for alpha particles.

The **electron volt** (eV) is the unit of energy of radiation; one eV equals 1.60×10^{-12} erg. The kilo-electron-Volt (keV, $= 1.6 \times 10^{-9}$ erg) and the mega-electron-Volt (MeV, $= 1.6 \times 10^{-6}$ erg) are more frequently used in nuclear-radiation measurement.

The **curie** is a unit of radioactivity; it equals the quantity of any radioactive material in which the number of disintegrations per seconds is 3.7×10^{10}.

Neutron flux density is usually expressed in **neutrons per square centimeter-seconds** (*n/cm²-s*).

The *window thickness* (of a radiation transducer) is commonly expressed in **milligrams per square centimeter** (*mg/cm²*).

9.2 Transduction Methods

There are two basic methods of converting nuclear radiation into usable electrical output signals. Both methods employ the interaction of the incident radiation with a material contained in the transducer. The first method relies upon the production of an ion pair in a gaseous or solid material and the separation of the positive and negative charges by an electric field to produce an electromotive force. The second method uses a scintillator material to generate light radiation upon incidence of nuclear radiation and a light sensor to provide an output proportional to this light.

9.2.1 Ionizing transduction

Charged particles, such as α-and β-particles and protons, can exert sufficient electromagnetic forces on the outer electrons of atoms which they pass at high velocity to separate one of the electrons. When this occurs, an ion pair is produced which consists of an ion and a secondary electron. The particle can cause this ionization a number of times before its energy is expended and it comes to rest. Additional ionization can be produced by some of the secondary electrons having sufficient energy to cause it. Neutrons and other uncharged particles can produce occasional ionizing particles by transferring some of their energy to nuclei with which they collide. Similarly, X rays and gamma rays can remove secondary electrons from atoms with which they interact. These secondary electrons can then produce ion pairs.

Ionization due to nuclear radiation occurs in certain gases and solids to various extents. Figure 9-1 (a) shows ion pairs formed in a gas such as argon or krypton. The separation of positive and negative ions is due to the electric field between two electrodes (cathode and anode) connected to a d-c power supply. In the commonly used current mode of operation, the increase in current caused by the flow of charges to the electrodes of opposite polarities (the *ionization current*) can be monitored as the average IR drop across the load resistor (R_L). In the pulse mode of operation, the ionization is measured as a single event. The output is a train of voltage pulses, each pulse generated by the ionization due to one particle. Since the anode voltage is quite high, the output pulses are conveniently taken through a coupling capacitor.

Different particles cause different amounts of ionization; hence, the output-pulse amplitude can be indicative of the types of incident particles. The ionization current will also be greater when some of the same type of particles have higher energy than others. The pulse amplitude then becomes a measure of particle energy. The average ionization current resulting from a steady radiation flux can be used as a measure of the average magnitude of this flux.

Ionization in solid crystals, such as diamond and silver-chloride, has also been used for the transduction of nuclear radiation. Charge separation and flow of ionization current are effected by two electrodes of opposite polarity on opposite crystal surfaces connected to a d-c power supply [Figure 9-1 (b)] Only a perfect or near-perfect crystal structure will provide usable transduction characteristics. Commercial

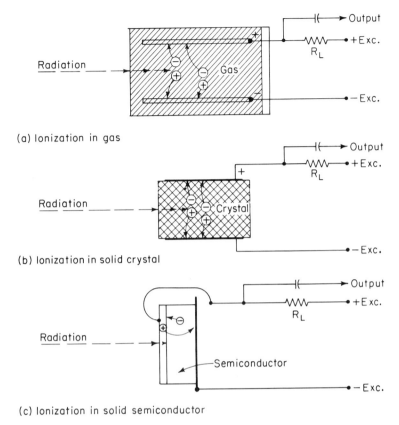

(a) Ionization in gas

(b) Ionization in solid crystal

(c) Ionization in solid semiconductor

Fig. 9-1. Ionizing transduction of nuclear radiation.

applications of crystal ionization detectors have become relatively rare since the advent of semiconductor detectors.

Semiconductor materials, such as germanium and silicon, also produce a positive-negative charge pair as a result of incident nuclear radiation [Figure 9-1 (c)]. The charge separation is accomplished by the permanent electric field established by the junction between "p" (positively charged) and "n" (negatively charged) material.

9.2.2 Photoelectric transduction

Certain materials have been found to convert nuclear radiation into light with a relatively high conversion efficiency. Such *scintillator materials* are used as sensing elements in a number of radiation transducers. Photoelectric transduction, usually provided by a photomultiplier tube (see 7.3.4), is then used to convert the photon flux from the scintillator into an electrical output (Figure 9-2). The anode current of

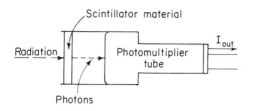

Fig. 9-2. Photoelectric transduction of nuclear radiation.

the photomultiplier tube is proportional to the illumination at the photocathode of the tube. The scintillator is optically coupled very closely to the photocathode. Due to their light-emitting action scintillator materials act as *phosphors*. These phosphors can be organic or inorganic, gaseous, liquid, or solid.

A different effect occurs in pure, transparent, nonluminescent materials whose refractive index is such that the velocity of the radiation through the material is greater than the velocity of light. This effect, known as the *Cerenkov effect* (since it was first observed by Cerenkov in 1934 in the U.S.S.R.), also results in the emission of light which can be transduced by a light detector. It is unrelated, however, to the effect occurring in fluorescent materials. Materials such as glass have been typically used as Cerenkov radiators.

9.3 Design and Operation

Emphasis in this section is placed on those types of nuclear-radiation sensors which are most commonly available commercially. Some additional types are described briefly.

The ionization chamber, proportional counter, and Geiger-Mueller tube are gas-filled ionizing transducers. Semiconductor radiation detectors are solid-material ionizing transducers. Scintillation detectors, usually employing a solid-material scintillator, are photoelectric radiation transducers.

9.3.1 Ionization chambers

The usual configuration of an ionization chamber ("ion chamber") is cylindrical (Figure 9-3). The outer cylinder is the cathode. An internal rod or wire running along the axis of the cylinder is the anode. The outer surface of the cathode need not be insulated since it is the negative electrode and can be grounded. The configuration of Figure 9-3 (a) mates with a socket which provides spring contacts for the cathode as well as a female connector for the anode pin. The electric field is highest close to the anode. The "pancake" style [Figure 9-3 (d)] is a special configuration. The cathode is made from a corrosion-resistant metal such as stainless steel. Ionization chambers with glass envelopes containing cathode, anode, and gas are still in use, however.

Although all gases will produce some ionization, the gases used in ionization chambers are selected on the basis of ionization potential and energy per ion pair. Typical fill gases are argon, krypton, neon, xenon, helium, hydrogen, nitrogen, and certain compounds such as barium trifluoride (BF_3) and methane (CH_4). Favorable

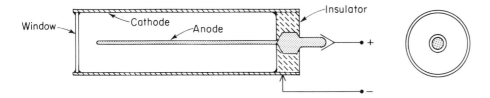

(a) With end window

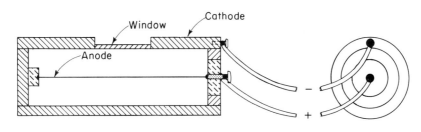

(b) With side window

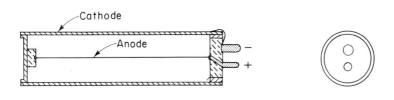

(c) Without window

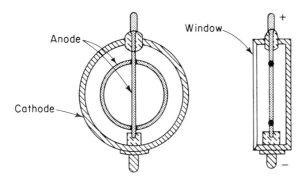

(d) Pancake style, with window

Fig. 9-3. Typical ionization chamber configurations.

operation is obtained by using mixtures, e.g., neon with a small amount of argon, xenon or argon with a small percentage of nitrogen. The pressure in the chamber can be below, at, or above the ambient atmospheric pressure. Atmospheric-pressure fill gas is necessary when very thin windows are used.

Windows are critical parts of ionization chambers. Their material and thickness control the types of radiation to which the chamber will respond with the necessary output. If the mean range of a heavy particle or a particle of very low energy is exceeded by the window thickness, the particle will be stopped by the window and will not be detected by the transducer. Very thin windows are required for α-particle measurements. Mica, or less commonly, nylon windows having a thickness of 4 mg/cm^2 or smaller are used for this purpose. Thin metal windows are used for β-, γ-, and X rays when no α-radiation is to be measured. Windowless thin-wall chambers [Figure 9-3 (c)] exist in a number of designs. Their cylindrical case, typically 30 to 50 mg/cm^2 thick, acts as cathode as well as window. Such a thin-wall ionization chamber will have a near-spherical field of view. When the field of view must be narrower, the chamber can be placed in a thick-walled enclosure (*shield*) with a small opening. For the best conversion of γ-rays into the high-speed electrons needed for ionization, the window should be made from a material having a large γ-ray absorption coefficient, e.g., tantalum or tungsten. Stainless steel has been used for windows and for thin-wall tube cathodes when both β- and γ-rays are to be measured. Anodes are typically made of thin tungsten wire.

9.3.2 Proportional counters

A proportional counter is an ionization chamber whose excitation voltage is above a certain limiting value so that gas amplification occurs within the chamber. The output pulse is proportional to the amount of ionization created by the incident particle which in turn is proportional to the energy of the radiation. Gas amplification (gas multiplication)is the production of additional ions by the electrons produced by the primary ionizing event. The additional ions multiply themselves by the avalanche effect, each secondary electron ionizes additional gas molecules, producing more secondary electrons, etc. The avalanche effect starts at a threshold value of anode-to-cathode potential and increases when the potential is raised above this value. The gas amplification factor can increase from 1 (at the voltage threshold) to over 10,000 (at a voltage several times the threshold value). Threshold values and amplification-factor variation with excitation voltage are different for different fill gases or gas mixtures. Mixtures such as xenon/nitrogen, argon/carbon dioxide, or hydrogen/ methane are most frequently used. The admixture of gases such as CO_2 or CH_4 greatly reduces the production of ionization by ultraviolet radiation created during the avalanche process. The absorption of this radiation by the gas additives allows higher anode voltages to be used and larger gas amplification factors to be realized (see Figure 9-4) while maintaining stable counter operation.

Since a proportional counter produces a pulse for each ionizing event and the

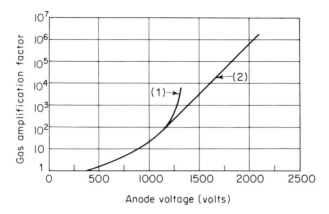

Fig. 9-4. Variation of gas amplification factor with anode-to-cathode potential for (1) pure argon and (2) 95% argon/5% CO_2.

amplitude of the pulse is proportional to the energy of the incident radiation, this type of transducer can be used to differentiate between various particles, e.g., to measure α-particles in the presence of β- and γ-radiation. This is accomplished by gating circuitry associated with the transducer, adjusted to pass only pulses of a certain range of amplitudes.

The windowless *gas-flow counter* is a special version of the proportional counter or of the Geiger counter (see 9.3.3), depending on the type of gas utilized. It is used primarily for the measurement of α-radiation and for weak β-radiation emitted by isotopes such as C^{14}, S^{35}, Ca^{45}, and H^3 (tritium). It consists of an ionization chamber, open at one end and equipped with two hose connections which serve as inlet and outlet for the counting gas. The chamber is mated to a cavity, containing the sample to be counted, by a sliding or compression seal (Figure 9-5). A gas (-mixture) supply is connected to the chamber, and the required exci-tation voltage is applied while the gas flows through the chamber. The quick make-and-break seal feature is the essential design characteristic of a gas-flow counter that differentiates it from other (perma-nently sealed) proportional or Geiger counters. Its advantages are the capability of direct insertion of radiation-emitting material into the chamber and the continuous replacement of the gas which makes the life of the counter virtually infinite. Very thin replaceable windows (0.25 mg/cm² or thinner) are sometimes used in gas-flow counters to reduce possible effects of static charges, accidental contami-nation from loosely packed materials, and vapor effects from moist samples.

Fig. 9-5. Gas-flow counter.

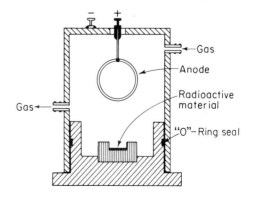

9.3.3 Geiger counters

The *Geiger-Mueller tube* (*Geiger counter*) can be considered as a proportional counter containing a *quenching agent* and whose excitation voltage (and gas amplification factor) is higher than for a proportional counter of the same design. Each primary ionizing event causes a discharge within the fill gas that spreads along the entire length of the anode. The amplitude of the output pulse is therefore independent of the type and energy of the particle initiating the ionization. The quenching vapor, typically alcohol or a halogen such as bromine or chlorine, quenches the discharge by preventing the production of secondary electrons by positive ions at the cathode. The gas mixture is frequently at less than atmospheric pressure.

Since some of the organic quenching fluid is dissociated during each discharge, alcohol-quenched Geiger counters have a limited life (e.g., 10^8 total counts). Halogen quenching gas, however, tends to recombine after dissociation, and there appears to be no limitation on the life of a tube whose counting gas has a halogen admixture. Figure 9-6 illustrates a windowless, thin-wall (30 mg/cm^2), halogen-quenched Geiger-counter tube intended for beta and gamma radiation measurements.

Gas-flow proportional counters can be operated as Geiger counters with the admixture of a quenching gas to the counting gas and with the application of anode potentials in the Geiger region.

A critical characteristic of Geiger counters is the plateau of the anode voltage-vs-counting rate curve (Figure 9-7). This is the portion of the curve, with a small positive slope, between the Geiger threshold voltage ("minimum voltage") and the voltage at which the quenching action begins to fail ("maximum voltage"). It is desirable to make the nominal anode voltage equivalent to the middle of the plateau so that power-supply variations will not cause operation of the tube beyond either end of the plateau. It should be noted that the absence of such a plateau characteristic in ionization chambers and proportional counters requires that well regulated power supplies be used with these devices to assure calibration stability.

The output pulse of a Geiger-Mueller ("G.M.") tube has a fast rise time (less than 1 μs) and a duration of several microseconds before its decay due to quenching

Fig. 9-6. Thin-wall, halogen-quenched Geiger counter tube (courtesy of EON Corp.).

following the discharge. The decay is followed by a *dead time* of 50 to 150 μs. During a subsequent recovery period (*recovery time*), usually of shorter duration than the dead time, the counter gradually recovers before becoming capable of providing a full-height output pulse. The counter is inoperative during the dead time, and any particle that could produce an ionizing event will not be detected during this time. A particle causing an ionizing event during the recovery time will create an output pulse of less than full height. The pulse height of the output of a Geiger-Mueller tube is independent of the type of particle or its energy level. Each ionizing event produces a pulse of the same height whenever the tube is fully operative since a complete discharge occurs.

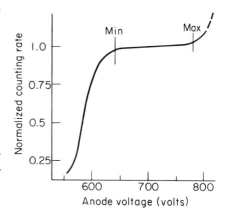

Fig. 9-7. Plateau of typical Geiger counter.

Dead time can be reduced by pulsing the anode supply voltage above a constant voltage level at which almost no ionization occurs or above a slightly negative bias voltage. The counter is then operative only while the Geiger-region voltage pulse is applied. The pulse on and off times can be substantially shorter than the normal dead time and recovery time. As a result, the tube will respond to a larger number of particles per unit time.

9.3.4 Semiconductor nuclear-radiation sensors

Semiconductor ionizing radiation transducers differ from gas-filled ionizing transducers primarily in the following areas: the semiconductor material is about 1000 times denser; its average threshold energy for electron-hole-pair production is roughly ten times lower; its carrier mobilities are higher; and the difference between the mobilities of its electrons and holes (negative and positive charges) is less.

Intrinsic semiconductors are pure crystals, i.e., those having a negligible concentration of impurities. They have been used for the measurement of electromagnetic radiation including light, gamma rays, and X rays only to a limited extent since more favorable transduction characteristics are obtained with *extrinsic semiconductors*, which contain controlled impurities. Charge carriers (electrons and holes) are produced in intrinsic semiconductor material by the interaction between incident photons and the atoms of the material as results of (a) the elastic scattering of photons by electrons which sets electrons into motion (*Compton effect*), (b) the ejection of a bound electron from an atom with absorption of all of the photon's energy (*photoelectric effect*), and (c) the conversion of a photon traversing a strong nuclear electric field into an electron and a positron (*pair production*). When an excitation voltage is applied across two electrodes on the semiconductor crystal (usually silicon or germanium), the electrons and holes flow toward their corresponding electrodes and an IR drop appears across the load resistor (R_L) as output voltage (see Figure 9-8). The output pulse produced by

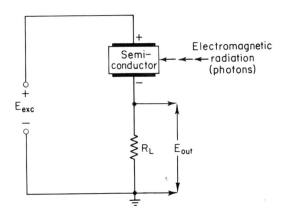

Fig. 9-8. Intrinsic-semiconductor radiation sensor.

each event is characterized by a short rise time and a slower decay time because the electrons have a mobility about three times that of the holes. The thermal noise level of an intrinsic semiconductor is high enough to affect its room-temperature operation as radiation transducer substantially. The noise current is reduced with increasing bulk resistivity of the material. Effective operation is best obtained by immersion of the transducer in a cryostat so that its operating temperature is in the liquid-nitrogen region.

Extrinsic-semiconductor radiation sensors have greater sensitivity and less noise current than intrinsic-semiconductor designs. They exist in two major types. The *surface-barrier* type [Figure 9-9 (a)] usually consists of n-type single-crystal silicon on one surface of which a p-type layer of silicon dioxide is formed. This thin (often monomolecular) layer is covered with a very thin film of vacuum-evaporated gold. Only a few designs use an n-type layer on p-type silicon. The *diffused-junction* type [Figure 9-9 (b)] is commonly made by a very shallow diffusion of n-type material into a base of p-type single-crystal material, although some sensors use a p-type diffusion in an n-type base.

N-type silicon has an excess of electrons (negative charges), and p-type silicon has an excess of holes (positive charges). An electric field (potential gradient) exists in the *depletion layer* (space charge region) close to the junction where the net charge density of donors and acceptors is not neutralized by the mobile-carrier density and is hence substantially different from zero. It is in this depletion region that electron-hole pairs are produced by incident radiation in the same manner in which ion pairs are

Fig. 9-9. Extrinsic-semiconductor radiation sensors (courtesy of ORTEC Inc.).

Evaporated gold contact
←P—type oxide layer
←N—type single crystal silicon
←Ohmic contact
○ + Bias voltage

←Phosphorus—doped N—type layer
←P—type single crystal silicon
←Ohmic contact
○ — Bias voltage

(a) Surface barrier type

(b) Diffused junction type

produced in the gas of gas-filled radiation sensors. The depth of the depletion region (the *depletion depth*) varies with the resistivity of the n- or p-type single-crystal material and with the reverse-bias voltage applied across the two terminals. The depletion layer is considered to extend primarily below the junction, and the depletion depth gives the extent of the sensitive region in which a nuclear particle of specified energy must normally be stopped.

The portion between the surface and the active portion, in which some of the incident energy is dissipated without producing electron-hole pairs collectable by the electrodes, is often referred to as the *dead layer*. The dead layer of surface-barrier type sensors is almost always thinner than that of the diffused-junction type. This advantage is some-

Fig. 9-10. Typical surface-barrier type radiation transducer (courtesy of ORTEC Inc.).

what offset by the location of the barrier at the surface, which makes it more subject to possible mechanical damage. A typical surface-barrier radiation sensor is shown in Figure 9-10.

In the *p-i-n junction* radiation sensor the n- and p-type semiconductors are separated by an intrinsic semiconductor to increase depletion depth. The best known p-i-n-junction type of sensor is the *lithium-drift* type detector (Figure 9-11). Lithium ions, which act as high-mobility donors in silicon, are diffused into a p-type silicon crystal to form a p-n junction at the crystal surface. An electric field is then applied across the crystal in the reverse-bias mode while the crystal is subjected to controlled elevated temperatures. This process causes the lithium ions to drift deeply into the p-region, where they migrate to and electrically compensate impurity acceptor sites, thus forming an intrinsic region at a small distance below the surface. The thickness of

Fig. 9-11. Configuration examples of lithium-drifted germanium radiation sensors (cross sections) (courtesy of ORTEC Inc.).

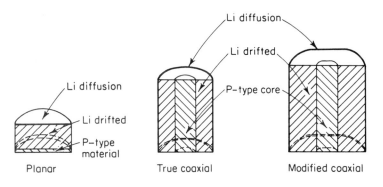

Planar   True coaxial   Modified coaxial

this intrinsic region governs the depletion depth of this detector. Depletion depths larger than in most other types of semiconductor devices have been achieved in lithium-drift p-i-n junction detectors. Lithium-drifted semiconductor radiation sensors, especially germanium types, should be operated and stored in vacuum at liquid-nitrogen temperatures.

Cadmium-sulphide photoconductive sensors (see 8.3.2) have been used for some types of radiation measurements. When furnished with a thin beryllium window, for example, this device has been found usable for the measurement of X rays below 50 keV.

9.3.5 Scintillation counters

This group of nuclear-radiation transducers includes all light sensors which have a scintillator placed in front of their sensing surface to convert incident radiation into light. The term "counter" for this category as well as for the ionizing transducers described in 9.3.2 and 9.3.3 should properly be applied only to a complete measurement system including transducer, signal-conditioning and display equipment. However, the application of the term "counter" to the transducer alone can be condoned because of common usage.

In a scintillation photoelectric radiation transducer the scintillator (body of scintillator material together with its container) acts as the sensing element, since it responds directly to the measurand, while the light sensor acts as transduction element because the transducer's output originates in it. Photomultiplier tubes (of the end-window type) are used most frequently as transduction elements. Of the two groups of scintillator materials—fluorescent and Cerenkov scintillators—the first is made up of inorganic crystals, organic crystals, solution scintillators (liquid, plastic and glass), and gaseous scintillators.

Inorganic crystals are impurity-activated. The impurity content is typically 0.1%. Thallium-activated sodium-iodide crystals, represented by "NaI (Tl)," are most common. ZnS (Ag) is useful for α-particle detection. Most inorganic crystals are thallium-activated alkali halides such as CsI (Tl), CsBr (Tl), KI (Tl), NaCl (Tl), NaBr (Tl), and RbBr (Tl).

Organic crystals are typically grown from coal-tar derivatives such as benzene (C_6H_6), naphthalene ($C_{10}H_8$), and anthracene ($C_{14}H_{10}$), which have conjugate double bonds in their benzene-ring structures. Other organic scintillator materials include naphthacene, pentacene, and trans-stilbene.

Solution scintillators are usually either solid plastics or solid materials dissolved in a liquid such as toluene. An example of a plastic scintillator is a polymerized solution of p-terphenyl and tetraphenylbutadiene in styrene. Gaseous scintillators, e.g., xenon, are infrequently used.

Glass and similar transparent solids are employed as Cerenkov scintillators.

Photomultiplier tubes are discussed in 7.3.4. The optical coupling between the tube's photocathode and the scintillator can be a thin layer of silicone oil or a glass or plastic light guide ("light pipe") which may be quite long. The assembly comprising scintillator, optical coupling, and photomultiplier tube is covered by an enclosure which frequently acts as magnetic shield and which is light-proof except at the scintillator sensing surface. The emission spectra of scintillator and photocathode should be matched closely to each other. Peak-emission wavelengths are most frequently in the blue region. Certain compounds (*wavelength shifters*) can be added to some scintillator materials to shift their peak toward the red region when required by photocathode characteristics. The photomultiplier output pulse height is proportional to the number of photons impinging on the photocathode. The conversion efficiency of scintillators depends on their material as well as on the type and energy of incident ionizing radiation.

9.3.6 Signal-conditioning and display equipment

A number of special design considerations apply to the ancillary equipment for nuclear-radiation transducers.

Ionization chambers produce output currents as low as 10^{-15} A which must be conditioned and indicated. In order to obtain a measurable IR drop across the load resistor (connected between anode and positive high-voltage terminal or between cathode and ground), these resistors can have values up to 10^{13} ohms. The design and manufacturing problems of stable high-value resistors have been largely overcome. *Electrometer tubes* having low grid current or quartz-fiber electrometers can provide the high input impedance required to avoid loading errors. Integration of the instantaneous ionization currents over time periods up to 30 s, by allowing the current to charge a capacitor, provides another means for measuring currents of less than one picoampere.

The output pulses of all types of radiation transducers usually require amplification. Special linear *pulse amplifiers* with large gain-adjustment ranges (e.g., 50 to 50,000) and the ability to amplify pulses with a variety of rise and decay times have been designed for use with transducers other than ionization chambers.

The usual indication expected from a radiation transducer, except for spectrometric purposes, is either the total number of counts or the count rate (typically counts per minute). Electromagnetic impulse counters (mechanical registers) are usable as *pulse counters* at counting rates up to 50 counts per second. Electronic counters can display total counts at rates up to about 1000 per second. Faster count rates require the use of a *scaler* which produces one output pulse for a prescribed number of input pulses. *Rate meters* (counting-rate meters) indicate the time rate of occurrence of input pulses averaged over a time interval. Selection of the time intervals can be accomplished by changing circuit constants of an RC integrating network and, hence, its time constant, in known discrete increments. The analog indication of counting rates is displayed on a panel meter.

The *pulse-height analyzer* furnishes displays of radiation energy spectra. It indicates either the number of pulses or the rate of occurrence of pulses falling within one or more specified amplitude ranges. Each channel of such an analyzer is adjusted to pass only those pulses whose amplitude is between a threshold voltage V and an upper limit $V + V_W$, where V_W is referred to as the "window width." A typical analyzer provides for adjusting the threshold voltage at any value between zero and 100 V and the window between zero and 10 V. A spectral plot can be obtained by adjusting the window to a fixed value and varying the threshold voltage linearly (*sweeping*), either manually or automatically at a known rate to provide a time correlation for each energy level analyzed.

The *coincidence circuit* is used to assure that a given indication is due only to a specific particle or several genetically related particles. The circuit obtains separate inputs from two or more transducers and uses logic components to produce an output pulse only when a specified number or a specified combination of input terminals receive pulses within a specified short time interval. A *delayed coincidence circuit* is actuated by two input pulses (due to genetically related nuclear occurrences) one of which is delayed by a specified time interval with respect to the other. An *anticoincidence circuit*, on the other hand, produces an output pulse only when one of two input terminals receives a pulse and the other receives no pulse. This circuit is useful in reducing unwanted *background counts* (counts caused by ionizing radiation other than the measurand).

9.4 Performance and Applications

9.4.1 Specification characteristics

The characteristics specified by users and manufacturers of nuclear-radiation transducers differ appreciably for the various transducer types.

Mechanical characteristics such as outline dimensions, dimensions and location of mountings, electrical connections, and gas connections (for gas-flow counters), case sealing, case material, identification markings, and weight are described just as for any other transducers. Additionally, window material and thickness are stated for radiation transducers. For gas-filled ionizing transducers the cathode material (if different from the case material), anode thickness and material, effective cathode dimensions, and type of fill gas and its pressure are shown, as well as the quenching fluid of Geiger-Mueller tubes. Scintillation counters can be specified by a description of the scintillator configuration, dimensions, and material, type of optical coupling, and either the model number of a commercial photomultiplier tube or the complete characteristics of the portion of the transducer constituted by an integrally packaged photomultiplier or other light sensor. The semiconductor material, sensitive area (active area), and total sensitive volume are stated for semiconductor-type transducers.

Electrical characteristics include the range of (operating) excitation voltages, the maximum allowable voltage, and the recommended range of external series resistances as well as the internal capacitance which together determine the effective output impedance of the transducer. Unless a commercial electrical connector is called out with its part number and manufacturer, the details of the electrical connections (primarily the conductor and insulator materials) should be shown. Excitation current (at a nominal voltage) is shown primarily for scintillation counters. The operating (voltage) plateau length, the averaged plateau slope over this length, and the output pulse amplitude are shown for Geiger-Mueller tubes. The essential electrical characteristics of semiconductor transducers are the ranges and nominal value of the resistivity, the bias voltage, depletion depth or width, and capacitance.

Performance characteristics list the types of radiation intended to be measured by the transducer and the characteristics of required as well as unwanted outputs. The latter comprise the maximum background counts (usually per minute), maximum leakage current (for semiconductors), and maximum dark current (for photomultipliers) at a nominal operating voltage. Required output pulses are described by rise time and decay time. Transfer characteristics have been expressed in various ways, such as detection efficiency for given types of radiation (in percent) or as counting efficiency (ratio of average number of measured particles to average number incident on sensitive area). The incident energy-vs-counting rate characteristic is often shown as "full width at half maximum (FWHM)." In the case of scintillation and proportional counters this refers to the width of an energy distribution curve (which somewhat resembles a Gaussian distribution curve) measured at one-half the height of the peak of the curve. Energy resolution is the smallest difference in energy between two particles that can be discerned by the counter. For semiconductors the energy resolution is usually defined as the FWHM height (expressed in energy units) of a monoenergetic peak.

Geiger-counter specifications should show dead time as well as recovery time. The effects or extent of photosensitivity and hysteresis, if any, should be stated. Geiger-counter life is shown (in total counts) whenever an organic quenching fluid (e.g., alcohol) is used. The wavelength of maximum emission and sometimes a curve descriptive of the complete emission spectrum are shown for scintillators.

Environmental characteristics are not often specified for radiation transducers since most of them (with the exception of cryogenic-operated sensors) are operated at room conditions. In some cases the levels of nonoperating shock, acceleration, and vibration are shown. When operation during such environments is required, their effects must usually be determined experimentally. Similarly, thermal effects on transducer operation must be verified by tests, although an operating temperature range is frequently specified. For certain transducer types it is possible to calculate the degradation of their performance characteristics with temperature. More complete environmental specifications for nuclear-radiation transducers are necessary when they are intended for operation at other than room conditions.

9.4.2 Applications

The determination of specific types of radiation emanating from a given material is facilitated by existing knowledge of the radiation-emitting characteristics of radioisotopes and other radioactive elements (see Tables 9-1 through 9-3). Table 9-4 is included to facilitate the identification of various elements mentioned in this chapter. Transducer design characteristics must be selected to optimize response to the radiation to be measured by maximizing the required type of interaction between radiation and transducer materials. In some applications it is possible to select a design which excludes one or more types of radiation simultaneously incident with the radiation to be measured.

Heavy charged particles such as alphas and protons of medium and low energy

Table 9-1 NATURALLY OCCURRING ALPHA EMITTERS†

Name	At. No.	At. wt.	α-energy (MeV)	β-energy (MeV)	γ-energy (MeV)	Half-life
Actinium C' (AcC') (Po)	84	211	7.434	6.34, 6.56, 6.88	—	.005 s
Polonium (RaF) (Po)	84	210	5.298	5.303	.800	138 days
Radium (Ra)	88	226	4.791	—	.188	1622 yr
Radium C' (RaC') (Po)	84	214	7.680	7.683	—	1.5×10^{-4} s
Radon (Rn)	86	222	5.486	—	—	3.825 days
Thorium (Io)	90	230	4.66	4.66	.068, .14, .24	80,000 yr
Thorium (Th)	90	232	3.98	3.98, 4.2	.055, .075	1.39×10^{10} yr
Thorium C' (ThC') (Po)	84	212	8.776	—	—	3.0×10^{-7} s
Uranium (UI)	92	238	4.80	4.77	.05, .093, .118	4.51×10^{9} yr
Uranium (UII)	92	234	4.763	4.21	.048	2.35×10^{5} yr

†Courtesy of EON Corp.

Table 9-2 AVAILABLE PURE BETA EMITTING RADIOISOTOPES†

Radioisotope	Half-life	Maximum energy (MeV)
Yttrium-90	64.2 hr	2.18
Phosphorus-32	14.3 days	1.701
Yttrium-91 (Gamma 1.22 MeV − 0.3%)	58.0 days	1.537
Strontium-89	50.4 days	1.463
Bismuth-210 (Alpha − 5×10^{-5}%)	5.02 days	1.17
Praseodymium-143	13.95 days	0.932
Thallium-204 (EC ∼ 2%)	4.1 yr	0.765
Chlorine-36	3.2×10^{5} yr	0.714
Strontium-90	25 yr	0.61
Technetium-99	2.12×10^{5} yr	0.29
Calcium-45	164 days	0.254
Promethium-147	2.5 yr	0.223
Sulfur-35	87.1 days	0.167
Carbon-14	5.57×10^{3} yr	0.155
Nickel-63	125 yr	0.067
Hydrogen-3	12.46 yr	0.01795

†Courtesy of EON Corp.

Table 9-3 AVAILABLE GAMMA EMITTING RADIOISOTOPES†

Radioisotope	Gamma (MeV)	Beta (MeV)	Half-life
Antimony 124	.605, .642, .72, .99, 1.38, 1.71, 2.11	.24, .61, .966, 1.602, 2.317	60 days
Arsenic 76	.549, .643, 1.20, 1.40, 2.05	.36, 1.76, 2.41, 2.96	26.6 hr
Barium 137m	.662	—	2.6 min
Bromine 82	.547, .608, .692, .766, .823, 1.031, 1.312	.465	35.87 hr
Cerium 144	.034, .054, .081, .100, .134	.17, .30	290 days
Cesium 134	.561, .567, .601, .794, 1.037, 1.164, 1.365	.09, .648	2.3 yr
Cobalt 60	1.17, 1.33	.306	5.27 yr
Europium 152	.122, .245, .445, .586, .692, .872, .969, 1.09, 1.12, 1.17, 1.42	.68, 1.00, 1.46	12.7 yr
Gallium 72	.63, .68, .84, 1.05, 1.20, 1.59, 1.87, 2.21, 2.51	.6, .9, 1.5, 2.52, 3.15	14.2 hr
Gold 198	.411, .68, 1.09	.29, .97	65 hr
Iridium 194	.293, .328, .466, .620, .643, .937, 1.149, 1.18	.430, .975, 1.905, 2.236	19 hr
Iron 59	.191, 1.098, 1.289	.271, .462, 1.560	45.1 days
Lanthanum 140	.093, .329, .487, .815, 1.60, 2.50, 3.0	1.32, 1.67, 2.26	40 hr
Niobium 95	.745	.160	35 days
Potassium 42	.320, 1.51	2.04, 3.58	12.7 hr
Praseodymium 142	1.6	.56, 2.166	19.3 hr
Rhodium 106	.513, .624, .87, 1.045, 1.85, 2.41	2.0, 2.44, 3.1, 3.53	30 s
Rubidium 86	1.08	.72, 1.82	18.6 days
Scandium 46	.89, 1.12	36, 1.2	85 days
Silver 110m	.116, .656, .885, .935, 1.399, 1.516	.087, .53, 2.12, 2.86	270 days
Sodium 24	1.368, 2.754	1.39	15.0 hr
Tantalum 182	.066, .068, .085, .100, .114, .116, .152, .156, .179, .198, .222, .229, .264, 1.121, 1.188, 1.223	.525	112 days
Tungsten 187	.072, .134, .480, .552, .686, .78	.63, 1.33	24 hr
Zinc 65	1.12	.325 $\beta+$	245 days
Zirconium 95	.722, .754	.364, .396, .833	63.3 days

†Courtesy of EON Corp.

lose energy rapidly in passing through matter without appreciable scattering. Their mean range decreases with increasing absorber density. Alpha particles can be detected with scintillation counters using silver-activated zinc sulphide as scintillator material,

Table 9-4 PROPERTIES OF ELEMENTS*

Name Note: Condition of state refers only to state in which density was measured.	Symbol	At. No.	International atomic weight 1949	Density gm/cc (at approximate atmospheric temperature)	Melting point °C	Boiling point °C
Actinium	Ac	89	227	—	—	—
Aluminum (hard drawn)	Al	13	26.97	2.699	659.7	2450
Americium	Am	95	(241)	—	—	—
Antimony (compressed)	Sb	51	121.76	6.691	630.5	1380
Argon	A	18	39.944	1.7837†	−189.2	−185.7
Arsenic (metallic)	As	33	74.91	5.73	814³⁶ᴬᵀᴹ	615 (Subld.)
Astatine	At	85	(210)	—	—	—
Barium	Ba	56	137.36	3.78	850	1140
Berkelium	Bk	97	(243)	—	—	—
Berylium, glucinum	Be	4	9.013	1.84	1278 ± 5	2970
Bismuth (electrolytic)	Bi	83	209.00	9.8	271.3	1560 ± 5
Boron (crystal)	B	5	10.82	3.33	2300	2550
Bromine (liquid)	Br	35	79.916	3.12	−7.2	58.78
Cadmium (cast)...............	Cd	48	112.41	8.648	320.9	767 ± 2
Calcium	Ca	20	40.08	1.54	842.8	1240
Californium	Cf	98	(244)	—	—	—
Carbon (crystal)	C	6	12.010	3.52	> 3500	4200
Cassiopeium, see *Lutetium* ...	—	—	—	—	—	—
Cerium (pure)	Ce	58	140.13	6.9	640	1400
Cesium	Cs	55	132.91	1.873	28.5	670
Chlorine.................	Cl	17	35.457	3.214†	−103 ± 5	−34.6
Chromium	Cr	24	52.01	6.52—73	1890	2200
Cobalt.................	Co	27	58.94	8.71	1495	2900
Columbium, see *Niobium*......	—	—	—	—	—	—
Copper (hard drawn)	Cu	29	63.54	8.89	1083	2336
Curium	Cm	96	(242)	—	—	—
Dysprosium	Dy	66	162.46	—	—	—
Erbium	Er	68	167.2	4.77(?)	—	—
Europium	Eu	63	152.0	—	—	—
Fluorine.................	F	9	19.000	1.696†	−223	−188
Francium	Fr	87	(223)	—	—	—
Gadolinium	Gd	64	156.9	—	—	—
Gallium	Ga	31	69.72	5.903	29.78	1983
Germanium	Ge	32	72.60	5.46	958.5	2700
Gold, aurum (cast)	Au	79	197.2	19.3	1063	2600
Hafnium, celtium	Hf	72	178.6	—	2207	> 3200
Helium	He	2	4.003	0.17847†	272.2²⁶ᴬᵀᴹ	−268.9
Holmium	Ho	67	164.94	—	—	—
Hydrogen	H	1	1.0080	0.08938†	−259.14	−252.8
Indium	In	49	114.76	7.28	156.4	1450
Iodine.................	I	53	126.92	4.94	113.7	184.35
Iridium	Ir	77	193.1	22.42	2454	> 4800
Iron, ferrum (pure)	Fe	26	55.85	7.85 − 8	1535	3000
Krypton.................	Kr	36	83.7	3.708†	−156.6	−152.9
Lanthanum	La	57	138.92	6.15	826	1800
Lead, plumbum (compressed)	Pb	82	207.21	11.347	327.43	1620
Lithium	Li	3	6.940	0.534	186	1336 ± 5
Lutetium	Lu	71	174.99	—	—	—
Magnesium	Mg	12	24.32	1.741	−651	1107
Manganese	Mn	25	54.93	7.42	1260	1900
Mercury, hydrargyrum	Hg	80	200.61	13.546	−38.87	356.9

Name Note: Condition of state refers only to state in which density was measured.	Symbol	At. No.	International atomic weight 1949	Density gm/cc (at approximate atmospheric temperature)	Melting point °C	Boiling point °C
Molybdenum	Mo	42	95.95	9.01	2620 ± 10	3700
Neodymium	Nd	60	144.27	6.96	840	—
Neon	Ne	10	20.183	0.90035†	−248.67	−245.9
Neptunium.........................	Np	93	(237)	—	—	—
Nickel	Ni	28	58.69	8.60-90	1455	2900
Niobium, columbium	Nb	41	92.91	8.4	1950	2900
Niton, see *Radon*			—	—	—	—
Nitrogen	N	7	14.008	1.25055†	−209.86	−195.8
Osmium	Os	76	190.2	22.5	2700	> 5300
Oxygen	O	8	16.000	1.42904†	−218.4	−182.97
Palladium	Pd	46	106.7	12.16	1555	2200
Phosphorus (white)	P	15	30.98	1.83	44.1	280
Platinum...........................	Pt	78	195.23	21.37	1773.5	4300
Plutonium	Pu	94	(239)	—	—	—
Polonium	Po	84	210	—	—	—
Potassium, kalium	K	19	39.096	0.87	62.3	760
Praseodymium	Pr	59	140.92	6.475	940	—
Promethium	Pm	61	(147)	—	—	—
Protactinium	Pa	91	231	—	—	—
Radium	Ra	88	226.05	—	960	1140
Radon, niton	Rn	86	222	9.73†	−110	−61.8
Rhenium	Re	75	186.31	—	3167 ± 60	—
Rhodium	Rh	45	102.91	12.44	1985	> 2500
Rubidium	Rb	37	85.48	1.532	38.5	700
Ruthenium	Ru	44	101.7	12.06	2450	> 2700
Samarium	Sm, Sa	62	150.43	7.7-8	> 1300	—
Scandium	Sc	21	45.10	—	1200	2400
Selenium...........................	Se	34	78.96	4.3-8	220	688
Silicon (crystal)	Si	14	28.06	2.42	1420	2600
Silver, argentum (cast)	Ag	47	107.880	10.42-53	960.8	1950
Sodium, natrium	Na	11	22.997	0.9712	97.5	880
Strontium	Sr	38	87.63	2.50-58	757 FREEZE	1150
Sulfur	S	16	32.066	2.0-1	112.8	444.6
Tantalum	Ta	73	180.88	16.6	3027	> 4100
Technetium	Tc	43	(99)	—	—	—
Tellurium (crystal)	Te	52	127.61	6.25	452	1390
Terbium	Tb	65	159.2	—	—	—
Thallium...........................	Tl	81	204.39	11.86	303.5	1457 ± 10
Thorium	Th	90	232.12	11.3-11.7	1845	> 3000
Thulium	Tm	69	169.4	—	—	—
Tin (white cast)	Sn	50	118.70	7.29	231.89	2260
Titanium	Ti	22	47.90	4.5	1800	> 3000
Tungsten, see *Wolfram*			—	—	3370	5900
Uranium...........................	U	92	238.07	18.7	1133	—
Vanadium	V	23	50.95	5.69	1710	3000
Wolfram, tungsten	W	74	183.92	18-19.1	−112	−107.1
Xenon	Xe	54	131.3	5.851†	—	—
Ytterbium	Yb	70	173.04	—	1800	—
Yttrium	Y	39	88.92	3.80	1490	2500
Zinc (cast)	Zn	30	65.38	7.04-16	419.47	907
Zirconium	Zr	40	91.22	6.44	1900	> 2900

*Courtesy of EON Corp.

†gm/liter measured at 0°C, 760 mm pressure.

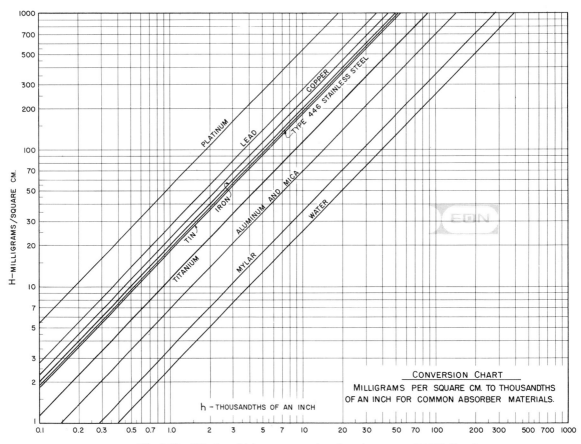

Fig. 9-12. Window thickness conversion chart (courtesy of EON Corp.).

with proportional and Geiger counters, with cryogenic-operated semiconductor transducers provided with a very thin window, and with windowless semiconductor transducers and gas-flow counters. A conversion chart for the effective thickness of typical window materials is shown in Figure 9-12. High-energy charged particles are best measured with fluorescent and Cerenkov scintillation counters. Beta and gamma radiation is detected by virtually all types of radiation transducers, although γ-rays must first interact with matter to produce ionizing particles. Weak beta radiation requires a very sensitive detector such as a windowless or ultra-thin-window gas-flow counter. Thin beryllium windows have been found usable in relatively low-energy X- and gamma-ray spectrometry. The window (or cathode-envelope) material and thickness of ionizing transducers can be selected to exclude beta particles and provide a transducer output only for gamma radiation when both are present. Most gamma sensors can also be used for X-ray measurements.

Neutrons are measured by measuring the charged particles emitted from a material in which a nuclear reaction is caused by the incident neutrons. The reaction is caused either by the interaction of neutrons with atomic nuclei or by recoils of charged particles from collisions with neutrons. The materials (*conversion materials*) used for this purpose are usually stable isotopes of various elements having a high absorption cross section in the energy range of interest. The neutron *cross section* (σ) is an area of such a magnitude that the number of reactions (N) occurring is equal to the product of the number of target nuclei or particles (n_t) and the number of incident neutrons (n_0), or $\sigma = N/n_0 n_t$. The unit of nuclear cross section is the *barn*, which has a magnitude of 10^{-24} cm^2 per nucleus.

The conversion material is used either as gas or internal coating or both in ionization chambers and proportional counters, as constituent in a scintillator material, or in a film or on a foil placed in front of the sensitive area of a charged-particle or gamma-ray transducer (depending on the reaction used and its products). Semiconductor transducers can be equipped with a foil on which the conversion material is deposited and which is clamped to the sensing surface by a cap (see Figure 9-13). In some cases the conversion material is vacuum evaporated onto the active surface of a semiconductor transducer, which is then placed face to face with another semiconductor transducer. The resulting "sandwich" geometry allows simultaneous counting of the two reaction products in the two transducers. The energy of the incident neutron is the sum of the energies of the reaction product plus the so-called "Q-value" of the reaction in question.

Typical conversion materials are B^{10} (solid) and B^{10}-enriched BF$_3$ (boron trifluoride, gas) for slow neutron measurements and He^3 (gas) and H^1 (as polyethylene

Fig. 9-13. Semiconductor radiation transducer for neutron detection. *Note:* A flat disc of polyethylene 1.5 mm thick is used for detection of fast neutrons instead of the steel disc (courtesy of ORTEC Inc.).

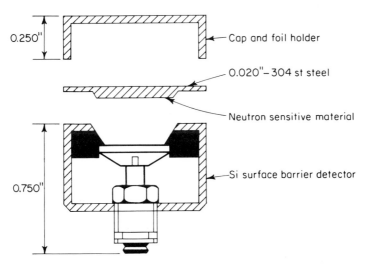

film) for fast neutrons. Li_6F (solid) is usable for both slow and fast neutron detection. Heavy charged particles are produced as fission fragments by neutron reactions with materials such as Pu^{239}, U^{235}, Np^{237}, and Th^{232}.

Applications of nuclear-radiation transducers are found in many sciences and industries either for the measurement of radiation as such or for other measurands such as liquid level (see 8.3.3) and thickness of materials. Radiation measurements are of particular importance to the medical, biophysical, space, and atmospheric sciences, to nuclear weaponry, and related civil defense activities, and to the nuclear power-generating industry. Archeological research received an important new tool when C^{14}-dating was introduced. Radiation spectrometry is used in chemical and metallurgical analysis. The petroleum industry has used gamma and neutron detection for oil-well logging. Many additional applications can be expected to arise from continued development and accuracy improvement of nuclear-radiation transducers.

9.5 Calibration and Tests

The calibration of nuclear-radiation transducers can be established by means of a standardized radiation source or by comparison with a reference standard instrument. Calibrated radiation sources are available for charged-particle calibrations. A typical *alpha source* uses isotopically pure Am^{241} with a radioactivity of 0.1 microcurie. The calibration uncertainty of the disintegration rate is within 1.0%. Monoenergetic *neutron generators* are available as calibrated neutron sources. Gamma- and X-ray calibrations can be performed by using a *free-air ionization chamber* for comparison with the test transducer by the substitution method. The radiation entering this chamber ionizes a precisely known volume of air. The free-air ionization chamber provides measurements directly in roentgens since its operation is based on the definition of the roentgen. Calibration uncertainties below 1.0% have been achieved with this instrument.

Other tests are similar to those briefly described for light sensors (see 7.5.2) except that radiation hazards and radiation effects on test equipment must be considered in setting up any test on nuclear-radiation transducers.

9.6 Related Measuring Devices

Various devices not classifiable as transducers have been employed for radiation detection. Most of these involve the use of photographic films or emulsions to display visible evidence of radiation. This visible record is then sometimes analyzed by photometric methods.

When a gas is supersaturated with a vapor, droplets of condensate will form on ions in the gas-vapor mixture. This effect, first described by C. T. R. Wilson in England in 1896, is utilized in the *cloud chamber*. A typical cloud chamber has a transparent window and a radiation window. Argon or air can be used as gas, and water or ethyl alcohol can be used as the vaporized liquid. The chamber floor is a membrane which when deflected downward by a suitable actuation mechanism causes expansion of the mixture and its resultant supersaturation. The water droplets which condense on ions are illuminated by a transverse beam of light and are photographed as small spots of light against a dark background. Particles penetrating the chamber through the radiation window cause characteristic tracks of bright spots which can then be analyzed visually from the photograph.

A similar instrument is the *bubble chamber* which contains a superheated liquid in which small bubbles form around any ions produced by incident particle radiation. The illuminated strings of bubbles will show as bright tracks when photographed.

The *spark chamber* also produces visible ion tracks which can be photographed. It consists of one or more pairs of electrodes in a sealed, at least partly transparent envelope filled with a gas such as argon or air. An electrical potential just below the sparking threshold is applied across the electrodes. An ionizing event traversing this gap results in the formation of a spark along the ion track which is bright enough to be photographed.

Photographic emulsion films containing silver bromide or other silver halides are are also used for display of particle tracks. Incident particles produce ion pairs in grains of the halide which appear as silver grains upon development of the film. Three-dimensional track displays can be obtained from a number of films stacked upon each other. After exposure each lamination is peeled off the stack and developed. The stack can then be reassembled for visual analysis using a binocular microscope of high power. Neutron reactions can be observed when a conversion material is included in the emulsion.

In the *film badge* the degree of blackening of a photographic film is a measure of the radiation absorbed by the film.

BIBLIOGRAPHY

1. Korff, S. A., *Electron and Nuclear Counters*. Princeton, N. J.: D. Van Nostrand Co., Inc., 1946.
2. Goldsmith, H. H., "Bibliography on Radiation Detection," *Nucleonics*, Vol. 4, No. 5, pp. 142–150, May 1949.
3. Rossi, B. B. and Staub, H. H., *Ionization Chambers and Counters*. New York: McGraw-Hill Book Company, 1949.
4. Curtiss, L. F., "The Geiger-Mueller Counter," *N.B.S. Circular 490*, Washington, D. C.: National Bureau of Standards, January 1950.
5. Pieper, G. F., "Instrumentation for Radioactivity," *Science*, Vol. 112, pp. 377–381, October 6, 1950.
6. Wilkinson, D. H., *Ionization Chambers and Counters*. New York: Cambridge University Press, 1950.
7. Cole, D. P., Duffy, T. A., Hayes, M. E., Lusby, W. S., and Webb, E. L., "The Phosphor-

Phototube Radiation Detector," *Electrical Engineering*, Vol. 71, No. 10, pp. 935–939, 1952.

8. Birks, J. B., *Scintillation Counters.* New York: McGraw-Hill Book Company, 1953.

9. Curran, S. C., *Luminescence and the Scintillation Counter.* London: Butterworths Scientific Publications, 1953.

10. Gillespie, A. B., *Signal, Noise, and Resolution in Nuclear Counter Amplifiers.* London: Pergamon Press, 1953.

11. McCreary, H. S., Jr. and Baynard, L. T., "Neutron Sensitive Ionization Chamber with Electrically Adjusted Gamma Compensation," *Review of Scientific Instruments*, Vol. 25, pp. 161–164, February 1954.

12. Batchelor, R., Aves, R., and Skyrme, T. H., "He³-Filled Proportional Counter for Neutron Spectrometry," *Review of Scientific Instruments*, Vol. 26, pp. 1037–1047, 1955.

13. Bell, P. R. and Siegbahn, K. (Eds.), *Beta and Gamma Ray Spectroscopy.* New York: Interscience Publishers, Inc., 1955.

14. Sharpe, J., *Nuclear Radiation Detectors.* New York: John Wiley & Sons, Inc., 1955.

15. Price, W. J., *Nuclear Radiation Detection.* New York: McGraw-Hill Book Company, 1958.

16. Salzberg, B. and Siegel, K., "Semiconductor p-n Junction Radiation Counter," *Proceedings of the IRE*, Vol. 46, p. 1536, August 1958.

17. Handloser, J. S., *Health Physics Instrumentation.* New York: Pergamon Press, 1959.

18. Nokes, M. C., *Radioactivity Measuring Instruments: A Guide to their Construction and Use.* New York: Philosophical Library, Inc., 1959.

19. Van Duuren, K., Jaspers, A. J. M., and Hermsen, J., "G-M Counters," *Nucleonics*, Vol. 17, No. 6, pp. 86–94, June 1959.

20. Whyte, G. N., *Principles of Radiation Dosimetry.* New York: John Wiley & Sons, Inc., 1959.

21. Sharpe, J., *Nuclear Radiation Measurement.* London: Temple Press, Ltd., 1960.

22. Stolyarova, Ye. L., Suchkov, G. M., and Nesterov, L. S., "Instruments and Methods of Radiation," *Collection of Scientific Works of MIFI*, No. 2 (Ye. L. Stolyarova, Ed.), *ATOMIZDAT*, Moscow, 1960.

23. Walter, F. J., Dabbs, J. W. T., and Roberts, L. D., "Large Area Germanium Surface Barrier Counters," *Review of Scientific Instruments*, Vol. 31, pp. 756–762, July 1960.

24. Yavin, A. I., "Detection of Alpha Particles with Commercially Available Transistors," *Review of Scientific Instruments*, Vol. 31, p. 351, 1960.

25. "IRE Standards on Nuclear Techniques: Definitions for the Scintillation Counter Field, 1960," *60 IRE 13.S1*, New York: Institute of Electrical and Electronics Engineers, 1960.

26. "Scintillators and Semiconductors—Symposium Report," *Nucleonics*, Vol. 18, pp. 85–100, May 1960.

27. Chase, R. L., *Nuclear Pulse Spectrometry.* New York: McGraw-Hill Book Company, 1961.

28. O'Kelley, G. D., "Detection and Measurement of Nuclear Radiation," *Report No. NAS-NS-3105*, Oak Ridge, Tennessee: Oak Ridge National Laboratory, December 1961. (Available from National Academy of Sciences.)

29. Shockley, W., "Problems Related to p-n Junctions in Silicon," *Solid State Electronics*, Vol. 2, pp. 35–67, January 1961.

30. Dearnaley, G., "Semiconductor Nuclear Radiation Detectors," *Journal of the British Institute of Radio Engineers*, Vol. 24, pp. 153–169, August 1962.

31. Williams, R. L. and Webb, P. P., "Silicon Junction Nuclear Particle Detectors," *RCA Review*, Vol. 23, pp. 29–46, March 1962.

32. Ammerlaan, C. A. J. and Mulder, K., "The Preparation of Lithium-Drifted Semiconductor Nuclear Particle Detectors," *Nuclear Instruments and Methods*, Vol. 21, pp. 97–100, January 1963.

33. Friedland, S. S., Katzenstein, N. S., and Ziemba, F. P., "Advances in Semiconductor Detectors for Charged Particle Space Spectrometry," *IEEE Transactions*, NS-10, pp. 190–201, January 1963.

34. Taylor, J. M., *Semiconductor Particle Detectors*. London: Butterworths Scientific Publications, Ltd., 1963.

35. Bartoli, A. L. and Missoni, G., "Precision Measurement with the Free-Air Ionization Chamber of the Instituto Superiore di Sanita," *Acta IMEKO 1964*, pp. 509–515, IMEKO, Budapest, Hungary, 1964.

36. Miller, F. M. and Kleppe, L. M., "Development, Testing and Calibration of the LRL Thermoluminescent Dosimetry System," *Report No. UCRL-11613*, Berkeley, California: University of California, Lawrence Radiation Lab., 1964.

37. Wallace, R., "Three Types of 4π Neutron Spectrometers," *Report UCRL-16311*, Berkeley, California: University of California, Lawrence Radiation Lab., 1965.

38. Dearnaley, G. and Northrop, D. C., *Semiconductor Counters for Nuclear Radiation*, 2nd Ed., New York: Barnes and Noble, Inc., 1966.

39. Dempsey, K. C. and Polishuk, P. (Eds.), *Radioisotopes for Aerospace*. New York: Plenum Press, 1966.

40. Kramer, G., Closser, W. H., and Mengali, O. J., "Study of Semiconductor Fast-Neutron Dosimeter for Range 0–50,000 rads," *Report NDL-TR-55*, Columbus, Ohio: Battelle Memorial Institute, April 1966.

41. "A Glossary of Terms in Nuclear Science and Technology," *U.S.A. Standard N1.1-1967*, New York: USA Standards Institute, 1967.

42. Polishuk, P. (Ed.), *Nucleonics in Aerospace*. New York: Plenum Press, 1968.

10   Pressure

10.1 Basic Concepts

10.1.1 Basic definitions

Pressure is force acting on a surface. Pressure is measured as *force per unit area*, exerted at a given point.

Absolute pressure is measured relative to zero pressure.

Gage pressure is measured relative to ambient pressure.

Differential pressure is the difference in pressure between two points of measurement. It is measured relative to a reference pressure or a range of reference pressures.

Static pressure is the pressure of a fluid, exerted normal to the surface along which the fluid flows. A *fluid* can be liquid or gaseous. The static pressure of a moving fluid is measured normal to the direction of flow.

Impact pressure is the pressure in a moving fluid exerted parallel to the direction of flow, due to flow velocity.

Stagnation pressure (also called **ram pressure** or **total pressure**) is the sum of the static pressure and impact pressure.

414

Head is the height of a liquid column at the base of which a given pressure would be developed.

Partial pressure is the pressure exerted by one constituent of a mixture of gases.

Vacuum is pressure reduced to a practically attainable minimum in a volume or region of space. Perfect vacuum is zero absolute pressure and the complete absence of any matter.

Standard pressure is a pressure of one normal atmosphere (see 10.1.3).

10.1.2 Related laws

Gas volume

Isothermal compression or expansion of an ideal gas (Boyle's Law)

$$pV = \text{constant, if temperature is constant}$$

where: p = pressure
V = volume

Adiabatic compression or expansion of gas when too rapid for heat flow during process

$$pV^k = \text{constant}$$

where: p = pressure
V = volume
k = adiabatic constant (typically 1.2 to 1.7)

Pressure in liquid

$$p_0 = wh = qgh$$

where: p_0 = pressure (at depth h)
w = specific weight of liquid
h = depth below surface
q = density ("mass density") of liquid
g = acceleration due to gravity

Liquid level

$$h = \frac{p_L - p_H}{w}$$

where: h = height of surface (above level at which p_L is measured)
p_L = pressure at sub-surface reference level
p_H = pressure on surface
w = specific weight of liquid

Pressure in vertically accelerating liquid mass

With upward acceleration

$$p_L = \frac{g + a}{g} wh$$

With downward acceleration

$$p_L = \frac{g - a}{g} wh$$

where: p_L = pressure at depth h
  h = depth below surface
  w = specific weight of liquid
  a = acceleration
  g = acceleration due to gravity

Velocity of fluid in motion

$$V_0 = \sqrt{\frac{2}{\rho}(p_s - p_0)}$$

V_0 = flow velocity
ρ = density of fluid
p_s = stagnation pressure
p_0 = static pressure

10.1.3 Units of measurement

Pressure is usually expressed in gravitational units of force per unit area: **pounds per square inch** (*psi*) when English units are used and **kilograms per square centimeter** (*kg/cm²*) in metric units. In order to avoid confusion between absolute-, gage-, and differential-pressure measurements, it is good practice to use the more specific units: **pounds per square inch, absolute** (*psia*); **pounds per square inch, gage** (*psig*); or **pounds per square inch, differential** (*psid*). At sea level, 0 psig = 14.7 psia (approximately).

Linear units of liquid head are frequently used to express relatively low pressures in terms of the height of a column of water or mercury. Vacuum is always measured in **millimeters or microns of mercury** at 0°C (*mm Hg, μ Hg*). One mm Hg is the pressure indicated by a column of mercury 1 mm high, at 0°C, and at standard gravity. The **torr,** equal to one mm Hg at 0°C, is frequently used in vacuum work. Other commonly used "head" units are **inches of water** at 4°C (*in. H_2O*), **centimeters of water** at 4°C (*cm H_2O*), and **inches of mercury** at 0°C (*in. Hg*). Absolute units of force per unit area, *dynes/cm²* (or *microbars*), are used primarily in acoustics and in work involving very low pressures.

A unit sometimes used in high pressure measurement is the *atmosphere.* One (normal) atmosphere, essentially representing the ambient atmospheric pressure at earth sea level, is the pressure indicated by a 760-mm high column of mercury, at 0°C, at a density of 13.595 gm/cm³, and an acceleration due to gravity of 980.665 cm/s².

Table 10-1 UNITS OF PRESSURE MEASUREMENT

	psi	kg/cm²	in. H₂O	cm H₂O	in. Hg	mm Hg	dyne/cm²	Atm.	See Note
1 psi =	—	.0703	27.67	70.13	2.036	51.715	68,947	.0680	—
1 kg/cm² =	14.22	—	393.7	1,000	28.96	735.56	980,665	.9678	(1)
1 in. H₂O =	.03613	.00254	—	2.540	.07355	1.868	2,491	.002458	(2)
1 cm H₂O =	.0142	.001	.3937	—	.02896	.7356	980.64	.0009678	(2)
1 in. Hg =	.4912	.0345	13.59	34.53	—	25.40	33,864	.03342	(3)
1 mm Hg =	.01934	.001359	.5352	1.35	.03937	—	1333	.001316	(3) (4)
1 dyne/cm² =	1.45×10^{-5}	1.02×10^{-6}	4.015×10^{-4}	1.02×10^{-3}	2.953×10^{-5}	7.5×10^{-4}	—	9.87×10^{-7}	(5)
1 atmosphere =	14.696	1.033	406.8	1,033	29.92	760	1,013,250	—	(6) (7)

Notes:
(1) 1 kg/m² = 10,000 kg/cm²
(2) in. H₂O and cm H₂O are referenced to a temperature of 4°C (39.2°F)
(3) in. Hg, mm Hg, and μ Hg are referenced to a temperature of 0°C (32°F)
(4) 1 mm Hg = 1000 μ Hg
(5) 1 dyne/cm² = 1 microbar = 1×10^{-6} bar
(6) 1 bar = .987 Atmospheres
(7) Atmospheres are more precisely called normal atmospheres

In the *Système International d'Unitès* (*SI*) the unit for pressure is the *newton per square meter* ($= 1.45 \times 10^{-4}$ psi). The newton per square centimeter (N/cm^2), equal to 1.45 psi, could be considered of more practical value.

Table 10-1 shows conversion factors for units of pressure measurement.

10.1.4 Pressure above and below sea level

Pressure transducers are frequently used to measure altitude and water depth. Atmospheric pressure decreases with increasing altitude. The relationship is not linear since air density and temperature decrease with increasing altitude. Accepted values for pressure-vs-altitude are shown in Table 10-2. Measurements taken by scientific satellites indicate that the barometric-pressure-vs-altitude relationship follows the curve given in this table until a pressure in the vicinity of 10^{-4} torr is reached. Approximate pressures beyond this point are: 10^{-5} torr at 180 km, 10^{-6} torr at 250 km, 10^{-7} torr at 360 km, and 10^{-8} torr at 500 km altitude.

Ambient pressure increases with water depth but is also somewhat affected by density changes, such as those caused by temperature variation, by compressibility effects, and by the salinity of ocean water. When corrections for these additional variants are not considered, the average *seawater* depth can be calculated from pressure measurements on the basis of *0.445 psi per foot of depth*, and the average *freshwater* depth can be calculated on the basis of *0.434 psi per foot of depth*.

10.2 Sensing Elements

Virtually all electrical-output pressure transducers sense the pressure to be measured by means of a mechanical sensing element. These elements are relatively thin-walled elastic members, such as plates, shells, or tubes, which offer the pressure (*force*) a surface (*area*) to act upon. When this pressure is not balanced by an equal pressure acting on the opposite surface, the element is caused to deflect. The *deflection*, frequently translated into a secondary mechanical displacement, is used to produce an electrical change in a transduction element.

The most commonly used sensing elements and the nature of their deflection are illustrated in Figure 10-1.

Although all sensing elements shown measure a differential pressure, transducers can be designed to measure either differential, gage, or absolute pressure depending on the *reference pressure* maintained in, or admitted to, the reference side of the sensing element. The basic pressure-reference configurations of sensing elements, using diaphragms as a typical example, are shown in Figure 10-2.

The reference side of an *absolute-pressure* sensing element is usually evacuated and sealed. *Gage pressure* is measured when ambient (atmospheric) pressure is admitted to the reference side.

Table 10-2   Pressure vs Altitude
(per N.A.C.A. Report No. 538; based on U.S. Standard Atmosphere)

ALTITUDE (Feet)	In. Hg.	Mm. Hg.	P. S. I.	ALTITUDE (Feet)	In. Hg.	Mm. Hg.	P. S. I.
−1,000	31.02	787.9	15.25	32,500	7.91	201.0	3.89
− 500	30.47	773.8	14.94	33,000	7.73	196.4	3.80
0	29.921	760.0	14.70	33,500	7.55	191.8	3.71
500	29.38	746.4	14.43	34,000	7.38	187.4	3.63
1,000	28.86	732.9	14.18	34,500	7.20	183.0	3.54
1,500	28.33	719.7	13.90	35,000	7.04	178.7	3.46
2,000	27.82	706.6	13.67	35,332	6.93	175 9	3.40
2,500	27.31	693.8	13.41	35,500	6.87	174.5	3.375
3,000	26.81	681.1	13.19	36,000	6.71	170.4	3.296
3,500	26.32	668.6	12.92	36,500	6.55	166.4	3.22
4,000	25.84	656.3	12.70	37,000	6.39	162.4	3.14
4,500	25.36	644.2	12.45	37,500	6.24	158.6	3.067
5,000	24.89	632.3	12.23	38,000	6.10	154.9	2.994
5,500	24.43	620.6	12.00	38,500	5.95	151.2	2.925
6,000	23.98	609.0	11.77	39,000	5.81	147.6	2.852
6,500	23.53	597.6	11.56	39,500	5.68	144.1	2.798
7,000	23.09	586.4	11.34	40,000	5.54	140.7	2.72
7,500	22.65	575.3	11.12	40,500	5.41	137.4	2.66
8,000	22.22	564.4	10.90	41,000	5.28	134.2	2.595
8,500	21.80	553.7	10.70	41,500	5.16	131.0	2.535
9,000	21.38	543.2	10.50	42,000	5.04	127.9	2.47
9,500	20.98	532.8	10.30	42,500	4.92	124.9	2.415
10,000	20.58	522.6	10.10	43,000	4.80	122.0	2.36
10,500	20.18	512.5	9.91	43,500	4.69	119.1	2.304
11,000	19.79	502.6	9.73	44,000	4.58	116.3	2.25
11,500	19.40	492.8	9.53	44,500	4.47	113.5	2.195
12,000	19.03	483.3	9.35	45,000	4.36	110.8	2.14
12,500	18.65	473.8	9.15	45,500	4.26	108.2	2.094
13,000	18.29	464.5	8.97	46,000	4.16	105.7	2.042
13,500	17.93	455.4	8.81	46,500	4.06	103.2	1.997
14,000	17.57	446.4	8.63	47,000	3.97	100.7	1.948
14,500	17.22	437.5	8.46	47,500	3.873	98.38	1.90
15,000	16.88	428.8	8.28	48,000	3.781	96.05	1.858
15,500	16.54	420.2	8.13	48,500	3.693	93.79	1.813
16,000	16.21	411.8	7.96	49,000	3.605	91.57	1.772
16,500	15.89	403.5	7.81	49,500	3.52	89.41	1.729
17,000	15.56	395.3	7.64	50,000	3.436	87.30	1.689
17,500	15.25	387.3	7.49	51,000	3.276	83.22	1.610
18,000	14.94	379.4	7.34	52,000	3.124	79.34	1.533
18,500	14.63	371.7	7.19	53,000	2.978	75.64	1.463
19,000	14.33	364.0	7.04	54,000	2.839	72.12	1.395
19,500	14.04	356.5	6.90	55,000	2.707	68.76	1.33
20,000	13.75	349.1	6.75	56,000	2.581	65.55	1.269
20,500	13.46	341.9	6.61	57,000	2.460	62.49	1.208
21,000	13.18	334.7	6.48	58,000	2.346	59.58	1.152
21,500	12.90	327.7	6.34	59,000	2.236	56.80	1.098
22,000	12.63	320.8	6.21	60,000	2.132	54.15	1.048
22,500	12.36	314.1	6.08	61,000	2.033	51.63	1.000
23,000	12.10	307.4	5.94	62,000	1.938	49.22	0.952
23,500	11.84	300.9	5.82	63,000	1.847	46.92	0.906
24,000	11.59	294.4	5.70	64,000	1.761	44.73	0.865
24,500	11.34	288.1	5.58	65,000	1.679	42.65	0.825
25,000	11.10	281.9	5.45	66,000	1.601	40.66	0.786
25,500	10.86	275.8	5.33	67,000	1.526	38.76	0.748
26,000	10.62	269.8	5.22	68,000	1.455	36.95	0.714
26,500	10.39	263.9	5.11	69,000	1.387	35.23	0.681
27,000	10.16	258.1	4.99	70,000	1.322	33.59	0.649
27,500	9.94	252.5	4.88	71,000	1.261	32.02	0.619
28,000	9.72	246.9	4.78	72,000	1.202	30.53	0.590
28,500	9.50	241.4	4.67	73,000	1.146	29.10	0.562
29,000	9.29	236.0	4.56	74,000	1.093	27.75	0.536
29,500	9.08	230.7	4.46	75,000	1.041	26.45	0.512
30,000	8.88	225.6	4.36	76,000	0.993	25.22	0.488
30,500	8.68	220.5	4.27	77,000	0.946	24.04	0.465
31,000	8.48	215.5	4.17	78,000	0.902	22.92	0.443
31,500	8.29	210.6	4.07	79,000	0.860	21.85	0.423
32,000	8.10	205.8	3.98	80,000	0.820	20.83	0.403

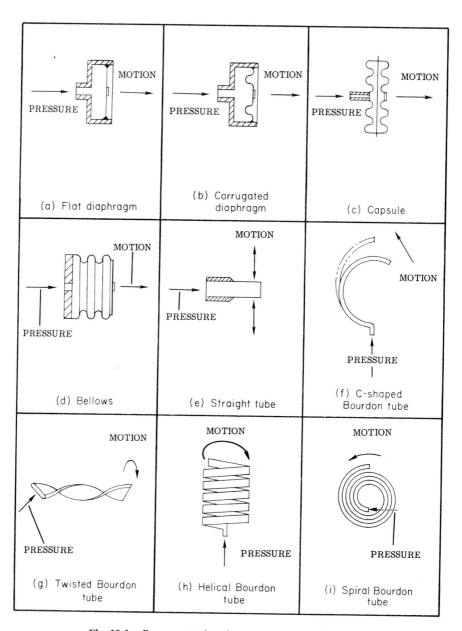

Fig. 10-1. Pressure sensing elements (courtesy of Bourns, Inc.).

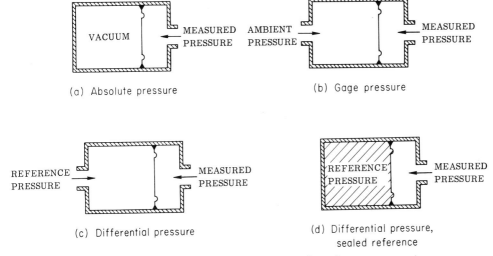

Fig. 10-2. Pressure reference configurations.

A differential-pressure sensing element deflects with an increasing difference between two pressures. It can be considered, therefore, to measure one pressure (measured pressure) relative to another pressure (reference pressure). In a special version of this element configuration, a fixed pressure which is greater than zero is permanently maintained on the reference side.

The measured pressure applied to (hollow) sensing elements other than diaphragms is traditionally shown as entering the sensing element. Equivalent operation can usually be obtained, however, by using the inside of such an element as the reference side and applying the measured pressure to the cavity surrounding the element. An example of this, showing a capsule used as an absolute-pressure sensing element, is shown in Figure 10-3. With the exception of their absolute-pressure configurations, all sensing elements can be used for unidirectional pressure measurement as well as for bidirectional measurement.

Fig. 10-3. Introduction of measured pressure to sensing element.

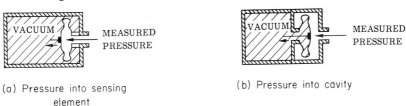

10.2.1 Diaphragms

A diaphragm is essentially a circular plate fastened continuously around its edge. Two basic diaphragm configurations are used in pressure transducers: the flat diaphragm [Figure 10-1 (a)] and the corrugated diaphragm [Figure 10-1 (b)]. Noncircular diaphragms for which such usage has been reported are quite rare and will not be considered further here.

10.2.1.1 Flat diaphragms (Figure 10-4) deflect in accordance with theories generally applicable to circular plates under conditions of symmetrical loading. In its basic configuration, the flat diaphragm presents, in cross section, an uninterrupted straight web supported only at its edge [Figure 10-5 (a)]. In the *spherical diaphragm* [Figure 10-5 (b)], the web is slightly concave. The *catenary diaphragm* [Figure 10-5 (c)] is additionally supported by the edge of an inner ring or tube concentric with the structure to which the web is fastened. The term "catenary" (Latin: *catena*, chain) suggests the curve formed by a chain suspended at two points at equal height. *Drum-head diaphragms* [Figure 10-5 (d)] are bent around and fastened to the outside of their supporting structure while radial tension is applied to stretch the diaphragm.

Fig. 10-4. Flat diaphragm, before (left) and after (right) assembly in transducer end-cap (courtesy of Data Sensors, Inc.).

Fig. 10-5. Flat diaphragms.

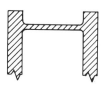

(a) Flat diaphragm

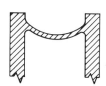

(b) Spherical diaphragm

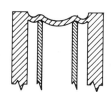

(c) Catenary diaphragm

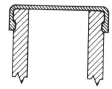

(d) Drum-head diaphragm

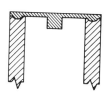

(e) Annular diaphragm
(flush)

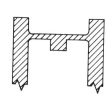

(f) Annular diaphragm
(recessed)

Flat diaphragms with a central circular reinforcement, such as a boss to facilitate the translation of their deflection into a secondary mechanical displacement, are known as *annular diaphragms* (Latin: *annulus*, ring), since the deflecting portion of the web is essentially ring-shaped. Figures 10-5 (e) and 10-5 (f) show flat annular diaphragms in the flush and recessed configurations.

10.2.1.2 Corrugated diaphragms contain a number of concentric corrugations (Figure 10-6). These increase the stiffness as well as the effective area of the diaphragm. The useful deflection is therefore larger than that of flat diaphragms. The amount of deflection and its linearity with a change in applied pressure are dependent on the shape, location, and number of corrugations.

The stiffness of a corrugation varies with its depth. Bending of a corrugated diaphragm is maximum near its periphery and minimum near its center. The corrugations of most diaphragms will therefore become progressively shallower from the periphery to the center of the diaphragm.

Fig. 10-6.   Corrugated diaphragm.

10.2.1.3 Membranes. Membranes are soft diaphragms sometimes made either from thin metal or from nonmetallic materials such as neoprene or from plastics, including especially Teflon® and Kel-F®. When used as pressure sensing elements, they are backed by a spring which provides the necessary elastic properties. They are also used to isolate an elastic sensing element or its adjacent cavity from fluids considered detrimental to the transducer mechanism and to the transduction element itself. When membranes are used as isolation diaphragms, a liquid such as silicone oil is used to transfer the force to the sensing element.

10.2.1.4 Deflection characteristics. The deflection of a diaphragm varies inversely to the 1.2 to 1.6 power of its thickness. It is proportional to the applied pressure (pressure differential across the diaphragm), and it varies with approximately the fourth power of its diameter. Other factors influencing the amount of useful deflection include: the number, shape, and spacing of corrugations (if any), the material used and its preparation, the diameter of the central reinforcement (if any), and the manner of attachment to its peripheral supporting wall.

The usable deflection is commonly taken as that amount of deflection which varies linearly with pressure. Diaphragms can also be designed to produce a specific nonlinear variation, if required.

10.2.1.5 Materials and manufacture. Sensing diaphragms are made from elastic metal alloys such as brass, bronze, phosphor bronze, beryllium copper, stainless steel, and such proprietary alloys as Monel, Inconel-X, and Ni-Span-C (a ferrous nickel alloy). They are pressed, stamped, or spun from sheet stock, or are integrally machined, with their supporting wall, from bar stock. In the latter case, one of the more critical details is the fillet radius between diaphragm and wall. Similar problems can be encountered in the attachment, by clamping, brazing, soldering, or welding, of separate diaphragms.

An important consideration in the selection of diaphragm materials is the chemical nature of the measured fluid which is expected to come in contact with the diaphragm. Corrosive media demand the use of special alloys. Other considerations involve the range of temperatures of fluid and ambient environment, the possible effects on the diaphragm of shock and vibration, and the requirements for response time, i.e., the time it takes the diaphragm to deflect in proportion to a step change in applied pressure.

Temperature effects are minimized in some alloys, notably Ni-Span-C. It is common good practice, however, to cycle diaphragms of any material a number of times over the expected range of temperatures. Proper manufacturing methods also include heat-treating and pressure cycling to reduce elastic aftereffects (drift) and hysteresis in the diaphragm.

10.2.2 Capsules

The pressure-sensing capsule [Figure 10-1 (c)] is frequently referred to as an *aneroid*. It consists of two annular corrugated metal diaphragms, formed into shells of opposite curvature and fastened together with a hermetically sealing joint around their peripheries (Figure 10-7). The seal can be made by brazing or soldering, but the preferred method is tungsten-inert-gas welding. The capsule is usually heat-treated after welding to relieve stresses built up during welding and subsequent rapid cooling.

One diaphragm is provided with a central reinforced opening (port) to admit the pressure to be measured. The other diaphragm contains a centrally located external boss from which the mechanical displacement originates. In another version, one diaphragm is provided with an internal boss. The mechanical member (pushrod) transmitting the displacement passes through a port in the opposite diaphragm through which pressure (usually reference pressure) can be applied to the capsule.

Fig. 10-7. Capsule.

Factors influencing the deflection of capsules are those which apply to diaphragms (see 10.2.1.4). The use of two diaphragms in a capsule nearly doubles the deflection obtained from each diaphragm. A typical value for capsule deflection is 0.04 in. for a $\frac{7}{8}$ in. diameter capsule. Multiple capsule elements consist of two or more capsules fastened together so that each capsule sees the same internal and external pressure (Figure 10-8). The deflection of such multiple elements is equal to the deflection produced by a single capsule multiplied by the number of capsules.

Fig. 10-8. Dual and quadruple capsules.

10.2.3 Bellows

The bellows sensing element [Figure 10-1 (d)] is usually made from a thin-walled tube formed into deep convolutions and sealed at one end which moves axially when pressure is applied to a port in its opposite end (Figure 10-9). The number of convolutions vary from less than ten to over twenty, depending on pressure and displacement (*stroke*) requirements and on outside diameter. Since inside diameters of commonly used bellows range between 50% and 90% of outside diameters, the effective area of a convolution is substantially less than that of a capsule. Due to the large number of their convolutions available, however, bellows are frequently desirable as sensing elements when a large stroke is required, when the pressure to be measured is relatively low, and when no severe vibration levels are encountered in their operating environments.

Fig. 10-9. Bellows.

The materials used in bellows are similar to those used in diaphragms (see 10.2.1), with brass, stainless steel, and nickel alloys predominating.

A typical bellows element has an open fitting (port) attached to one end and a sealed closed fitting at the opposite end. The latter is equipped with a boss to transmit the mechanical displacement obtained when the bellows deflects axially.

Bellows elements can be furnished with a restraining spring which opposes their axial deflection. This allows their use for measuring higher pressure (to about 300 psia) or limits their stroke to increase their cycling life.

There are two other applications of bellows in pressure transducers, where they are not used as sensing elements. The *isolation* bellows is used to isolate a transduction element or a sensing element from the measured fluid. The *expansion* bellows, sealed at both ends and containing an inert gas at low pressure, compensates for changes in the volume of damping oil due to temperature changes in oil-filled pressure transducers.

10.2.4 Bourdon tubes

The Bourdon tube [see Figures 10-1 (f), (g), (h), and (i)] is a curved or twisted tube, oval or elliptical in cross section, which is sealed at one end (tip). When pressure is applied internally to the tube, the tube tends to straighten. This results in an angular tip deflection in a twisted tube and a curvilinear tip deflection (*tip travel*) in curved tubes. The Bourdon tube is named after the French inventor Eugene Bourdon who patented it in June 1849.

10.2.4.1 The C-shaped Bourdon tube (Figure 10-10) has a total angle of curvature of 180 to 270 degrees. Variations of the C-shaped configuration are: the *U-shaped* tube (Figure 10-11), which has the pressure port at the center of its length instead of at one end so that internally applied pressure causes the two tips to deflect away from each other; and the *J-shaped* tube, which is straight near its port and curved only over a portion of its length near the tip.

10.2.4.2 The helical Bourdon tube (Figure 10-12) is similar in deflection behavior to a C-shaped tube. Since it is coiled into a multi-turn helix so that its total angle is

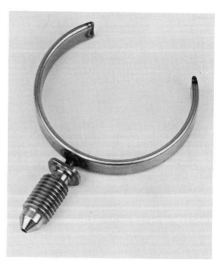

Fig. 10-11. U-shaped Bourdon tube.

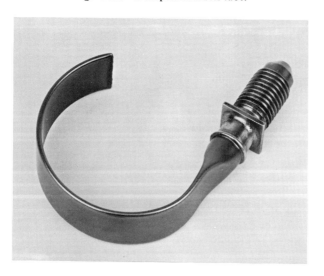

Fig. 10-10. C-shaped Bourdon tube.

Fig. 10-12. Helical Bourdon tube.   **Fig. 10-13.** Spiral Bourdon tube.

typically between 1800 and 3600 degrees, its tip travel is proportionally greater than that of a tube with less than 360 degrees curvature.

10.2.4.3 The spiral Bourdon tube (Figure 10-13) also amplifies tip travel due to its multi-turn configuration (usually 4 to 8 turns). Its radius of curvature increases with each turn, from port to tip. Its tube centerline is in one plane. The helical tube centerline is not in one plane, but its radius of curvature is constant. Performance of both tubes is very similar, and the choice of either tube is primarily dictated by limitations on overall sensing element diameter and length. A partly helical, partly spiral tube design has also been developed.

10.2.4.4 The twisted Bourdon tube (Figure 10-14) is a flattened tube, twisted along its length. Whereas its twist rate (twists per inch of length) varies, the total number of twists is usually between 2 and 6, but it can be less than one full turn. The centerline of the tube is straight throughout its length.

A special prestressed twisted-tube design shows no appreciable deflection until a pressure near the upper design limit of its measuring range is reached. Deflection then occurs only over the narrow remainder of the range so that scale expansion (zero suppression) can be obtained for the transducer utilizing this tube design.

Fig. 10-14. Twisted Bourdon tube.

10.2.4.5 Deflection characteristics. The deflection of a Bourdon tube varies with the ratio of its major to minor cross-sectional axes, the tube length, the difference between internal and external pressure, the radius of curvature and total angle (tip to port) of a curved tube measured from its center of curvature, and the rate of twist of a twisted tube. It also varies inversely to the wall thickness of the tube and to the modulus of elasticity of the tube material after processing.

10.2.4.6 Materials and manufacture. Bourdon tubes are almost invariably metallic. Alloys used include cold-worked brasses, bronzes, "300"-series stainless steels, precipitation-hardened beryllium copper, proprietary nickel alloys such as Monel and Inconel-X, and quenched and tempered ferrous alloys with carbon, chromium, molybdenum, and nickel content. Ni-Span-C is one of the most popular materials because it can be heat-treated to minimize the effects of temperature changes on tube-deflection characteristics. With the exception of the brasses and the bronzes (including phosphor bronze), the alloys used require heat treatment.

The tubing can be machined from bar stock, or it can be drawn. It is then flattened and formed into the desired configuration. Machined tubing can have one or both end fittings integrally machined from the same bar stock. Drawn tubes are welded or brazed to their fittings or to the transducer's machined base and port section, and the tip is sealed, welded, or brazed at the opposite end with great care taken to assure leak-proof joints. These operations are usually followed by the repeated pressure-and-temperature cycling performed on all mechanical pressure-sensing elements.

Fig. 10-15.   Expansion bellows in liquid-damped Bourdon tube pressure transducer (courtesy of Bourns, Inc.).

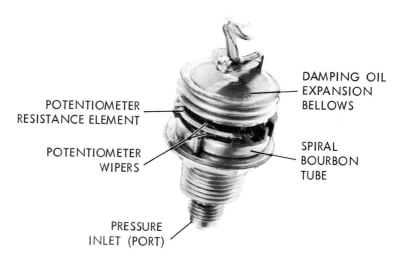

POTENTIOMETER
RESISTANCE ELEMENT

POTENTIOMETER
WIPERS

DAMPING OIL
EXPANSION
BELLOWS

SPIRAL
BOURBON
TUBE

PRESSURE
INLET (PORT)

10.2.4.7 Vibration damping. Bourdon tubes, particularly multiturn tubes, are more sensitive to shock and vibration than diaphragms, although they are less sensitive than bellows. C-shaped tube elements may need to be balanced for vibration effects in the most sensitive axis by a counterweight in the mechanical linkage to the transduction element unless tip travel requirements are so low that the tube can be made very stiff. Helical and spiral tubes, however, are often surrounded with a liquid such as silicone oil to dampen vibration effects. Damping is optimized when the space around the tube is as small as possible. Expansion of the damping oil with temperature and ensuing degradation of damping and apparent change of reference pressure may have to be counteracted by an expansion bellows (Figure 10-15).

10.2.5 Straight tubes

The straight-tube sensing element [Figure 10-1 (f)] is utilized in a limited number of transducers. Its only similarity to the Bourdon tube is in its end fittings—the sealed tip on one end, the port on the other, and the materials used. The straight tube has a circular cross section. A pressure differential across its wall causes a very slight expansion or contraction of the tube diameter. The deformation is maximum near the center of the tube's length. One type of transducer uses the change in the tube's resonant frequency caused by the wall deformation (hoop strain) produced by a change in applied pressure (Figure 10-45).

10.3 Design and Operation

Since the area of pressure measurement is the largest in the entire field of instrumentation in which transducers with electrical output are utilized, a great variety of pressure transducers have been developed for the industry. A considerable number of different operating principles are used in these designs. Only some of them, however, have attained widespread use. Others have limited or specialized applications; some are too difficult to manufacture profitably; some are too strongly affected by one or more environmental conditions; and some have not yet been fully developed for general use.

A general listing of practical pressure transducers is shown in Table 10-3. The various types of transducers are grouped primarily on the basis of their transduction principle. Design and operation of the more commonly used transducers are explained in the following portion of this section.

10.3.1 Capacitive pressure transducers

The capacitive transduction principle (see II.2.1) is utilized in pressure transducers primarily in either of the following two basic designs:

Table 10-3   Pressure Transduction

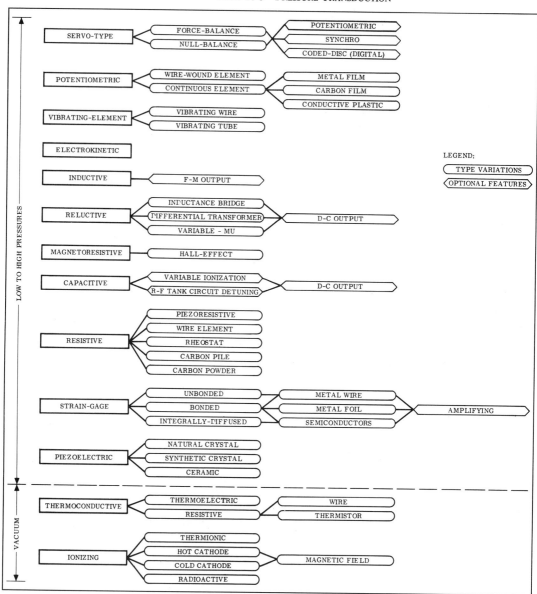

1. Wherein pressure is applied to a diaphragm which moves to and from a stationary electrode (*stator*) or
2. Wherein the pressure is applied to a diaphragm supported between two stationary electrodes (stators).

10.3.1.1 Dual-stator design. Since the dual-stator version provides two arms of a four-arm a-c bridge circuit (Figure 10-16), whereas the single-stator version provides only one arm, it is used in most capacitive pressure transducers.

The sensing element is a thin, metallic, flat diaphragm. It can be integrally machined with its supporting rim, but it is more frequently prestressed and clamped. Applying and maintaining tension by prestressing (stretching) increases the diaphragm's resonant frequency above the value usually encountered in vehicles and machinery. This reduces vibration error in the transducer. Prestressing also tends to shorten the transducer's time constant and lower its hysteresis.

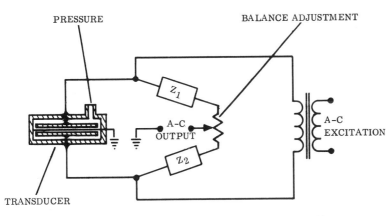

Fig. 10-16. Bridge connection schematic for dual-stator capacitive pressure transducer.

The stators are either metallic and mounted to the case through an insulating material or ceramic with a metal film deposited on the surface facing the diaphragm.

Since the measured fluid is admitted directly to the gap between the stators and the diaphragm, the dielectric constant of this fluid becomes significant to the calibration characteristics of the transducer, especially when liquids are measured. Most pure gases have negligible differences in dielectric constant.

A typical dual-stator transducer design is illustrated in Figure 10-17. The diaphragm is supported on one end of a narrow thin-walled cylinder whose opposite end is part of the main mounting rim. This supporting method tends to isolate the diaphragm from case strains. Temperature error is reduced by using the same relatively corrosion-resistant nickel alloy material for case and diaphragm.

10.3.1.2 Single-stator design. An entirely different design is shown in Figure 10-18. The integrally-machined grounded diaphragm moves to and from an insulated metallic stator which is connected to one side of the secondary winding of an air-core transformer. The other side of this winding is grounded. An a-c voltage, its frequency

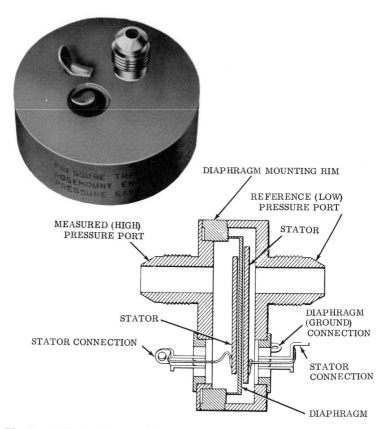

Fig. 10-17.   Dual-stator capacitive pressure transducer (courtesy of Rosemount Engineering Co.).

in the megahertz range, is applied across the primary winding from an associated oscillator-amplifier unit. The changes in diaphragm-stator capacitance which result from pressure variations cause changes in the tuning of the LC-tank circuit formed by the secondary winding in parallel with the sensing element. The changes in tank-circuit tuning are detected by the oscillator-amplifier and appear as a varying voltage or current in its output.

10.3.1.3 Ionization-tube utilization. Among special circuitry developed to utilize the capacitive transducer is the ionization tube circuit [Figure 10-19 (b) and (c)], in which a dual-electrode ionization tube is closely coupled to a single- or dual-stator capacitive pressure transducer. Since the 250 to 300 kHz voltage applied across the anodes (P_1, P_2) can have a large amplitude (about 250 V rms), very small variations in capacitance can cause several volts of d-c output to appear across the cathodes (C_1, C_2)

SCHEMATIC

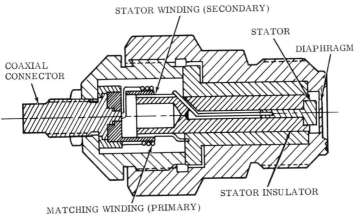

STATOR WINDING (SECONDARY)

STATOR

DIAPHRAGM

COAXIAL
CONNECTOR

STATOR INSULATOR

MATCHING WINDING (PRIMARY)

Fig. 10-18. Single-stator capacitive pressure transducer with integral fixed inductor (courtesy of Omega Dynamics Corp.).

for relatively small pressure variations. The capacitive pressure transducer in the "differential-pressure sensor" module, which includes the ionization tube [Figure 10-19 (a)], is of the dual-stator type with center diaphragm. The associated oscillator/output unit contains a voltmeter whose readings (in "Pressure Units") can be correlated with pressure by a suitable calibration curve.

The existence of a separate case containing the associated circuitry, with which the transducer must be calibrated as a "matched pair," is typical for virtually all capacitive pressure transducers.

10.3.2 Inductive pressure transducers

In an inductive pressure transducer the self-inductance of a single coil is varied by pressure-induced changes in displacement of a magnetic member in close proximity

(a) Transducer module and oscillator/output unit
with integral read-out

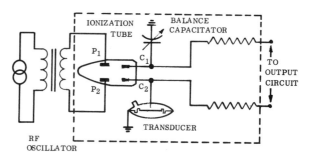

(b) Single-stator pressure
transducer circuit

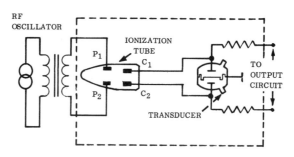

(c) Dual stator pressure
transducer circuit

Fig. 10-19. Capacitive pressure transducer used with ionization tube (courtesy of Decker Corp.).

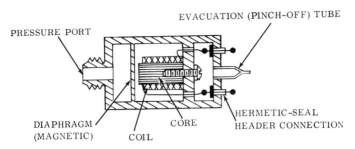

Fig. 10-20. Inductive absolute pressure transducer.

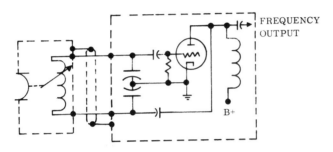

Fig. 10-21. Inductive pressure transducer in oscillator tank circuit.

to the coil. A typical design is shown in Figure 10-20. Motion of the diaphragm to and from the iron core changes the coil's inductance. In the absolute configuration shown, the cavity between diaphragm and header-sealed electrical connections is evacuated.

In other designs inductance changes are produced by the displacement of the iron core within a coil. The displacement can be created by the motion of a bellows or the tip travel of a Bourdon tube.

This type of transducer is generally difficult to compensate for environmental effects, especially temperature, and requires extensive external circuitry. It is, therefore, much less frequently used than a related type, the reluctive transducer (see 10.3.5).

10.3.2.1 Frequency-output circuit. One circuit combination with the inductive pressure transducer has found limited application where frequency output is required. In this version the transducer is used in conjunction with two fixed capacitors in the tank circuit of an LC oscillator such as the Colpitts circuit shown in Figure 10-21 or with one fixed capacitor and a tap on the coil winding as in Hartley oscillator circuits. The oscillator unit is usually installed in the telemeter package. Transducers with integral oscillator circuitry, usually transistorized, are still used in some applications.

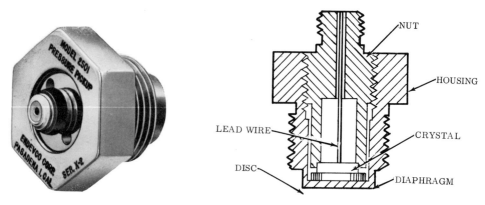

Fig. 10-22. Piezoelectric pressure transducer (courtesy of Endevco Corp.).

By careful adjustment of coil inductance and range of inductance variations, and of the fixed capacitor(s), the output frequency range of a standard IRIG telemetry channel (e.g., 675 to 785 Hz, "Channel 3") can be duplicated for the pressure range of the transducer.

10.3.3 Piezoelectric pressure transducers

Pressure transducers containing a piezoelectric transduction element are used for high-frequency pressure and sound-pressure measurements. In the latter application they have been referred to as *microphones*. Most transducer designs use a flat diaphragm whose inside surface is directly in contact with the crystal or ceramic element. Others use a disc, a column, or some similar mechanical link to transmit deflection forces to the element. The diaphragm motion can cause tension/compression, bending, or shear stresses in the element, depending on design. A typical transducer design is shown in Figure 10-22.

Some models use an annular diaphragm with a large central boss. One-piece elements are most common. Different element constructions, including a dual semi-cylindrical element, have been used in some designs.

10.3.3.1 Crystals. The piezoelectric element is commonly referred to as a *crystal* regardless of its natural or artificial origin. A variety of crystal materials is used. Quartz and tourmaline crystals are selected from those found in their natural state. Rochelle-salt and ADP (ammonium dihydrogen phosphate) crystals are usually grown artificially from aqueous solutions under controlled conditions. The required piezoelectric elements are made from slabs cut from the crystal with careful attention to the existing crystallographic axes. Presently available transducers rarely use the early developed tourmaline materials or the Rochelle-salt crystal, which has a low melting-point and cannot withstand moisture.

Ceramic elements are made from finely powdered materials which are pressed into the required shape (bisque) and then sintered by firing at high temperatures. After cooling, the top and bottom surfaces of the usually cylindrical or annular elements are machined flat and parallel. Silver foil is applied, or a silver film is sprayed and fired on the surfaces to form the required electrodes. Connecting leads can then be attached to the electrodes. The elements gain their piezoelectric characteristics when they are polarized by exposure to an orienting electric field during cooling after they are first heated in silicone oil.

Barium titanate ($BaTiO_3$) was the first material used in commercial units. Addition of controlled impurities, such as calcium titanate, was found to improve some of the characteristics. Later developments showed favorable performance from the use of lead zirconate, lead niobate, and lead zirconate-barium zirconate mixtures and resulted in the most promising mixtures, those of lead titanate-lead zirconate ("PZT"®). Most of these basic materials are also mixed with controlled small amounts of one or more additives. Piezoelectric ceramics reach their *Curie point* when sufficiently heated. At the Curie-point temperature the crystalline structure changes; polarization is lost; and the element becomes useless until it is again subjected to a polarization process. Curie points vary from about 300°F ($BaTiO_3$) to over 750°F (PZT®).

10.3.3.2 Minimization of environmental effects.

Temperature effects on piezoelectric pressure transducers can be minimized by selection of proper crystal materials. The crystalline structure of virtually all elements is affected by temperature. At low and cryogenic temperatures the crystal is still polarized, but its sensitivity can be substantially different from that at room temperature. Quartz crystals are usually limited to a −65°F to +200°F range, except PZT® which has a considerably higher Curie point. Several other ceramic mixtures have more recently been developed to extend the operating temperature range in both directions.

Acceleration effects can be minimized by proper mechanical design, reducing as much as possible the action of diaphragm and mechanical link as seismic mass. One ceramic-element transducer design (Figure 10-23) incorporates a piezoelectric accelerometer which compensates the output of the pressure transducer portion for acceleration effects to a considerable degree. This transducer type is useful for measuring rapidly varying ambient pressures and high-intensity sound-pressure levels in the presence of vibration.

Since the source impedance of a piezoelectric pressure transducer is very high, even a slight conduction path across the electrodes or between pin and shell of the coaxial electrical connector can cause erratic output readings. Sealing of the transducer is therefore quite important, and potting or sealing of the mating connector and cable after connection is very desirable.

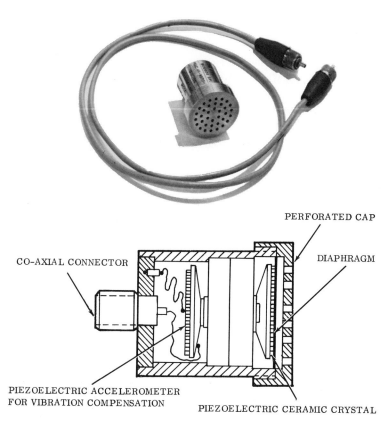

CO-AXIAL CONNECTOR

PERFORATED CAP

DIAPHRAGM

PIEZOELECTRIC ACCELEROMETER
FOR VIBRATION COMPENSATION

PIEZOELECTRIC CERAMIC CRYSTAL

Fig. 10-23. Vibration-compensated piezoelectric pressure transducer (courtesy of Gulton Industries, Inc.).

10.3.3.3 Amplifiers. The transducer's high source impedance and the necessary minimization of charge leakage rate across the crystal requires, in most cases, the use of an associated emitter-follower or amplifier with an impedance of at least 100 megohms across its signal-input terminals. The output of this signal-conditioning device can then have a source impedance of 500 ohms or less, suitable for use with a long transmission line and direct connection to indicating, storage, or transmission equipment. The two types of amplifiers commonly used, the voltage amplifier and the charge amplifier, are described in more detail in the portion of this book dealing with piezoelectric acceleration transducers.

10.3.4 Potentiometric pressure transducers

Pressure transducers with potentiometric transduction elements are widely used throughout the industry because of their simple operation, choice of high-level output,

Fig. 10-24. High-pressure potentiometric transducer with twisted Bourdon-tube sensing element (courtesy of Colvin Laboratories, Inc.).

relatively low cost, and ready availability. A large number of designs and configurations have been developed to meet different measurement requirements.

The most commonly used sensing elements are capsules (single or multiple) and Bourdon tubes (C-shaped, U-shaped, helical, spiral, and twisted). Capsules are normally used for pressure ranges up to 0 to 300 psi, and Bourdon tubes are used for

Fig. 10-25. Low-pressure transducer with lever linkage between capsule and wiper (courtesy of Fairchild Controls Corp.).

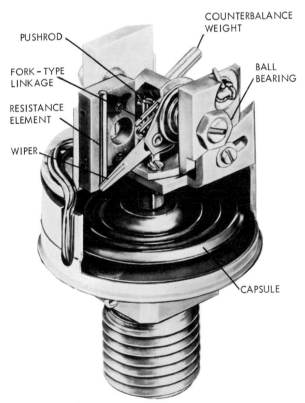

ranges over 0 to 500 psi. Either type can be used in the 300 to 500 psi region. Twisted and U-shaped Bourdon tubes have been used for ranges as low as 0 to 100 psi. A twisted-tube design is illustrated in Figure 10-24.

10.3.4.1 Amplification linkages. Potentiometer resistance elements vary in length from 0.2 to 0.8 in. in most transducers. Since the average full-scale travel of a capsule is approximately 0.04 in. the sensing element motion must be amplified mechanically to move the wiper over the full length of the resistance element. This amplification is achieved by a lever linkage (see Figure 10-25) or by a flexure linkage. The free end of the linkage arm can be provided with an adjustable counterbalance weight to offset the effect of acceleration transmitted from the capsule to the pushrod end of the linkage arm.

Pivotal jewel or ball bearings are selected for low friction, cycling endurance without excessive wear, and resistance to vibration. Flexure linkages use thin spring-metal bands instead of bearings. Combinations of flexure bands and bearing-equipped pulleys have been used successfully in capsule-type transducers. Flexures are also popular for the amplification of tip travel of C-shaped or U-shaped Bourdon tubes.

Use of helical or spiral Bourdon tubes obviates the necessity for amplification linkages. The wiper, always on an insulated supporting member, can be assembled directly onto the tube tip (Figure 10-26).

10.3.4.2 Overpressure stops. Most pressure transducers are designed to withstand a limited amount of overpressure (*proof pressure*) without damage or excessive change in performance. In potentiometric transducers the capsule boss to which the pushrod is attached frequently acts as overtravel stop. In deflating capsules it bottoms against the boss in the opposite capsule half. In inflating capsules it butts against a machined collar which also acts as pushrod support in some designs. Bourdon-tube travel can be similarly limited, either by providing stops for the tube tip or by machining the inside of the case to such a close tolerance that motion is stopped throughout most of the tube length.

10.3.4.3 Vibration damping. Potentiometric pressure transducers are widely used on vehicles or machinery where they are subjected to high vibration levels. Linkages can be compensated for vibration effects by counterbalance weights, damping grease between the pushrod and a retaining sleeve, or by a pneumatic dashpot. The vibration of helical and spiral Bourdon tubes is usually damped by surrounding the tube with silicon oil or, if required, with a liquid-oxygen-compatible fluid. Damping is improved by minimizing the free space around the tube. An expansion bellows (Figure 10-27) can be used to compensate for changes in oil volume due to environmental variations and resulting changes in internal pressure. Oil damping is effective at temperatures from $-100°F$ to about $+500°F$. Since the viscosity of the damping oil changes with temperature, more damping occurs at lower than at higher temperatures. At very low temperatures the transducer is usually damped to such an extent that its

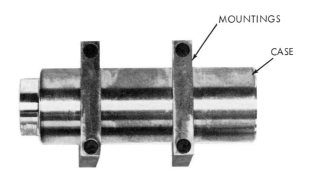

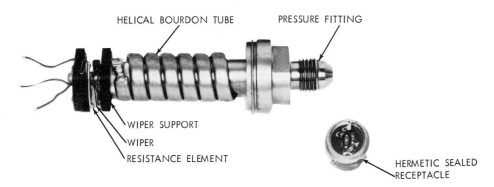

Fig. 10-26. Potentiometric high-pressure transducer with helical Bourdon tube (courtesy of Bourns, Inc.).

Fig. 10-27. High-pressure transducer with bellows for damping-oil temperature compensation (courtesy of Bourns, Inc.).

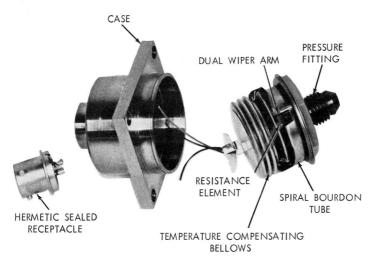

response time is lengthened measurably as compared to its response time at room temperature. Lack of adequate damping can contribute to vibration-induced wiper lift-off from the resistance element, producing noise spikes in the transducer output as well as erratic output levels. These problems can be further prevented by proper wiper design.

10.3.4.4 Wiper design. Wipers are usually stamped from spring alloy and are grooved or dimpled to provide a narrow contact surface. Frequently, a precious-metal wire is fastened along the bottom of the wiper arm to reduce contact width and optimize the resolution of the instrument. A poorly designed wiper can cause troublesome lift-off and lateral-excursion noise when excited at its own resonant frequency. Increasing the pressure of the wiper against the resistance element will tend to overcome this problem, but it can also increase friction errors beyond acceptable limits and greatly reduce wiper life.

Wiper shapes with protrusions resonating at different frequencies have been used with a fair amount of success. A double wiper construction with one contact below, and one above, the element has yielded satisfactory results because one-half of the wiper assembly always remains in contact with the element; thus noise due to discontinuities is reduced.

10.3.4.5 Resistance elements. The resistance element usually consists of a length of precious-metal or nickel-alloy wire, either insulated or bare, wound on a flat, oval, or round mandrel. Platinum or platinum-alloy wire, 0.3 to 2 mils thick, is used most commonly. Between 300 and 600 turns are wound around the mandrel, which can be an insulating material (such as ceramic, phenolic, or glass) or a coated metal (such as anodized aluminum). Epoxy or ceramic materials are sometimes used to fix the wire turns in position after winding. The mandrel material is selected to match the temperature coefficient of expansion of the wire. If insulated wire is used, the insulation is removed after winding along the area to be contacted by the wiper arm.

Because the resolution of transducers using wirewound elements is finite (normally between 0.15 and 0.35% FSO), other types of elements have been developed with varying degrees of success and have been found useful in some applications. These include carbon-film, conductive plastic, cermet (conductive ceramic), and metal-film elements which all share almost infinitely small resolution characteristics.

10.3.4.6 Dual-output transducers. Because of their modular construction, potentiometric pressure transducers lend themselves to special design modifications. Some applications, for example, call for more than one output from the transducer. The two outputs can be of identical nature, or one output can be linear, and the other can follow a specified nonlinear curve, or the second output can indicate one or more discrete levels (*switch output*). Potentiometric transducers have been designed for these special needs. A single sensing element and two transduction elements can be used, or,

Fig. 10-28. Dual-tube, dual-element high-pressure transducer (courtesy of Servonic Instruments, Inc.).

as illustrated in Figure 10-28, separate sensing and transduction elements can be assembled in one configuration.

10.3.4.7 Transduction-element isolation. The design of potentiometric differential-pressure transducers presents no unusual problems as long as the measured fluid is fed into the sensing element and the fluid on the reference side has no harmful effects on the resistance element, wiper, their electrical connections, and any linkage in the transducer. When the fluid can affect the transduction element, the latter must be isolated. When the measured fluid is incompatible with the reference fluid, they should be well isolated from each other. Figure 10-29 shows a transducer which provides isolation between measured ("high") and reference ("low") pressure ports and between either port and the transduction element. A plastic membrane is used for isolation, and silicone oil is used as pressure transfer medium. One disadvantage of using oil for isolation purposes is the difficulty of balancing acceleration-induced forces acting on the mass of the oil itself, and the resulting acceleration error.

Isolation of the transduction element may also be necessary in a gage-pressure transducer. When the transducer case is vented to the ambient atmosphere, a filter or porous plug in the *gage vent* affords limited protection against contaminants. One

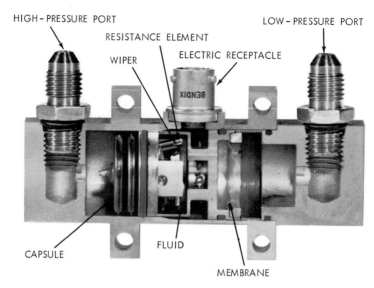

Fig. 10-29. Differential-pressure transducer with port isolation (courtesy of Colvin Laboratories, Inc.).

solution to the problem of obtaining complete isolation is the use of a second sensing element, such as a capsule, linked to the measuring element so as to provide the necessary displacement reference. This reference capsule is vented to the atmosphere while all other internal portions of the transducer remain sealed.

10.3.5 Reluctive pressure transducers

The reluctance changes in the magnetic circuit necessary to produce output-voltage changes in reluctive pressure transducers are commonly created by one of the following methods:

1. The deflection of a magnetically permeable diaphragm supported between two ferric-core coils which are connected as a two-arm inductance bridge. Diaphragm motion decreases the gap in the magnetic flux path of one coil while simultaneously increasing the gap in the other coil's flux path (Figure 10-30). The output voltage is proportional to the ratio between the two inductances.

2. The rotation of a twisted Bourdon tube with a strip of magnetically permeable material (*armature*) fastened to and protruding radially from its tip. Angular deflection of the tube tip causes deflection of the armature. Armature motion decreases the flux gap of one coil and increases the gap of the other coil. The two coils are wound around opposite arms of an E-shaped ferric core (*E-core*) (Figure 10-31). Coil connection and production of output voltage are the same as for method 1 above.

3. The motion of a core within a triple-coil assembly connected as a differential transformer. Displacement of the core toward either of the secondary windings

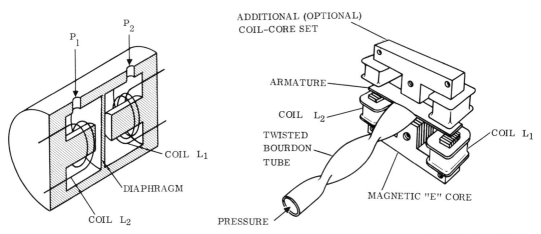

Fig. 10-30. Reluctance change by diaphragm deflection.

Fig. 10-31. Reluctance change by angular deflection of Bourdon tube (courtesy of Wiancko Engineering Co.).

decreases the reluctance path and increases the magnetic coupling between that winding and the centrally located primary winding. This produces an output-voltage change in a manner similar to that of the inductance-bridge circuits of methods 1 and 2. Core motion is caused by the deflection of the sensing element, e.g., capsule, bellows, or Bourdon tube (Figure 10-32), frequently with the assistance of a mechanical amplification linkage.

Fig. 10-32. Reluctance change by core motion in differential transformer.

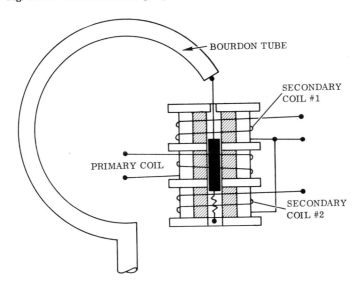

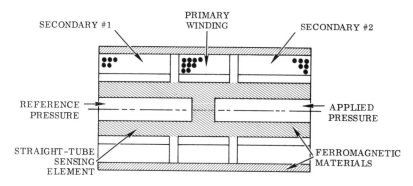

Fig. 10-33. Reluctance change by stress-induced permeability change in straight tube (courtesy of International Resistance Co.).

4. The change in permeability due to a change in stress of a straight-tube sensing element used as fixed core in a differential-transformer circuit (Figure 10-33). Circuit operation is similar to that described in 3 above, except that the decrease in reluctance path occurs in the fixed tubular core rather than by displacement of a moving core.

The most commonly used designs, 1 and 2, are further discussed below.

10.3.5.1 Diaphragm-type transducers. Reluctive pressure transducers using a diaphragm as sensing element are popular for the measurement of low pressures, especially differential pressures below 0 to 5 psid, but are equally suited for higher pressure ranges, e.g., up to 6000 psi. The thin metallic diaphragm is either integrally machined with its supporting ring or is clamped, soldered, brazed, or welded to its annular support. Figure 10-34 illustrates a model which is assembled, by sandwich build up, from the end cap with pressure port, the seal disk with central opening, and the coil assembly with matching central opening, diaphragm, and the corresponding opposite pieces. The central holes in the seal plate and coil assembly admit pressure to the diaphragm.

Earlier models used O-rings as compression seals. In later designs the possibility of leakage around O-rings and their deterioration with time was prevented by welding or brazing the coils and diaphragm into an integral module and admitting pressure to it through braze-sealed internal tubing sections. It is desirable to use the same metal for diaphragm, support, and other portions of the sensing assembly to avoid differential expansion of the various pieces due to temperature variations.

One design uses a shell around the sensing module to equalize pressures inside and outside the diaphragm-supporting cylinder or "inner case." This presents one solution to the problem of error-causing stresses introduced by varying high reference pressures, relative to which small differential pressures must be measured.

Coils are wound over ferric or ferroceramic cores of various shapes. A shape resembling an "E" in cross section (*E-core*) is frequently used. The windings are completely potted and sometimes even covered with a nonmagnetic metal foil so that the coil assembly itself does not respond to pressure variations and is not damaged by chemical interaction with a large variety of measured fluids. Best stability is obtained by temperature cycling and aging of the completed transducer after assembly.

Output voltages up to 20% of applied excitation can be obtained with full-scale diaphragm motion of a few thousandths of an inch. The small deflection requirement permits diaphragms to be relatively stiff, even for low-pressure ranges. The stiffness sometimes obtained by prestressing the diaphragm and the low mass of the thin diaphragm tend to minimize vibration effects in the transducer. A dual-diaphragm design has been developed in which vibration effects are further reduced by connecting the pressure inlets and the electrical outputs of two separate sensing-transduction modules in such a manner that no net output results from simultaneous deflection of both diaphragms in the same direction (caused by acceleration), whereas a larger output results from diaphragm deflection in opposite directions (caused by pressure variations). The two sets must be mutually balanced within close limits.

Fig. 10-34. Diaphragm-type reluctive differential-pressure transducer before assembly (courtesy of Mitchell Camera Corp.).

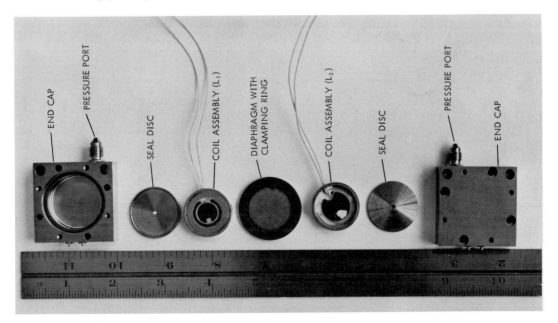

10.3.5.2 Twisted-Bourdon-tube type transducers. Pressure transducers using the twisted Bourdon tube as sensing element have been developed for gage-, absolute-, and differential-pressure measurements over ranges from below 0 to 5 psi to over 0 to 10,000 psi. Figure 10-35 illustrates a basic transducer of this type. The pressure is admitted into the Bourdon tube. In the gage-pressure version the case is vented to the ambient atmosphere. In the absolute-pressure version the case is evacuated and sealed. In other models the Bourdon tube is evacuated and sealed, and the pressure is admitted to the case. This reduces the frequency response of the transducer due to the relatively large cavity formed by the case. Differential-pressure versions introduce one pressure into the tube and the other into the case.

The transducer's output can be increased by adding a second E-core coil assembly so that it faces the upper surface of the armature. The four coils can then be connected as a four-active-arm a-c bridge.

Temperature effects are minimized by using special alloys such as Ni-Span-C® as Bourdon-tube material. Vibration effects are reduced by supporting the tube tip in a crosswire flexure, in which radial supporting wires are rigidly attached to the armature center near its joint with the tube tip, and, at their opposite ends, to thick supporting studs.

Although the coils are sealed, the volume around armature and coils is suffi-

Fig. 10-35.   Reluctive pressure transducer using twisted Bourdon tube (case removed) (courtesy of Wiancko Engineering Co.).

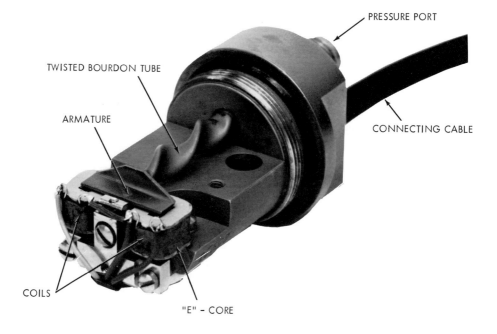

ciently large, and the internal materials used are such that conductive or highly corrosive fluids can cause performance change and possible damage when admitted into the case. An isolation membrane with silicone oil or Fluorolube® as transfer medium can be used when the pressure of such fluids must be measured. Among the disadvantages of membrane isolation are increased acceleration effects due to the mass of the oil, possibility of leaks into the measured system, and damping of response.

A very useful design variation incorporates a prestressed Bourdon tube which does not begin to untwist until a pressure near the upper-range end point is reached. Armature motion required for full-scale output is obtained between this pressure and the upper-range limit pressure. This permits zero suppression and range expansion. The expanded range can be as small as 10% of the upper-range limit pressure, e.g., 900 to 1000 psi for zero to full-scale output.

10.3.5.3 D-C conversion. The increasing use of d-c measuring systems required manufacturers of reluctive transducers to develop dc-ac-dc conversion circuitry so that obsolescence of this transducer type could be averted and its beneficial features made more generally useful.

Since these d-c output transducers were designed primarily for aerospace instrumentation systems, they usually operate from a 28 V d-c excitation source (typical for aircraft and missile batteries) and produce the 0 to 5 V d-c output required by a majority of airborne telemetry systems.

A typical pressure transducer with integral conversion circuitry is illustrated in Figure 10-36. As shown in the associated block diagram, the excitation voltage is applied to a d-c to a-c converter (oscillator) which creates a sinusoidal, square-wave, or sawtooth-shaped a-c voltage. A voltage regulator is frequently incorporated in the input circuit in order to assure that the a-c voltage amplitude remains independent of excitation-voltage variations. The a-c voltage (carrier) generated by the oscillator stage is applied to the sensing/transduction module through a "carrier supply" (conditioner). The a-c voltage resulting from pressure variations may require amplification before conversion to d-c in the demodulator. Amplification can be accomplished by use of a voltage step-up transformer or an amplifier stabilized by feedback control. Some designs use a d-c amplifier after the demodulator. An output limiter ("clamp circuit") can be added when output limiting is required during pressure overloads. Transducers whose range is bidirectional are frequently equipped with an additional output-biasing circuit which offsets the d-c output in such a manner that a specific positive output voltage (e.g., +2.500 V d-c) corresponds to zero differential pressure.

10.3.6 Resistive pressure transducers

A number of resistive pressure-transducer designs have been developed with varying amounts of success. Most of these rely on resistance variations in a conductive material due to variations in applied pressure.

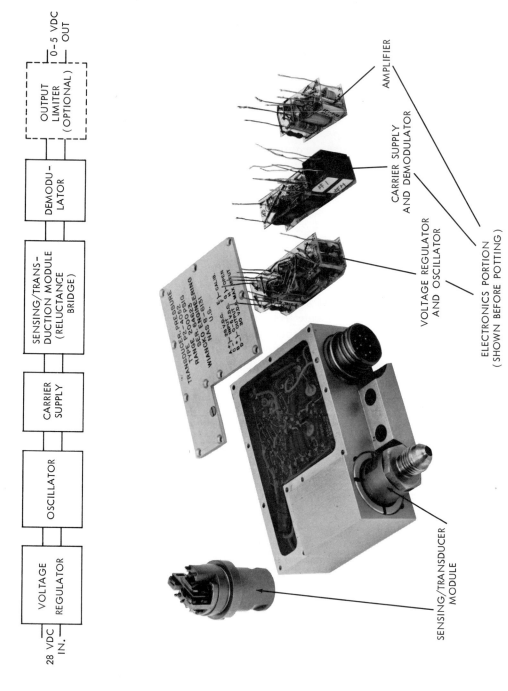

Fig. 10-36. D-C output, reluctive pressure transducer (courtesy of Wiancko Engineering Co.).

450

10.3.6.1 Carbon resistance. The compression of carbon powder or a stacked set of carbon discs results in a decrease of this material's resistance. A diaphragm, capsule, or bellows can provide this compression. The resistance change appears across two internal conducting surfaces, one of which can be the metallic diaphragm. This principle has been used in carbon microphones for many years. Its application to pressure transducers has not been too successful.

More satisfactory performance appears to have been obtained with transducers whose sensing element consists of two carbon resistors of a commonly available type, connected in two opposite arms of a Wheatstone-bridge circuit. The remaining two arms are also carbon resistors which are, however, sealed off from the pressure applied to the two active resistors. Since carbon resistors incur a variation in their resistance proportional to pressure ambient to such resistors, the bridge circuit produces a change in output voltage proportional to pressure when the bridge-excitation voltage is held constant. The absence of a primary mechanical sensing element is significant. The active resistors constitute the sensing element as well as a portion of the transduction element which is the bridge network itself. The operation of this type of transducer is comparable to that of a two-active-element strain-gage bridge.

10.3.6.2 Rare-earth-mixture resistance. Resinous mixtures of rare earths and zirconium tetrachloride have been found to undergo a decrease in electrical resistance when pressure is applied to such mixtures contained between two conductive surfaces. The rare earths are typically combinations of Lanthanide-series elements—those elements having atomic numbers from 58 through 71.

Transducers incorporating rare-earth mixtures, either as a resinous layer or as a dried paint layer between two electrodes, have been used in a number of experimental applications.

10.3.7 Strain-gage pressure transducers

The conversion of pressure changes into changes of resistance due to strain in two or, more commonly, four arms of a Wheatstone bridge has been used in commercial pressure transducers for many years. Numerous strain-gage pressure-transducer designs are widely used, differing to some extent in their sensing elements and to a large extent in their transduction elements.

10.3.7.1 Sensing elements. The usual pressure-sensing element is a flat or corrugated diaphragm. Catenary and other more specialized diaphragms are used sometimes. Another group of designs incorporates versions of the straight-tube sensing element.

Most transducers use a secondary sensing element (auxiliary member) as deforming member. It produces the necessary strain to be sensed by the gages when acted upon by the primary sensing element. Commonly used auxiliary members are: a thin rectan-

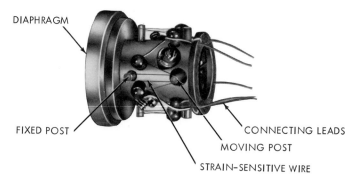

DIAPHRAGM

FIXED POST

CONNECTING LEADS

MOVING POST

STRAIN-SENSITIVE WIRE

Fig. 10-37. Basic elements of unbonded-strain-gage pressure transducer (courtesy of Statham Instruments, Inc.).

gular strip, usually thinner near its longitudinal center than at its ends (*beam*), a ring of rectangular cross section with two diametrically opposite radial bosses (*force ring*), and a thin-walled cylinder (*strain tube*).

10.3.7.2 Transduction elements. The variety of strain-gage elements used in pressure transducers can be classified as follows:

1. Unbonded metal-wire gages.
2. Bonded metal-wire gages.
3. Bonded metal-foil gages.
4. Bonded semiconductor gages.
5. Integrally diffused semiconductor gages.

Unbonded metal-wire elements are stretched and unsupported between a fixed and a moving end (Figure 10-37). The wire is usually looped one or more times over supporting posts, and friction between wire and post is minimized by proper selection of post materials.

Bonded gages are permanently attached over the length and width of their active metal-wire, metal-foil, or semiconductor element. Figure 10-38 illustrates the assembly of a typical absolute-pressure transducer utilizing: (1) a corrugated diaphragm spotwelded around its edge to an annular base plate; (2) a beam to which four active metal-foil gages are bonded; (3) a stop bar with adjusting screw and locknut; (4) a hermetically sealed header, solder-sealed to the sensing element under vacuum; (5) an end cap with pressure port; and (6) an end cap with electrical connector. When pressure is applied to the diaphragm, its deflection is transmitted to the beam through the pushrod. The stop bar prevents further beam motion when excessive pressure, beyond the rated range, is applied to the diaphragm. Compensation resistors (connected to header pins) and wire are not shown.

Typical for a different type of bonded-strain-gage transducer is the model illus-

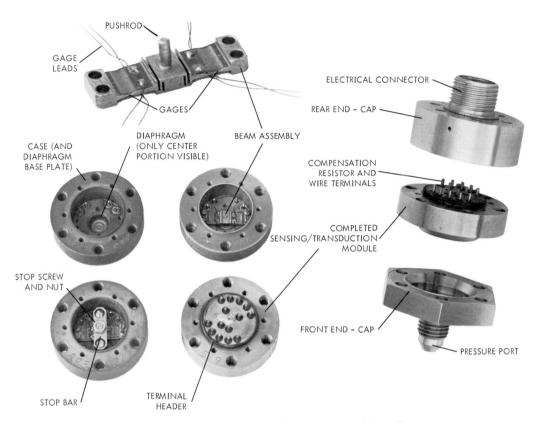

Fig. 10-38. Bonded-strain-gage pressure transducer (courtesy of Data Sensors, Inc.).

trated in Figure 10-39. This *flush-diaphragm* unit transmits diaphragm deflection to a strain tube to which circumferential and longitudinal wire windings are bonded.

Another proprietary design utilizes two concentric strain-tubes with gages bonded to the outside of the outer tube and the inside of the inner tube. Both tubes are sealed-off at one end and sealed to an end cap at the other end. Pressure is applied at the end cap into the space between the two tubes so that the outer tube expands and the inner tube contracts.

An entirely different manner of obtaining a bonded-strain-gage element is the vacuum deposition, or similar application, of a thin metallic film through a four-gage-pattern mask upon an insulating substrate which is first deposited on a metal diaphragm or other sensing element. Related to this design is the strain-gage bridge pattern, etched from metal (e.g., Constantan) foil and cemented onto a diaphragm sensing element, with the cement acting as insulating layer.

Semiconductor gages are bonded to a primary or secondary sensing element by

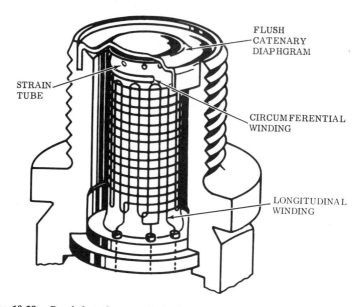

STRAIN
TUBE

FLUSH
CATENARY
DIAPHGRAM

CIRCUMFERENTIAL
WINDING

LONGITUDINAL
WINDING

Fig. 10-39. Bonded-strain-gage, flush-diaphragm pressure transducer (courtesy of Advanced Technology Labs.).

use of special bonding techniques to provide adhesion to the usually polished crystal surfaces. Silicon with controlled impurities ("dopants") is the crystal material most commonly used in these applications. Although these semiconductor elements have inherent temperature-sensitivity problems, considerable development efforts have resulted in doping variations useful to match temperature sensitivity between gages as well as to exert control over resistance and resistance-change characteristics over a relatively wide range of temperatures.

Among the more promising semiconductor-strain-gage pressure-transducer developments are diaphragms made of silicon with doping material diffused directly into their surface in a controlled pattern, so that doped portions form the required piezoresistive area connected by lines of diffused or deposited conductive material between them into the typical strain-gage bridge.

10.3.7.3 Compensation. The full-scale output of a basic four-active-element strain-gage bridge, as used in a pressure transducer incorporating metal-wire or foil gages, is approximately 50 to 60 mV for bonded gages and 60 to 80 mV for unbonded gages, with 10 V excitation applied. All such transducers contain compensating and adjusting resistors (Figure 10-40) which reduce the full-scale output to a nominal 30 mV for bonded gages, and 40 mV for unbonded metallic gages at 10 V excitation.

These resistors allow zero-balance adjustment, sensitivity (full-scale-output) adjustment, thermal-zero-shift compensation, and thermal-sensitivity-shift compensation. The number and type of these resistors are determined by the amount of adjustment and compensation required, which varies between transduction elements. Hence, one or more of the resistors may be omitted in a transducer and replaced by a jumper.

10.3.7.4 Shunt calibration. External *shunt calibration* of the transducer is used in some pressure-measuring systems. Simulation of pressure application is obtained, for system-calibration purposes, when a resistor is connected in parallel with one arm of the strain-gage bridge. The resulting bridge unbalance can be made equal to the known transducer output at a desired pressure level by careful selection of the shunt calibration resistor. Some systems provide for simulation of several pressure levels (e.g., 25, 50, and 75% of range) by sequentially switching-in several external calibration resistors of different values.

10.3.7.5 Output standardization. Some measuring systems require complete interchangeability of all strain-gage transducers used. The full-scale output of each transducer must then be held to a fixed value within narrow tolerances. When the normally used sensitivity-adjustment resistors are insufficient for this "output standardization," additional resistors can be added to the internal circuit, such as a series network of two resistors, one temperature-sensitive, one temperature-insensitive, placed across the transducer's output (terminals *2* and *3* in Figure 10-40).

10.3.7.6 Integral amplification. The full-scale output of metal-wire and metal-foil strain-gage type transducers is on the order of 30 to 40 mV at 10 V excitation. The maximum excitation for such transducers is usually limited to between 12 and 18 V. Semiconductor strain-gage transducers yield a much higher output, on the order of 500 mV for 10 V excitation. Most semiconductor designs have limitations on excitation voltage more stringent than metal-gage types and produce a full scale output of 250 to 500 mV, using an internal voltage-dropping circuit to reduce the excitation voltage to the level required for proper bridge operation.

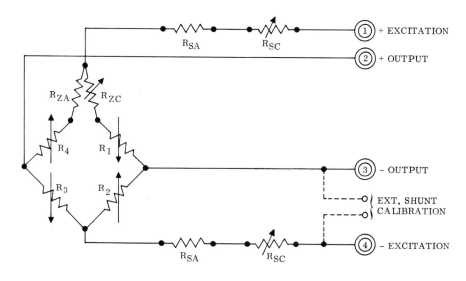

SYMBOL	FUNCTION
R_{ZA}	ZERO BALANCE ADJUSTMENT
R_{ZC}	COMPENSATION FOR THERMAL ZERO SHIFT
R_{SA}	SENSITIVITY ADJUSTMENT
R_{SC}	COMPENSATION FOR THERMAL SENSITIVITY SHIFT
$R_1 - R_4$	STRAIN GAGE BRIDGE (TRANSDUCTION ELEMENT)

Fig. 10-40.   Typical compensation and adjustment network for strain-gage transducer.

In order to obtain the popular 5 V output (required for many telemetry and recording systems), an amplifier must be used for metal-gage, as well as for most semiconductor-gage, transducers. Many designs now have an integral amplifier. These amplifiers provide additional functions in most cases. They are designed to operate with the 28 V d-c excitation typically available in aerospace instrumentation systems. Some amplifiers provide optional excitation regulation. The next stage of the amplifier reduces the excitation to a value optimized for the strain-gage bridge used or converts the d-c excitation to a reduced a-c excitation for the bridge. The output of the bridge is then amplified by a d-c differential amplifier or, if a-c was used, by an a-c amplifier followed by a demodulator. Additional circuitry may be incorporated to provide low source impedance, to limit output even when overpressure is applied to the transducer, or to permit use of a common excitation and output ground connection.

10.3.8 Servo-type pressure transducers

This group of transducers incorporates a closed servo loop and is better known by this special feature than by its primary transduction element, which is usually reluctive. The two most common servoed transducer types are the null-balance and the force-balance transducers.

10.3.8.1 Null-balance pressure transducers. In this transducer type [Figure 10-41 (a)] the sensing element is allowed to deflect freely. The displacement, such as by a pushrod, is detected by a coil system arranged in such a manner as to produce a minimum absolute value of output (*null*) when a condition of configurational balance exists. When the sensing element deflects, an unbalance occurs in the coil system resulting in an output increase. This error signal is applied to a servo amplifier which energizes a servo motor. The shaft rotation of the servo motor causes a positioner to drive the coil system into a new null position while simultaneously operating an output device (e.g., moving a wiper arm over a potentiometric resistance element).

10.3.8.2 Force-balance pressure transducer. In the more frequently used version the sensing element is not allowed to deflect freely but is restrained by a force generator so that no displacement occurs [Figure 10-41 (b)]. When the sensing element tries to deflect in response to applied pressure, the balance of a coil system is disturbed so that an output signal is generated. This output signal is fed into a servo amplifier which drives a servo motor. The servo motor operates an output device (secondary transduction element) while simultaneously causing a force generator to apply sufficient force to the sensing element to maintain balance in the coil system. No null condition is created in this system; instead, the transducer output is proportional to the force required to restrain the sensing element so that no displacement occurs.

In a less frequently used version the servo motor is omitted and the output from the servo amplifier is directly applied to the forcing coil or other displacement-restraining device [Figure 10-41 (c)]. The amplifier output current is proportional to the pressure applied to the sensing element by virtue of being proportional to the force required to keep the sensing element from causing a displacement. A resistor is then placed in series with the servo-amplifier output, and the IR drop across this resistor is the output voltage (E_{out}) of the transducer. An amplifier or buffer stage is frequently used to isolate the circuitry following the transducer from the servo amplifier and to minimize loading effects on the transducer. This concept is used more commonly in acceleration transducers.

Figure 10-42 illustrates a force-balance pressure-ratio transducer incorporating two sensing elements connected mechanically by a linkage. The linkage is prevented from displacing by the restraining force proportional to the ratio between the pressures applied to the two sensing elements.

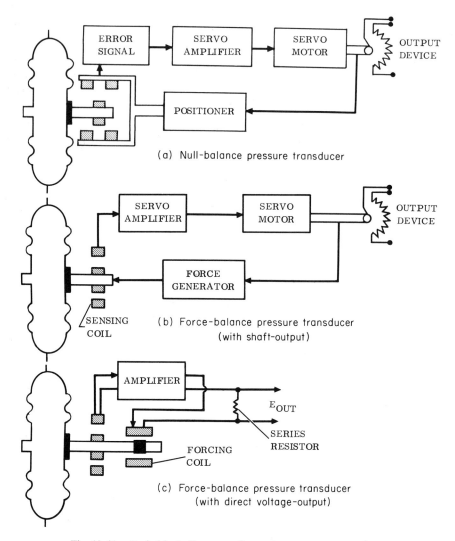

Fig. 10-41. Basic block diagrams of servo-type pressure transducers.

10.3.9 Vibrating-element pressure transducers

This type of transducer is capable of providing very close data-system accuracy. Its output is the fundamental or a harmonic of the frequency at which its sensing element or its mechanically coupled transduction element vibrates. This frequency output or frequency-modulated output (frequency deviation from a given center frequency)

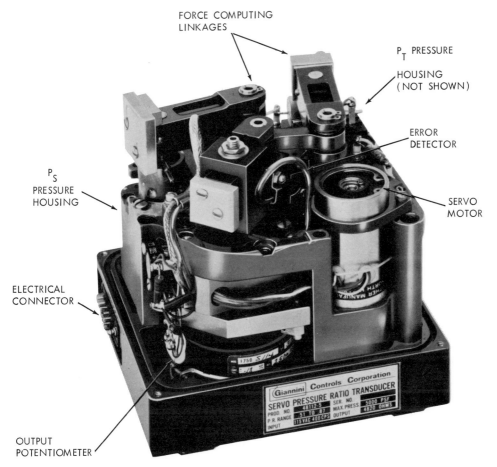

Fig. 10-42. Force-balance pressure-ratio transducer (courtesy of Conrac Corp.).

can be transmitted through a complex data-transmission system with virtually no deterioration since system errors affect only the amplitude, not the frequency, of the transducer output signal. These transducers can be adjusted so as to provide a full-scale frequency output equal to that of subcarrier oscillators used in FM telemetry systems.

10.3.9.1 Vibrating-wire pressure transducers. This version uses a diaphragm sensing element. One end of a very thin wire (usually tungsten) is attached to the center of the diaphragm. The other end is mechanically fixed and electrically insulated. The wire is located in a magnetic field usually obtained from a permanent magnet.

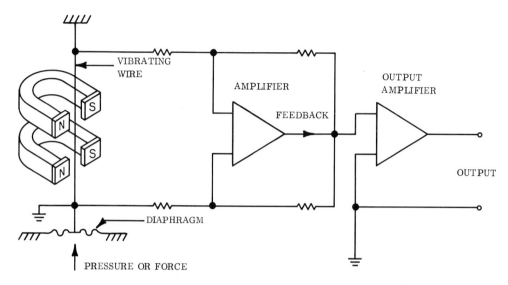

Fig. 10-43. Simplified schematic of vibrating-wire type transducer (courtesy of White Avionics Corp.).

The wire also forms a portion of one leg of a Wheatstone bridge (Figure 10-43). When a current is passed through the wire, as occurs when excitation is first applied to the transducer, the wire moves across the magnetic field. Although this motion is very small, it is sufficient to cause a current to be induced in the wire by the field. This current appears as a potential across the bridge. Sustained oscillation is produced when this voltage is amplified and fed back. The bridge-oscillator output is then further amplified by an output amplifier which also acts as buffer between load and oscillator.

Vibrating-wire pressure transducers are normally so designed that increasing pressure results in decreasing wire tension which in turn causes a reduction in the output frequency of the transducer. The frequency of vibration of a stretched wire is given by the relationship

$$f = \frac{n}{2L}\sqrt{\frac{F}{Ad}}$$

where: n = any integer expressing mode of oscillation
          (1 for the fundamental, 2 for the second mode, etc.)
     L = wire length
     F = tension force acting on wire
     A = cross-sectional area of wire
     d = density of wire material

From the above equation it can be seen that a plot of frequency-vs-pressure could never be linear since output frequency is proportional to the square root of wire tension. Over the frequency range of a typical IRIG-standardized telemetry channel—$\pm 7.5\%$ of the center frequency—the independent linearity of the "bow"-curve representing frequency-vs-pressure is within approximately $\pm 1.5\%$ FSO. The end-point linearity would be close to $+3, -0\%$ FSO.

The diaphragm deflection required for full-scale output is extremely small and is measured in microinches. This allows use of a stiff sensing element with minimized hysteresis and very good repeatability. However, a slight deformation due to thermal effects in the wire-supporting frame or in the wire itself can cause relatively large errors in the transducer output. Optimum performance of this type of transducer can only be achieved by close matching of thermal coefficients of expansion of all

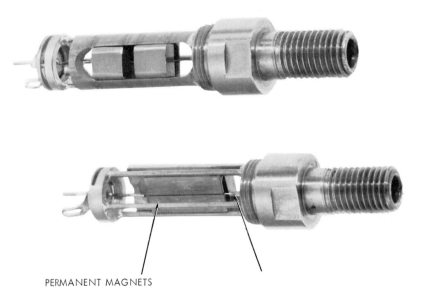

PERMANENT MAGNETS

Fig. 10-44. Sensing transduction module of vibrating-wire pressure transducer (courtesy of White Avionics Corp.).

materials used in wire and frame assembly and by use of a constant-temperature heater jacket around the sensing/transduction module. The temperature must be maintained within such close limits that thermostatic control is usually inadequate and a proportional-control system must be used in which one or more suitably located temperature sensors control the heater current through a control amplifier.

The sensing/transduction portion of a vibrating-wire pressure transducer is illustrated in Figure 10-44. The diaphragm is contained within the end cap (at the pressure fitting end). Two opposing magnetic fields are established by the two permanent magnets (upper picture) to obtain second-mode wire oscillation in the design shown.

10.3.9.2 Vibrating-cylinder pressure transducers. This type of transducer utilizes a straight-tube sensing element—a thin-walled cylinder capped at one end. This cylinder is caused to vibrate at its mechanical resonant frequency by a driving coil and a velocity pick-up coil mounted on a supporting cylinder within the sensing element (Figure 10-45). Vibration of the sensing cylinder at its resonant frequency is maintained by an oscillator/amplifier which has the coil set connected in its feedback circuit. When the pressure within the vibrating cylinder is constant and the pressure applied externally to the cylinder increases, the effective stiffness of the system is reduced and its resonant frequency is lowered. The frequency depends on the material, length, diameter, and wall thickness of the cylinder, the vibration mode, the pressure differential across the cylinder wall, and the physical characteristics, including density, of the measured fluid and the reference (internal) fluid. Effects of the latter can be minimized by proper sensing/transduction-element design. The variation of frequency with pressure follows the relationship

$$f = f_0 \sqrt{1 - AP}$$

where: f = frequency at pressure P
$\quad\quad f_0$ = frequency at lower limit of pressure range
$\quad\quad A$ = cylinder constant
$\quad\quad P$ = measured pressure

The full-scale output frequency can be made to match the frequency ranges of standard telemetry subcarrier oscillators. Sinusoidal oscillation and output are assured by use of automatic gain control in the amplifier. Temperature errors can be minimized by use of cylinder material relatively free of thermal effects, matching of materials as to thermal coefficients of expansion, and use of a heater jacket if further reduction of temperature errors is required.

10.3.9.3 Vibrating-diaphragm pressure transducers. This design utilizes a diaphragm as sensing and secondary-transduction element. Pressure is admitted to one side of a diaphragm so that it is deflected toward the transducer's reference cavity. An electromagnetic drive coil on the reference side excites the diaphragm into mechanical resonance. A stator plate, placed close to the diaphragm on its reference side, forms a capacitor with the diaphragm itself acting as the rotor or moving plate. The secondary sensing capacitive element and the driving coil are connected into a feedback circuit, an oscillator/amplifier which maintains the diaphragm in oscillation and from which the transducer's frequency-modulated output is obtained. Stress build up in the diaphragm, due to increasing applied pressure, changes the resonant frequency of the diaphragm and, hence, the frequency of the transducer's output voltage.

Transducers of this type have been used for low-pressure measurements where an overall system accuracy of ± 0.25 to 0.5% FSO is required.

SENSING/TRANSDUCTION
MODULE

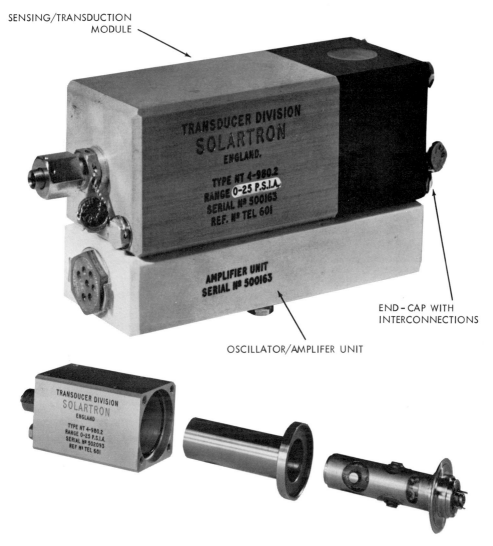

END – CAP WITH
INTERCONNECTIONS

OSCILLATOR/AMPLIFER UNIT

Fig. 10-45. Vibrating-cylinder pressure transducer (courtesy of Solartron Electronic Group, Ltd.).

10.3.10 Photoelectric pressure transducers

Among the recently developed photoelectric or photoconductive pressure transducers, one relatively successful design uses a straight-diaphragm sensing element linked, through a short pushrod, to a double-knife-edge shutter. Shutter motion varies the incidence of light emanating from a miniature ruggedized lamp on two photo-transistors connected in a feedback-controlled bridge circuit. This design is capable of producing a 5 V d-c full-scale output with little or no internal signal amplification.

10.4 Performance and Applications

10.4.1 Specification characteristics

Specifications for pressure transducers should include most of the design and performance characteristics listed and explained below with appropriate values and their tolerances shown for each specified characteristic.

10.4.1.1 Mechanical design characteristics. The mechanical design characteristics of pressure transducers, as listed in Table 10-4, include primarily those pertaining to configuration, materials, weight, construction, and identification.

A drawing should accompany every specification to show outline dimensions, type of mountings and their dimensions (unless the transducer is mounted by its pressure port), sizes of any screw threads used, the pressure port or ports, their locations, type, size, and thread, and the type and location of all electrical connections. The location of the nameplate and other identification markings on the transducer's case should be included on the drawing.

The written specification usually starts with the nature of the pressure to be measured, which can be absolute, gage, unidirectionally or bidirectionally differential, or sealed-reference differential (i.e., the reference pressure has a fixed value and is sealed within the transducer). For specialized applications, a secondary measurand in terms of which the transducer is to be calibrated and read out may be shown instead of the primary measurand (pressure), e.g., "seawater depth" or "altitude."

The specified measured fluids should include those which may be used for cleaning the transducer (e.g., alcohol, carbon tetrachloride, trichloroethylene), and the specification may have to show concentrations when these are significant (e.g., "90% hydrogen peroxide").

Material and finish of the transducer's case should be specified when operation in a damp, salty, or other deleterious atmosphere is required. To assure such operation further it is often desirable to state requirements for case sealing in detail. Some transducers use O-rings or gaskets to form a seal between portions of the case. When such elastic seals may be damaged by atmospheric elements, cleaning solvent, or one of the measured fluids, a different type of seal (e.g., welding or brazing) should be specified. An explosion-proof or waterproof enclosure may be required by the transducer's usage or by an applicable electrical code.

Gage-pressure and differential-pressure transducers usually expose their transduction element to the atmosphere or to a measured fluid, respectively. Corrosive or salty air can penetrate the usual gage-vent filter, enter the inside of the case, and cause

Table 10-4 Design Characteristic for Pressure Transducer
Specifications

Mechanical Design
a. Specified by User and Manufacturer

Configuration and dimensions (shown on drawing)
Mountings (shown on drawing)
Type and location of pressure ports (shown on drawing)
Type and location of electrical connections (shown on drawing)
Nature of pressure to be measured, including range
Measured fluids
Case sealing (explosion-proof, burst-proof, or water-proof enclosure)
Isolation of transduction element
Mounting and coupling torque
Weight
Identification
Nameplate location (shown on drawing)

b. Stated by Manufacturer

Sensing element
Transduction element details
Materials in contact with measured fluids
Dead volume
Type of damping (including type of damping oil if used)

Electrical Design

Excitation (nominal and limits)
Power rating (optional)
Input impedance (or element resistance)
Output impedance
Insulation resistance (or breakdown-voltage rating)
Wiper noise (in potentiometric transducers)
Output noise (in d-c output transducers)
Electrical connections and wiring diagram
Integral provisions for simulated calibration (optional)

---

eventual malfunction of the transduction element. Contact between some measured fluids and portions of a transduction element can not only cause output errors, output discontinuities, and partial or complete short circuits, it may even cause explosive hazards. A clearly stated requirement for transduction-element isolation from these media may then be indicated.

Some transducers are so sensitive to externally applied stresses that output shifts occur when the coupling nut is tightened over the pressure port or when the mounting hardware is tightened during installation. It is desirable to specify the mounting torque as well as the coupling torque whenever there is any doubt as to their effects on the transducer's operation.

Manufacturers' specifications usually show the actual weight, whereas user-prepared specifications show the maximum allowable weight of the transducer. The latter should not be chosen until a realistic appraisal of the state of the art is made.

The identification of the transducer can be either shown on a nameplate or etched or engraved directly on the case. Specifications should always list those items which must be included in the identification. Recommended requirements include:

1. Nomenclature (e.g., "Capacitive Pressure Transducer," "D-C-Output, Reluctive Pressure Transducer").
2. Range and units (e.g., "0 to 1000 psia," "±5 psid," "−5 to +10 psig").
3. Manufacturer's name, part number, and serial number.
4. Customer's specification and part number (optional).
5. Identification of electrical connections (connector pins by number or letter, leads by color).
6. Essential electrical characteristics (e.g., potentiometer winding resistance, maximum excitation, etc.).
7. Sensitivity or full-scale output at specified excitation (optional).
8. Reference pressure or range of reference pressures and identification of "high" and "low" pressure ports (on differential-pressure transducers).

Additional mechanical design characteristics are required primarily in manufacturers' specifications. Most of them, in fact, are superfluous in user-prepared specifications which should emphasize performance requirements. These additional characteristics include:

1. Type of sensing element used (e.g., dual capsule, flat diaphragm, helical Bourdon tube, etc.).
2. Transduction element details (e.g., type, location and number of active strain-gage elements, type of crystal in piezoelectric transducers, length and number of turns of potentiometer windings, etc.).
3. Identification of those materials which come in contact with measured fluids (to facilitate determination of compatibility). In the case of gage-pressure transducers it also is desirable that materials in contact with the environment are shown.
4. Size of the "dead volume" (total volume of the cavity enclosed by the pressure port and sensing element) of a nonflush-mounted transducer and the change in dead volume over the pressure range.
5. Type of artificial damping, if any, and identification of any damping oil used.

10.4.1.2 Electrical design characteristics. Excitation, impedance, and output characteristics constitute the electrical design characteristics to be specified (see Table 10-4). The nominal excitation voltage or current should be shown with its tolerances indicating the amount of regulation provided (e.g., "10.00 ± 0.01 V d-c," "28 ±3 V d-c," "5.00 ± 0.05 mA rms at 400 Hz"). The maximum excitation rating should include a statement regarding the time duration over which it is applicable.

Transducers incorporating integral amplifiers or d-c/a-c converters may require an additional specification to limit their susceptibility to reverse excitation and to high-amplitude transients or a-c voltages superimposed on the excitation voltage.

The ripple in the output of such transducers should also be limited by a specification for maximum ripple amplitude and minimum ripple frequency. The excitation specification can be associated with a rating for minimum power dissipation (without self-heating) or maximum power or current drain. This is not necessary if the input impedance (across the excitation terminals) or the element resistance of a potentiometric transducer is shown in conjunction with maximum excitation rating. The output impedance (across the output terminals) should be specified for all except potentiometric transducers. Tolerances on output noise should be shown for d-c output transducers.

When showing the insulation resistance or, less commonly, the breakdown-voltage rating of the transducer, it is desirable to call out the connector pins or pigtail leads from which a measurement should be taken to case ground as well as the test voltage at which the insulation resistance test is to be performed.

Limits on noise in potentiometric transducers are specified in terms of equivalent maximum contact resistance, usually measured as noise voltage produced when the wiper arm is cycled over the winding at a specific rate while the potentiometer element is connected to a constant-current supply.

The transducer's external electrical connections are best defined by a simple wiring diagram showing the internal electrical wiring of a simple transducer or a "black-box" outline, with the function of each connection properly labeled, for a transducer containing more than just the basic transduction element. Callouts must include the exact receptacle (and recommended mating plug), solder lugs, leads or pigtail cable (and color code) used for electrical connection of the transducer.

When provisions for simulated calibration are required to be contained within the transducer, the integral shunt- or series-calibration resistor or network and their connector-pin or solder-lug numbers or lead colors must be shown. If the resistor or network is external to the transducer, the internal auxiliary wiring must be shown and the additional external connections identified.

10.4.1.3 Static performance characteristics. Static characteristics (see Table 10-5) are applicable at room conditions. They can be verified by a static calibration during which pressure is applied to the transducer slowly and in discrete amplitude intervals.

The transducer's range can be unidirectional (e.g., 0 to 100 psia) or bidirectional (symmetrical, e.g., ± 5 psid, or asymmetrical, e.g., -5 to $+10$ psig). Suppressed-zero ranges are unidirectional (e.g., 70 to 100 psig).

Unless theoretical end points are specified (as for an ideal transducer of a given design), tolerances should always be specified for end points (e.g., "0.000, -0.000, $+0.020$ V and 5.000 ± 0.025 V"). If end-point tolerances are specified (or theoretical end points are used), the full-scale output becomes a nominal value (no tolerances). If end points are not specified, the full-scale output must carry a tolerance, and zero-balance as well as zero-shift tolerances must be shown in the specification so that the allowable outputs at the range limits can be defined. When the transducer's output varies with its excitation (such as in strain-gage type transducers), the end points and full-scale output specifications must state the excitation voltage or current at which

Table 10-5 STATIC PERFORMANCE CHARACTERISTICS FOR PRESSURE
TRANSDUCER SPECIFICATIONS

A. Individual characteristics spec.	B. Error-band spec.
Range	Range
End points	Full-scale output (nominal)
Full-scale output	End points (defining reference line)
Creep (optional)	Resolution (where applicable)
Resolution (where applicable)	Static error band
Linearity	Reference-pressure range†
Hysteresis	Warm-up period (optional)
Repeatability	
Friction error	
Zero balance	
Zero shift	
Sensitivity shift	
Warm-up period (optional)	
Reference pressure range†	
Reference pressure effects†	

†Applies to differential-pressure transducers only

these characteristics are to be measured (e.g., "Full-scale output: 30 mV, at 10.00 V d-c excitation"). Output voltage is always assumed to be "open-circuit" output unless otherwise specified (e.g., "with a resistive load impedance of $3,500 \pm 100$ ohms").

Zero balance is the output at rated excitation with zero pressure (psia, psig, or psid) applied to the transducer. Zero shift is the change in zero balance over a specified period of time; sensitivity shift is the change in full-scale output at rated excitation over a specified period of time. The latter two characteristics are difficult to verify if the time period chosen is relatively long.

For improved accuracy of static measurements it may be desirable to specify limits on *creep*, a change in output occurring over a specific time period while the measurand and all environmental conditions are held constant.

Resolution defines the allowable magnitude of output step changes. A specification for resolution is important primarily for potentiometric transducers and can be omitted for most other types. The same holds true for friction error, except that a relatively smaller friction error can also occur in any transducer containing a mechanical linkage between the sensing and transduction elements.

Specifications for differential-pressure transducers must contain the range of reference pressures over which the transducer must operate and the maximum allowable effect such reference pressure variations may have on the transducer's performance. An example of such a specification would be: "Reference pressure: 0 to 1000 psig; Reference pressure effects: additional zero shift—0.5% FSO/100 psig, max., additional sensitivity shift—0.1% FSO/100 psig, max., referred to output with zero psig."

Among the most important static performance characteristics are either the static error band or individual specifications for a specific type of linearity, for hysteresis, and for repeatability.

An "error-band" specification of static performance characteristics must call out range, resolution, full-scale output, end points, and the reference-pressure range of a differential pressure transducer. The full-scale output, however, is a nominal value only and carries no tolerances. The specified end points, which may or may not carry tolerances, are used to define the reference line of the error band (unless a nonlinear output-vs-pressure relationship is required, in which case the error band is referred to a specified curve). The static error band includes linearity, hysteresis, repeatability, friction error, zero balance (unless included in tolerances on the lower range end point), zero shift, sensitivity shift, and, usually, reference-pressure effects (when no separate tolerances are allowed for them).

Specifying the required warm-up period is necessary for heater-jacketed pressure transducers; it may also be desirable for other transducers in which measurable effects of self-heating can be expected.

10.4.1.4 Dynamic performance characteristics. Dynamic performance characteristics are specified to establish minimum transducer performance in measuring rapidly fluctuating or step changes in pressure.

Frequency response is specified as a maximum percentage of a reference output-amplitude ratio over a given frequency range, e.g., "within $\pm 5\%$ from 0 to 500 Hz, as referred to the amplitude at 50 Hz."

When phase shift is significant to the application of the transducer, it should be similarly specified as maximum degrees of shift over the same frequency range.

An alternate manner of specifying the dynamic pressure response of the transducer is given by specifications for minimum resonant frequency and damping ratio (percent of critical damping).

In a specification for a transducer known to be overdamped or near-overdamped, the terms explained above may be replaced by a callout for time constant and maximum allowable overshoot.

10.4.1.5 Environmental performance characteristics. When a transducer is to be used during (or after) exposure to environmental conditions other than room conditions, its environmental characteristics (Table 10-6) must be specified.

Limitations on thermal effects can be specified in terms of individual characteristics or as an error band. In either case, the operating temperature range, over which the transducer must perform within specified tolerances, must be shown. For some types of transducers which incorporate temperature compensation, it has been customary to specify an additional "compensated temperature range," narrower than the operating temperature range. In such cases temperatures between the limits of the "compensated" range and the broader "operating" range are merely considered as "safe" temperatures without any tolerances imposed on temperature errors.

Parallel shifts of the calibration curve (thermal-zero shift) and slope changes of the calibration curve (thermal-sensitivity shift) due to thermal effects are usually specified as a maximum allowable percentage of full-scale output per degree Fahrenheit

Table 10-6 Environmental Performance Characteristics for
Pressure Transducer Specifications

1. Basic

A. Individual characteristics spec.	B. Error-band spec.
Operating temperature range	Operating temperature range
Temperature error *or*	Temperature error band
Thermal zero shift *and*	
Thermal sensitivity shift	
Temperature-gradient error	Temperature-gradient error
Ambient-pressure error	Ambient-pressure error band
Acceleration error	Acceleration error band
Vibration error	Vibration error band

Performance after exposure to:
  Shock (triaxial)
  Humidity
  Salt spray or salt atmosphere

2. Special

Performance during and after exposure to:
  High sound-pressure levels
  Sand and dust
  Ozone
  Nuclear radiation
  High-intensity magnetic fields
  Etc.

(or Celsius) between the limits of either the operating or the compensated temperature range. An alternate specification for temperature error combines thermal zero and sensitivity shifts into a single value. The temperature error band is always referred to the same reference line or curve used to define the static error band. A typical error-band specification may read: "Temperature Error Band: $\pm 3.5\%$ FSO (from $-80°$ F to $+200°$F)," unless an operating temperature range is separately specified.

A related characteristic is temperature gradient error, specified as the maximum allowable output change following exposure of the transducer to a step change in temperature of either the measured fluid applied to the sensing element (which requires a difficult test) or of the ambient atmosphere. The error resulting from the temperature gradient within the transducer should be observed for a specified length of time, usually 3 to 5 min.

Errors due to changes in ambient pressure are more significant in gage-pressure than in absolute-pressure transducers, unless the latter are poorly sealed. Most specifications state that no additional error is allowed for ambient-pressure changes between values existing at sea level and at high elevations or very high altitudes, sometimes even the vacuum condition of interplanetary space.

The relationship between error band (total maximum output deviation from a constant reference line or curve) and error (changes from the room-condition calibration curve) is the same for ambient pressure, steady-state acceleration, and pro-grammed vibration.

Vibration errors in transducers having ranges larger than about 15 psi are usually introduced more by internal mechanical resonances than by the vibration amplitude itself.

Since pressure transducers are frequently exposed to some mishandling or rough usage (shock) and are required to operate in a humid or salt-laden atmosphere, it is customary to specify that little or no performance changes must be observable in the transducer after exposure to these environmental conditions. Other environmental conditions which may be applicable to some transducers, as given by the intended usage, are listed in Part 2 of Table 10-6 and deserve special consideration in a specifica-tion.

10.4.1.6 Performance reliability characteristics. The final group of pressure transducer specification characteristics limit overpressure effects and define life expectancy. The maximum pressure which may be applied to the transducer's sensing element without changing the transducer performance beyond stated error tolerances is the proof pressure, usually specified as a precentage of the range which is taken as 100%, e.g., "Proof Pressure: 150% of range." The burst-pressure rating, the maximum pressure which can be applied to the sensing element without rupture of the case (or of the sensing element, if so stated) is also shown in terms of percent of range.

Before a transducer is used in its intended application, it is frequently stored "on the shelf" and in the installed condition for a considerable length of time. If the environmental conditions during this storage period ("storage conditions") can be defined, a specification for storage life can be written in terms of the number of months or years over which the transducer can be stored without performance change beyond stated tolerances. The operating life of a pressure transducer is stated as cycling life, the minimum number of full-range (or specific partial-range) excursions over which the transducer must operate and remain within specified performance tolerances.

10.4.2 Considerations for selection

Although various groups of characteristics can be of major importance for any given applications, the characteristics of pressure transducers which most frequently govern selection are primarily range, accuracy, output, and frequency response, and secondarily, environmental conditions and nature of measured fluids. The following guidelines are only generally applicable since developmental efforts are constantly directed at extending the capabilities of all transducer types so that a competitive position can be maintained by their manufacturers.

10.4.2.1 Range. Pressure transducers of almost all types are designed to measure pressures over specified ranges. Stated performance tolerances apply only over a specific pressure range. An exception to this is the self-generating piezoelectric pressure transducer for which only a maximum range is normally stated. The actual measuring range is determined by the setting of associated signal-conditioning and readout equipment for a given transducer output corresponding to the upper limit of the pressure range of interest.

The following grouping of pressure ranges has been arbitrarily established and is sometimes useful in general performance descriptions:

$<$15 psi—very low
15–400 psi—low
400–10,000 psi—high
$>$10,000 psi—very high

Of the most commonly used transducer types, *capacitive* transducers are used primarily for very low and some low ranges, *inductive*, *resistive*, and *potentiometric* transducers for low and high ranges, *reluctive* transducers for very low through high ranges, *strain-gage* and *piezoelectric* transducers for low through very high ranges, and *vibrating-element* and *servo-type* transducers for low ranges.

Pressure-range reference is a major design criterion. Absolute-pressure transducers require an internal vacuum or partial vacuum, maintained by a seal which normally also provides a hermetic seal for the transduction element and isolates it from any atmospheric elements. Gage- and differential-pressure transducers, except sealed-reference types, usually expose the transduction element to the ambient atmosphere or to the reference-pressure fluid and provide transduction-element isolation only when specified, and only in a limited number of designs.

When close accuracy is required within a narrow range of interest, special transducer designs can be used which incorporate zero-suppression or high proof-pressure capability, or both.

Resistance to shock and vibration is largely dependent upon sensing-element stiffness and is therefore proportional to pressure range. However, a number of transducer designs exist in which very small sensing-element deflections yield adequate output. This allows the design of a vibration-resistant pressure transducer for very low ranges. The required reference-pressure range of a differential-pressure transducer can affect its design considerably and should always be specified, especially when the reference pressure is expected to vary over a large portion of its range during the time of interest of the differential-pressure measurement itself.

Bidirectional pressure ranges are only specified for differential- and, sometimes, for gage-pressure transducers, but never for absolute-pressure transducers, since negative absolute pressures do not exist. Differential-pressure transducers can always be used for gage-pressure measurements by venting the reference port to the ambient atmosphere.

10.4.2.2 Accuracy. Pressure-transducer accuracy capability and accuracy requirements are strongly related to signal-conditioning, signal-transmission, and readout system capabilities. Servo-type pressure transducers with digital output and vibrating-element transducers which have frequency or FM output offer the closest accuracies, since they are unaffected by transmission-system errors which affect only signal amplitude. Reluctive pressure transducers and, to a slightly lesser extent, strain-gage pressure transducers can provide very close accuracies in conjunction with a low-error transmission and readout system adjusted, compensated, and balanced to optimize the accuracy capability of these transducers. Other types of pressure transducers can often be designed to provide very close accuracy under certain conditions but can usually be relied upon to yield the degree of accuracy normally required for a large majority of pressure measurements.

10.4.2.3 Output. Although any pressure transducer can be made to furnish any required output to transmission and readout equipment by use of suitable signal-conditioning circuitry, certain types, notably potentiometric transducers and servo-types with a potentiometric secondary transduction element, can provide high output amplitude without signal conditioning. The output of strain-gage pressure transducers is large enough to drive the commonly used millivolt recorders. Where output levels must be higher than provided by the transduction element, integral or separate amplifiers can be used. Capacitive and inductive pressure transducers always require some signal conditioning to yield usable outputs. Piezoelectric transducers are almost invariably used in conjunction with a voltage or charge amplifier since their output impedance is very high. Reluctive transducers, which require a-c in their transduction element, can be equipped with demodulators to provide d-c output. Digital output can be obtained from analog-output transducers by use of an analog-to-digital converter which can be built into the same case with the transducer. The relationship of load impedance to transducer output impedance is critical in its effect on output levels during actual use.

10.4.2.4 Frequency response. The ability to reproduce rapid pressure fluctuations in their output is most pronounced in piezoelectric and certain types of capacitive pressure transducers and to a lesser extent in strain-gage pressure transducers. Reluctive transducers usually have less frequency-response capability than the above three types, and potentiometric transducers are normally not used where a large frequency range must be monitored. Servo-types have very low frequency response. Vibrating-element transducers have about the same frequency-response capability as reluctive transducers. However, this capability is normally not utilized since the associated readout counters average the output over relatively long time intervals so as to optimize measurement accuracy.

Pressure-transducer frequency response is dependent upon the natural frequency of the sensing element, as dictated mainly by its stiffness and by a number of other transducer design factors. As a rule, stiffness, natural frequency, and frequency range of flat response increase with increasing pressure ranges for any one transducer design. Transducer frequency response is also inversely proportional to the dead volume of the transducer and the length of tubing from point of measurement to transducer pressure port.

When tubing is used, the frequency range obtainable in liquid systems, especially where the liquid extends into the transducer sensing element, is considerably higher than in gaseous systems where the fluid in the tubing is also gaseous. Improved frequency response through tubing has been obtained in gas-pressure measurements by use of "grease lines," isolated from the measured fluid by a membrane at the point of measurement. Some pressure transducers are equipped with "bleed ports"—resealable openings in the transducer cavity just upstream of the sensing element. This permits venting of any air entrapped in the cavity before measurements are made of rapid pressure fluctuations in liquid systems or where grease lines are used, so that only liquid contacts the sensing element.

Some measurements call for limitations on the ability of the transducer to respond to rapid pressure variations, so that only the very low-frequency components of the variations are reflected in the transducer's output. Certain design features can be added to mechanical linkages within the transducer, or small-bore orifices or "snubbers" (capillary, baffle-type, etc.) can be attached to the pressure port to accomplish this. Integral electrical filters can also be used for this purpose.

10.4.2.5 Environmental temperature. The ability of pressure transducers to perform with little additional error while exposed to very high or very low ambient temperatures has long been an area of strong developmental efforts by transducer manufacturers. Almost identical design considerations apply to transducers which must measure the pressures of very hot or very cold fluids while exposed to relatively normal ambient temperatures. The major difference is the direction of heat flow and of the ensuing temperature gradient across the external and internal portion of the transducer. The consideration of exposure time is second in importance only to the temperature level itself.

No general rule can be established relating design simplicity to ease of compensation for thermal effects. Minimization of temperature error through selection of materials with either very small or properly matched temperature coefficients is usually superior to accomplishing this by use of temperature-sensitive components in the transduction or integral signal-conditioning circuitry. In the latter case, temperature-compensation reliability is dependent upon proper execution of production processes rather than being inherent in the basic design.

Some piezoelectric pressure transducers can perform with acceptable error limits at temperatures from -400 to $+600°F$. Potentiometric pressure transducers have been developed to operate at temperatures between -320 and $+350°F$ with temperature errors about double the magnitude of static errors. Neither of these types uses compensation in their circuitry.

Strain-gage pressure transducers always contain compensation resistors in their transduction circuitry. With careful compensation such transducers can be built so that they have little additional temperature error—equal to or less than static errors—between room temperature and −450°F and between room temperature and +250°F. They can also be compensated to over +500°F, but temperature errors will then be larger. The addition of provisions for water cooling (see example, Figure 10-46) enables such transducers to measure the pressure of fluids whose temperature is over 1000°F (liquids) and over 4000°F (gases).

Fig. 10-46. Water-cooled strain-gage pressure transducer (courtesy of Advanced Technology Labs.).

Reluctive pressure transducers show very promising temperature characteristics on the basis of design. Their sensing elements and coil windings can be made relatively insensitive to temperature variations and magnetic characteristics of armature (or diaphragm), and coil cores can be matched to a considerable extent.

Capacitive and inductive transducers can be designed to minimize thermal effects by matching of materials as to dimensional variations for the former, and as to magnetic characteristics for the latter type. Their usual applications, however, are in environments at or near room temperature.

Resistive and servo-type transducers can be temperature-compensated to some extent but are commonly used in fairly normal temperature environments.

Vibrating-element pressure transducers, with their very small sensing-element deflections, must be carefully designed to minimize thermal effects on dimensions and stiffness of internal components. They must also be provided with a well-controlled heater jacket to optimize their accuracy when exposed to temperature variations. The effectiveness of heater jackets is limited for this transducer type to an operating temperature range of about 0 to 140°F. Heater jackets can also be used to minimize temperature errors in all other types of pressure transducers.

Transducers with integral signal conditioning (amplifiers, dc-ac-dc converters, analog-to-digital converters, etc.) always require temperature compensation within this circuitry.

The possibility of obtaining temperature-gradient errors in response to rapid changes in temperature, when those are expected to occur, should also be considered. Such errors are usually most prevalent in low-output-level transducers and in those types using temperature-compensation components not located directly on the sensing element or made integral with the transduction element.

10.4.2.6 Environmental acceleration and vibration. The effects of static and vibratory acceleration on pressure transducers are largely dictated by sensing-element stiffness and support and design and complexity of any mechanical-amplification linkages. Acceleration forces act on the sensing element in the same manner as pressure forces. The relative magnitude of acceleration-induced output errors is proportional to sensing-element mass and deflection. Very low-pressure transducers are affected to a greater degree by acceleration forces, especially those acting on the sensing element in the same direction as pressure forces, than high-pressure transducers. Even the acceleration due to gravity can cause output shifts (*attitude error*). Transducers so affected must be oriented in their installation so as to minimize these effects, by assuring that their acceleration-sensitive axis does not coincide with the axis of gravity forces. In transducers using oil as transfer medium, acceleration can act upon the mass of the internally contained oil. Such acceleration effects can be considerably larger than those given by the mass of the sensing element alone.

Vibration affects pressure transducers not only by applying low-frequency acceleration forces to the sensing element which "look like" pressure, but also by exciting internal components into higher-frequency "sympathetic" vibration at their resonant frequencies (*resonances*). At these resonances the applied vibration amplitude is often greatly amplified by the vibrating components, and large output errors can be observed within narrow vibration-frequency bands. One of the primary design factors for a vibration-resistant pressure transducer is the selection of sensing elements and any linkage components having natural frequencies appreciably higher than the highest significant frequency of vibration of the structure on which the transducer is to be installed.

Reluctive and capacitive transducers, whose only moving part is a diaphragm element, can be designed to be relatively insensitive to vibration. The diaphragm can be quite stiff and still deflect sufficiently for the desired full-scale output even when pressure ranges are very low. Reluctive transducers using twisted Bourdon tubes are made vibration-resistant by a stiff crosswire support at the tube tip.

Strain-gage transducers for high and very high pressure ranges usually incorporate stiff diaphragms with sufficiently high natural frequencies to minimize vibration effects. The vibration resistance of transducers with range limits below 100 psi can be marginal unless special design precautions are taken. This is very difficult in the very low pressure ranges. One attempted solution to this problem is the use of high-output semiconductor gages so that acceptable output levels can be obtained with less diaphragm deflection, as obtained with stiffer diaphragms, than could be used with metal-wire or foil gages.

Potentiometric pressure transducers, with their relatively large sensing-element deflections, amplifying linkages, and cantilevered wiper arms, were originally quite sensitive to vibration. Continuous development for aerospace applications resulted in new designs with vibration errors well within acceptable limits in environments of over 50 g at frequencies ranging above 2000 Hz. In high-pressure transducers, vibration problems were overcome by viscous damping of the Bourdon tube, a simple counterbalanced linkage—or no linkage at all—and wiper arms with high resonant frequencies and sufficient tension to prevent lift-off. A forklike dual-wiper-arm system having one arm above and one arm below the resistance element proved quite successful. Low-pressure transducers were made relatively insensitive to vibration by an adjustable counterbalance weight in their linkage; by using two capsules, each with a slightly different resonant frequency; by using a stiff flexure linkage instead of a pivoted lever linkage; by doing away with the linkage entirely; and by damping the pushrod motion.

Piezoelectric pressure transducers are most frequently used for high-pressure measurements where diaphragm stiffness, low mass, and high system natural frequency combine to minimize vibration effects. Low-pressure designs need special design considerations to prevent their response to vibratory acceleration, such as reducing as much as possible the action of the diaphragm and any link to the crystal as a seismic mass. One design incorporating a compensating piezoelectric accelerometer has proved quite successful.

Vibrating-wire, servo-type, and resistive transducers are usually designed to operate only in moderate or low-vibration environments but can be designed to minimize the effects of slowly varying acceleration. Inductive transducers using diaphragms compare in acceleration and vibration resistance with reluctive transducer if large inductance changes are not required.

10.4.2.7 Environmental atmospheric elements. Like most other transducers, pressure transducers can suffer deterioration of performance and reliability when exposed to corrosive and electrically conductive constituents in their ambient atmosphere. Incorrect case finishes and poorly made seals in the case and in the electrical and plumbing connections allow corrosion, leaks, and partial short circuits between conductors due to exposure to salt, acids, and other corrosives, and due to high relative humidity in the atmosphere in which the transducer must operate.

More severe effects can be encountered in gage-pressure transducers and in differential-pressure transducers whose reference port is vented to the atmosphere. The metal-mesh inserts in the vent openings will usually stop only coarse-grained solids from entering internal operating portions of the transducer, such as transduction elements, allowing them to be severely contaminated by finer-grained solids, droplets of corrosive or conductive liquids, or corrosive gases. Even fine-pore sintered-metal or ceramic filter plugs may not be sufficient to prevent internal contamination. A special transducer design may then have to be selected which isolates internal components whose operation may be affected by the ambient atmosphere.

10.4.2.8 Environmental nuclear radiation. Certain design precautions are necessary when pressure transducers must operate while exposed to nuclear radiation, especially gamma rays and high-energy (fast) neutrons. These can cause deterioration and eventual breakdown of:

1. Plastic and elastomer inserts in electrical connectors.
2. Teflon® standoffs and terminals.
3. Teflon® wire insulation.

4. Silicone O-rings and damping fluid.
5. Organic and silicone lubricants.
6. Organic-material potentiometer-element mandrels.
7. Adhesives used to bond strain gages.
8. Semiconductors such as silicon transistors and diodes.
9. Semiconductor strain gages.
10. Acrylic castings.
11. Organic potting material.

Piezoelectric-crystal characteristics can degenerate so that their frequency response and output amplitude are reduced. Certain wire materials, such as those used in strain-gage bridges, can change their resistance.

Effects of high-energy nuclear radiation on pressure transducers are minimized when the following design precautions are taken:

1. "Use ceramics instead of organics."
2. Do not use O-rings.
3. Use no lubricants if possible.
4. For potentiometer elements use ceramic mandrels, ceramic-coated metallic mandrels and ceramic-coated or bare wire, or use metal-film elements on ceramic or glass mandrels.
5. Obtain vibration resistance without use of damping fluid.
6. Do not use organic or silicone fluids as force-transfer media.
7. Use ceramic-filled or mineral-filled epoxy potting and insulating materials.
8. Do not use Teflon®.
9. Take extreme care in selecting the proper adhesives for bonding strain gages or use unbonded strain-gage bridges when feasible.
10. Minimize use of piezoelectric transducers—select quartz-crystal transducers when they are required for a high-frequency-response measurement.
11. Minimize use of transducers with integral transistorized circuitry.

10.4.2.9 Measured fluids. The transducer's pressure port, cavity behind the port, if any, and the sensing element must all be compatible with the liquids and gases expected to come in contact with them before and during use of the transducer. Additionally, compatibility must be established between measured fluids and any other internal portions of some transducers in which the sensing element does not provide a complete seal for them.

This compatibility is usually obtained by proper selection of metals used for the pressure port, case, and sensing element. Attention must also be paid to soldering or brazing materials, finishes, and any components such as O-rings or gaskets, utilized as seals. Because the same consideration must apply to all components and materials used in the reference (or "low") side of differential-pressure transducers, they are much more difficult to design for measured-fluid compatibility than absolute-pressure transducers.

Transducers which must be cleaned, passivated, or treated before use must also be compatible with fluids used for such processing. The type, state, temperature, and concentration of all fluids, and the estimated duration of exposure to them must be clearly established before a pressure-transducer design is selected.

Certain highly corrosive fluids may limit the choice of compatible transducer materials in such a manner that excessive performance compromises would be incurred. An isolating membrane can then be used, with a pressure-transfer liquid between membrane and sensing element. A membrane and transfer fluid may also be used in the reference side of differential-pressure transducers.

The membrane must not only be compatible with measured fluids and pre-use cleaning fluids, it must also be sufficiently compliant ("soft") to avoid introducing additional hysteresis and other errors in the transducer's operation. The membrane may be contained within the transducer's case or can be remotely located at the point of measurement with the entire length of connecting tubing filled with the pressure-transfer fluid. The term "grease line" is sometimes used to describe such sealed-off, liquid-filled tubing although silicone fluids or high-temperature radiation-resistant liquid metals (e.g., NaK) have largely replaced the previously used organic or mineral oils and greases.

10.5 Calibration and Tests

10.5.1 Calibration

The basic performance test on a pressure transducer is the determination or verification of its output over its measuring range. This usually takes the form of a static calibration, performed at room conditions and in the absence of shock, vibration, or acceleration, during which known pressure levels are applied to the transducer and corresponding output readings are recorded.

10.5.1.1 Calibration equipment. The excitation and readout equipment used for calibrations depends upon the type of pressure transducer to be tested and is covered in Chapter III. The pressure levels are obtained either from a calibrated pressure source or from a supply of pressurized gas or liquid monitored by an indicator.

The two most commonly used calibrated pressure sources are the air- and oil-dead-weight testers (or "piston gages"). A typical dead-weight tester (Figure 10-47) is essentially a piston of known area, closely fitted within a cylinder with provisions for placing weights on the piston. Since pressure is defined as force per unit area, the pressure applied by the piston to a fluid in the cylinder is known within the accuracy that the piston area and the weights are known. The current state of the art permits dimensional and weight determinations within very close tolerances.

In operation, the dead-weight tester is made a part of a closed fluid system containing the transducer to be calibrated and a means of varying the pressure within

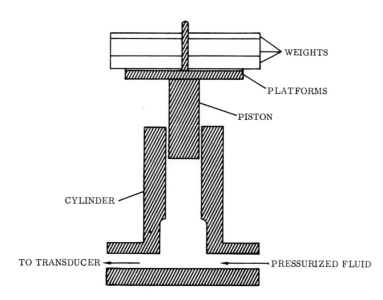

Fig. 10-47. Dead-weight tester.

the closed system. The pressure is raised until it balances the pressure exerted by the weighted piston, so that the piston floats freely. The weights are usually disc-shaped and are placed concentrically on a circular platform on the protruding end of the cylinder. Static friction within the cylinder is removed by spinning the weights, causing the piston to spin within the cylinder.

Air-dead-weight testers are used to calibrate very low- and low-pressure transducers, and oil-dead-weight testers are to used calibrate high- and very high-pressure transducers.

Typical accuracies are within $\pm 0.015\%$ FS for air-dead-weight testers and within $\pm 0.010\%$ FS for oil-dead-weight testers.

Manometers, gages, and transducer-indicator systems require an external supply of pressurized gas or liquid. Gas—air, nitrogen, or helium—is used most frequently. Manometers are used for low- and very low-pressure measurements. The vertical-tube types illustrated in Figure 10-48 are commonly used for pressures up to 50 psi. Inclined-tube manometers are used for very low-pressure measurements, usually below 1 psi. Mercury-filled manometers are most frequently used. Other liquids used in manometers include water, ethyl alcohol, and others with densities either higher or lower than that of water, depending upon the pressure range required. Manometer readings are normally in inches of mercury, inches of water, or of other liquids. The scale indicates the difference in height between liquid levels in the two legs. Their relationship, for static balance, is

$$P_m - P_r = dh$$

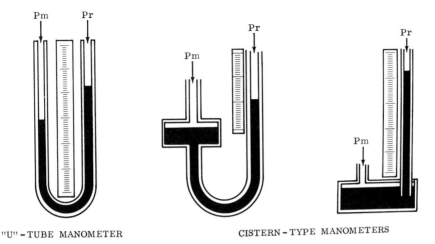

"U" – TUBE MANOMETER   CISTERN – TYPE MANOMETERS

Fig. 10-48. Manometers.

where: P_m = measured pressure (absolute, gage, or differential)

P_r = reference pressure (zero, with tube sealed, for absolute; ambient, for gage; as existing at reference point, for differential)

d = density of manometer liquid

h = difference in height of columns (between cistern zero line and level in vertical tube of base-cistern type)

Typical manometer accuracies are within ±0.01 to 0.02% FS. Gages (dial indicators) use diaphragms, capsules, bellows, and Bourdon tubes similar to those found in transducers. The mechanical deflection of the sensing element is used to move a pointer over a circular dial face.

A number of special pressure indicators have been developed, many of which utilize a transducer connected to a readout device. Reluctive, strain-gage, capacitive, piezoelectric, and servo-type transducers have been used for this purpose. The transducers are carefully selected and adjusted for minimum error and for long-term stability. They are frequently enclosed in a constant-temperature chamber to prevent introducing errors due to room-temperature variations. Figure 10-49 shows a portable "secondary pressure standard" being calibrated against a dead-weight tester. The electronic counter is used as readout device.

10.5.1.2 Calibration methods. The static calibration of a pressure transducer is accomplished by applying pressure to the transducer in discrete steps, e.g., at 0, 10, 20, 30, 40, 50, 60, 70, 80, 90, and 100% of range, then reducing the pressure in steps of the same interval, in reverse order. The transducer's output at each level is read and recorded on the calibration record. At least one additional calibration cycle is then

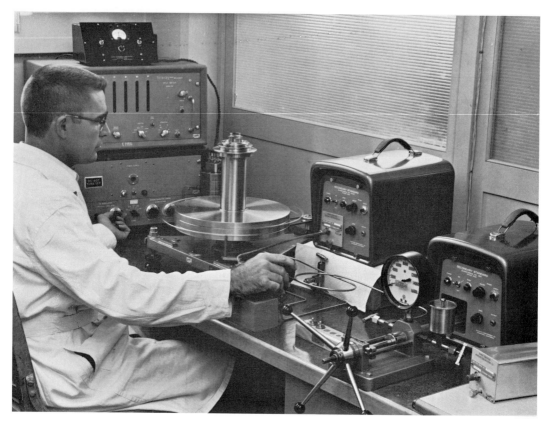

Fig. 10-49. Calibration of transducer-indicator system against dead-weight tester (courtesy of Wiancko Engineering Co.).

performed to verify the transducer's repeatability and absence of short-term shifts and drifts.

A pneumatic diagram of a typical setup for the calibration of absolute-pressure and gage-pressure transducers is shown in Figure 10-50. The gaseous-nitrogen bottle supplies slightly over 2000 psi when fully charged. The pressure regulator is used to limit pressures in the system to the highest level required, as given by the transducer's range. The vacuum pump is used to reduce absolute-pressure levels in the system below ambient atmospheric pressure (approximately 14.7 psia). The gage is used to indicate system pressure. It is first exercised by opening the pressure valve—with the needle valve open, gage valve open, and other valves closed—then closing the pressure valve and opening the vent valve to bleed off the pressure. The transducer valve is then opened, and pressure is applied in discrete intervals using the pressure valve to admit pressure and the needle valve for fine control. After the highest pressure is reached,

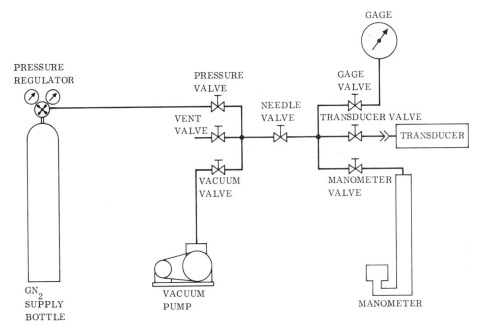

Fig. 10-50. Calibration set-up for absolute- and gage-pressure transducers.

the pressure valve is kept closed; the vent valve is opened; and decreasing pressure levels are obtained by bleeding the pressure down by operation of the needle valve. The manometer is used only for low-pressure indications within its range. Each pressure level is approached gradually, without overshoot, to obtain the correct measure of hysteresis within the transducer. Gages are often tapped, during the final phase of obtaining a reading, to reduce internal static friction.

A more complex setup is required for the calibration of differential-pressure transducers. It must be designed to provide low differential pressures in the presence of high reference pressures. It must also be capable of applying reverse pressure differentials to transducers having a bidirectional range. A practical setup, shown in Figure 10-51, allows differential-pressure transducer calibration in the following manner:

1. Calibration of Unidirectional-Range Differential-Pressure Transducers:
   A. When the reference pressure is zero psig:
      (i) Open valves 2, 5, 6, 8, 10, 13, and 14.
      (ii) Close valves 1, 3, 7, 9, 11, and 12.
      (iii) Increase pressure to the transducer's "high" port, in discrete intervals, to the range limit by operating valve 4 to pressurize.
      (iv) Close valve 2.
      (v) Open valve 1.

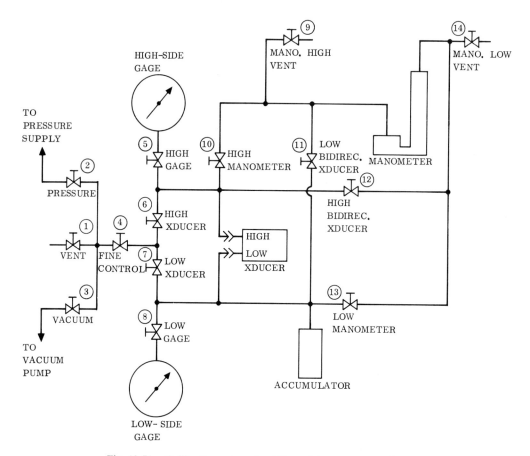

Fig. 10-51. Calibration set-up for differential-pressure transducers.

(vi) Reduce pressure at the transducer's "high" port, in discrete intervals, to ambient pressure by operating valve 4 to bleed off.

B. When the reference pressure is at a high (specified) level:

(i) Open valves 2, 5, 6, 7, 8, 10, and 13.

(ii) Close valves 1, 3, 9, 11, 12, and 14.

(iii) Increase pressure, applied equally to the transducer's "high" and "low" ports, to the specified reference-pressure level by operating valve 4 to pressurize.

(iv) Close valve 7.

(v) Increase pressure on the transducer's "high" ports, in discrete intervals, by operating valve 4 to pressurize further.

(vi) Close valve 2.

(vii) Open valve 1.

    (viii) Reduce pressure at the transducer's "high" port, in discrete intervals, by operating valve 4 to bleed down pressure until the specified reference pressure is again reached.

    (ix) Open valve 7 to assure equalization of pressure on both transducer ports.

    (x) Reduce pressure to ambient pressure by operating valve 4 to bleed off.

2. Calibration of Bidirectional-Range Differential-Pressure Transducers:

  A. When the reference pressure is zero psig:

    (i) Open valves 3, 6, 8, 9, 11, and 12.

    (ii) Close valves 1, 2, 5, 7, 10, 13, and 14.

    (iii) Using vacuum pump, reduce pressure at the transducer's "high" port below ambient pressure, in discrete intervals, to the negative range limit, by operating valve 4.

    (iv) Close valve 3 and turn off vacuum pump.

    (v) Open valve 1.

    (vi) Increase pressure at the transducer's "high" port back to ambient pressure, in discrete intervals, by operating valve 4.

    (vii) Close valve 1, 9, 11, and 12.

    (viii) Open valves 2, 10, 13, and 14.

    (ix) Increase pressure at the transducer's "high" port, in discrete intervals, to the positive range limit, by operating valve 4 to pressurize.

    (x) Close valve 2.

    (xi) Open valve 1.

    (xii) Reduce pressure at the transducer's "high" port, in discrete intervals, back to ambient pressure, by operating valve 4 to bleed off.

  B. When the reference pressure is at a high (specified) level:

    (i) Open valves 2, 5, 6, 7, 8, 10, and 13.

    (ii) Close valves 1, 3, 9, 11, 12, and 14.

    (iii) Increase pressure, applied equally to the transducer's "high" and "low" ports, to the specified reference-pressure level by operating valve 4 to pressurize.

    (iv) Close valves 2, 7, 10, and 13.

    (v) Open valves 1, 11, and 12.

    (vi) Reduce pressure at the transducer's "high" port, in discrete intervals, to the negative range limit (below the reference-pressure level), by operating valve 4 to bleed down.

    (vii) Close valve 1.

    (viii) Open valve 2.

    (ix) Increase pressure at the transducer's "high" port, in discrete intervals, back to the reference-pressure level, by operating valve 4 to pressurize.

    (x) Open valves 10 and 13 to equalize any pressure differential between "high" and "low" sides.

    (xi) Close valves 11 and 12.

(xii) Increase pressure at the transducer's "high" port, in discrete intervals, to the positive range limit (above the reference-pressure level), by operating valve 4 to pressurize.

(xiii) Close valve 2.

(xiv) Open valve 1.

(xv) Reduce pressure at the transducer's "high" port, in discrete intervals, back to the reference-pressure level, by operating valve 4 to bleed down.

(xvi) Open valve 7 to assure equalization of pressure at both transducer ports.

(xvii) Reduce pressure to ambient pressure by operating valve 4 to bleed off.

Several types of automatic and semiautomatic pressure calibration equipment have been developed, some of which automatically cycle the pressure over preset intervals and print out the transducer output reading at each level.

A number of essential performance characteristics can be determined by the static calibration. The two (or more) consecutive calibration cycles show the transducer's end points, full-scale output, zero balance, sensitivity, linearity, hysteresis, repeatability, or the static-error band in lieu of these characteristics. Measuring the transducer's output before and during the application of dithering to the transducer, at each calibration point over one calibration cycle, shows friction error.

Applying the specified coupling and mounting torque or force to the transducer and then repeating one calibration cycle shows mounting error. Calibration cycles repeated over a specified period of time (usually anywhere from several hours to several weeks) shows zero shift and sensitivity shift (or full-scale-output shift). Measuring output changes after rapid application of full-range pressure shows any creep the transducer may have. Measuring both end points over a period of time, starting with the first application of excitation voltage to the transducer, until output changes are no longer measurable, establishes the transducer's warm-up time.

Since reference-pressure effects are often significant in differential-pressure transducers, at least one calibration cycle should always be performed with the maximum specified reference pressure applied to both ports at the starting point of the calibration cycle (zero psid at high reference pressure); the other calibration cycles are normally performed with ambient pressure as reference pressure (zero psid at zero psig), unless an absolute pressure less than atmospheric is specified as the low end of the reference-pressure range.

A static calibration performed after completion of an environmental test shows any permanent effects on the transducer's performance due to such environmental effects.

10.5.2 Proof-pressure tests

The purpose of the (positive) proof-pressure test is to verify the performance of the transducer after pressure in excess of its range has been applied to it, e.g., after a high-pressure surge in the system whose pressure is being measured. During this test the specified proof pressure is applied to the transducer for the specified length of time and a calibration check is performed immediately after this pressure has been bled off so that even short-term elastic aftereffects can be detected.

A negative proof-pressure test is performed, additionally, on differential-pressure transducers. These transducers are frequently used, connected across an orifice plate or Venturi tube, for flow measurement. When flow is in the reverse direction, a negative differential pressure is applied to the transducer, which it must be able to withstand. Other applications will normally also require the performance of the negative proof-pressure test. Further tests of general nature, often performed immediately before or after a calibration, are explained in Chapter III.

10.5.3 Response-time tests

This test is performed to verify the transducer's performance when the pressure applied to it varies with time in a specified manner; it is also known as *frequency-response test* or, sometimes, as a *dynamic calibration*.

10.5.3.1 Sinusoidal-pressure application. Sinusoidal-pressure generators have been developed to provide a direct measurement of frequency response. The amplitude of the pressure step is held constant and only the frequency of the pressure variation over the selected step is varied. The transducer's output amplitude can then be plotted against frequency, and response can be determined on the basis of output attenuation at the higher frequencies. This method is considered by many engineers as ideal, primarily because data requirements are usually stated in terms of accuracy and frequency response, rather than accuracy and response time. Sinusoidal-pressure generators are relatively rare and costly and can have shortcomings which limit their usability, such as waveshape impurities and inability to provide sufficiently high frequencies and large enough pressure steps.

Among the relatively successful types of sinusoidal-pressure generators are those using the following operating principles: a quick-acting servo valve driven by an audio oscillator through a power amplifier to vary the pressure from a hydraulic supply; a rotating disc to alternately decrease and increase the pressure of a gas supplied from one tubing section, through the perforated disc, to the other tubing section; a sonar-type sound emitter (or a strong loudspeaker voice-coil element) acting on a test fluid in a closed system; and acoustic calibration devices, usable only for low- and very low-pressure transducers.

10.5.3.2 Pressure-step application. Pressure-step generators are more commonly used. The transient response of the transducer is usually displayed on a camera-equipped oscilloscope to which the transducer's output is connected. A switch in the step-function generator can be used to trigger the oscilloscope for best presentation.

The ideal pressure step is a square pulse with zero rise time and of sufficient duration to cause proper response of the transducer.

Shock tubes come closest to meeting these requirements by providing pulses with rise times of between 1 ns and 1 μs, pulse durations of about 1 ms, and pressure steps from 3 to over 600 psi in amplitude, depending on their design. The shock wave is originated by a rapid displacement of air as caused by a high-voltage spark or by rupture of a diaphragm used as seal on a volume of pressurized gas.

Simpler test devices without shock tubes provide pressure steps with higher rise time (0.15 to over 3 ms), utilizing a high-voltage spark, a burst-diaphragm mechanism, or a special quick-opening valve. Some of these devices provide only negative pressure steps. Burst-diaphragm devices usually incorporate replaceable plastic diaphragms. Their rupture is initiated by a mechanically activated knife edge which perforates the diaphragm. The subsequent rupture action is comparable to the bursting of a balloon which was punctured by a pin.

Careful attention must be paid to close coupling of the transducer's sensing element to the point in the test device at which the pressure variation occurs when only the transducer's transient response is to be determined. When tubing (sensing line) is used in a given transducer installation, the length, diameter, straightness, and even internal smoothness of the tubing will affect the transient response of the transducer tubing system. If system response must be measured, the transducer must be coupled to the test device through an exact replica of the tubing used in the actual installation.

The analysis of a transducer's response to a step function of pressure allows the approximate determination of frequency response, resonant frequency, damping ratio, and ringing period as explained in Chapter III.

10.5.4 Environmental tests

In addition to the general test methods described in Chapter III, a number of specific methods and precautions apply to environmental tests of pressure transducers.

10.5.4.1 Temperature tests. A calibration consisting of at least two consecutive calibration cycles is usually performed after the transducer has been stabilized at the required temperature. On differential-pressure transducers one of these calibration cycles should be performed at the lowest, the other at the highest, specified reference pressure. This test may have to be repeated at several temperature levels when thermal shifts are to be determined in terms of percent of full-scale-output error per degree of temperature. Among other tests frequently performed during exposure to specified high and low temperatures are proof-pressure and insulation-resistance tests.

During temperature stabilization the transducer should be vented, through tubing, to the ambient atmosphere outside the temperature chamber. If this is not done, and the fluid system is kept closed, expansion or contraction of the fluid, due to temperature, can damage the transducer by under- or over-pressurizing it.

Response-time tests at specific temperatures may be required for viscous-damped transducers when thermal effects on damping are considered critical for a given transducer application.

10.5.4.2 Temperature-gradient tests.

This test is usually performed with ambient pressure applied to the transducer. The output is recorded on a strip-chart recorder and is monitored for the maximum zero shift obtained from the time the transducer is suddenly exposed to a step change of temperature until the output indicates temperature stabilization. This test is relatively simple to perform on a flush-diaphragm type transducer which can simply be inserted into a hot or cold liquid to just slightly over its diaphragm end.

It is more difficult to perform this test on a cavity-type transducer. Attempts to apply a hot or cold fluid through the pressure port usually do not result in proper temperature stabilization of the entire transducer. The actual temperature of the fluid in contact with the sensing element is very difficult to determine, unless the transducer's cavity is equipped with a bleed port which can be opened to permit the fluid to flow. Acceptable results can be obtained by connecting a length of tubing to the pressure port and immersing the entire transducer into a hot or cold liquid, with the open end of the tubing protruding well beyond the liquid's surface so that the sensing element sees only the ambient atmosphere as measured fluid.

For temperature-gradient tests on very small transducers with fast response, such as piezoelectric pressure transducers, the energizing of a photographer's flash-bulb in close proximity to the sensing element has been suggested as an acceptable method.

10.5.4.3 Vibration, acceleration, and shock tests.

Calibration cycles are usually not performed during these tests. Acceleration and shock tests are too brief to allow such a complete performance determination. During vibration tests the vibration frequency is normally varied continuously, and it is poor testing practice to vary more than one parameter during any one test.

Instead, the pressure applied to the transducer is held at a constant level by closing the fluid system after a desired pressure level has been reached, keeping it closed during the test, and ascertaining its constancy at frequent intervals. The level selected should be different for each portion of each test to avoid undue wear and fatigue of the sensing and transduction elements. The general range of these pressure points, however, should be in the vicinity of the pressure most likely to be measured by the transducer at the time when it is expected to see these environments at their most severe level in its actual application.

For each test setup, care must be taken to minimize environmental effects on the pneumatic or hydraulic connections as well as on the electrical connections, tubing, and cabling to the transducer. Accelerators (centrifuges) can be equipped with a hollow central shaft incorporating a sealed swivel fitting to which rigid tubing from the transducer can be radially installed and connected. Flexible tubing or hose can be used to link the pressure port of the transducer, installed on the platform of a vibration exciter or shock-test machine, to the pressure supply.

10.5.4.4 Ambient-pressure tests. These tests, also known as altitude tests, are only necessary for gage-pressure transducers with vented cases. In such transducers a near-vacuum ambient atmosphere can cause outgassing of internal components as well as self-heating of the transduction element. The latter may introduce undesirable thermal effects. One or two calibration cycles should be performed on such a transducer. The tubing from the pressure port is carried through a seal in the altitude-chamber wall to the pressure supply and monitoring equipment.

10.5.4.5 Humidity, salt-atmosphere, and immersion tests. These tests are primarily performed to verify case-sealing, case-material, and case-finish suitability. Gage-pressure transducers with vented cases, i.e., without provisions for transduction-element isolation, are frequently not expected to pass these tests. Other transducers should have tubing connected tightly to their pressure ports. The tubing should lead to the ambient laboratory atmosphere outside the environmental chamber, except if one or more of these severe environments are considered to be measured fluids for the transducers. Electrical connections to the transducer should be made so that they duplicate those specified for its actual end-use application, with special attention paid to all sealing and potting requirements.

10.5.5 Life tests

Pressure-transducer life requirements are usually specified in terms of cycling life (e.g., "25,000 full-range increasing and decreasing pressure cycles"). The cycling-life test is normally performed by sequentially pressurizing and depressurizing the transducer until the specified number of cycles has been reached. The test can be automated by using cam-driven or timer-programmed pressurization and vent valves. Pressure steps with excessively fast rise times should be avoided. If necessary, a small orifice can be added between valves and transducer to smoothen the pressure variation.

Special performance requirements may call for additional types of life tests. These include a partial-range life test in which the pressure is cycled rapidly over only a small part of the test specimen's measuring range within a specific portion of this range (e.g., between 70 and 80 % of its full range). Another special life test is the single-level operating life test, the application of pressure at a specific point within the range for a certain number of hours. This test requires the complete absence of any leaks in the pressurizing system.

The effect of a life test is most commonly determined by interrupting it periodically and performing a static calibration on the transducer during the intermissions, as well as after completion of the life test.

10.5.6 Burst-pressure tests

This test must necessarily be performed after the completion of all other tests since, by definition of "Burst-Pressure Rating," changes in the transducer's performance beyond specified tolerances are allowed. The transducer is considered as having passed the test when no rupture occurs as a result of applying to it the specified rated burst pressure a specified number of times. An additional specification may also prohibit leakage from the transducer in excess of a certain leakage rate.

During the burst-pressure test the transducer is best installed behind a safety barrier since the possibility of explosive rupture always exists. When a burst-pressure rating for the transducer's case is additionally specified (normally only the sensing element is burst-rated), the sensing element must be punctured so that during the next burst-pressure applications the pressurized fluid is applied to the inside walls of the transducer's case.

A *rupture-point test*, which is related to burst-pressure tests, is rarely performed because it is usually more important to verify that a transducer can safely withstand a certain overpressure than to determine at which pressure the transducer will actually rupture. During such a test it is important that the overpressure be gradually applied so that the rupture point can be determined with reasonable accuracy. Gaseous pressurizing fluids should not be used for this test. The use of liquids, such as hydraulic oil, minimizes the force of the explosion at the rupture point.

10.6 Specialized Measuring Devices

10.6.1 Vacuum transducers

Although vacuum is, theoretically, a space devoid of matter, the term is used in the field of pressure measurement to denote, generally, pressures below one standard atmosphere. Vacuum transducers, however, are primarily those pressure transducers whose measuring ranges are below one torr. The *torr* is the most frequently used unit of measurand in vacuum measurement. It is equal to 1/760 standard atmosphere. It is, for all practical purposes, equivalent to one millimeter of mercury (*mm Hg*), differing from it by only one part in seven million.

10.6.1.1 Thermoconductive vacuum transducers. This type of transducer measures pressure as a function of heat transfer by a gas. Heat flow originates at the surface of an electrically heated wire (filament) and is transferred by the gas in an enclosed vessel to the inside wall surface of this vessel. The quantity of heat transferred is proportional to the number of gas molecules and, hence, to the pressure in the vessel. A decrease in pressure causes the filament to get hotter, because less gas molecules transfer less heat away from the filament and to the wall of the vessel. At low pressures the heat conductivity of a gas decreases linearly with pressure.

Thermoconductive vacuum transducers require a well-regulated filament supply for their excitation. The filament, typically of tungsten or platinum wire, and its end supports must be carefully selected and designed to minimize heat loss due to radiation from its surface as well as heat loss due to conduction through the supports and connecting leads. Improved operation can be obtained by cooling the vessel, preferably to cryogenic temperatures.

The normal measuring range of these transducer types is from 10^{-3} to 1 torr.

10.6.1.1.1 Resistive Thermoconductive Vacuum Transducers. In these transducers (Figure 10-52) the heating of the filament is measured in terms of the change of filament resistance due to heating. An increase in filament temperature causes an increase in filament resistance. The transducer is usually connected so that the hot filament forms one arm of a Wheatstone bridge. A second transducer of the same configuration, but sealed so as not to respond to any pressure changes, can then be connected as a second arm of the bridge, primarily to compensate for changes in the temperature of the wall of the vessel or chamber.

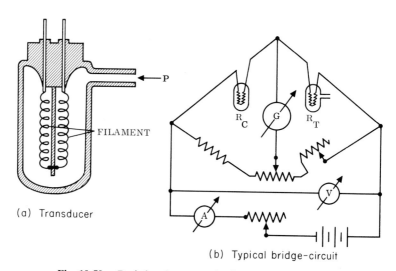

(a) Transducer

(b) Typical bridge-circuit

Fig. 10-52. Resistive thermoconductive vacuum transducer.

This type of transducer is commonly referred to as the *Pirani gage*. It operates from a constant-voltage filament supply. The filament is normally heated to a temperature of about 400°C at the lower end of the measuring range. A design variation uses a thermistor instead of a filament.

10.6.1.1.2 Thermoelectric Thermoconductive Vacuum Transducers. This type, sometimes called the *thermocouple gage*, uses a thermocouple as transduction element (Figure 10-53). The thermocouple's sensing junction is welded to the center portion of

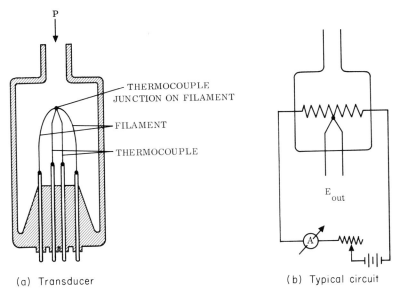

(a) Transducer   (b) Typical circuit

Fig. 10-53. Thermoelectric thermoconductive vacuum transducer.

the filament and measures the filament temperature. The transducer operates from a constant-current filament supply. It is somewhat less accurate than the resistive type. Design variations use a thermopile instead of a single thermocouple for increased output voltage. One design uses a center-tapped thermopile both as heat source and as transduction element.

10.6.1.2 Ionizing vacuum transducers. In this general category of vacuum transducers pressure is measured as a function of gas density by measuring ion current. The ion current results from positive ions which are collected at a negatively charged electrode when the gas is ionized by a stream of electrons or other particles. The ion current is proportional to gas density (molecular density) and, hence, proportional to pressure, when the number of electrons and their average path length are constant and all ions are collected. Because these transducers measure density, however, their calibrations are different for different gases. Nitrogen is most frequently used as reference gas.

10.6.1.2.1 Thermionic Ionizing Vacuum Transducers. The basic thermionic vacuum transducer, often referred to as the *ion gage* or *ionization gage*, resembles a triode-type radio tube and differs from it primarily by having an opening in the bulb. In a typical design [Figure 10-54 (a)] the filamentary cathode is surrounded by a helical grid around which a cylindrical anode is placed. Positive ions are collected at the anode, which is kept at a low negative voltage with respect to the filament.

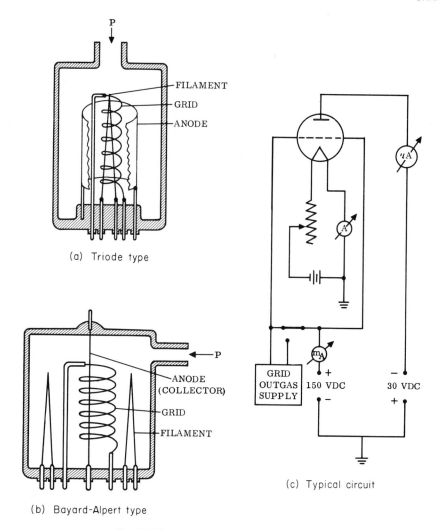

(a) Triode type

(b) Bayard-Alpert type

(c) Typical circuit

Fig. 10-54. Thermionic vacuum transducers.

Commonly used accessories include an ion-current amplifier which replaces the microammeter, a feedback-controlled filament supply which replaces the ammeter, battery, and rheostat, and a grid-outgas power supply inserted in series with one of the two grid connections, in the typical circuit illustrated in Figure 10-54 (c).

Tungsten or iridium filaments are used in most designs, and thoriated tungsten is used in some designs to allow operation at low temperatures and to prolong filament life. The pressure port for tubing connection (the *tubulation*) is kept short and of relatively large (typically 0.75 in.) diameter. To further reduce tubing effects, some thermionic ionizing vacuum transducers are furnished without a glass envelope, for direct installation in vacuum vessels. Such a bulbless transducer is referred to as a *nude gage*.

The normal measuring range of the triode-type transducer is from 10^{-8} to 10^{-3} torr. At pressures higher than this the space charge effect, as well as recombination of ions due to a reduction in the mean free path of the ions, cause extreme nonlinearity and effectively limit the range to that shown. A design variation, the *Shulz-Phelps ion gage*, incorporates a separate electron collector plate in addition to the usual ion collector plate to minimize space charge effects and yield a usable range from 10^{-5} to 1 torr.

The lower end of the triode-type's measuring range is limited to 10^{-8} torr because of X-ray effects. The electron stream produces soft X rays at the grid. When these X rays strike the collector electrode, they drive electrons from it by secondary emission. The resulting current is of the same polarity as the normally produced ion current. At about 10^{-8} torr the calibration curve of the triode-type vacuum transducer becomes asymptotic to this residual X-ray current; hence, gas pressures below this value cannot be measured.

This difficulty is overcome in the *Bayard-Alpert gage*. In this design variation [Figure 10-54 (b)], the arrangement of filament, grid, and anode are inverted. The anode is a centrally mounted thin wire, surrounded by the helical grid. A filament is placed outside the grid. A second filament is frequently added. It can be switched-in for continued transducer operation after the first filament has burnt out. Because of the small size of the collector, only a relatively small number of the X rays produced at the grid strike the collector. The resultant X-ray current is much lower than in the normal triode type. This feature extends the lower range limit and yields a measuring range of 10^{-10} to 10^{-3} torr for the Bayard-Alpert type thermionic ionizing vacuum transducer.

Because the Bayard-Alpert transducer does not contain the large outer anode plate of the normal triode type, more electrons and ions strike the inside wall of the glass envelope, where they build up an electrostatic charge. The existence of such a charge can cause erratic operation of the transducer. This difficulty is overcome in a further design variation, the *Nottingham ion gage*, in which an electrically conductive coating, typically a platinum film, is applied to the inside of the glass bulb and connected to the operating transducer circuit. In this configuration the grid is provided with end shields. This allows a measuring range of 10^{-11} to 10^{-3} torr for the Nottingham modification of the Bayard-Alpert type transducer.

Further design modifications intended to virtually eliminate the X-ray current, such as the *Schuemann modification*, have resulted in an extension of the Bayard-Alpert gage's lower range limit to almost 10^{-12} torr.

In the *photomultiplier ionizing vacuum transducer*, an electron multipier is used in place of the usual anode to amplify the ion current. This allows much lower electron currents to be used for the formation of positive ions, with a resulting decrease in X-ray production. The electrons are emitted from a tungsten filament surrounded by a cylindrical electron collector grid. Two accelerator grids, positioned between the collector grid and the first photomultiplier dynode, boost the flow of positive ions to the dynode. With anode area minimized and a relatively large electron path length, the lower range limit of the design is given by the "dark current" of the photomultiplier, which can be as low as 10^{-20} A. This current is equivalent to about 10^{-18} torr.

10.6.1.2.2 Magnetic-Field Ionizing Vacuum Transducers. At very low pressures, where the mean molecular free path is very large compared to electrode spacing, ionization in an ionizing vacuum transducer increases with electron path length because of the increasing probability of electron collision with gas molecules. This increase in electron path length is accomplished in the magnetic-field ionizing vacuum transducer, commonly called the *magnetron gage*, by a magnetic field which forces electrons accelerated by an electric field to travel in a helical path. This increase in ionization is obtained without an increase of X-ray current.

Magnetic-field vacuum transducers can be used for very low pressures ranges. Two basic versions of such transducers exist, one using hot-cathode, the other cold-cathode, electron emission.

A typical hot-cathode, magnetic-field, ionizing vacuum transducer—*hot-cathode*

Fig. 10-55. Hot-cathode, magnetic-field, ionizing vacuum transducer.

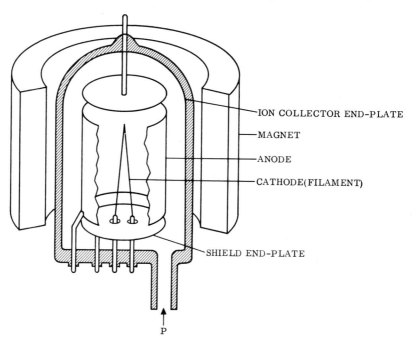

magnetron gage—is the *Lafferty gage*, illustrated in Figure 10-55. The electric and magnetic fields are crossed in this design so that electrons emitted from the filamentary cathode and accelerated radially toward the anode are also subjected to an axially acting magnetic field which forces them into a helical path. The shield and collector end plates are negative, and the anode is positive with respect to the cathode. Positive ions are collected at the collector end plate, and ion current can be measured between collector and cathode. Magnetic field strength must be maintained above the magnetron cutoff value. This type of transducer has a measuring range of 10^{-14} to 10^{-5} torr. The range can be extended by adding an electron multiplier within the same envelope so that ions are focused on the first dynode of the multiplier rather than on a collector plate. The addition of the multiplier can extend the measuring range downward to 10^{-17} torr and perhaps as low as 10^{-18} torr.

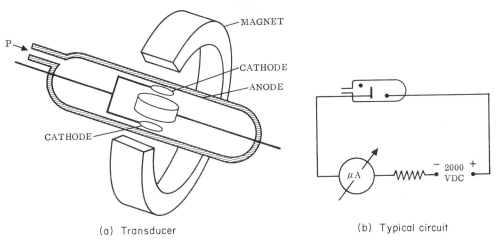

(a) Transducer   (b) Typical circuit

Fig. 10-56. Cold-cathode, magnetic-field, ionizing vacuum transducer.

An example of the basic design of a cold-cathode, magnetic-field, ionizing vacuum transducer—*cold-cathode magnetron gage*—is the *Philips gage* (named after its first manufacturer) or *Penning gage* (named after its developer), shown in Figure 10-56. When the cathode surfaces of this transducer are bombarded by high-energy ions, the surfaces emit electrons which join the total electron stream and produce additional ions which, in turn, produce additional electrons. This *avalanche effect* is primarily responsible for the high sensitivity of the transducer. The electrons are accelerated by the high-voltage electric field between cathodes and anode. By the crosswise action of the magnetic field the electrons are forced to travel along a helical path. The total current between cathodes and anode is the sum of the ion current and the electron current. The current, as measured by a microammeter, is therefore not linear with pressure.

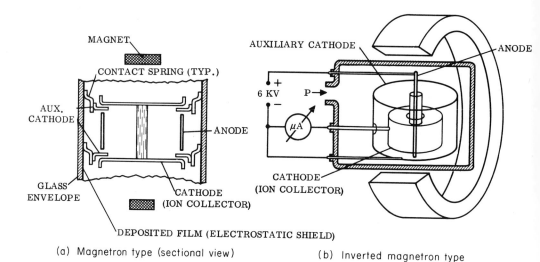

Fig. 10-57. Cold-cathode, magnetic-field, ionizing vacuum transducer with auxiliary cathode.

The measuring range of this transducer design is 10^{-7} to 10^{-3} torr. The sensitivity over this range is about 0.5 A per torr. The range can be extended downward, to as low as 10^{-12} torr, by the addition of a flash filament which triggers the discharge at very low pressures when the filament is briefly activated. The main advantage of cold-cathode designs over hot-cathode designs is the absence of a hot filament which can burn out. Cold-cathode designs need large tubulation because of their relatively high ionic pumping speed.

A very successful design variation of the basic cold-cathode, magnetic-field, ionizing vacuum transducer is the *Redhead gage* (named after its developer), in which field emission and ion collection are separated by use of an auxiliary cathode. Two versions of this design variation are illustrated in Figure 10-57. The normal magnetron type [Figure 10-57 (a)] uses a magnetic field of about one kilogauss. The ion-current-vs-pressure relationship is a straight line on a log-log plot. The inverted magnetron type [Figure 10-57 (b)] uses a magnetic field of about two kilogauss. The auxiliary cathode acts as electrostatic shield and prevents field emission from the edges of the circular opening in the ion-collector cathode. The field-emission cathode is connected directly to the negative side of the high-voltage supply so that field emission is not measured by the electrometer (picoammeter). The ion-current-vs-pressure relationship is linear down to about 5×10^{-10} torr and exponential below that level. The sensitivity is approximately 4.5 A per torr over the range. The range of the Redhead-type transducer is 10^{-13} to 10^{-4} torr. Its lower range end is limited primarily by a limitation on the ability to measure currents smaller than those corresponding to the 10^{-13} torr level.

10.6.1.2.3 Radioactive Ionizing Vacuum Transducers. The use of particles other than electrons to ionize gas is exemplified (Figure 10-58) in an alpha-particle ionizing vacuum transducer, the *Alphatron*®. The radioactive-particle source is a small thin plaque usually containing radium, tritium, polonium, or a similar radioactive material which is in equilibrium with its daughter products. In a typical design a radioactive source of Ra^{226} with a strength of 200 microcuries is used to provide a relatively constant alpha-particle flux. When the average path length is fixed, the average energy is constant, and all ions which are formed are collected. The ion-current-vs-pressure relationship is linear up to about 50 torr. One design variation contains an auxiliary smaller ionization chamber for use at higher pressures. Although this transducer type is most commonly used over the range from 10^{-1} to 100 torr, its range capability is from 10^{-5} to 1000 torr. When the higher range is available on a given design, the ambient atmospheric pressure level can be used as a convenient calibration checkpoint. Ion current can be amplified and read out in terms of voltage, current, or pulse frequency. Output readings can differ for different gases or gas compositions which have different ionization cross sections for alpha particles. It should be noted, however, that calibrations for different gases are closer to each other for this type of transducer than for any other type of ionizing vacuum transducer.

Use of a beta source, such as tritium absorbed in titanium foil, in place of an alpha source has also shown promising results and appears to allow an extension of the lower range limit to 10^{-8} torr.

Fig. 10-58.   Alpha-particle ionizing vacuum transducer.

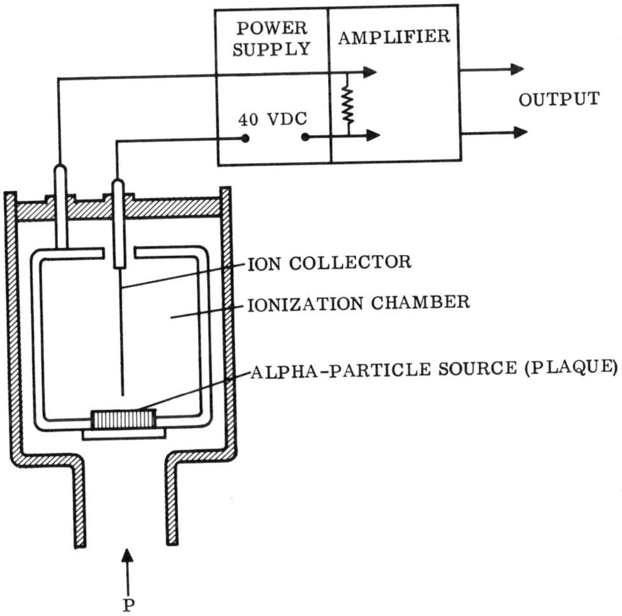

10.6.1.3 Vacuum-transducer calibration equipment. Equipment used for vacuum-transducer calibration below those pressure levels (about 10 torr) which can be measured more conveniently with standard manometers includes primarily the McLeod-type compression manometer (*McLeod gage*) for measurements between 10^{-3} and 10 torr and the Knudsen radiometric vacuum indicator (*Knudsen gage*) for measurements between 10^{-8} and 10^{-3} torr.

10.6.1.3.1 McLeod Gage. A typical McLeod gage, as shown in Figure 10-59, consists essentially of a vertical tube with a bypass section whose vertical portion is a capillary, a bulb with an upper sealed capillary extension, a connecting tube from the bulb downward to the vertical tube, and a mercury reservoir connected to the bottom end of the vertical tube. The two capillaries have the same cross-sectional dimensions (same diameter).

In operation the McLeod gage is connected to the system whose pressure (vacuum) is to be measured. The gas in the measured system fills the vertical tube, the bypass section, and the bulb since the mercury level is below the "Y" junction between bulb connection and vertical tube. The mercury level is then raised in the measuring system by one of several means: by physically lifting a reservoir connected to the bottom end of the vertical hose by means of a flexible U-shaped hose; by pushing a piston into the reservoir to displace the mercury upward into the measuring system; by admitting pressurized gas to the reservoir to force the mercury upward; by raising the reservoir

Fig. 10-59. McLeod-type manometer.

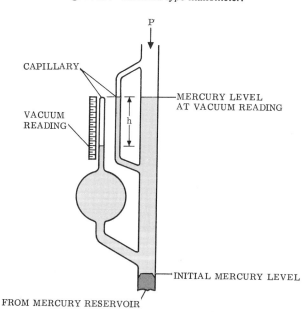

to force mercury up a coaxial tube protruding into the reservoir; or by admitting air into a previously evacuated reservoir.

As the mercury rises in the system, it seals off the spherical bulb. As it rises further in the bulb, it compresses the gas in the bulb's upper capillary extension. At the same time the mercury will have risen in the vertical tube and its capillary bypass section. The mercury is allowed to rise until it reaches a level mark corresponding to the upper end of the capillary extension on the bulb. At this point the pressure is read on the square-law calibrated scale at the level reached by the mercury in the capillary bulb extension. The use of the square-law scale is predicated upon the relationship between pressure (p), the difference in mercury level between the two capillaries of equal cross section (h), and a constant (k) given by the ratio of capillary cross-sectional area (A) to bulb volume (V). This relationship is:

$$p = \frac{A}{V}h^2 = kh^2$$

A number of adaptations of the McLeod-type manometer have been designed, including swivelling or tilting manometers, multirange manometers, and manometers with optical or electronic accessories intended to improve reading accuracy.

10.6.1.3.2 Knudsen Gage. Since the principle employed in this instrument has been used in rotating-vane type radiant-energy detectors (radiometers), the Knudsen gage has been described as a "radiometric" type of device. It measures pressure as a function of the force produced by the impact of heated gas molecules upon a surface.

The operation of a simple Knudsen gage is shown in Figure 10-60. Two upright heated vanes are mounted on a circular base. A moving vane, not heated, is suspended (e.g., by a thin wire or a quartz fiber) at its center so as to allow angular motion of the vane tips to and from the fixed vanes such that the torsion of the suspension wire acts as restoring force. The entire mechanism is enclosed in an envelope with tubulation for connection to a vacuum system.

As gas molecules come in contact with the fixed heated vanes, they rebound from these vanes with an increased velocity, impinge upon the cooler movable vane ends, and cause this vane to rotate so that its tips move away from the fixed vanes.

In the relatively simple indicator version illustrated, a light beam emanating from a light source, reflected by a mirror attached to the center of the moving vane, and intercepted by a translucent scale is used to indicate vane rotation and, hence, pressure in the vacuum system to which the indicator is connected.

The Knudsen gage is essentially linear with pressure below 10^{-3} torr, but is usable, with increasing nonlinearity, to about 1 torr. However, the McLeod gage is more practical over the 10^{-3} to 1 torr range. The Knudsen gage is usually calibrated against a McLeod gage at one point or over a narrow portion of its operating range. An absolute-zero reading, very useful as lower-end calibration point, can be obtained on the Knudsen gage by simply turning off the vane heaters.

A number of design variations of the Knudsen gage include transducers with

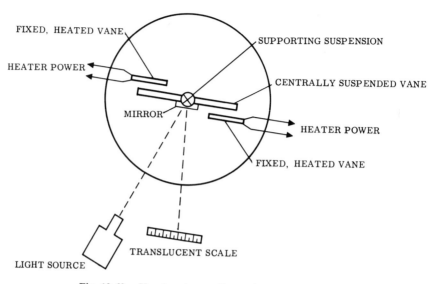

Fig. 10-60.   Knudsen-type radiometric vacuum indicator.

capacitive or reluctive null-balance-type transduction elements and instruments with an electromagnetic restoring force in place of the wire or fiber suspension. With such additional refinements, and extreme care during measurement, the range of the Knudsen gage can be extended downward to as low as 10^{-9} torr.

10.6.2 Very-high-pressure transducers

The measuring ranges of virtually all commonly used transducer designs are no higher than 15,000 psi. A few types have ranges up to 20,000 psi. Some designs exist, using potentiometric, reluctive, piezoelectric, or strain-gage transduction, with ranges up to 100,000 psi, and, in a few rare cases, up to about 500,000 psi.

The special sensors usable for measuring pressures in the 100,000 to 1,000,000 psi ranges include primarily the following: (1) resistive devices, such as metal alloys whose resistance increases with pressure or polymorphic elements whose resistance increases and decreases in accordance with a known pattern; (2) inductive devices in the form of coils bonded to but insulated from a solid core so that a change of inductance occurs when pressure changes cause changes in the volume of the core; and (3) solid-electrolyte ionic cells whose output voltage into a relatively high load-resistance decreases after having reached an initial peak as pressure increases.

10.6.3 Pressure-altitude transducers

Pressure-altitude transducers are absolute-pressure transducers whose output variation is linear with altitude variation rather than with absolute-pressure variation.

The relationship between atmospheric pressure and altitude is nonlinear; it is fixed by international agreement and stated in published tables as "normal," "standard," or "model" atmosphere. One of the commonly accepted model atmospheres is shown in Table 10-2.

In order to provide an output which is linear with altitude, pressure-altitude transducers incorporate transduction elements which compensate for the nonlinear altitude-vs-pressure relationship by responding to pressure applied to their sensing elements in an exactly matched oppositely nonlinear manner. The most commonly used transduction elements are potentiometric, reluctive, and servo-type.

The range of such transducers, always stated in altitude units (feet or meters), is based on the operating ceiling of the vehicle they are intended for. It always exceeds that ceiling, however, by some amount. Ranges of most existing designs have lower limits as low as −1000 ft and upper limits to 300,000 ft.

10.6.4 Airspeed transducers

Airspeed transducers are essentially differential-pressure transducers which measure the difference between stagnation pressure (impact pressure) and static pressure as sensed by either a pitot-static tube or a pitot tube and separate static tap. Since airspeed transducers measure air-flow velocity (see 10.1.2), this combination of pitot-static tube and differential-pressure transducer is more properly classified as a flow transducer and is described further in Chapter 4.

Airspeed, however, is the rate of motion of a vehicle relative to an air mass. The output of an airspeed transducer is a measure of the *indicated airspeed* (*IAS*), which is normally equivalent to the calibrated airspeed, applicable at sea level where calibrations are derived. IAS differs from calibrated airspeed primarily by the amount of position error (error due to the location of the transducer on the vehicle) and the various errors in the transducer. To obtain *true airspeed* (*TAS*), additional corrections must be made in relatively slow-moving vehicles for air density and pressure altitude. In relatively high-speed vehicles (traveling at Mach 0.3 and greater) further corrections must be made for the compressibility of air, which increases with Mach number.

10.6.5 Mach-number transducers

The Mach number—named after the Austrian scientist Ernst Mach—is the ratio of the speed of a vehicle to the speed of sound in the medium in which the vehicle is traveling. This definition is applicable for an undisturbed medium (such as free-stream air), and corrections must be made for configurational factors at the location at which the pitot-static tube or separate pitot tube and static tap are installed. Usually, these installations are made where virtual free-stream conditions can be obtained so that no such corrections are necessary.

The output of a Mach-number transducer increases with an increasing ratio of differential pressure (stagnation pressure minus static pressure) to static pressure.

For most practical applications the relationship between Mach number (M), stagnation pressure (P_t) and static pressure (P_s) can be approximated over the range Mach 0.3 to 5.0 by

$$M \simeq 1.04\left(\frac{P_t - P_s}{P_s}\right)^{0.45}$$

A typical Mach-number transducer can provide an output proportional to Mach number by incorporating a mechanical computing linkage between a differential-pressure sensing element and an absolute-pressure sensing element such that the linkage motion acts upon the transduction element. It can also incorporate a transduction element for each of the two sensing elements and feed the two outputs into a small integral electronic computer from which the output of the entire Mach-number transducer is obtained. Servo-type transduction is most frequently used.

10.6.6 Pressure-depth transducer

Just as pressure transducers can be used to measure height above sea level (pressure-altitude), they can also be used to measure height below sea level (pressure-depth). Transducers used for pressure-depth measurement are frequently used for the determination of water depth, especially in oceanographic work. In the latter field they are known as *seawater-depth transducers*. Other applications include those in the geophysical area, such as oil-well and mine-shaft depth measurements. Pressure-depth transducers are usually absolute-pressure transducers.

Whereas the pressure-vs-altitude and atmospheric-pressure-vs-depth relationships are nonlinear, the underwater-pressure-vs-depth relationship is considered linear for all practical purposes (see 10.1.4) since water is much less compressible than air.

Special design considerations apply to water-depth transducers because of the nature of the fluid in which they must operate continuously, often for long periods of time, and because of the long cables usually required between transducer and either telemetry transmitter or readout equipment. Water, especially seawater, is highly corrosive. This indicates a need for a corrosion-proof liquid acting as pressure-transfer medium.

The problem of transducer-output attenuation in long interconnecting cables has been solved most satisfactorily by the use of vibrating-element, FM-output pressure transducers. Some other transducer types capable of providing relatively high output amplitudes, such as potentiometric transducers, have also been found satisfactory for oceanographic applications.

A recurring problem in seawater-depth transducers is their exposure to marine life. Life-forms of small size but relatively fast growth can cover all portions of the transducer and effectively seal the pressure port. Larger life-forms, notably fish such as sharks, can cause severe damage to cables and destruction and loss of portions of the cables and of the transducers themselves.

BIBLIOGRAPHY

1. Wildhack, W. A., "Pressure Drop in Tubing in Aircraft Instrument Installations," *NACA Technical Note TN 539*, Washington, D.C.: Government Printing Office, 1937.

2. Vennard, J. K., *Elementary Fluid Mechanics*. New York: John Wiley & Sons, Inc., 1940.

3. Pike, E. W. and Gibbs, N. E., "A Study on Aneroid Capsules," *Journal of Applied Physics*, Vol. 19, pp. 106–108, 1948.

4. Fryburg, G. C. and Simons, J. H., "Precision Vacuum Gage," *Review of Scientific Instruments*, Vol. 20, pp. 541–548, 1949.

5. Guthrie, A. and Wakerling, R. K., *Vacuum Equipment and Techniques*. New York: McGraw-Hill Book Company, 1949.

6. Taback, I., "The Response of Pressure Measuring Systems to Oscillating Pressures," *NACA Technical Note TN 1819*, Washington, D.C.: Government Printing Office, February 1949.

7. Bayard, R. T. and Alpert, D., "Extension of the Low Pressure Range of the Ionization Gage," *Review of Scientific Instruments*, Vol. 21, pp. 571–572, 1950.

8. Evans, E. C. and Burmaster, K. E., "A Philips-Type Ionization Gauge for Measuring of Vacuum from 10^{-7} to 10^{-1} mm of Mercury," *Proceedings of the IRE*, Vol. 38, p. 651, 1950.

9. Iberall, A. S., "Attenuation of Oscillatory Pressures in Instrument Lines," *N.B.S. Research Paper RP 2115, Journal of Research*, N.B.S., Vol. 45, July 1950.

10. Wagner, S. and Johnson, C. B., "Calibration of Ionization Gauges for Various Gases at Low Pressures," *Journal of Scientific Instruments*, Vol. 28, p. 278, 1951.

11. Humphreys, J. D., "Pressure Sensing Calculations for Aircraft and Guided Missiles," *Tele-Tech*, April 1953 and *Instrument Note No. 25*, Los Angeles: Statham Instruments, Inc., June 1953.

12. Li, Y. T., "High-Frequency Pressure Indicators for Aerodynamic Problems," *NACA Technical Note TN 3042*, Washington, D.C.: Government Printing Office, November 1953.

13. Sibley, C. B. and Roehrig, J. R., "Wide Range Vacuum Gage," *Electronics*, Vol. 26, No. 11, p. 176, 1953.

14. Van der Pyl, L. M., "Bibliography on Bourdon Tubes and Bourdon Tube Gages," *ASME Paper No. 53-IRD-1*, 1953.

15. Grey, J., "Pressure Transducers," *Product Engineering*, June 1954.

16. Philips, K., "Some Experiments with a Cold-Cathode Vacuum Gauge," *Journal of Scientific Instruments*, Vol. 31, pp. 110–111, 1954.

17. Barry, F. W., "Determination of Mach Number from Pressure Measurements," *ASME Paper 55-SA-28*, 1955.

18. Brombacher, W. G. and Lashof, T. W., "Bibliography and Index on Dynamic Pressure Measurement," *N.B.S. Circular 558*, Washington, D.C.: Government Printing Office, 1955.

19. Leggat, J. W., "Applications of a Capacitor Type Pressure Indicator," *I.S.A. Journal*, Vol. 2, p. 290, 1955.

20. Presson, A. G., "Analytical Study of Frequency Response of Pressure Transducer Systems," *J.P.L. Report 20-91*, Pasadena, Calif.: Jet Propulsion Laboratory, August 1955.

21. Sandell, R. P. and Ceaglske, N. H., "Frequency Response in Pneumatic Transmission Lines," *I.S.A. Journal*, Vol. 3, December 1956.

22. Stedman, C. K., "Alternating Flow of Fluid in Tubes," *Instrument Note No. 30*, Los Angeles: Statham Instruments, Inc., 1956.

23. Brombacher, W. G., "Force-Balance Systems for Measuring Static Pressure, Pressure Altitude and Mach Number," *N.B.S. Report No. 5296*, Washington, D.C.: Government Printing Office, 1957.

24. Leck, J. H., "Pressure Measurement in Vacuum Systems," London: Institute of Physics, 1957.

25. Lederer, P. S. and Smith, R. O., "Performance Tests on Two Piezoelectric Quartz Crystal Pressure Transducers and Calibrator," *N.B.S. Report No. 4973*, Washington, D.C.: Government Printing Office, 1957.

26. Smith, R. O. and Lederer, P. S., "The Shock Tube as a Facility for Dynamic Testing of Pressure Pickups," *N.B.S. Report No. 4910*, Washington, D.C.: Government Printing Office, 1957.

27. Stedman, C. K., "The Characteristics of Flat Annular Diaphragms," *Instrument Note No. 31*, Los Angeles: Statham Instruments, Inc., 1957.

28. Barton, J. R., "A Note on the Evaluation of Designs of Transducers for the Measurement of Dynamic Pressures in Liquid Systems," *Instrument Note No. 27*, Los Angeles: Statham Instruments, Inc., 1958.

29. Brombacher, W. G. and Jenny, C. J., "Some Factors Affecting the Performance of Corrugated Diaphragm Capsules Having a Deflection Nonlinear with Pressure," *ASME Paper No. 58-A-169*, 1958.

30. Li, Y. T., "Pressure Transducers for Missile Testing and Control," *I.S.A. Journal*, November 1958.

31. Gillis, R. A. and Rice, C. F., Jr., "Performance Tests on Two Potentiometer Type Pressure Pickups," *N.B.S. Report No. 6299*, Washington, D.C.: Government Printing Office, 1959.

32. Hausner, A., "A Guide to the Selection and Use of Dynamic Pressure Transducers," *Report TR-814*, Diamond Ordnance Fuze Laboratories, September 1959.

33. Lederer, P. S. and Smith, R. O., "Performance Tests on Two Strain-Gage Pressure Transducers," *N.B.S. Report No. 6336*, Washington, D.C.: Government Printing Office, 1959.

34. Perino, P. R., "The Effect of Transmission Line Resistance in the Shunt Calibration of Bridge Transducers," *Instrument Note No. 36*, Los Angeles: Statham Instruments, Inc., 1959.

35. Spencer, N. W. and Boggers, R. L., "A Radioactive Ionization Gage Pressure Measurement System," *A.R.S. Journal*, Vol. 29, No. 1, 1959.

36. Baba, A. J. and Rice, C. F., Jr., "Performance Tests on Two Variable Inductance Type Pressure Gages," *N.B.S. Report No. 6633*, Washington, D.C.: Government Printing Office, 1960.

37. Barnes, G., "New Type of Cold Cathode Vacuum Gage for the Measurement of Pressure below 10^{-3} mm Hg," *Review of Scientific Instruments*, Vol. 31, No. 6, pp. 608–611, 1960.

38. Harris, C. J., Kaegi, E. M., and Warren, W. R., "Pressure and Force Transducers for Shock Tunnels," *I.S.A. Journal*, August 1960.

39. Hyman, C. and Knapp, F. M., "A Simple Inexpensive Pressure Transducer," *Journal of Applied Physiology*, Vol. 15, p. 727, 1960.

40. Lederer, P. S., "A Simple Pneumatic Stepfunction Pressure Calibration," *N.B.S. Report No. 6981*, Washington, D.C.: Government Printing Office, 1960.

41. McQuarrie, R. A. and Lederer, P. S., "Performance Tests on Two Unbonded Strain-Gage Pressure Pickups," *N.B.S. Report No. 6942*, Washington, D.C.: Government Printing Office, 1960.

42. Norton, H. N., "Potentiometers Measure Pressures in Advanced Flight Tests," *Space/Aeronautics*, December 1960.

43. Van der Pyl, L. M., "Bibliography on Diaphragms and Aneroids," *ASME Paper No. 60-WA-122*, 1960.

44. Bradspies, R. W., "Bourdon Tubes," *Giannini Controls Technical Note*, Duarte, California: Conrac Corp., 1961.

45. Neubert, H. K. P., "Pressure Transducers," *Control*, Vol. 4, No. 32, p. 114, February 1961.

46. Norton, H. N., "Potentiometric Pressure Transducers," *Instruments and Control Systems*, Vol. 34, February 1961.

47. Norton, H. N., "Reluctive Transducers," *Instruments and Control Systems*, Vol. 34, December 1961.

48. Streeter, V. L. (Ed.), *Handbook of Fluid Dynamics*. New York: McGraw-Hill Book Company, 1961.

49. Aronson, M. (Ed.), "Capacitive Pressure Sensors," *Instruments and Control Systems*, Vol. 35, May 1962.

50. Clark, D. B., "Rare Earth Pressure Transducers," *Instruments and Control Systems*, Vol. 35, November 1962.

51. Schweppe, J. L., "Calibration of Pressure Transducers with Aperiodic Input-Function Generators," *I.S.A. Paper No. 46.1.62*, October 1962.

52. Rubinstein, E., "Pressure Transducers," *Product Engineering*, pp. 102–106, November 1962.

53. Norton, H. N., "Specification Characteristics of Pressure Transducers," *Instruments and Control Systems*, Vol. 36, December 1963.

54. Schweppe, J. L. *et al.*, "Methods for the Dynamic Calibration of Pressure Transducers," *N.B.S. Monograph 67*, Washington, D.C.: National Bureau of Standards, 1963.

55. Newton, G. P. *et al.*, "Response of Modified Redhead Magnetron and Bayard-Alpert Vacuum Gauges Aboard Explorer XVII," *NASA Technical Note TN D-2146*, NASA-Goddard Space Flight Center, February 1964.

56. "Guide for Specifications and Tests for Strain-Gage Pressure Transducers," *I.S.A. Tentative Recommended Practice RP37.3*, 1964.

57. Lloyd, C. and Giardini, A. A., "Measurement of Very High Pressures," *Paper No. 21-USA-267, Acta IMEKO 1964*, IMEKO, Budapest, Hungary, 1964.

58. Roehrig, J. R., "High Vacuum Measuring Instrumentation and Methodology," *Report No. FDL-TDR-64-68 (AFSC-R&TD-AFFDL)*, Washington, D.C.: Office of Technical Services, U.S. Dept. of Commerce, 1964.

59. Simons, J. C., Jr., "Measurement and Calibration Techniques for Very Low Gas Densities," *Paper No. 21-USA-259, Acta IMEKO 1964*, pp. 295–302, IMEKO, Budapest, Hungary, 1964.

60. Cohen, M. M. and Horn, L., "Tunneling Junction Diode Pressure Gage," *Report No. TR-1282*, Washington, D.C.: U.S. Army Materiel Command, Harry Diamond Laboratories, March 1965.

61. Horn, L., "The Response of Flush Diaphragm Pressure Transducers to Thermal Gradients," *Preprint No. 13.3-4-65*, Pittsburgh: Instrument Society of America, October 1965.

62. Jones, H. B., Jr., Knauer, R. C., Layton, J. P., and Thomas, J. P., "Transient Pressure Measurements in Liquid Propellant Thrust Chambers," *ISA Transactions*, Vol. 4, No. 2., pp. 116–132, April 1965.

63. "Specifications and Tests of Potentiometric Pressure Transducers for Aerospace Testing," *ISA Standard S 37.6*, Pittsburgh: Instrument Society of America, 1967.

11   Sound

11.1 Basic Concepts

11.1.1 Basic definitions

Sound is an oscillation in pressure, stress, particle displacement, particle velocity, etc., in an elastic or viscous medium; it is also the superposition of such propagated oscillations. *Note:* Primary consideration in this chapter is given to pressure oscillations in fluid media.

Sound (sensation) is the auditory sensation evoked by the oscillations associated with sound.

Sound energy of a portion of a medium is the total energy in that portion minus the energy which would exist in the same portion with no sound waves present.

Sound pressure is the total instantaneous pressure at a given point, in the presence of a sound wave, minus the static pressure at that point.

Peak sound pressure is the maximum absolute value of the instantaneous sound pressure within a specified time interval.

Effective sound pressure is the root-mean-square value of the instantaneous sound pressure over a specified time interval at a given point.

Sound pressure level (SPL or L_p) is 10 times the logarithmic ratio of the mean-square sound pressure to a mean-square reference pressure. It is normally expressed in decibels as 20 times the logarithm to the base of 10 of the ratio of the rms sound pressure to an rms reference pressure, or

$$SPL = 20 \log_{10} \frac{p(\text{rms})}{p_{\text{ref}}(\text{rms})}$$

The reference pressure must be stated. It is usually taken as 2×10^{-4} dynes/cm^2 (2×10^{-4} microbar), and sometimes as 1 dyne/cm^2 (10^{-1} N/m^2). *Note:* This chapter deals primarily with sound-pressure transducers whose range is expressed in terms of sound-pressure level.

Sound level is a weighted sound-pressure-level reading obtained with a meter complying with U.S.A. Standard S 1.4-1961, "Specification for General-Purpose Sound Level Meters"; the reference pressure is 2×10^{-4} dyne/cm^2.

Propagation velocity is a vector quantity which describes the speed and direction with which a sound wave travels through a medium.

Sound intensity is the average rate of sound energy transmitted in a specified direction through a unit area normal to this direction at a given point.

Sound power of a source is the total sound energy radiated by the source per unit of time.

Sound absorption is the process by which sound energy is diminished by being partially changed into some other form of energy, usually heat, while passing through a medium or striking a surface.

A **simple sound source** is a source which radiates sound uniformly in all directions under free-field conditions.

A **sound field** is a region containing sound waves.

A **free sound field** is a sound field in a homogeneous medium free of any acoustically reflecting boundaries.

Free-field frequency response (of a sound-pressure measuring transducer) is the ratio, as a function of frequency, of the output of the transducer in a sound field to the free-field sound pressure that would exist at the transducer location were the transducer not present.

Free-field normal incidence response is the free-field frequency response (of a sound-pressure measuring transducer) when sound incidence at a specified sensing surface of a transducer is from the direction normal to that surface.

Free-field grazing incidence response is the free-field frequency response (of a sound-pressure measuring transducer) when sound incidence at a specified sensing surface of the transducer is from the direction parallel to that surface.

Pressure frequency response is the ratio, as a function of frequency, of the output to sound pressure which is equal in phase and amplitude over the entire sensing-element surface of a sound-pressure measuring transducer.

Random incidence response is the diffuse-field frequency response (of a sound-pressure measuring transducer) where sound incidence at a specified sensing surface of a transducer is from random directions.

Directivity is the solid angle, or the angle in a specified plane, over which sound incident on a transducer's sensing element is measured (within specified tolerances) at a specified measurand frequency or in a specified band of measurand frequencies.

The **directivity factor** is the ratio of the square of the transducer output produced in response to sound incident from a specified direction to the mean-square output that would be produced in a perfectly diffused sound field of the same frequency or band of frequencies and of the same mean-square sound pressure.

A **directional response pattern (directivity pattern, directivity characteristics,** "beam pattern") of a sound transducer is a description, usually in graphical form, of the transducer's response as a function of direction of incidence of sound waves in a specified plane and at a specified frequency or band of frequencies.

The **equivalent volume** of a sound-pressure transducer is its acoustical input impedance expressed in terms of the acoustical impedance of an equivalent volume of a gas enclosed in a rigid cavity.

A **sound-pressure transducer** is a device which provides a usable (electrical) output in response to sound pressure which is to be measured. *Note:* A *microphone* is a device which provides electrical signals in response to sound waves which are not necessarily associated with a measurand.

11.1.2 Related laws

Sound pressure (considered as pressure, in general terms)

$$p = \frac{f}{S}$$

where: p = sound pressure
$\quad\quad f$ = force due to sound acting on a surface
$\quad\quad S$ = surface area

Acoustic impedance

$$Z_a = \frac{p}{Su} \quad\quad\quad Z_a = R_a + jX_a$$

where: Z_a = acoustic impedance
$\quad\quad R_a$ = acoustic resistance
$\quad\quad X_a$ = acoustic reactance

p = rms sound pressure
S = area of surface (through which the sound waves act)
u = rms particle velocity (of an infinitesimal portion of the medium)
Note: $Su = U$ = rms *volume velocity*

Sound-energy flux

$$J = \frac{p^2 S}{\rho c} \cos \theta$$

Sound intensity

$$I = \frac{p^2}{\rho c}$$

where: J = sound-energy flux (for one period)
   I = sound intensity (in the direction of propagation)
   p = rms sound pressure
   ρ = density of medium
   c = velocity of propagation of free (plane or spherical) sound wave
   S = area (of surface through which the flux acts)
   θ = angle between the normal to area S and the direction of travel of the sound wave

Total acoustic power radiated from a point source

$$W_p = 4\pi r^2 I$$

where: W_p = acoustic power
   r = distance from point source
   I = sound intensity

Spectrum level (for a given band of frequencies)

$$S(f) = L_p - 10 \log_{10} \Delta f$$

where: $S(f)$ = Spectrum level at center (of band) frequency (in db)
   $L_p = SPL$ = sound pressure level (in db)
   Δf = bandwidth (in hertz)

11.1.3 Units of measurement

Sound pressure, as measurand for a transducer, is expressed in terms of sound pressure *level*. Sound pressure level is expressed in **decibels** (*db*). The decibel is one-tenth of a **bel**. The bel, which is not commonly used, is a unit of level when the base of the logarithm of a ratio (of a quantity to a reference quantity of the same kind) is 10. Hence, for quantities proportional to power, the decibel is a unit of level when the base of the logarithm is the tenth root of ten. A logarithm to the base the tenth root of ten is equal to ten times the logarithm to the base of ten. Hence, for sound pressure level

$$L_p = 10 \log_{10} \left(\frac{p}{p_{\text{ref}}}\right)^2 = 20 \log_{10} \frac{p}{p_{\text{ref}}}$$

where: $L_p = SPL$ = sound pressure level, expressed in db

p = rms sound pressure

p_{ref} = rms reference pressure

The reference pressure must be stated. Its most commonly used value is 2×10^{-4} μbar (= 0.0002 dyne/cm²). A reference pressure of 1 μbar has also been used. Both values are rms values.

Sound pressure is also expressed in **newtons per square meter** (N/m^2), **newtons per square centimeter** (N/cm^2), or **dynes per square centimeter** $(dyne/cm^2)$. Sound power is expressed in **watts** (W) or **ergs**.

Acoustic impedance, resistance, or reactance are expressed in **newton-seconds per (meter)**⁵ (Ns/m^5), a unit which is also referred to as the **mks acoustic ohm**. Alternately, they are expressed in **acoustic (cgs) ohms**. They have a value of 1 acoustic (cgs) ohm when a sound pressure of 1 μbar produces a volume velocity of 1 cm³/s.

11.2 Sensing Elements

The flat diaphragm is used as sensing element in virtually all sound-pressure transducers. It is a circular flat plate supported continuously around its edge. The thin diaphragm is usually attached to its supporting rim by welding, preferably while tension is applied ("prestressed"). Diaphragms are discussed in more detail in the sensing-element section of the chapter dealing with pressure transducers. In some transducer designs where the transduction element itself acts as the sensing element, the diaphragm is used only as a membrane.

Sound-pressure sensing elements are usually constructed in a gage-pressure configuration; i.e., ambient pressure is admitted to their reference side (inside of case) so that sound pressure is measured with respect to ambient static pressure while static pressures acting on the outer and inner diaphragm surfaces are equalized. The gage vent (case opening) also acts as "low-pass-filter" acoustic leak which prevents access of the sound to the reference side of the diaphragm. Some sensing elements exist in sealed-reference differential-pressure configurations where their reference side is sealed and sometimes partly evacuated.

11.3 Design and Operation

Sound-pressure transducers are, essentially, special-purpose pressure transducers. Their design optimizes certain features of general-purpose pressure transducers while neglecting other features not considered necessary for their proper operation. Considered as pressure transducers, the upper limit of their pressure range is usually quite low, whereas the frequency range over which flat response is obtained is very large. The latter requirement limits the transducer designs usable for sound-pressure measurement to the few types having stiff low-mass sensing elements with small deflection and high natural frequency and transduction elements capable of following high-frequency changes in sensing-element deflection.

11.3.1 Capacitive sound-pressure transducers

Transducers in which changes in applied sound pressure are converted into changes in capacitance are popular for sound-pressure measurement applications because of their inherently high frequency-response characteristics. Such transducers are sometimes referred to as "condenser microphones." A basic transducer of this type (Figure 11-1) consists of a diaphragm which acts as the "rotor" and a back plate which acts as the "stator" of a variable capacitor. The two electrodes are spaced very close to each other. The thin metallic diaphragm is supported around its edge by the housing (case). The stator plate is rigidly supported by insulating material within the case. A narrow port is drilled through the case wall to admit ambient pressure to the inside of the case. When a well-regulated d-c polarization voltage (typically 200 V) is applied through a high resistance across the two electrodes, a constant charge is maintained on the electrodes. Changes in capacity due to diaphragm deflection produce a change in voltage across the electrodes. This output is typically fed to a cathode-follower or emitter-follower circuit before it is fed to an amplifier.

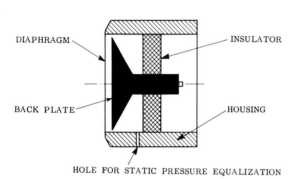

Fig. 11-1. Basic capacitive sound-pressure transducer (courtesy of A/S Bruel & Kjaer).

The cathode follower, a one-stage amplifier with a gain less than unity, provides the high-load impedance required by the transducer. Its output impedance, however, is relatively low. It should be located as close as possible to the transducer to avoid capacitive loading by the connecting cable. Preferably, the emitter or cathode follower is packaged integrally with the transducer. Its low output impedance then permits use of a fairly long connecting cable to an amplifier or other equipment.

Figure 11-2 shows details of construction of a typical capacitive sound-pressure transducer. This transducer, which has an outside diameter of about 0.5 in., is provided with a mechanical-protection cap ("grid") to protect the 0.16 mil-thick nickel diaphragm from damage in handling. The cap also affects the directivity of the transducer. The gap between diaphragm and stator (back plate) is slightly less than 1 mil. At a polarization voltage of 200 V d-c the steady-state capacitance of the transducer is approximately 20 pF. Transducers of this type exist in various sizes and configurations as exemplified by those shown in Figure 11-3.

Among developmental efforts aimed at improving characteristics of capacitive

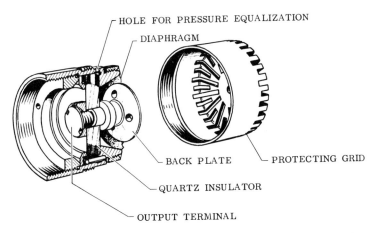

HOLE FOR PRESSURE EQUALIZATION
DIAPHRAGM
BACK PLATE
PROTECTING GRID
QUARTZ INSULATOR
OUTPUT TERMINAL

Fig. 11-2. Internal construction of typical capacitive sound-pressure transducer (courtesy of A/S Bruel & Kjaer.).

Fig. 11-3. Capacitive sound-pressure transducers (with integral cathode followers) (courtesy of A/S Bruel & Kjaer).

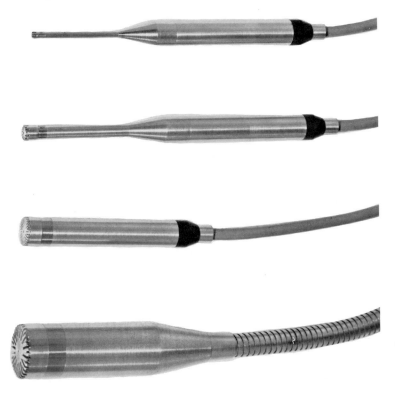

sound-pressure transducers are those directed towards eliminating the necessity for a high polarization voltage and increasing the steady-state capacitance. One approach uses a metallized 0.5 mil plastic foil which is permanently prepolarized by processing it through a high-voltage electrostatic field at high temperature. The use of a solid dielectric also permits closer spacing between the moving electrode (metallized surface on outside of film) and a perforated stator plate.

A different output-circuit design is used in the transducer shown in Figure 11-4, a water-cooled version of a capacitive sound-pressure transducer which incorporates a fixed inductance so connected to the capacitor represented by the transduction element that a tuned circuit is formed. The circuit constants are selected for resonance of the tuned ("tank") circuit at a radio frequency of about 700 kHz. Changes in the capacitance due to sound-pressure changes cause corresponding variations in the impedance of this circuit due to detuning from a resonance condition. A link-coupled low-impedance coaxial cable connects the transducer to an oscillator/detector unit which converts the impedance change into a d-c output.

Fig. 11-4. Water-cooled RF tank-circuit type capacitive pressure and sound-pressure transducer (courtesy of Photocon Research Products).

The sensing diaphragm of the transducer, which acts as the "rotor" plate of the capacitive transduction element, is grounded by its rim attachment to the case. The stator plate is insulated from the case. The water-cooled version, usable for sound-pressure measurement in the presence of ambient temperatures to 450°F, uses two parallel diaphragms connected together with a short pin to permit flow of cooling water between the diaphragms. The inner diaphragm is the grounded electrode. Its deflection for full-range sound pressure is approximately 0.3 mil. Mounting errors due to mounting-thread stretching are reduced in an alternate version of this transducer design which is constructed in two sections with the transducer proper threaded into a bushing which provides the mounting thread. This allows the strain-sensitive diaphragm portion of the transducer to protrude below the mounting thread.

11.3.2 Electromagnetic sound-pressure transducers

Devices incorporating this transduction principle are constructed with a moving coil concentric to a permanent magnet. Often referred to as "dynamic microphones," they are not commonly used for sound-pressure measurements because of their inherently limited frequency range.

11.3.3 Inductive sound-pressure transducers

This type of transducer uses a ferromagnetic sound-pressure-sensing diaphragm to change the self-inductance of a coil when the gap between diaphragm and coil is changed by deflection of the sensing element. The construction of one typical design incorporating a compensating coil as the second arm of an a-c bridge circuit is similar to that of the displacement transducer illustrated in Figure 3-7, except that a diaphragm is permanently mounted to the case in front of the sensing coil. The diaphragm in turn is protected by a cap which has a small opening (port) to admit sound pressure to it. The port can be plugged with a porous (sintered) filter plug to protect the diaphragm from dust and thermal radiation while not substantially reducing the frequency range of the transducer.

11.3.4 Piezoelectric sound-pressure transducers

The design and operation of sound-pressure transducers utilizing piezoelectric crystal transduction elements is essentially the same as that of the piezoelectric pressure transducers as described in 10.3.3. Ceramic as well as quartz crystals are used in modern sound-pressure transducers.

A typical quartz-crystal transducer is illustrated in Figure 11-5. This transducer is intended for flush-mounted installation in the wall of a duct, vessel, or chamber. It is provided with a miniature coaxial cable connector. Natural quartz crystals provide less output but better high-temperature performance than most ceramic crystals. However, some ceramics have been especially developed for use in transducers which have to operate over a wide range of ambient and measured-fluid temperatures.

Fig. 11-5. Quartz-crystal piezoelectric sound-pressure transducer (courtesy of Kistler Instrument Corp.).

Piezoelectric (and other) sound-pressure transducers are somewhat sensitive to vibration since the sensing element can also respond to acceleration forces. One manner of partially cancelling such vibration effects in a piezoelectric transducer is to mechanically connect a second piezoelectric element to the sound-pressure element in the same case. The two elements are back-to-back and are so interconnected that output due to vibration of the sensing element will be partially cancelled by the output due to the same vibration of the compensating element. A transducer incorporating this feature is shown in Figure 11-6. The transducer is provided with a perforated cap mounted over the diaphragm.

Similar caps are used on most sound-pressure transducers not only as mechanical protection for the diaphragm but also to provide some control over the transducer's directivity characteristics.

Many piezoelectric sound-pressure transducers have sealed cases, primarily to avoid reducing internal impedances by leakage paths due to moisture. This tends to increase their ambient-pressure error when used, e.g., at varying high altitudes.

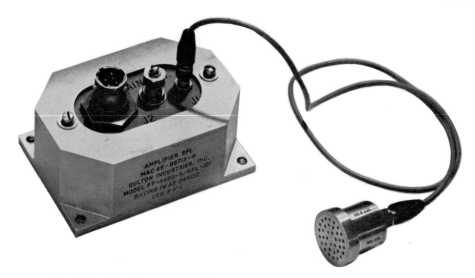

Fig. 11-6.   Vibration-compensated piezoelectric sound-pressure transducer with amplifier (courtesy of Gulton Industries, Inc.).

11.3.5 Reluctive sound-pressure transducers

A few reluctive-transducer designs, similar to those discussed in 10.3.5, have been used for sound-pressure measurement. Diaphragm as well as twisted-Bourdon-tube sensing elements are utilized in such transducers. The frequency range for flat response of reluctive sound-pressure transducers is limited by the natural frequencies of the sensing elements employed. These frequencies are usually lower than those inherent in most capacitive and piezoelectric transducers. Frequency response is also limited by the carrier (excitation) frequencies commonly provided for reluctive transducers. The upper limit of flat response is normally below the excitation frequency.

11.4 Performance and Applications

11.4.1 Specification characteristics

The characteristics which should be specified for sound-pressure transducers are quite similar to those specified for pressure transducers. However, some important differences exist, especially in the manner of specifying performance characteristics and in the type of design characteristics which must be emphasized for transducers having the relatively low output amplitude and high output impedance typical of sound-pressure transducers.

11.4.1.1 Mechanical design characteristics. An outline drawing should be used to define overall case configuration and dimensions as well as details of mounting provisions, mounting threads or flanges, and all necessary mounting dimensions. If a cap is used over the sensing element, its overall dimensions should be shown and its construction, including its perforations, should be described. It should be noted that the transducer's configuration, size, and mountings seriously affect its frequency response.

The location and type of electrical connector and any pneumatic or cooling-water connecting fittings should be stated. If the transducer is furnished with an integral connecting cable, the connector at the end of the cable should be defined. A description of the necessary mating connector should be added.

Specifications should include descriptions of materials used for case, sensing element, and transduction element, and of the manner of case sealing if employed. Limitations on constituents and contaminants of the atmosphere ambient to case and sensing element should be stated. A statement of equivalent volume (due to the compliance of the sensing element) is desirable. Any enclosed or semienclosed volume associated with the sensing element should be described in detail because of its effect on transducer frequency response.

Rated mounting force or torque on case mountings and any pressure connections as well as transducer weight and its tolerances should be shown. Minimum nameplate information should be detailed, including nomenclature (e.g., "Inductive Sound-Pressure Transducer"), manufacturer's name, address, and part number, serial number, range, sensitivity, and identification of all external electrical connections as to their function.

If the transducer is to be used with a separate connecting cable, this cable should be described as to type, length, maximum operating temperature, and connectors.

11.4.1.2 Electrical design characteristics. Specifications should include the excitation voltage, frequency (if a-c), and current or power of the transducer (including any integrally packaged circuitry). If the transducer is only usable with one type of signal-conditioning and excitation equipment, the specifications of such equipment should be included. Other requirements to be shown include insulation resistance, output impedance (transducer capacitance and shunting resistance for piezoelectric transducers), output noise, triboelectric (mechanical-friction-induced) cable noise, internal grounding of any output or excitation connections, and the load impedance (with tolerances) at which specified performance characteristics apply.

11.4.1.3 Performance characteristics comprise dynamic characteristics at room conditions and under specified environmental conditions. Since sound pressure is inherently a dynamic pressure, static characteristics as applicable to most other transducers are not shown. All characteristics should be shown without reference to any separately packaged signal-conditioning equipment and separate connecting cable, unless a statement is included that the transducer must be used only with such equipment and cable.

Range is expressed in terms of sound-pressure level (*SPL* or L_p), e.g., "Range: 140 to 180 db SPL re 0.0002 dyne/cm²." Range has sometimes been expressed in psi, microbars, dyne/cm², or newtons/m².

Output is usually expressed in terms of sensitivity (at a specified excitation if not a "self-generating" transducer). The sensitivity has often been shown as "open-circuit" sensitivity. This is unrealistic because any meters used to measure transducer output have an input impedance (representing the load impedance for the transducer) less than infinity. Performance characteristics must always be shown in such a manner that they can be verified by use of available test equipment. It is, therefore, usually necessary to specify a load impedance (with tolerances) as part of the electrical design characteristics (e.g., "100 megohms, minimum, resistive, shunted by 50 pF, maximum"). Output characteristics are then applicable only at this specified load impedance and readout equipment can be selected on this basis. When specifications do show an "open-circuit" sensitivity and no load impedance values are shown, the loading effect of the impedance presented by the associated measuring system must be calculated and applied.

Sensitivity of sound-pressure transducers is normally stated as *sensitivity level* (even when called "sensitivity," as has been customary). Sensitivity level is expressed in decibels referred to a reference sensitivity of 1 V per dyne per square centimeter (or 1 V per μbar), e.g., "Sensitivity: −60 db re 1 V per dyne/cm²" or "Sensitivity: −60 db re 1 V/μbar." This sensitivity level is determined by use of the following equation.

$$\text{Sensitivity Level in db} = 20 \log_{10} \left(\frac{\text{output in volts rms}}{\text{effective sound pressure in dyne/cm}^2} \right)$$

$$= 20 \log_{10} \left(\frac{\text{output in volts}}{1 \text{ volt}} \right) - (\text{applied SPL in db}) + 74 \text{ db}$$

The +74 db is the level of the ratio 1 dyne/cm² over 0.0002 dyne/cm².

The sensitivity of a piezoelectric sound-pressure transducer can be stated as *voltage sensitivity* as above or as *charge sensitivity*, e.g., "Sensitivity: _____ db re 1 picocoulomb per dyne/cm²."

Sensitivity has also been expressed (as ratio rather than level) in millivolts per microbar, picocoulombs per microbar, or either millivolts or picocoulombs per psi, if range was specified in these terms. Both quantities in each ratio of units should be shown as either rms, zero-to-peak, or peak-to-peak values. The full-scale output can be shown in lieu of sensitivity. A nomograph suitable for correlation of various output characteristics is shown in Figure 11-7.

Most sensitivity specifications represent a "nominal" sensitivity, and transducers with the same part number but different serial numbers will have sensitivities slightly

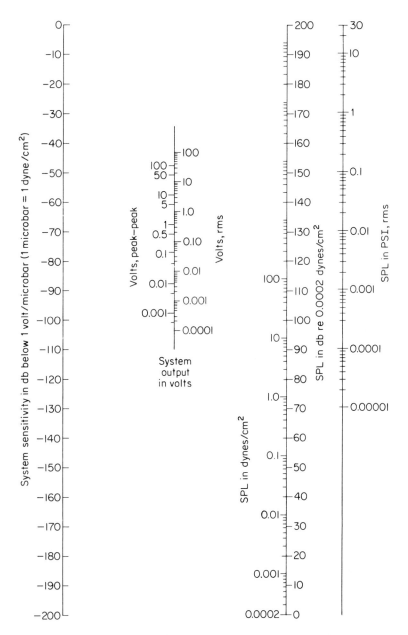

Fig. 11-7. Nomograph for sound-pressure-level calculations.

different from the specified value. This necessitates the inclusion of tolerances with the nominal value, e.g., "Sensitivity: -60 ± 5 db re 1 V/μbar." Sensitivity specifications are best shown as applicable at a specific sound-pressure level and at a specific measurand frequency and direction of sound incidence. Further specifications for linearity and frequency response (at the same direction of sound incidence) then show how sensitivity can vary over the measurand amplitude and frequency range.

Linearity, such as end-point, independent, or least-squares linearity is usually expressed in decibels, e.g., "Independent Linearity: within ± 3 db," as a tolerance on the actual sensitivity (which must be within the sensitivity tolerance of the nominal sensitivity). This is why sensitivity should be stated as applicable at a specific sound-pressure level. Linearity is always "amplitude linearity" over the specified range. It should be expressed in terms of output, e.g., as tolerances on the specified sensitivity level. A well-defined statement of linearity should include the measurand frequency at which it is applicable. Nonlinearity in an acoustic system generally produces harmonic distortion.

Frequency Response specifications should state whether pressure response or free-field response is meant and, for the latter, should show the direction of incidence (grazing, random, or normal-incidence response). Preferably, the frequency response should be referred to a specific amplitude and frequency. The flatness of frequency response is expressed in db as a tolerance on the sensitivity, over a specified frequency range, e.g., "Free-Field Normal-Incidence Frequency Response: within ± 3 db from 25 to 8000 Hz (as referred to the response at 100 Hz and at 160 db SPL)."

The pressure frequency response is generally equal to the free-field random- or grazing-incidence frequency response at wavelengths which are long compared to the maximum dimension of the transducer. Frequency response can sometimes be calculated from the transducer's response to transient sound pressure, from its mechanical properties, or from its geometry, when the latter is very simple. If determined in this manner it is stated as "Calculated Frequency Response." If the transducer is intended for use with a measured fluid other than air, and frequency response as well as other characteristics are stated as applicable for this fluid, a note to this extent must be included in the specification.

Overload. The maximum sound-pressure level in a certain band of frequencies which can be applied to the transducer without causing permanent changes in its performance is sometimes specified, expressed in units of range, e.g., "Overload (without subsequent performance change): 190 db, 50 to 10,000 Hz."

Directivity can be specified as directivity factor or as solid angle symmetrical about (bisected by) the transducer's principal axis. It is frequently shown in manufacturer's literature as one or more typical directional response patterns (see Figure 11-8 for examples).

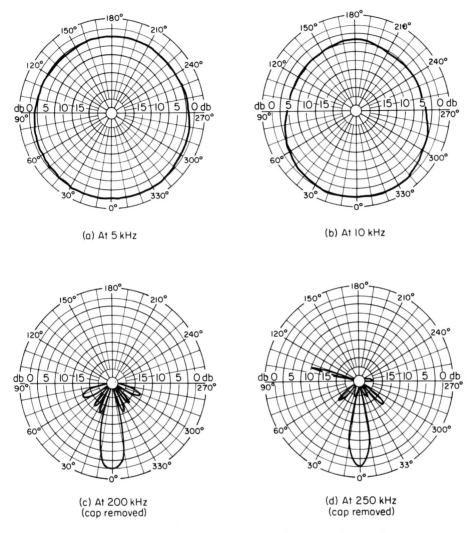

(a) At 5 kHz

(b) At 10 kHz

(c) At 200 kHz
(cap removed)

(d) At 250 kHz
(cap removed)

Fig. 11-8. Typical directional response patterns for a capacitive sound-pressure transducer (courtesy of A/S Bruel & Kjaer).

Threshold is sometimes specified. Resolution is not shown for sound-pressure transducers since all existing designs have continuous resolution.

Thermal Effects. The operating temperature is stated for the transducer's ambient environment and, if it is different (e.g., for a water-cooled transducer), for the measured fluid. A limit on exposure time may have to be included. Where output is stated as sensitivity, the only steady-state temperature error to be specified is *thermal sensitivity shift*, expressed in db, e.g., "Thermal Sensitivity Shift ($-40°$C to $+110°$C): ± 5 db."

Where transducer output is shown in other units of sensitivity or as full-scale output, thermal sensitivity shift must be expressed in the same units and any thermal zero shift that may occur in certain transducer designs should additionally be specified, also in the same output units. Alternately, one or both types of shift can be expressed as a percentage of sensitivity (ratio-type only, not level) or of full-scale output. Thermal zero and sensitivity shifts can be combined as temperature error.

Requirements for a limit on temperature gradient error, also expressed in units used for output specifications, should be considered when the transducer can be exposed to step changes in ambient or measured-fluid temperature (within the operating temperature range) in its application.

Ambient-Pressure Error is expressed as allowable sensitivity shift with varying ambient pressure, e.g., "Ambient-Pressure Sensitivity Shift (15 to 0.5 psia): ± 3 db, maximum." Sound-pressure transducer performance can be affected by changes in ambient pressure regardless of case sealing. Normal variations in ambient pressure at the same altitude above sea level usually do not cause measurable performance changes; however, output shifts can be encountered at high altitudes, especially in airborne transducers.

Other Atmospheric Effects. Limitations may have to be specified on the effect on transducer performance of constituents and contaminants of the ambient air or of measured fluids other than air. Allowance should be made for exposure, of specified duration, to humidity of various levels, salt atmosphere (when the application is near an ocean), soot, sand, dust, particles of corrosive material, etc.

Vibration Error tolerances have been specified as output equivalent to a certain (apparent) sound-pressure level. The tolerances are stated as equivalent SPL (expressed in db) per rms g of the vibration applied along a specified axis over a specified vibration-frequency range, e.g., "Vibration Error: 0.04 db, maximum, per g, any axis, 5 to 4000 Hz." A maximum acceleration level is sometimes added, e.g., "up to 50 rms g." Vibration error can also be expressed as maximum equivalent sound-pressure level in db, with a specific, separately described vibration-vs-frequency program applied along a specific axis. A better way to express vibration error may be to specify it as output due to equivalent SPL per g shown as sensitivity level equal to

$$20 \log_{10} \left(\frac{\text{apparent rms sound pressure due to vibration}}{\text{rms applied vibration amplitude}} \right)$$

Other Environmental Effects. The specification of tolerances should be considered for effects on performance after ("nonoperating") exposure to other environmental conditions such as shock and nuclear radiation and during or after exposure to electromagnetic fields and interference.

Environmental Effects on Frequency Response. During the discussion of environ-mental-performance specifications, emphasis was placed on sensitivity shift or output at a reference frequency. Environments such as temperature can also affect the frequency response. When such requirements are considered important enough to warrant a test setup for environmental frequency-response tests, additional specification requirements can be shown accordingly.

Performance Reliability. Stability can be specified by tolerances on maximum allowable sensitivity shift after a specified storage period (storage life) and after a specified number of temperature cycles between two specific temperatures (thermal stability).

11.4.2 Considerations for selection

Since a limited number of transducer designs are available for remote and unat-tended measurement of sound pressure (the principal application of such transducers), the selection of a specific design is much less laborious than for most other measur-ands.

Reluctive transducers, which usually have limited frequency response, have been used primarily in measurement systems where transducers for other measurands are also reluctive and suitable excitation (carrier) supplies and signal-conditioning mod-ules are more or less standardized and available and where their frequency range is considered adequate. High-performance inductive transducers are of fairly recent origin and require excitation and signal-conditioning equipment sufficiently com-plex to make the procurement of the entire measuring system from the same manufac-turer advisable.

Capacitive transducers have been available from several suppliers for a long enough time to build up a considerable level of experience in their use among per-sonnel involved with sound-pressure measurements. Such transducers are usually furnished with an integral cathode follower which allows relatively long cables and low impedance. When the sound-pressure range, required frequency response, and directivity have been established for a given measurement, a specific design can usu-ally be found to satisfy the measurement requirements. Some capacitive transducer designs have been found sufficiently stable and error-free to serve as laboratory standards.

Typical performance characteristics for capacitive sound-pressure transducers include very high frequency response (flat to more than 50 kHz for some designs), upper range limits slightly less than 180 db SPL, and lower range limits below 40 db SPL.

The associated power supply must be capable of providing the necessary polari-zation voltage (about 200 V d-c) for the transducer, as well as filament and plate power for the cathode follower. Hence, the interconnecting cable typically has six or seven conductors including output and ground wires. The presence of a 200-V potential limits applications of these transducers at very high altitudes where corona discharge may occur at any exposed electrical connection (e.g., inside the vented transducer case).

Piezoelectric transducers are somewhat more versatile and easier to apply than other types of sound-pressure transducers. Their performance characteristics include a range capability from a lower limit of about 80 db SPL to an upper limit near 200 db SPL and frequency response flat to 15 kHz. Many of their design variations are usable in adverse environments.

Since piezoelectric transducers are "self-generating," they do not require a power supply. Their output impedance, though usually lower than that of capacitive types, is very high (greater than 20,000 megohms), and variations in cable capacitance and insulation resistance can have substantial effects on transducer output. A number of such transducers are available with an integrally packaged emitter follower (using a transistor) or source follower (using a field-effect transistor). Use of such a device can lower the output impedance of the transducer package to around 100 ohms. If the transducer is used with a separate cathode, emitter, or source follower or just a separate amplifier, the interconnecting cable should be a rugged, flexible, low-noise, miniature coaxial cable with reliable connectors securely attached. Longer cables can be used with charge amplifiers than with voltage amplifiers. After mating, the connectors can be covered with a moisture-proofing compound to prevent any lowering of insulation resistance by moisture.

11.5 Calibration and Tests

11.5.1 Calibration

A number of different calibration methods involving different types of equipment are used on sound-pressure transducers. All calibrations are dynamic calibrations in that the sound pressure varies with time in a specified manner. In virtually all calibration methods sound pressure is made to vary sinusoidally at one or more specific frequencies.

11.5.1.1 The reciprocity method is a primary technique; i.e., it does not rely on a reference transducer to determine the measurand level applied to the transducer being calibrated. The basis of the reciprocity method is the reciprocity theorem for linear, bilateral, passive four-terminal electrical networks. This theorem states that the ratio of a voltage applied across any two terminals to the resulting short-circuit current flowing through the other two terminals (the "transfer impedance") is equal in magnitude and phase to the ratio that would exist were the positions of voltage source and current meter interchanged. Hence, the reciprocity calibration method involves operation of the transducer as a bilateral device, i.e., as emitter as well as receiver of sound.

The *self-reciprocity method* has been used at times, primarily as a research procedure. It requires a large rigid plane wall and an electronic switch capable of alternately connecting the transducer under test to a constant-current a-c supply and an amplifier. Switching time must be extremely short. The transducer is first used as an emitter by connecting it to the constant-current generator. The quasi-steady-state current is measured while the transducer generates a wave train at a specific frequency. The wave travels to the wall and is reflected by it toward the transducer. While the wave train travels in this manner, the transducer is switched into the receiving mode by disconnecting it from the generator and connecting it to the amplifier. When the reflected wave train arrives at the transducer, the transducer's output voltage is measured. The input impedance of the amplifier must be high enough to allow measurement of a near-"open-circuit" output voltage. The arriving sound pressure is the same as the emitted sound pressure. The sensitivity of the transducer, the logarithmic ratio of rms output (in volts) to rms sound pressure (in dyne/cm^2), can then be calculated from knowledge of the ratio of measured output voltage to measured generator current, the distance between wall and transducer, the frequency, and the characteristic impedance of the air.

The *auxiliary-transducer reciprocity method* does not require a reflecting wall or electronic switch. The transducer under test is used as sound emitter and the current from a constant-current source is measured. An auxiliary transducer, a known distance away from the first transducer, is used as sound receiver whose "open-circuit" output voltage is measured. By placing the two transducers successively in the same sound field their two sensitivities are next compared. The sensitivity of the test transducer can then be calculated from knowledge of the ratio of the two sensitivities, the ratio of the output voltage to generator current first measured, the distance between the two transducers in the first measurement, the frequency, and the characteristic impedance of the air. Neither reciprocity method is commonly used for the routine calibration of sound-pressure transducers. Either is used, at times, for very precise calibrations, on a sampling basis, or for the calibration of reference transducers used in comparison calibrations.

11.5.1.2 Pressure calibrations are considerably simpler. A commonly used method requires a *pistonphone* as pressure-application equipment. This is a small chamber equipped with an oscillating piston whose displacement is measurable. The pistonphone establishes a known sound pressure at a specific frequency in the chamber to which the transducer is closely coupled. Similar operation for somewhat less precise calibrations is provided by an "artificial voice" (or "artificial mouth") containing an oscillating source of sound pressure comparable to a loudspeaker and equipped with a coupler for the transducer to be calibrated. A device of this type is illustrated in Figure 11-9. It is best used after its characteristics have been established by using it on a reference transducer.

The relative pressure frequency response of capacitive sound-pressure transducers can be obtained by use of an *electrostatic actuator* (Figure 11-10), a device which

Fig. 11-9. "Artificial voice" for sound-pressure transducer calibration (courtesy of A/S Bruel & Kjaer).

Fig. 11-10. Electrostatic actuator for pressure-response tests on capacitive sound-pressure transducers (courtesy of Bruel & Kjaer).

provides an auxiliary external electrode that permits the application of known electrostatic forces to the transducer diaphragm so that a simulated sound pressure is established for calibration purposes.

When an accurate pressure calibration has been performed on one of a limited number of laboratory-standard type sound-pressure transducers, its free-field calibration can be established on the basis of a comparison calibration in an anechoic chamber or by referring to published data, such as U.S.A. Standard S1.10-1966 (American Standard for the Calibration of Microphones) rather than by the reciprocity method.

11.5.1.3 Secondary calibration techniques (comparison calibrations) are used more frequently than primary methods for routine calibrations of sound-pressure transducers. Such techniques use a reference transducer system, whose performance characteristics have been sufficiently well-established for this purpose, to measure and indicate the sound pressure applied to the transducer undergoing tests. A laboratory-standard microphone, with associated amplifier and readout equipment, is commonly used for this purpose.

Free-field (grazing or normal-incidence) comparison calibrations are performed in an *anechoic room* (anechoic chamber), whose boundaries absorb all the sound incident on them. Diffuse-field (random incidence) comparison calibrations are accomplished in a *reverberation room* (reverberation chamber) which is designed to produce such a field.

It should be noted that the free-field response usually approximates the pressure response at low and medium frequencies when the sound wavelength (in the measured medium) is large compared to the dimensions of the transducers. The two types of response differ only at higher frequencies because of diffraction of the sound field by the transducer and its associated mounting hardware.

Calibration methods using response to a step change in pressure, such as obtained in a shock tube, are rarely used for sound-pressure transducers. They find an occasional application in the testing of transducers having low sensitivity.

11.5.1.4 Performance verification. Dynamic calibrations are performed on sound-pressure transducers for determination of sensitivity, linearity, and frequency response and for the verification of specification compliance of these characteristics.

Sensitivity is determined at one sound-pressure level and one frequency. If nominal sensitivity or sensitivity with tolerances was specified at a reference frequency and reference measurand level, these parameters are used during calibration. If no such reference parameters were stated, a frequency is arbitrarily selected within the lower portion of the frequency range where flat response can be assumed, usually between 100 and 250 Hz, and a sound-pressure level is chosen within the lower half of the range.

Frequency response is determined at the previously chosen measurand level over a frequency range slightly larger (at both ends) than the frequency range over which flatness tolerances were specified. The frequency-response and sensitivity tests constitute those calibration tests normally performed on each transducer. Most other tests are performed on a sampling or qualification basis only. Random-incidence tests are often performed when no specific incidence was specified. Grazing- and

Fig. 11-11.   Typical frequency-response plot for sound-pressure transducer.

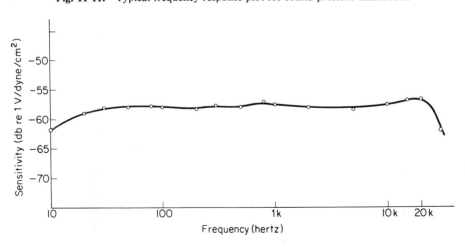

normal-incidence response, however, are either specified or need to be determined (in the absence of specifications) quite frequently. Figure 11-11 illustrates typical frequency-response test results.

Linearity can be determined at a specific frequency by application to the transducer of sound-pressure levels at discrete increments over the full specified range. Frequency-response tests at two or more sound-pressure levels can be performed instead of linearity tests.

11.5.1.5 Sound-pressure transducer calibration records should always show the calibration method used, a description of the reference transducer system in a comparison calibration and of the auxiliary transducer(s) used for a reciprocity calibration (except when the self-reciprocity method is used), barometric pressure, temperature, and humidity, the characteristics of the cable, cathode- or emitter-follower (if not integral with the transducer), amplifier, and readout device used with the transducer, the location of the transducer and the sound-source in a reverberation or anechoic chamber, the sound-pressure level and type of incidence for frequency-response tests, the frequency for linearity test, the frequency and sound-pressure level for sensitivity tests, and model and serial number of any ancillary equipment used.

11.5.2 Other tests at room conditions

11.5.2.1 Visual and electrical inspections are normally performed prior to calibration. The former consist of dimensional checks, nameplate verification, visual determination of acceptable workmanship, connector-mating check of any associated connecting cable, dimensional verification of mountings and any pressure connections and perforated cap, and, on a sampling basis, a weight determination. Electrical inspections include an insulation-resistance test (unless the signal-return lead is grounded internally to the case), determination of capacitance and shunting resistance of some transducers and of any nonintegral connecting cable, and a verification of correct external electrical connections (if more than two).

11.5.2.2 Directivity tests are special free-field calibrations sometimes performed during a qualification test by varying the transducer attitude (or source position) in such a manner that angle of incidence in two planes is varied in discrete increments.

11.5.2.3 Mounting-error tests are perfomed (if applicable) on a sampling basis by measuring transducer sensitivity before and after using the specified mounting force or torque to simulate its typical installation.

11.5.2.4 Case-seal tests can be performed (usually during qualification) by immersing a transducer for which a sealed case is specified into a beaker of water and checking for any air bubbles emanating from a leak in the case.

11.5.2.5 Cable-noise tests are sometimes performed on cables associated with piezoelectric transducers. One method utilizes a small weight clamped over the middle of the cable which is held fixed at two points so that it is suspended limply between them. After the cable is connected to a shunt capacitor and an emitter follower or amplifier, the weight is lifted, then dropped, and any (triboelectric) noise created by the resulting cable bounce is measured.

11.5.3 Environmental tests

Whenever environmental tests on sound-pressure transducers are comparison tests, the characteristics of the reference transducer while exposed to this environment must be accurately known. Environmental tests are only performed when corresponding environmental performance is defined in a specification. Test records should include those data which were listed as necessary on a calibration record.

11.5.3.1 Ambient-pressure tests. During such tests the transducer and the reference transducer are inserted simultaneously or alternately into a pressure-calibration coupler cavity whose static pressure is varied over the ambient-pressure range at discrete intervals. By use of this technique, sensitivity, linearity, and frequency response (often not over the full frequency range) can be determined at various ambient pressures.

11.5.3.2 Temperature tests are performed while either the transducer alone or the entire transducer system is stabilized at the specified limits of the operating temperature range. The transducer and reference transducer are inserted, preferably simultaneously, into the pressure-calibration coupler cavity. When feasible, only the transducer in test is heated or cooled. However, it is almost impossible to keep the coupler and the measured fluid from stabilizing at the same temperature. It is difficult to thermally isolate the reference transducer and even the pressure source sufficiently well from the transducer. Hence, all or most of the calibration setup is placed in the temperature chamber, or temperature is only applied to the transducer while the temperatures of the other devices are monitored. Tests in which only the measured-fluid temperature is changed are still more difficult to perform properly.

Temperature tests can be used to determine thermal sensitivity and thermal frequency response (as a pressure response). Certain other characteristics can be measured individually, such as temperature effects on transducer capacitance and shunting resistance, on cable capacitance, and on insulation resistance.

Temperature-gradient tests can be performed by rapid immersion of the (sealed) transducer in a beaker filled with hot water or oil and by monitoring transducer output in the absence of applied sound pressure. Such tests are more difficult when the transducer case is vented. Some success has been obtained by switching on a source of radiant heat placed near the transducer and radiating this heat upon the sensing element (e.g., by discharging a photographic flashbulb in front of the transducer).

11.5.3.3 Vibration tests are performed to determine transducer output due to vibration. They do not involve a comparison calibration. Care must be taken not to report any air-pressure variations ("windage" effects) seen by the transducer sensing element due to rapid back-and-forth motion of the transducer as vibration error. Equal care must be used not to confuse shaker noise sensed by the transducer with vibration error. During vibration tests such erroneous indications can be minimized by placing a cover over the sensing element (but not in contact with it).

11.5.3.4 Other environmental tests on sound-pressure transducers, such as shock tests, salt-atmosphere, and nuclear-radiation tests, are usually nonoperating tests. The transducer's performance is verified only after exposure to these environments. Electromagnetic interference tests, such as prescribed in military specifications, are normally performed on the entire transducer system. Humidity tests are often performed as operating tests.

11.5.4 Life tests

Transducer stability over specified temperature cycles (thermal stability) can be determined by repeated calibrations at high, room, low, and again room temperature.

11.6 Special Measuring Devices

11.6.1 Underwater sound detectors

Devices very similar to some of the sound-pressure measuring transducers previously described are widely utilized in the detection of sound under water. Some of these are used for pure measurements of sound pressure. Others, however, have as their primary purpose the detection of objects in large bodies of water and the determination of the location and nature of these objects. The equipment or method used for such purposes is usually designated as sonar, an acronym derived from "*so*und *na*vigation *a*nd *r*anging."

The *hydrophone* is an underwater microphone, commonly used as a listening device. Analysis of the frequencies and waveshapes of the signals it provides can yield information about the nature of the object from which the sound emanates. Two or more hydrophones can also be used for obtaining information about the location of the object (*ranging*) by comparison of their electrical output signals.

Ranging is more commonly accomplished by using an *underwater sound projector*, a device used to produce sound waves under water, in conjunction with a hydrophone. Frequently, the functions of transmission and reception are combined in a single (reciprocal or bilateral) device which could be referred to as an underwater sound transceiver. This equipment is used as *echo-ranging sonar*. Its usual operating mode involves the generation of a pulse of sound energy ("*ping*") whose travel time to and from an object (*target*) is measured by determining the difference in time between projection of the "ping" and reception of its "echo" from the target. The travel time, combined with knowledge of the directional characteristics of the transducer, provides locational (range) information about the target. Further analysis of the "echo" can yield additional information about the nature of the target. When an underwater sound transceiver is used for echo ranging, it is switched continuously between its "transmit" and "receive" modes of operation. Piezoelectric and magnetostrictive devices are most commonly used as hydrophones and underwater sound projectors and transceivers.

In addition to detection of objects such as submarines and fish under water, sonar is used for ocean-floor mapping and other oceanographic research.

11.6.2 Sound-level meters

Sound level is, by definition, a *weighted* sound-pressure level—an attempt to approximate a subjective quantity, i.e., a measure of loudness (of pure tones) as perceived by the human ear. A sound-level meter consists of a microphone, an amplifier, standard (per U.S.A. Standard S1.4-1961) weighting networks, a graduated attenuator, and an output-indicating meter. The directional response of the microphone is chosen as close to spherical (omnidirectional response) as possible for the usual measurement of diffuse sound fields and as less than hemispherical for frontal sound fields. The weightings, referred to merely as A, B, and C, denote different frequency-response characteristics of the instrument. Sound-level range is always referred to a sound pressure of 10^{-4} μbar. A typical range of a sound-level meter is 24 to 150 db. This measuring equipment is widely used in noise measurements, primarily those having to do with human comfort. It is usually portable and battery-powered so that it can be moved rapidly to various locations by its operator. Typical applications of sound-level meters involve measurements of the loudness of noise near operating machinery, in industrial areas, in residential and commercial buildings, in and near airports, in ground-transportation vehicles and aircraft, in entertainment halls, and on city streets.

BIBLIOGRAPHY

1. *Bruel & Kjaer Technical Review*, Naerum, Denmark: Bruel & Kjaer (quarterly periodical).

2. Cook, R. K., "Absolute Pressure Calibrations of Microphones," *N.B.S. Research Paper RP1341*, Washington D.C.: National Bureau of Standards, 1940.

3. Beranek, L. L., *Acoustic Measurements.* New York: John Wiley & Sons, Inc., 1949.

4. Randall, R. H., *Introduction to Acoustics.* Cambridge, Massachusetts: Addison-Wesley Press, 1951.

5. Beranek, L. L., *Acoustics*. New York: McGraw-Hill Book Company, 1954.

6. Fischer, F. A., *Fundamentals of Electroacoustics*. New York: Interscience Publishers, Inc., 1955.

7. "Acoustical Terminology," *U.S.A. Standard S1.1-1960*, New York: U.S.A. Standards Institute, 1960.

8. "Preferred Frequencies for Acoustical Measurements," *U.S.A. Standard S1.6-1960*, New York: U.S.A. Standards Institute, 1960.

9. "Specification for General-Purpose Sound Level Meters," *U.S.A. Standard S1.4-1961*, New York: U.S.A. Standards Institute, 1961.

10. "Method for Physical Measurement of Sound," *U.S.A. Standard S1.2-1962*, New York: U.S.A. Standards Institute, 1962.

11. Sessler, G. M. and West, J. E., "Self-Biased Condenser Microphone with High Capacitance," *Journal of the Acoustical Society of America*, Vol. 34, No. 11, p. 1787, November 1962.

12. Ziemer, R. E. and Lambert, R. F., "Shock-Wave Transducer Calibration," *Journal of the Acoustical Society of America*, Vol. 34, No. 7, pp. 987–988, July 1962.

13. "Method for the Calibration of Microphones," *U.S.A. Standard S1.10-1966*, New York: U.S.A. Standards Institute, 1966.

14. Koidan, W., "A New Standard for the Calibration of Microphones," *The Magazine of Standards*, pp. 141–144, May 1966.

15. Ohme, W. E., "Loudness Evaluation," *Hewlett-Packard Journal*, Vol. 19, No. 3, November 1967.

16. "Laboratory Standard Microphones, Specifications For," *U.S.A. Standard S1.12-1967*, New York: U.S.A. Standards Institute, 1967.

17. Keast, D. N., *Measurements in Mechanical Dynamics*. New York: McGraw-Hill Book Company, 1967.

18. "Specifications and Tests for Piezoelectric Pressure and Sound-Pressure Transducers," *ISA Standard S37.10*, Pittsburgh: Instrument Society of America, 1969.

12 Speed and Velocity

12.1 Basic Concepts

This chapter deals primarily with two types of transducers: linear-velocity transducers and angular-speed transducers (*tachometers*). The output of the former contains information about direction as well as magnitude, whereas the output of a tachometer usually contains only magnitude information.

12.1.1 Basic definitions

Speed is the magnitude of the time rate of change of displacement; it is a scalar quantity.

Velocity is the time rate of change of displacement with respect to a reference system; it is a vector quantity.

Average speed is the magnitude of the average velocity vector.

Average velocity is the total displacement divided by the total time taken by this displacement.

Instantaneous velocity is the first derivative of displacement.

12.1.2 Related laws

Displacement, velocity, acceleration

Linear (Translational)   Angular (Rotational)

$$v = \frac{dx}{dt}$$ $$\omega = \frac{d\theta}{dt}$$

$$a = \frac{d^2x}{dt^2}$$ $$\alpha = \frac{d^2\theta}{dt^2}$$

where: x = linear displacement
θ = angular displacement
v = linear velocity
ω = angular velocity
a = linear acceleration
α = angular acceleration
t = time

Translation

$$v = \omega r$$

where: r = radius (of rotating member)
ω = angular velocity
v = linear velocity (translational)

Electromagnetic induction

$$e = -N\frac{d\phi}{dt}$$

where: e = induced electromotive force (in abvolts, 1 abvolt = 10^{-8} volt)
N = number of turns in coil
$\frac{d\phi}{dt}$ = change in magnetic flux (ϕ) per unit time (t)

Note: The negative sign in the equation indicates that the direction of induced emf opposes the change of flux that produced it.

12.1.3 Units of measurement

Linear velocity is expressed in **inches per second** or **centimeters per second.** It can also be shown in other units of length per unit of time.

Angular speed should be expressed, in accordance with the International System of Units, in **radians per second** (rad/s). However, it is normally shown in **revolutions per minute** (rpm).

12.2 Sensing Elements

12.2.1 Linear-velocity sensing elements

Linear-velocity transducers—usually referred to as just "velocity transducers"—are almost invariably of the electromagnetic type, in which a change of magnetic flux induces an electromotive force in a conductor. The flux change results from relative motion between a coil and a permanent magnet. In designs having a stationary magnet, the *moving coil* is the sensing element. Where the coil is fixed in the transducer case, the *moving magnet* is the sensing element. In either case the conductor cutting the flux has an electromotive force generated in it.

12.2.2 Angular-speed sensing elements

The sensing element of a tachometer is the *sensing shaft*. The shaft end can be provided with a coupling, but it can also be plain (a solid cylinder), with or without a radial hole, splined, serrated, square, slotted, threaded, or conical. Some of the shaft ends illustrated in Figure 3-1 are typical for angular-displacement as well as for angular-speed transducers. In a few noncontacting angular-speed-measuring systems a rotating member other than a shaft can act as sensing element.

12.3 Design and Operation

12.3.1 Electromagnetic linear-velocity transducers

The simplest form of this category of transducers consists of a coil in a stainless-steel housing and a coaxial cylindrical permanent magnet (core) attached to a shaft with a threaded end (Figure 12-1). The magnetic core moves freely within the coil

Fig. 12-1.   Coupled electromagnetic velocity transducer.

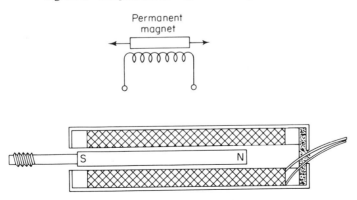

with the motion of the object to which the shaft end is mounted. The output from the coil in this coupled transducer is proportional to the velocity of the core.

A different fixed-coil design is shown in Figure 12-2. A permanent magnet is supported between two springs. Gold-palladium-alloy bearing rings are pressed on the ends of the cylindrical magnet to allow its vertical motion within a chrome-plated stainless-steel sleeve with a minimum of friction. The mechanical assembly is sealed within the coil-supporting bobbin by threaded retainers in the end caps of the transducer case. When the frequency of the magnet's oscillatory motion exceeds the rela-

Fig. 12-2. Fixed-coil electromagnetic velocity transducer (courtesy of Consolidated Electrodynamics Corp.).

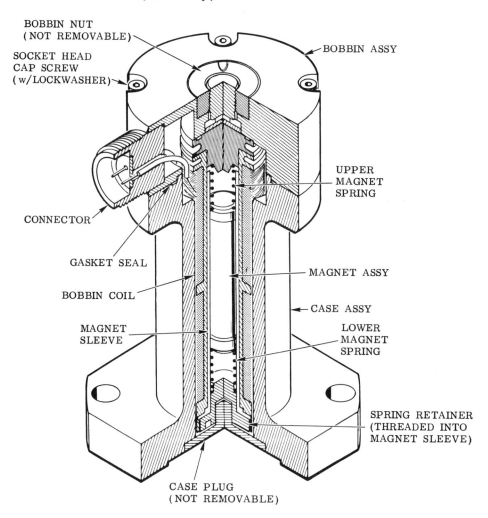

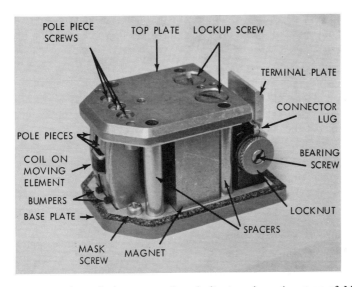

Fig. 12-3. Moving-coil electromagnetic velocity transducer (courtesy of MB Electronics).

tively low natural resonance (about 15 Hz) of the suspended system, the magnet remains in an essentially undisturbed position. The fixed coil, however, moves with the transducer case at the velocity of the measured object to which it is mounted. This results in relative motion between coil and magnet and causes the flux change necessary to obtain an output whose magnitude is proportional to velocity. Although the entire transducer case with its internally fixed coil moves about the stationary (suspended) magnet, this transducer appears as a moving-magnet design when the eye of the observer moves with the case.

A moving-coil design is shown in Figure 12-3. The coil is part of a moving structure pivoted at bearings at the opposite end of the transducer assembly. The coil moves within a magnetic field established by the pole pieces of a fixed permanent magnet.

12.3.2 Electromagnetic angular-velocity transducers

This type of transducer is sometimes used for attitude-control purposes, rarely for measurement. It is exemplified by a moving-magnet design with the magnet suspended on low-friction bearings permitting it to oscillate over a peak-to-peak arc of about 12 degrees within a fixed-coil assembly. A voltage proportional to the flux change due to magnet velocity is induced in the coil.

12.3.3 Electromagnetic angular-speed transducers

This category of widely used transducers includes tachometers whose output is either a varying d-c voltage or an a-c voltage varying in amplitude or frequency or both.

12.3.3.1 The D-C tachometer generator uses either a permanent magnet (*d-c magneto*) or a separately excited winding as its stator and a conventional generator winding on the commutator-equipped rotor. The output of a d-c magneto is between 3 and 7 V, that of the stator-winding type is between 10 and 20 V per 1000 rpm. The brushes necessitated by the commutator require maintenance. An advantage for certain applications is the indication of direction of shaft rotation since output polarity is dependent on it. This feature makes the d-c tachometer generator an angular-velocity transducer.

12.3.3.2 The A-C induction tachometer is essentially a variable-coupling transformer in which the coupling coefficient is proportional to rotary speed. When the primary (input) winding is excited by an a-c voltage, an a-c voltage at the excitation frequency is provided by the secondary (output) winding. The output amplitude varies with rotor speed. A squirrel-cage rotor is used in some of the devices. Others (*drag-cup tachometers*) use a cup-shaped rotor made of a high-conductance metal such as copper, copper alloy, or aluminum. Shaft rotation produces the shift of flux distribution on which the operation of the induction generator is based.

12.3.3.3 The A-C permanent-magnet tachometer uses the magnetic interaction between a permanent-magnet rotor and a stator winding to provide an a-c output voltage. The amplitude as well as the frequency of this *a-c magneto* are proportional to rotor speed. In some applications it is more desirable to convert the frequency of the generated voltage into a d-c output, such as by use of a discriminator which also allows making this d-c output bipolar from a null at a preselected frequency. Another advantage of operating on the frequency, rather than the amplitude, of the generated voltage is the relative freedom from effects of loading, temperature variations including those due to self-heating, and armature misalignments caused by vibration.

A different design uses a shaded-pole motor operated as a generator whose output originates from the shading winding.

12.3.3.4 Toothed-rotor tachometers. Tachometers using a ferromagnetic toothed rotor in conjunction with a transduction coil wound around a permanent magnet are the most commonly used group of frequency-output angular-speed transducers. The operating principle is illustrated in Figure 12-4. When the ferromagnetic (e.g., magnetic steel) tooth on the rotating shaft passes in proximity of the permanent magnet, the lines of flux of this magnet shift and cut across the coil winding thereby inducing an electromotive force in the coil. An output pulse is created once per each shaft revolution [Figure 12-4(a)]. The elapsed time between pulses is inversely proportional to shaft speed. With increasing speed the number of pulses per unit time increases and can be conveniently displayed as such on an EPUT (events-per-unit-time) meter, a type of frequency counter.

When the shaft rotates, e.g., at 600 rpm, ten pulses per second will be generated. Rotors with continuous teeth around their circumference are used more frequently than single-tooth rotors because they produce a higher output frequency as well as a greater voltage amplitude at low shaft speeds. At a shaft speed of 600 rpm the output frequency of the transducer shown in Figure 12-4(b), which has eight teeth, will be $(8)(600 \div 60) = 80$ pulses per second. The rms output-voltage amplitude increases with decreasing clearance between the coil assembly and the teeth, with increasing shaft speed, and with increasing tooth size.

Fig. 12-4. Operating principle of electromagnetic, frequency-output tachometer.

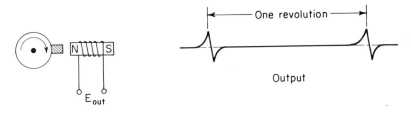

(a) Single tooth

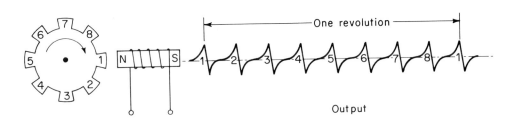

(b) Continuous teeth

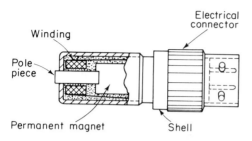

Fig. 12-5.   Typical transduction-coil
assembly.

A typical transduction-coil assembly is shown in Figure 12-5. The coil is wound on an insulating bobbin form which is slipped over a pole piece attached to the permanent magnet. The assembly is potted within a hermetically sealed sheath. An external thread on the sheath or an internal thread in the knurled or serrated shell is used to adjust the gap between pole piece and rotor teeth by rotating the coil in a mating thread which is part of the transducer housing. A locknut or safety wiring is then used to keep the coil from rotating further. A number of commercially available transduction-coil assemblies are illustrated in Figure 12-6.

Rotors have been designed in a multitude of configurations. Existing shafts can be converted into tachometer rotors by attaching to them a single ferromagnetic key [Figure 12-7(a)] or a multitoothed disc and mounting a transduction coil in suitable proximity. The rotor of the instrument shown in Figure 12-7(b) is provided with an internal spline for coupling to the drive shaft and an external spline for coupling to the driven shaft of a rotary device. The tachometer is sandwiched between driving and driven equipment. The rotor teeth are so shaped as to produce a quasi-sinusoidal voltage from the transduction coil. Two or more transduction coils can be installed in a transducer housing when outputs must be furnished simultaneously to two or more systems which are isolated from each other.

Fig. 12-6.   Various transduction-coil configurations (courtesy of Potter Aeronautical Corp.).

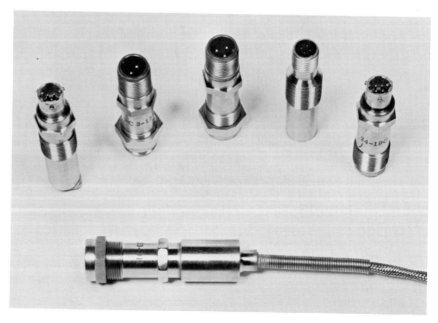

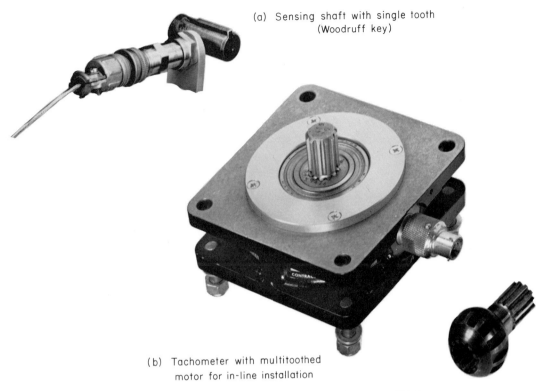

(a) Sensing shaft with single tooth
(Woodruff key)

(b) Tachometer with multitoothed
motor for in-line installation

Fig. 12-7. Electromagnetic, frequency-output tachometers (courtesy of Potter Aeronautical Corp.).

The output of multitoothed-rotor tachometers is typically of an amplitude between 2 and 10 V rms (into a load impedance of about 10K ohms) and of a frequency between 4 and 360 cycles per revolution at speeds between 100 and 40,000 rpm. Below a minimum speed and above a maximum speed, which are dictated by design characteristics for each transducer model, the output amplitude drops below a usable level (Figure 12-8). The additional load impedance presented by a long (shielded) cable between transduction coil and signal-conditioning or readout equipment tends to reduce the output voltage level. This level is adjustable, however, by varying the gap between the coil pole piece and the surface of the rotor teeth (Figure 12-9). Care must be taken not to make the gap so small that thermal expansion can cause binding of the rotor and damage to teeth and coil assembly.

The output waveshape is not critical for most applications as long as the "on" time occupies a significant portion of the width of each pulse. For certain telemetry systems, however, the input received from a tachometer must be as sinusoidal as possible, particularly when this signal replaces that of a standard subcarrier oscillator in an FM/FM telemeter. This objective can be accomplished by proper transducer design with special attention paid to tooth shape and exact equality of the shape of

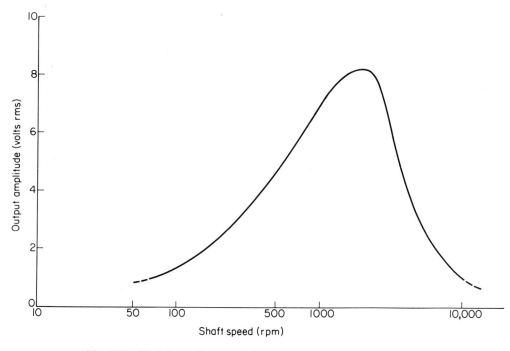

Fig. 12-8. Variation of output voltage amplitude with angular speed for a typical toothed-rotor electromagnetic tachometer design.

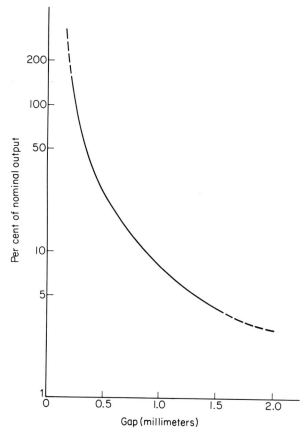

Fig. 12-9. Variation of output amplitude of toothed-rotor tachometer with gap between transduction-coil pole piece and rotor teeth surfaces.

all teeth. Frequency-to-dc converters are available for converting the output into a voltage analog of angular speed.

12.3.4 Capacitive angular-speed transducers

A number of designs have been developed using capacitance changes to produce an output proportional to angular speed, usually from an a-c bridge periodically unbalanced by capacitance variations in one or two of its arms. Rotor plates are either attached to or machined directly into the sensing shaft. One, two, or four stator plates can be used, shaped to comply with the rotor electrodes. Capacitive tachometers are rarely available as commercial hardware.

12.3.5 Photoelectric angular-speed transducers

The main advantage of this design over the electromagnetic frequency-output types is the essentially constant amplitude of its output pulses regardless of speed. It consists of a shutter disc, mounted to the sensing shaft, which rotates between a light source and a light detector (Figure 12-10). The disc has alternating opaque and trans-

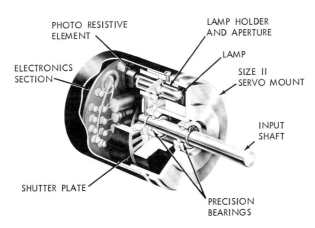

Fig. 12-10.   Photoelectric tachometer (courtesy of White Avionics Corp.).

parent sectors. The opaque sectors interrupt the light beam. A pulse, whose shape is determined by sector configuration, is produced in the output of the light detector for each transparent sector passing between it and the light source.

The sectors can be formed by stamping slots into the disc or by applying a pattern on a transparent disc by photographic means. A rugged miniature lamp is used as light source, a photoconductive or photovoltaic sensor as light detector. Circuitry can be added, within the transducer case or separately packaged, to provide a square or a sinusoidal waveshape of well-regulated constant amplitude, typically between 2 and 8 V zero-to-peak. The output frequency is determined by shaft speed and the number of transparent sectors in the disc.

12.3.6 Reluctive angular-speed transducers

Toothed rotors can be used in conjunction with a reluctive, rather than an electromagnetic, transduction element. Reluctive designs use a C-core transformer with an excitation winding on one leg of the core and an output winding on the other leg. A toothed rotor is placed in the gap of the core so that the reluctance path is varied by the alternating amounts of ferromagnetic material in the gap. A differential-transformer version uses an E-core with the excitation primary winding on the center leg and the two secondary windings on the outer legs of the transformer core. A toothed rotor in one of the two gaps varies the reluctance path between the primary and one secondary winding. The excitation (carrier) frequency should be higher than the highest frequency of its modulation, due to reluctance changes, expected in the transducer's operation. This type of transducer is rarely produced commercially in the U.S.A.

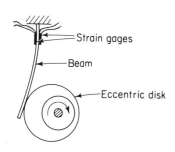

Fig. 12-11. Strain-gage tachometer.

12.3.7 Strain-gage angular-speed transducers

Strain-gage transduction can be used in conjunction with a deflecting beam and an eccentric disc attached to the sensing shaft, as shown in Figure 12-11. By proper mechanical design a sinusoidal output can be obtained from the strain-gage bridge, two arms of which (a compression gage and a tension gage) are on the bending beam. Some transducers of this type have been manufactured in countries other than the U.S.A.

12.3.8 Switch-type angular-speed transducers

Although rotary switches can be used as frequency-output tachometers, the more common designs incorporate a simple differentiating circuit so that a lower-cost analog readout device (e.g., a damped milliammeter) can be utilized. In the schematic circuit of Figure 12-12(a), rotating contacts on each end of a capacitor C permit the capacitor to charge to the voltage E first with one polarity, then with the opposite polarity. R represents the internal resistance of the milliammeter I as well as any

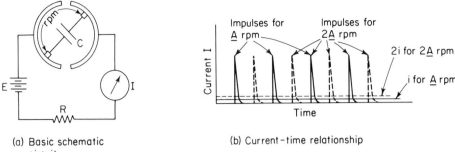

(a) Basic schematic circuit

(b) Current–time relationship

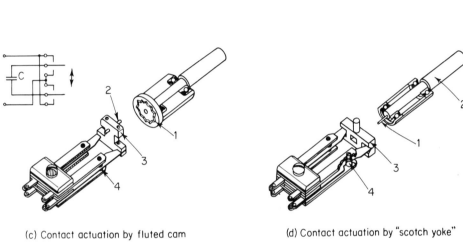

(c) Contact actuation by fluted cam

(d) Contact actuation by "scotch yoke"

Fig. 12-12. Switch-type differentiating tachometer (courtesy of Metron Instruments, Inc.).

resistance in interconnecting wiring (the meter can be located remotely). The meter is sufficiently damped to respond to the average of the individual current pulses [Figure 12-12(b)]. The average current $i = kCE\omega$, where k is a constant depending on units chosen and ω is the angular speed. An actual transducer uses a double-pole double-throw reed switch for reversal of capacitor polarity. The contacts of a 2 to 10 cycle-per-revolution transducer [Figure 12-12(c)] can be actuated by a cam track (1) having between 2 and 10 flutes into which a pin (2) projects to actuate a cam follower (3) and polarity-reversing contacts (4).

Figure 12-12(d) shows an alternate design in which one switching cycle per revolution is produced by an eccentric pin (1) attached to the sensing shaft (2) which moves a "scotch yoke" (3) sinusoidally up and down to actuate the polarity-reversing contacts (4). Shunt resistors across the indicating milliammeter, whose scale is calibrated in rpm, can be used to change or adjust the meter range.

12.4 Performance and Applications

12.4.1 Specification characteristics

12.4.1.1 Mechanical design characteristics of velocity and speed transducers include case dimensions, material and finish, mounting provisions and dimensions, electrical connector or cable type and location, nameplate contents and location, and weight. The sensing shaft of some linear-velocity and all angular-speed transducers should be defined as to material, dimensions, and coupling means. Materials of brushes and commutators, if used, and lubrication requirements, if any, of tachometer bearings should be shown. Most bearings are of the self-lubricating or sealed-lubricant type. Specifications of toothed-rotor electromagnetic tachometers should contain additional callouts for sealing, location, adjustability, and locking provisions of their transduction coil(s). Their cases should be of nonferrous metal, and rotor material should be of homogeneous crystalline structure. Fail-safe provisions, e.g., a shear pin between rotor and shaft, may have to be incorporated in order to prevent damage to the equipment driving the tachometer shaft if the rotor binds.

12.4.1.2 Electrical design characteristics comprise insulation resistance and breakdown-voltage rating (between any winding and the case), receptacle pin or cable conductor identification (including polarity if applicable), allowable load impedance, output impedance at a nominal frequency for a-c output, output resistance (for continuity checks), and any excitation voltage and power requirements for transducers which are not self-generating.

12.4.1.3 Performance characteristics at room conditions. For a linear-velocity transducer the range, sensitivity or full-scale output, hysteresis, repeatability, and linearity, frequency response, transverse sensitivity, and damped natural frequency are stated. Specifications for angular-speed transducers show the required range (e.g., "800 to 6000 rpm"), starting and running torque, and sometimes moment of inertia. Linearity and repeatability, sensitivity (in millivolts per rpm or volts per 1000 rpm), and output noise are specified for d-c tachometers; sensitivity, phase, and allowable harmonic content are called out for a-c tachometers, and the following are specified for frequency-output tachometers: output amplitude and its allowable variation over the range, uniformity of amplitude over one revolution, amplitude adjustment range, output frequency in cycles or pulses per revolution, pulse characteristics for nonsinusoidal output, and allowable harmonic content for sinusoidal output.

12.4.1.4 Environmental characteristics. Limitations on effects of operating or nonoperating environmental conditions, as demanded by the transducer application, are shown for a specified temperature range, for stated levels of shock, vibration, and acceleration (usually only in the transverse axes for linear-velocity transducers), for electromagnetic interference, high sound-pressure levels, magnetic fields, and nuclear radiation, and (unless the transducer is hermetically sealed) for ambient-pressure variations, humidity, and atmospheric contaminants.

12.4.2 Selection and applications

12.4.2.1 Linear-velocity transducers are most commonly used to measure vibration as vibratory velocity. Their outputs can be differentiated if acceleration data are desired and integrated if displacement data are desired. Selection of a required model is based primarily on frequency response and velocity (sometimes acceleration) range and secondarily on operating temperature range, sensitivity, weight, and mounting position.

12.4.2.2 Angular-speed transducers are widely used in measurements of rotational speed of engines, pumps, and all types of machinery. When a "measuring wheel" is attached to their shaft, they are usable for monitoring of (translational) linear speed, e.g., of moving vehicles, conveyor belts, and paper or textiles in mills. Selection is based primarily on the type of available readout or signal-conditioning and telemetry equipment, on range, accuracy characteristics, and on cost and secondarily on sensitivity and output characteristics, number of separate outputs needed, excitation requirements, if any, type of installation, weight, and environmental conditions.

Most tachometers can be designed with any type of required mounting. For in-line installations special attention must be directed at any potential hazards the tachometer, taken as a shaft link, may present to the operation of driving and driven devices.

12.5 Calibration and Tests

12.5.1 Calibration

12.5.1.1 Linear-velocity transducers are calibrated on a vibration exciter by methods similar to those used on acceleration transducers (see 1.5). The velocity applied to the transducer being calibrated can be determined by use of a reference transducer installed on the exciter so that it sees exactly the same velocity. The reference transducer used for such a comparison calibration must have well-established velocity-vs-output characteristics. An absolute method of determining the velocity is sometimes used. The peak-to-peak displacement of the transducer is measured optically; the frequency is known (read off the exciter-control console or a frequency counter); and the velocity is calculated as the time derivative of the displacement.

12.5.1.2 Angular-speed transducers are calibrated by a comparison method using a variable-speed drive and a reference transducer of the frequency-output type so that an EPUT-meter (electronic counter) can display the reference output with very close accuracy.

12.5.2 Performance verification at room conditions

After verification of essential specified mechanical and electrical characteristics, the calibration is used to determine transducer performance at room temperature.

The linearity, hysteresis, and repeatability of a linear-velocity transducer at a given frequency can be established from the data of at least two calibration cycles, each with discrete levels of velocity applied in an increasing and decreasing direction. Of somewhat greater significance is a frequency-response test, a dynamic calibration during which velocity amplitude is held constant but frequency is varied and output readings are recorded at preselected frequencies. If a reference transducer is used for such a dynamic calibration, it must have a wider frequency range of flat response than the test specimen.

The output characteristics of angular-speed transducers can also be determined from a calibration which is usually simpler than that of linear-velocity transducers. When hysteresis and repeatability are negligible, as is the case in frequency-output transducers, only half of one calibration cycle is needed. The application of three levels of angular speed normally suffices for the determination of linearity and other characteristics such as output amplitude and harmonic content. Figure 12-13 shows a typical test data form for frequency-output tachometers. It is intended for toothed-rotor transducers having either one or two transduction coils which are locked in position after their gaps are adjusted for a certain output amplitude at midrange.

Special structural-integrity tests of in-line installed tachometers may include X-raying of shaft and rotor to detect potentially failure-prone material.

Life tests can be performed with the same measurand-application equipment as is used for calibration.

12.5.3 Environmental tests

These tests are normally performed only on a sampling or one-time qualification basis. Operating environmental tests on angular-speed transducers require relatively complex test setups and, for some tests, flexible drive shafts to assure that only the transducer is exposed to the environment. The performance of speed and velocity transducers is usually verified at only one point in their range during environmental tests. However, the vibratory frequency applied to linear-velocity transducers may be successively changed to two or three different values.

Environmental tests on linear-velocity transducers are essentially the same as those described in 1.5 for (dynamic) acceleration transducers. Tachometers can be driven at constant midrange speed through a reasonably well-sealed hole in the

VENDOR MODEL NO.		PART NO.
PURCHASE ORDER NO.	**INDIVIDUAL ACCEPTANCE RECORD & CALIBRATION** **FOR FREQUENCY-OUTPUT** **ANGULAR SPEED TRANSDUCER**	SERIAL NO. NOMINAL RANGE

1. VISUAL: Mechanical ☐   Finish ☐

                        Nameplate ☐   Receptacle ☐

2. ELECTRICAL: PIN CONNECTIONS ☐   SOURCE IMPEDANCE @ _____ CPS: COIL "A" _____OHMS
                                                                             COIL "B" _____ OHMS

                        INSULATION RESISTANCE: COIL "A" _____ MEGOHMS AT _____ VDC
                                               COIL "B" _____ MEGOHMS AT _____ VDC

3. CALIBRATION AND OUTPUT CHARACTERISTICS:

ANGULAR SPEED	VOLTAGE (VAC-RMS)		FREQUENCY (CPS)		HARMONIC CONTENT (%)	
	COIL "A"	COIL "B"	COIL "A"	COIL "B"	COIL "A"	COIL "B"
LOWER RANGE END _____RPM						
MID-RANGE _____RPM						
UPPER RANGE END _____RPM						

MID-RANGE OUTPUT FREQUENCY: _____ CPS ALLOWED: _____ ± _____ CPS

COMMENTS _____
_____
_____
_____

BY _____ DATE _____ APPROVED _____

Fig. 12-13. Data form for calibration and other acceptance tests of tachometers with sinusoidal frequency output.

chamber during temperature, nuclear-radiation, humidity, and contaminating-atmosphere (salt, dust, etc.) tests. During vibration and high-sound-pressure level tests they can be driven through a flexible shaft, with care taken to prevent this shaft from whipping.   During acceleration tests the driving motor must be installed on the centrifuge with the transducer unless a special angle drive can be provided through the hollow center shaft of the turntable. Electromagnetic-interference and magnetic-field tests require no special driving provisions. Ambient-pressure and shock tests are performed as nonoperating tests.

12.6 Special Measuring Devices

12.6.1 Seismometers

The seismometer is an instrument used to measure earth tremors. Although various design approaches have been used, many of the commonly used seismometers are most closely related to linear-velocity transducers. Their large proof mass (Figure 12-14) with its suspension system has a natural frequency of less than one

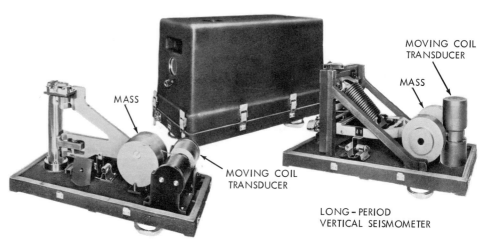

Fig. 12-14. Horizontal and vertical seismometers (courtesy of Geotech Div., Teledyne Industries, Inc.).

hertz. Artificial (e.g., electromagnetic) damping is employed. The mass is linked to a moving coil (of several hundred turns) in a strong magnetic field. A sensitivity of about 170 V/m/s is typical for transducer output. A complete seismographic installation requires two horizontal seismometers (one for east-west, the other for north-south, sensing) and a vertical seismometer. Some seismometers have been designed with special provisions to extend their period to as long as 3000 s.

12.6.2 Vertical-speed transducers

Transducers used to measure vertical speed of aircraft are not related to other velocity and speed transducers since this measurement is not based on displacement. Instead, a pressure transducer is used to measure pressure altitude (see 10.6.3), and the transducer output is differentiated to indicate vertical speed as the time rate of change of pressure-altitude.

12.6.3 Noncontacting angular-speed transducers

Angular speed can be measured by two types of optical systems without installing any measuring device on the rotating object. One is the stroboscopic system in which brief periodic flashes are produced from a special lamp which illuminates the object. The flash frequency is controlled by a variable-frequency generator such as an audio oscillator. When the light flashes at the frequency equal to the angular speed, the rotating object appears to stand still. The frequency setting of the generator is then used to determine angular speed. Since the object, however, will also appear to stand still at harmonics and subharmonics of the correct frequency, the approximate speed will have to be known unless special automatic synchronization provisions are included in the measuring equipment.

The other system is related to the photoelectric tachometer. A light beam and a light detector are aimed at the rotating object which must have a reflective portion (or two or more equally spaced portions) on its surface. The detector is adjusted to respond only to the reflected light so that a frequency output is generated which can be displayed on an electronic counter.

12.6.4 Sound velocimeters

The term "velocimeter" has been applied to a device intended for measurements of the velocity of sound in liquids. Such devices are related to the underwater sound transceivers of sonar systems briefly discussed in 11.6.1,

BIBLIOGRAPHY

1. Frazier, R. H., "Analysis of the Drag Cup A-C Tachometer," *AIEE Transactions 1951*, pp. 1894–1906, 1951.

2. Cook, G. W., "A Sensitive Vertical Displacement Seismometer," *Report No. 814*, Washington, D.C.: The David W. Taylor Model Basin, 1952.

3. Davis, S., "Using a Two-Phase Servo Motor as an Induction Tachometer," *Control Engineering*, pp. 75–76, November 1955.

4. Viskanta, V. Z., "A Study of Rotary Speed Measuring Techniques," *Report WADD TR 60-210*, Armour Research Foundation, February 1960.

5. Trickey, P. H., "Calculating No Load Output of Drag Cup Tachometers," *Electromechanical Design*, pp. 34–38, November 1961.

6. Savard, R. F., "Acceptance Testing of Precision Motor Tachometer Generator," *AIEE Conference Paper CP62-1054*, New York: Institute of Electrical and Electronics Engineers, June 1962.

7. Frey, J., "The A-C Tachometer," *Electro-Technology*, pp. 88–93, August 1963.

8. Miller, W. F., "Feedback Stabilized Seismometers," *Preprint No. 164-5-64*, Pittsburgh: Instrument Society of America, 1964.

13   Strain

13.1 Basic Concepts

13.1.1 Basic definitions

Strain is the deformation of a solid resulting from stress. It is measured as the ratio of dimensional change to the total value of the dimension in which the change occurs. *Note:* Strain can also be produced thermally.

Stress is the force acting on a unit area in a solid.

The **modulus of elasticity** is the ratio of stress to the corresponding strain (below the proportional limits).

The **elastic limit** is the maximum unit stress not causing permanent deformation of a solid.

Poisson's ratio is the ratio of transverse to longitudinal unit strain.

A **strain gage** is generally a resistive strain transducer.

The **gage factor** is the sensitivity of a resistive strain transducer.

13.1.2 Related laws

Gage factor

$$GF = \frac{\Delta R/R}{\epsilon} = \frac{\Delta R/R}{\Delta L/L}$$

where: GF = gage factor
$\Delta R/R$ = unit change in resistance
ϵ = unit strain
$\Delta L/L$ = unit change in length

Hooke's law

$$\frac{s}{\epsilon} = (\text{constant}) = E$$

where: E = modulus of elasticity
s = stress
ϵ = strain

Elasticity of length

$$E_t = Y\left(= \frac{s}{\epsilon}\right) = \frac{F_t/a}{\Delta L/L} \qquad E_c\left(= \frac{s}{\epsilon}\right) = \frac{F_c/a}{\Delta L/L}$$

where: F_t = force (tension)
F_c = force (compression)
a = cross-sectional area (normal to direction of force application)
ΔL = elongation or contraction (of solid along the direction of force application)
L = original length of solid
Y = Young's modulus ("stretch modulus of elasticity")
E_t = tensile modulus of elasticity
E_c = compressive modulus of elasticity

13.1.3 Units of measurement

Strain is expressed as the (dimensionless) ratio of two length units, usually in **microinches per inch** or, generally, in percent (for large deformations) or in **microstrain** (for medium and small deformations). One microstrain ($\mu\epsilon$) is the ratio of 10^{-6} of a length unit to this length unit.

Stress is expressed in units of force per units of area (e.g., lb/in^2, kg/mm^2, kg/cm^2, $dynes/cm^2$).

Modulus of elasticity is usually expressed in lb/in^2, kg/cm^2 or kg/mm^2.

13.2 Sensing and Transduction Methods

Although strain has been measured by mechanical indicators (extensometers), optical methods, and reluctive transducers, the device used for the vast majority of strain measurements is the resistive strain transducer or *strain gage*. It consists essentially of a conductor or semiconductor of small cross-sectional area which is mounted to the measured surface so that it elongates or contracts with that surface. This deformation of the sensing material causes it to undergo a change in resistance. Hence, a strain gage senses strain by its own deformation and transduces the deformation into a resistance change.

The change in resistance of a material due to an applied stress has been termed the *piezoresistive effect*, although this term has been used primarily in discussions of semiconductors. The change in resistance of a conductor when subjected to mechanical strain was first reported by Lord Kelvin (W. Thomson) in 1856, further elaborated by H. Tomlinson in 1877, both working in England, and demonstrated in 1881 by the Russian physicist O. D. Khvol'son. The first commercial application of this principle occurred 75 years after Lord Kelvin's announcement when an unbonded-wire strain gage was developed. This was followed in more rapid succession by commercial developments of the bare-wire bondable strain gage, the bondable wire strain gage premounted on a paper or plastic carrier base, the metal-foil bondable strain gage, the semiconductor strain gage, and the deposited-metal (thin-film) strain gage.

The resistance change of a strain gage is usually converted into voltage by connecting one, two, or four similar gages as arms of a Wheatstone bridge (*strain-gage bridge*) and applying excitation to the bridge. The bridge output voltage is then a measure of the strain sensed by each strain gage. Each arm of such a bridge containing a strain-sensing gage is referred to as an *active arm*. Strain transduction, then, can be said to be performed by the active arms of a strain-gage bridge.

13.3 Design and Operation

13.3.1 Metal-wire strain gages

Resistive metal-wire strain gages are manufactured in bondable, surface-transferable ("free-filament"), and weldable configurations. The unbonded strain gage, consisting of a wire stretched between two posts and unsupported between its ends, is used in some transducers for other measurands but rarely for strain measurements.

The bondable strain gage is commonly referred to as "bonded" strain gage since it is bonded, usually with cement, to the measured surface in its installation. It is the most widely used strain-gage configuration. A basic flat-wire-grid bondable strain gage is shown in Figure 13-1. It consists of a thin wire (*filament*) arranged in a zigzag pattern and cemented to a base. Lead wires are soldered or welded to the ends of the

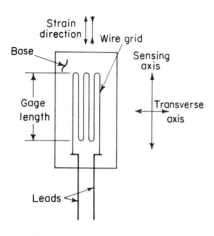

Fig. 13-1. Bondable wire-grid strain gage.

filament to facilitate external electrical connections. The length of the grid, exclusive of lead connections, is the *gage length* (active length). A thin cover plate is sometimes cemented over the wire grid.

The choice of filament material is based on five criteria: (1) the material should provide as high a gage factor as possible; i.e., its resistance change with strain should be large; (2) the temperature coefficient of resistance should be as low as possible to minimize temperature error; (3) the resistivity of the material should be high, so as to allow winding a grid of the required gage resistance within the smallest possible area; (4) the filament should have a high mechanical strength to minimize damage due to wire breaks and to allow high stresses to be applied to the filament without its undergoing plastic deformation; and (5) the material should generate the smallest possible thermoelectric potential at its junctions with the lead wires.

Copper-nickel alloys such as Constantan are most frequently used as filament materials in general-purpose strain gages. Wire diameters are typically between 0.02 and 0.04 mm. Temperature coefficients of resistance in the absence of any specifications can vary between -50×10^{-6} and $+50 \times 10^{-6}$ per degree Celsius over the range 0 to 100°C. However, the material manufacturer can usually supply such alloys with much smaller temperature coefficients. The strain-gage manufacturer can further select certain melts and rolls of wire in his possession to obtain either a low temperature coefficient or a specific temperature coefficient which matches that of the material to which the gage is intended to be bonded. Selection of wire material by temperature coefficient also allows the manufacture of split-element temperature-compensated gages using two half-grids connected in series and applied to the base as one whole grid where the temperature coefficient of each half-grid is essentially equal in magnitude but of opposite polarity. The gage factor of commercial copper-nickel wire gages can be as low as 1.7 but usually ranges between 1.9 and 2.1.

Other wire materials include nickel-chromium and platinum-iridium alloys for high-temperature applications and iron-chromium-aluminum as well as iron-nickel-chromium alloys when higher gage factors (2.8 to 3.5) are required and operating temperatures are moderate. Manganin wire (copper-manganese-nickel alloy) has been used in some applications, although it provides a gage factor of only 0.5.

Lead materials in common use are nickel-clad copper, stainless-steel-clad copper, nickel-clad silver, and nickel-chromium. The thermal emf generated by the various wire materials at their junction with copper varies from 2μV/°C for Manganin to $47\ \mu$V/°C for Constantan. The leads are frequently insulated by plastic coating or, for high-temperature applications, by glass sleeving.

Several types of base (carrier) materials are used to support the filament. A nitrocellulose paper is a popular carrier for general-purpose gages whose applications are at temperatures between about −100 and +150°F. Paper carriers having a thickness of 40 microns or less can be used for gages to be installed on flat as well as curved surfaces. Epoxy carriers are utilized when a larger operating temperature range (−320 to +250°F) is needed. Bakelite-impregnated cellulose or glass-fiber carriers extend the operating temperature range to +350°F. The filament is firmly embedded in the phenolic (bakelite) wafer. Paper and bakelite carriers are employed in the most common grid construction, the flat grid, as well as in the less frequently used wraparound or helical grid construction [Figure 13-2(a)] in which the wire is wound around a carrier card to double the wire length provided in a given area. The wound card is then cemented between two thin carrier plates. Strippable plastic (e.g., vinyl) backings are used for surface-transferable gages [Figure 13-2(b)] which are installed without any carrier. Among other carriers are polyester, poly-imide, epoxy/glass-fiber, and asbestos bases.

Special stress-relieving connections between filament and leads have been used in gages intended for fatigue and large-strain measurements. One design employs intermediate U-shaped wires with the filament end connected to the base of the U while both ends of the arms of the U are connected to the right-angle portion of the external connecting lead.

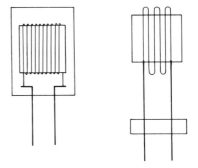

(a) Wrap–around　　(b) Surface–transferable

Fig. 13-2.　Special strain-gage windings.

Gages with multiple grids (*rosettes*) were developed for simultaneous measurement of strain in different directions. Rosettes are further described in 13.3.2. However, multiple wire gages are usually of sandwich construction with the grids superimposed over each other and insulated by cement.

Weldable strain gages are mounted within small flat or tubular envelopes made, preferably, of the same metal as the measured surface of their intended application. The base of the envelope is slightly larger than the main body of the gage providing a sufficient edge around the body for welding to the surface. Since the base is very thin and the edge is quite narrow, special microwelding techniques must be used for mounting weldable gages. Such gages have been used successfully in difficult applications, e.g., on inside and outside surfaces of tanks containing cryogenic liquids.

Special strain gages for strain measurements in concrete can be encapsulated in a waterproof brass container to protect the filament during installation when pouring concrete. Another special design is the "post-yield" strain gage usable for measurements up to 100,000 $\mu\epsilon$ (10% strain).

13.3.2 Metal-foil strain gages

Refinements in photoetching techniques permitted the development of low-cost metal-foil strain gages. Available manufacturing methods, which are related to those employed in the popular and mass-produced printed circuits, allow better utilization of a given area. Hence, foil gages can be made smaller than wire-grid gages. A full three-element rosette disc, for example, is available with an overall diameter of 0.35 in. (about 9 mm). Foil gages can generally be used to measure larger strains than wire gages, and they are not damaged as easily, especially in the delicate surface-transferable configuration. The relative merits of foil gages over wire gages are somewhat controversial. Some users feel that foil gages are superior in all essential respects. Others prefer wire gages for at least some of their applications.

Foil grids consist of a very thin (approximately 3 to 8 microns thick) layer of foil, part of whose metal is removed by etching so that the desired grid shape is obtained. The grid can be cemented to a carrier base. The foil can also be cemented to a base before etching so that the etching process results in a finished strain gage on a carrier. Surface-transferable gages are placed on a strippable carrier after completion. Weldable foil gages have been manufactured in several configurations, e.g., a foil grid prebonded and temperature cycled on a stainless-steel shim.

Foil, lead, and carrier materials are similar to those used in wire-grid strain gages, with Constantan and nickel-chromium alloys predominating as foil materials. A 0.05 mm thick 95% titanium-5% aluminum foil has been reported usable for high-strain (post-yield) measurements. Many foil gages are provided with relatively large integral foil portions (*tabs*) to which leads can be attached by the user.

Typical metal-foil strain-gage patterns are shown in Figure 13-3. The normal-width grid [Figure 13-3(a)] represents the most common configuration. The pattern is longer in the sensing axis to reduce transverse-strain effects. Soldering of external leads is facilitated by the long tabs. When strains are known to be mainly uniaxial, i.e., when transverse strains are expected to be negligible in the measured surface, the wide-grid design [Figure 13-3(b)] can be used. Because of its large size it can dissipate more power than the normal-width grid, thereby allowing greater excitation voltages and higher bridge output. Three-element rosettes with 45-degree [Figure 13-3(c)] or 60-degree angles between elements can be used to determine the direction as well as the magnitude of principal strains in the measured surface when neither is known.

Because of the versatility of the etched-foil technique, a number of special configurations have been manufactured to fill a variety of needs. Figure 13-4 shows some of the existing dual- and triple-element rosettes, additional to the rosette of Figure 13-3(c). The gages shown are normally supplied on epoxy bases, but rosette gages can have other backings as well. The biaxial gage [Figure 13-4(a)] consists of two identical grids 90 degrees from each other. It is used for measurements in which the direction of the orthogonal principal strains is known. The two elements of the biaxial gage shown in Figure 13-4(b) are also spaced 90 degrees from each other; however, they are spaced at 45-degree angles from an identifiable center line. This spacing makes them useful for torsional-strain determinations in torque measurements (see 5.3.6). The triaxial rosette [Fig. 13-4(c)], with elements I and II spaced at 60 degrees from the sensing axis of element III, is used when the direction of principal strains in the measured surface must be determined. The gage length of each of the three elements is

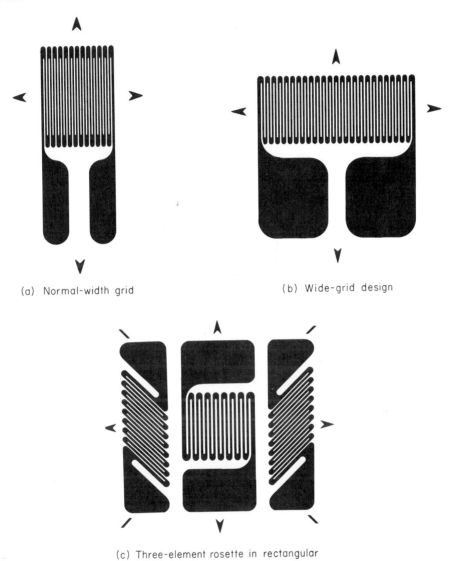

(a) Normal-width grid

(b) Wide-grid design

(c) Three-element rosette in rectangular
"V" Configuration

Fig. 13-3. Metal-foil strain gages (courtesy of The Budd Co., Instruments Div.).

2 mm. When using rosette gages, it is necessary to operate upon the output readings using Poisson's ratio to convert strain rates to stress.

Figure 13-5 shows a spiral foil gage usable in measuring tangential strain in a diaphragm. The foil pattern is bifilar to facilitate its construction and to cancel any inductive effects. A related full-bridge pattern is used more frequently.

(a) Biaxial
Rosette (90°)

(b) Biaxial
Rosette (45 + 45°)

(c) Triaxial
Rosette (60°)

Fig. 13-4. Metal-foil strain-gage rosettes (courtesy of Gulton Industries, Inc.).

Fig. 13-5. Spiral-foil strain gage (courtesy of Gulton Industries, Inc.).

Fig. 13-6. Four-element foil gage rosettes (courtesy of Dentronics, Inc.).

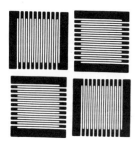

(a) Full-bridge
rosette

(b) Bending beam rosettes

Full-bridge rosettes are shown in Figure 13-6(a) (two opposing biaxial combinations, for structural applications) and Figure 13-6(b) (in-line configuration for bending-beam applications).

The *stress-strain gage* (Figure 13-7) is composed of two uniaxial strain-sensing elements oriented at 90 degrees to each other. The elements have a common connection so that either of them may be observed independently for conventional strain measurement or that the series combination of the two elements can produce an output proportional to stress along the principal gage axis. The latter is achieved by making the ratio between the resistances of element 2 to element 1 equal to Poisson's ratio of the material to which the gage combination is to be applied. For example, if the stress-strain gage is to be applied to aluminum, whose Poisson's ratio is 0.33, the resistances of element 2 and element 1 are made 115 and 350 ohms, respectively. Strain in either the principal axis (*a-a*) or the transverse axis (*n-n*) is measured in the usual manner using gage factors given for each element. Stress in the *a-a* axis (s_a) can be determined by using a manufacturer-supplied "stress-gage factor" (F_c), the

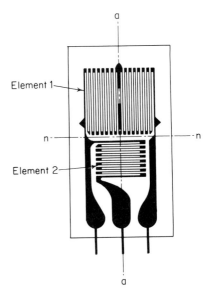

Fig. 13-7. Stress-strain foil gage (courtesy of BLH Electronics).

resistances of the two elements (R_1 and R_2), the resistance changes of the two elements (ΔR_1 and ΔR_2), and the modulus of elasticity of the material (E) in the following equation

$$s_a = \frac{(\Delta R_1 + \Delta R_2)}{R_1 + R_2} \cdot \frac{E}{F_c}$$

This gage design uses Constantan foil on a phenolic base. It is temperature-compensated for the specified end-use material (mild steel, stainless steel, or aluminum).

13.3.3 Deposited-metal strain gages

Thin-film techniques have been applied to the manufacture of strain gages applied directly to a measured surface. Such gages can be made substantially smaller than other equivalent metal strain gages. The measured surface is first coated with an insulating substrate upon which the gage or rosette can then be formed by evaporative or bombardment methods. Applications of thin-film metal strain gages have been limited to diaphragms of pressure transducers but can be expected to be extended to other mechanical elements, primarily other transducer sensing elements.

Flame-sprayed strain gages have been applied to structures which are to be exposed to extremely hostile environments. In such difficult applications they have been found a usable alternative to welded strain gages although the application techniques are more laborious and costly. They are, on the other hand, thinner than comparable welded strain gages. This is a substantial advantage in certain applications such as on exterior surfaces of rockets, which are subject to severe aerodynamic heating.

The strain-sensitive metal grid is applied to an insulating ceramic substrate on the measured surface. The substrate is formed by a process in which either the end of a ceramic rod is heated in a gas-oxygen medium and the molten particles projected onto the surface at high speed (Norton Co. "Rokide" process) or in which ceramic powder is suspended in an oxygen-acetylene gas mixture which is detonated to propel the particles at high speed from a long barrel (Linde Co. "Flame-Plating" process). "Rokide" substrates 0.05 to 0.1 mm thick have been used in typical installations, especially on aircraft engines, with satisfactory results. A surface-transferable metal strain gage can then be applied to the substrate and permanently bonded to the surface by a flame-sprayed ceramic coating. Earlier techniques required the flame spraying of the metal grid itself through a mask placed over the substrate.

Acceptable performance at temperatures to 2200°F has been reported for flame-sprayed strain sensors.

13.3.4 Semiconductor strain gages

When experiments during the early 1950's confirmed that the piezoresistive effect is much larger in semiconductors than in conductors, a number of commercial and governmental laboratories undertook the development of semiconductor strain gages. Since then a large number of such devices with satisfactory performance characteristics have become available. The gage factors of semiconductor strain gages are between 50 and 200, whereas those of metal strain gages are no greater than 6 and are most frequently around 2. However, semiconductor gages tend to be more difficult to apply to measured surfaces; their strain ranges are usually limited to $\pm 3000 \; \mu\epsilon$; their operating temperature range is more limited; and their temperature compensation within this range is more laborious.

The resistivity of a semiconductor is inversely proportional to the product of the electronic charge, the number of charge carriers, and their average mobility. The effect of an applied stress is to change both the number of carriers and their average mobility. The magnitude and the sign of the change depend on physical form, type of material, doping, carrier concentration, and crystallographic orientation. Hence the gage factor can be positive or negative. The gage factor (GF) of a semiconductor strain gage in its normal application—measuring longitudinal strain—is given by

$$GF = \frac{R}{R_0\epsilon} = 1 + 2v + Y\pi_e$$

where: R = change in resistance due to applied stress

R_0 = initial resistance (of the semiconductor)

ϵ = strain

$\nu = $ Poisson's ratio (for the semiconductor)

$Y = $ Young's modulus

$\pi_e = $ longitudinal piezoresistive coefficient

The gage factor then is dependent on resistance change due to dimensional change $(1 + 2\nu)$ as well as the change of resistivity with strain $(Y\pi_e)$. It is also inherently nonlinear with applied strain. This nonlinearity is usually minimized by appropriate strain-gage-bridge design. One of the simplest methods is to connect a pair of active gages, matched to each other, as adjacent bridge arms with one side of the output taken from the connection between the two gages. Other methods use cancellation of strain-gage nonlinearity by an opposite nonlinearity of an asymmetric bridge circuit, by using a four-active-arm bridge with two gages in tension and two in compression (provided the gages are matched), and by shunting a single active arm by a resistor of appropriate value. Another linearization technique is prestressing the gage in tension during mounting to shift its operation into a more linear portion of the range. The amount of doping and resulting carrier concentration can also be varied to obtain better linearity at a reduction in gage factor. The linearity of highly doped semiconductor gages approaches that of most metal gages.

Semiconductor strain gages are affected by temperature to a considerably larger extent than metal strain gages. The temperature coefficient of resistance of the semiconductors used as strain gages is 60 to 100 times greater than that of Constantan, the variation of gage factor with temperature is 3 to 5 times greater, and the Seebeck coefficient (thermoelectric potential) generated at connections can be 10 to 20 times larger. Differential thermal expansion between semiconductor and a metallic measured surface can also be greater. The linear expansion coefficient of the semiconductor strain gage is typically about one-half that of a metal strain gage.

Temperature compensation can be effected by a variety of techniques. A dummy gage can be connected into the conjugate bridge arm as described for metal strain gages. Resistors (e.g., thermistors) with controlled thermal resistance changes can be connected in series or parallel with the active semiconductor-gage bridge arm(s) or in series with one or both excitation leads. A p-type semiconductor (positive gage factor and temperature coefficient) can be connected in a conjugate active bridge arm with an active n-type gage (negative gage factor, positive temperature coefficient). The p- and n-type gages can be manufactured as a dual-element combination with a common terminal for the excitation connection. A thermocouple can be so used and connected that it provides an appropriate bucking voltage. For installation on a given material a gage can be selected such that its thermal expansion and temperature coefficient of resistance tend to counteract each other. Finally, the doping of the semiconductor material can be increased to obtain a lower resistivity and less thermal effects at the cost of a decreased gage factor.

Typical semiconductor strain gage configurations are shown in Figure 13-8. The gages can be surface-transferable (bare) or encapsulated in a carrier. Effective gage lengths vary from 0.01 to 0.25 in. Lead materials are gold, copper, or silver wire, or

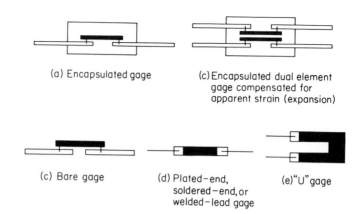

(a) Encapsulated gage

(c) Encapsulated dual element
gage compensated for
apparent strain (expansion)

(c) Bare gage

(d) Plated–end,
soldered–end, or
welded–lead gage

(e) "U" gage

Fig. 13-8. Typical semiconductor strain-gage configurations (courtesy of Kulite Semiconductor Products, Inc.).

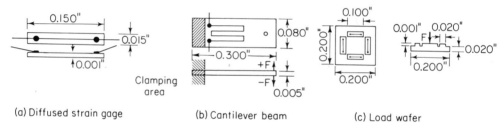

(a) Diffused strain gage

(b) Cantilever beam

(c) Load wafer

Fig. 13-9. Diffused-element single and multiple semiconductor strain gages (courtesy of Kulite Semiconductor Products, Inc.).

nickel ribbon. Silicon is used almost exclusively as a semiconductor material, although germanium has also been used at times. Special configurations have been obtained by diffusing the dopant directly into selected portions of silicon blocks, diaphragms, or wafers (Figure 13-9). Complete sensing/transduction elements made of silicon with integrally-diffused strain sensors have been used in pressure and force transducers. Because of the low density (2 grams/cm^3) and the high stiffness ($Y = 30 \times 10^6$ psi) of silicon, sensing elements made from it can have very high natural frequencies.

13.4 Performance and Applications

13.4.1 Specification characteristics

Resistive strain transducers are specified by a limited number of dimensional, electrical, and performance characteristics. Specifications for active-element rosettes should state the characteristics of one of the strain-gage elements and the degree of matching required from the other elements, as well as the internal geometry of the combination. Active/dummy combinations should be defined by specifications for both types of elements.

Dimensions are somewhat more critical for strain gages than for most other types of transducers. The gage length and width are always defined. Tolerances should be kept close, commensurate with the magnitude of these dimensions. Total grid length including terminals should also be shown in all cases. When the gage is furnished with leads, the diameter or thickness of the leads, lead material, insulation, and length and spacing between leads (if not given by the grid width) are stated. When the gage is bonded to a carrier or is encapsulated, the base length, width, thickness, and material should be specified. When the gage is surface-transferable, the size and material of the strippable carrier should be stated. The material of the grid or other active element should be defined. Grid-wire diameter or foil thickness is sometimes shown.

Recommended methods for bonding or other attachment of the strain gage to the measured surface or for its embedment in specific materials should be furnished by, or at least obtainable from, the manufacturer. Such information as type of cement, material compatibility, cure time, temperature and pressure, insulation resistance and its change at temperature extremes, resistance to humidity, and moisture proofing should be included with a description of the methods. When special bonding materials and techniques are intended to be employed by the user, the manufacturer should be consulted as to compatibility with the desired type of strain gage.

Electrical characteristics comprise gage resistance and its tolerances, power rating or maximum excitation current (which depend primarily on the heat sink provided by the measured surface and the bond to it), and, for some gages bonded to a carrier or encapsulated, the insulation resistance (at room temperature and at temperature extremes). The latter is important mainly for weldable gages since no insulation will be provided by any bonding cement. Gage resistances (at room temperature) have been largely standardized at the following values: 60, 120, 240, 350, 500, and 1000 ohms and, for semiconductor types only, 5000 and 10,000 ohms. The gage resistance tolerance can be stated either in ohms or in percent deviation from the stated nominal value. For a given lot a further tolerance may be imposed on the deviation from an average actual value. This type of lot specification can also be applied to strain-gage performance characteristics.

The primary room-temperature performance characteristic of all strain gages is their gage factor. It is shown as a nominal value with tolerances. Complete specifications should include the strain range over which linearity, hysteresis, and creep tolerances apply, as well as the allowable overload ("strain limit") which does not cause output errors during its application to exceed a specified amount. Linearity, when not otherwise stated, is intended to be the independent linearity and often applies only to a specified partial range. Hysteresis and creep are typically expressed in units of indicated strain. Drift, when specified for strain gages, refers to the change in gage resistance at a constant high or low temperature with no strain applied. Transverse sensitivity tolerances should be stated as applicable at a specific value of transverse strain. Life requirements are shown as cycling life ("fatigue life").

Temperature is the only environmental condition for which performance tolerances are normally specified. Other environments, such as shock and vibration, should not be stated for strain gages. They can, if necessary, be specified for a complete strain gage installation primarily as requirements for the integrity of the bond. Some applications, however, may require limitations on effects of nuclear radiation, illumination (for semiconductor strain gages), strong magnetic fields, and atmospheric conditions.

Thermal performance characteristics include resistivity or gage-resistance change and gage-factor change (equivalent to a thermal sensitivity shift) with temperature, expressed as percent per 100°F (or °C) over a specified operating temperature range. An additional statement of maximum temperature (which will not cause permanent performance changes in the gage) is desirable. Allowable overload at the temperature range limits is sometimes specified.

None of these thermal characteristics include any effects of thermal expansion of the strain gage and of differential thermal expansion between a strain gage and the specific material of the measured surface. Additional specifications are required for limits on these effects. The "apparent strain" as well as temperature gradient error ("transient temperature response") can be stated only for a strain gage installed by a specified method on a well-defined specimen. The different types of strains in an installation have been defined in some literature, such as NAS 942 (item 38 of bibliography), along the following lines: "Indicated strain" is the output of an installed strain gage as indicated on the readout equipment and as corrected for the errors introduced by all portions of the measuring system except the installed gage; "real strain" is the deformation in the measured surface (specimen) resulting from applied mechanical loads as well as thermal changes; "apparent strain" is the difference between indicated strain and real strain; "thermal strain" is the deformation of an unrestrained specimen due to a change in temperature alone; "mechanical strain" is the difference between real strain and thermal strain; "thermal output" is the indicated strain resulting only from thermal and apparent strains. Apparent strain has been defined more commonly as the indicated strain produced by a gage mounted on an unstrained specimen subjected to a change in temperature.

Thermal output tolerances, as applicable to a specific specimen material, are sometimes specified as deviation from a nominal and a possible additional average value.

13.4.2 Strain-gage attachment

With the exception of welded or embedded gages the usual method of attaching strain gages is bonding by means of an adhesive (cement).

The surface is first prepared by removing any paint, dirt, grease, and corrosion from it completely, then making it even and roughening it slightly by sandblasting or by use of fine-grit emery paper. If the surface is prepared too long before the strain gage is attached, it can be covered temporarily with vaseline which must later be removed with toluol and acetone or trichlorethylene. The prepared surface must never be touched by fingers. It is best prepared immediately before installation.

The cement is then applied to the surface, using an amount somewhat larger than required and squeezing out the excess later during clamping of the gage. The gage often receives a very thin coating of cement as well. The gage is then accurately positioned, lowered to the cement-carrying surface, pressed against it (through a plastic strip) manually or by clamping, and then left to air dry or temperature cure.

The type of cement and detailed procedures for applying it for gage installation and for curing depend on the strain gage selected and its carrier (if any), on the specimen material, and on the expected operating temperatures and humidity levels. Such detailed procedures are usually available from the manufacturer. Some manufacturers also offer cements and installation aids. Three categories of cement are in common use: acetone-celluloid and acetone-butyl-acetate cements which dry by solvent (acetone) release, polymerizing cements, and ceramic cements. Acetone-solvent cements cure at room temperature and require very little clamping pressure. Room-temperature-curing siliconitroglyptal, carbinol, and epoxy-polymerizing cements require a somewhat higher clamping pressure. Installations cured at room temperature are limited to operating temperatures of about 100°F. Heat-curing polymerizing cements such as bakelite and certain epoxies are usable at operating temperatures about 50°F below their cure temperature, which can be as high as 600°F. Required clamping pressures are between 20 and 90 psi. Ceramic cements are heat-cured or flame-sprayed without clamping pressure. The cure time of cements can be as short as 5 min and as long as 70 hr depending on cement used and type of gage and installation. Cure cycles can be quite complex and may involve controlling the rate of cure-temperature rise and holding this temperature constant for certain lengths of time at different levels. Heat curing can be accomplished by use of one or more infrared-heat lamps or an electric heater (e.g., of the film or blanket type).

Clamping can be performed by placing a metal plate over a strip of nonadhesive plastic (e.g., Teflon®) on the installation and, if more pressure is required, using a spring clamp or spring-loaded clamp to press against the plate. A rubbery material which is able to withstand the curing temperature (e.g., silicone rubber) should be placed between the plate and the plastic strip to assure an even pressure on the gage. The plate should be preformed to match the specimen or structure if the installation is on a curved surface.

After curing the installation should be moisture-proofed to avoid swelling of the cement, reduction of insulation resistance, and moisture condensation on the strain-sensing material. Moisture-proofing compounds include paraffin waxes, silicone rubber, and varnish, and certain resins. A thin layer of vaseline can be used for short-term moisture proofing at room temperatures.

13.4.3 Typical signal-conditioning circuits

The resistance changes in a strain gage are almost invariably converted into voltage changes. The conversion is performed by passive resistive signal-conditioning networks.

A simple voltage-divider circuit (Figure 13-10) can be used where only the variable component of dynamic strains needs to be measured. The d-c voltage drop across the strain gage is removed from the output by a coupling capacitor so that only the a-c variation of the voltage drop appears as output voltage.

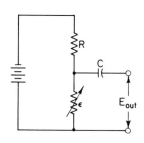

Strain-gage-bridge (Wheatstone-bridge) circuits are used in all other types of strain measurements. They are also usable in the measurement of dynamic strains. Either d-c or a-c excitation can be used. The latter is often preferred when the bridge output must be amplified and when interconnecting cables are relatively short. The bridge network is normally balanced (no output) for a condition of zero strain acting on the active gage(s) so that application of strain causes bridge output due to unbalance.

The one-active-arm strain-gage bridge (Figure 13-11) is useful only for metal strain gages at room temperature.

Fig. 13-10. Simple circuit for dynamic measurements.

Compensation for temperature effects can be achieved by using an active gage as well as a dummy gage of the same type and material and exposed to the same temperature as the active gage but isolated from mechanical strain. When connected in adjacent bridge arms as shown in Figure 13-12, the increase in resistance due to temperature alone is equal for both gages and is not reflected in the bridge output voltage since no bridge unbalance is effected. Only the additional resistance change due to strain in the active gage causes bridge unbalance.

It is important that the dummy gage be mounted on a plate of material having the same thermal characteristics as the structure under test. The dummy plate must be in good thermal contact with the test structure but this bond must not cause the gage to be strained when load is applied to the structure. The dummy-gage technique

Fig. 13-11. Strain-gage bridge with one active arm.

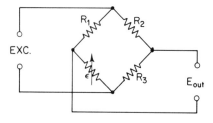

Fig. 13-12. Temperature compensation by dummy gage.

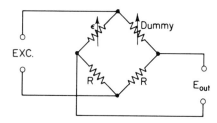

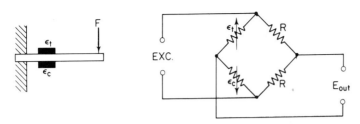

Fig. 13-13.   Two-active-arm strain-gage bridge.

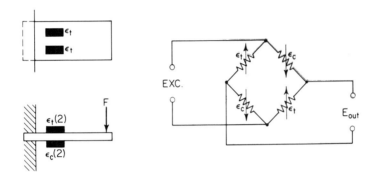

Fig. 13-14.   Four-active-arm strain-gage bridge.

is usually not effective in compensating for transient temperature conditions.

A two-active-arm strain-gage bridge (Figure 13-13) can be used in certain applications such as on bending beams, where a strain gage on the upper surface senses tensile strain while a gage of the same type and resistance on the lower surface senses compression strain. If the magnitude of the two strains is equal, the bridge output voltage will be twice that of a one-active-arm bridge. If both gages see the same temperature, they will also compensate each other to a large degree for thermal effects on gage resistance as well as for differential thermal expansion between each gage and the beam. Similarly, four closely matched gages mounted two on the upper, two on the lower, surface of a bending beam can be connected as a four-active-arm bridge (Figure 13-14) to provide four times the bridge output of a one-active-arm bridge as well as to compensate for thermal effects. Additional compensation provisions are required when the characteristics of the gages are not exactly matched in all active gages used.

The use of strain-gage-bridge connections for linearizing the output of two or four matched semiconductor strain gages (see 13.3.4) sensing strain in the same direction (all tensile or all compression) is shown by the examples in Figure 13-15 where N and P indicate gages having the same characteristics except for a negative or positive

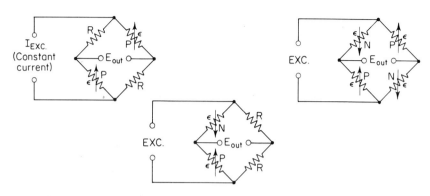

Fig. 13-15. Compensation for nonlinearity of semiconductor strain gages.

gage factor, respectively (as also indicated by the directions of the arrows in the strain-gage symbol).

13.4.4 Considerations for selection

The choice of a particular type of strain gage and its installation method depends primarily on the material and geometry of the measured structure or specimen, the expected temperature environment, the area available for gage installation, and whether static or dynamic strain measurements are intended. Other important criteria are gage factor, gage resistance, strain range, and expected overloads, the length of time over which the installation is planned to be used and the number of test operations during this time, and the relative availability of technician skill and installation tools.

Installations of single gages are much rarer than those of an active/dummy gage pair since some temperature compensation is usually necessary. It should also be noted that bondable strain gages are usually sold as a group of five or ten gages with similar characteristics. Two biaxially-installed active/dummy gage pairs or biaxial rosettes are used when strains are expected to occur in known orthogonal directions. Triaxial rosettes are useful in determining strains of unknown directions. Full-bridge rosettes provide maximum bridge output in certain applications. Separate temperature compensation by dummy gages or other means is usually required for rosette installations.

Before a specific strain gage is selected, its installation should be planned completely. Allowances should be made for proper wiring connections. Subminiature terminal strips are available for a connection between the delicate gage leads and the wires which connect to the signal-conditioning network. The resistance of these wires can be affected by the temperature environment so as to introduce measuring errors. When a-c excitation is used, the capacitance between the conductors and between conductors and ground can vary due to thermal effects and due to physical motions of the cable. The cable tie-down provisions as well as the gage installation itself must be chosen so as to minimize any effects on the strain that is to be measured. Excessively sturdy attachments can reinforce the test structure in the measuring area sufficiently to invalidate data from the measuring system.

Special sealing provisions must be incorporated in gage installations and associated cable connections used to measure strain under water. Gages used in a cryogenic-temperature environment must be of a strain-sensing material which still has sufficient resistance at such low temperatures and whose crystalline structure is not changed drastically. The conductors of connecting cables can be affected similarly. Insulating materials must not get brittle enough to fail at such temperatures. Many bonding cements are not usable at very low temperatures. Good results have been obtained with weldable gages provided with a thin stainless-steel-sheathed magnesium-oxide-filled cable whose conductors are made of Constantan. The cable is tied down to the structure by means of small welded clips. This type of installation has also been found satisfactory in high-temperature environments (to about 600°C). Satisfactory performance at both cryogenic and high temperatures has also been reported when surface-transferable gages were heat-bonded with ceramic cement. Flame-sprayed strain gages appear to be usable at temperatures exceeding 650°C.

13.5 Calibration and Tests

The calibration and test philosophy applicable to strain gages differs from that applied to most other transducers, for which each item manufactured is individually calibrated and some environmental tests can even be required on a 100% sampling basis. It has been said, with some justification, that a strain gage is not really a transducer until it is installed. The essential performance characteristics of a strain gage can only be determined when it has been installed to a well-defined specimen or structure by a specified procedure. After it has been so installed, it cannot normally be reused or sold. Hence it is not feasible to calibrate each gage individually. The only characteristic which can be tested without difficulty on an uninstalled gage is its resistance.

The *gage-resistance* test is usually performed by the manufacturer on an individual (100% sampling) basis. The *change of gage resistance with temperature* can also be determined and such tests are often performed on a 100% basis on semiconductor strain gages, where this resistance change is of significant magnitude, at one selected elevated-temperature level.

Dimensional checks could also be performed on each gage. However, this is made unnecessary by the closely controlled repetitive production methods which must be employed to keep the gages economically competitive.

All other characteristics can only be verified on a low-percentage sampling basis (2% to 0.01% depending on the characteristic). One or several gages are selected at random from a production run, mounted to a specimen and subjected to the tests. The results of these tests are then considered typical for the sampled production lot.

This test philosophy can be considered valid not only because too much testing would raise the price of strain gages excessively and because the nature of mass-production methods employed assure a fair degree of consistency in characteristics but also for another important reason: The methods and materials of the end-use installation can cause essential gage characteristics to be quite different from those reported by the manufacturer since his values are obtained from a limited number of installation types under closely controlled conditions, with the installations performed by specific skilled personnel. It is, therefore, a frequent and common practice to calibrate the actual installation or its facsimile by loading it in a testing machine, and in a temperature environment if necessary, so as to simulate as closely as possible the strains expected to occur in the measured structure during its intended use.

It should be noted that partial exceptions to this philosophy are made in the case of certain semiconductor strain gages the gage factor of which can be expected to vary appreciably within any given production lot. Removable bonds have been developed for test installations on bending beams so that 100% gage factor and overload tests can be performed. The critical lead bonds are also checked on each of such gages and visual inspection is more thorough than for metal-grid strain gages.

The procedure used to bond sample gages to a specimen for use in a test fixture should be well documented and it should be consistent for the same type of gage. Such information as surface preparation procedure, type of adhesive, bonding method, cure cycle (including time and temperature), clamping method and pressure, any lead wire attachment, and welding technique and equipment should be recorded for each test. This may be done by referencing the applicable paragraph in a general test document. Test equipment, test method, equipment certification, excitation voltage, frequency (if a-c), and current should be similarly recorded for each test.

A set of detailed test requirements intended for metal strain gages used in aerospace application is contained in National Aerospace Standard NAS 942.

13.5.1 Verification of gage factor, linearity, and hysteresis

These room-temperature-performance characteristics can be verified by mounting the gages to a constant-moment test beam which is part of a test fixture. Strain in the beam, at the location where the gages are mounted, is measured with a reference strain gage which has previously been calibrated against an optical strain indicator or with the optical strain indicator itself. Once the characteristics of the beam have been established with sufficient certainty, a reference strain gage is sufficient and occasional spot checks of actual strain in the beam can be made by means of a close-accuracy dial-indicator displacement gage. Force is applied to the beam in an increasing and decreasing direction so that strain at the gage location is varied at equal discrete increments over the strain range, typically at 500 $\mu\epsilon$ increments over a ± 3000 $\mu\epsilon$ range. The output of each strain gage is recorded at each point. When required by a specification, this calibration cycle is repeated twice and gage factors, hysteresis, and linearity determined from each of the three calibration cycles are averaged. Other specifications may require that the largest hysteresis and the poorest linearity of each of the three calibration cycles be reported as test results. If required, the data from the three cycles can also be used to determine repeatability which, however, is rarely specified.

13.5.2 Other room-temperature tests

Creep is determined by loading the specimen to which the gage samples are mounted in a loading device at a constant level of either tension or compression and recording the outputs for the prescribed length of time. Typical test requirements are for a one-hour application of a 1000 $\mu\epsilon$ strain level.

Transverse Sensitivity is verified, on a qualification basis, by mounting the gage to be tested on a specimen in a loading jig so that strain is applied at 90 degrees to the sensing axis of the gage. During this test the strain in both axes of the gage is monitored to assure that the strain along the sensing axis of the gage is negligible. The output of the tested gage is monitored, and transverse sensitivity is calculated by dividing the indicated strain by the applied transverse strain which is typically 1000 $\mu\epsilon$.

Fatigue Life tests are performed by mounting the gages to a constant-stress beam fixed at one end and deflected at the free end by a cam-driven linkage. Stress is applied unidirectionally at two or three different zero-to-peak levels or, preferably, bidirectionally at one peak-to-peak strain level, at a specified rate (frequency), for a specified number of cycles or until failure, while the outputs of the test gages are monitored for discontinuities on an oscilloscope. Typical test requirements call for a ± 1500 $\mu\epsilon$ application at a rate of 30 Hz. After each specified number of cycles the cycling is stopped and the output is recorded as indicated strain at positive and negative applied strain peaks and at zero applied strain. Test results either report the number of cycles after which failure occurred or that a specified number of cycles were applied without causing failure, depending on applicable specifications.

Strain Limit (allowable overload) can be determined by tensile loading of a specimen to which the strain gage is mounted at discrete increments until the error in the output of the tested gage reaches the tolerance specified for this characteristic. Since strain levels can be quite high, an extensometer can be used to monitor the actual deformation of the specimen. The test may be continued until failure of the gage or its bond.

13.5.3 Temperature tests

Thermal effects on strain-gage-performance characteristics are determined as gage factor change with temperature and as apparent or thermal strain. Additional tests or test cycles are used to verify thermal hysteresis, strain limit at operating temperature extremes, and temperature-induced drift.

For apparent strain or thermal strain and for drift measurements, the gages to be tested are mounted on a rectangular plate ("coupon"). For the determination of gage-factor change the gages are mounted to a constant-stress beam. A properly bonded gage should experience the same thermal strain as the specimen to which it is bonded.

Moderate variations from room temperature are obtained in a temperature chamber. Heat lamps are used for high temperatures. A cryostat can be used to expose the specimen to very low temperatures. The specimen temperature at the installed gages is always monitored by a thermoelectric or resistive temperature transducer attached to the specimen surface.

During *apparent strain* and *thermal strain* measurements the specimen remains unloaded while the temperature of the specimen is varied from room temperature to one or (successively) both limits of the operating temperature range while temperature and strain-gage outputs are recorded. For *drift* determination the temperature is maintained at a level less than the (usually upper) temperature extreme for a specified length of time while outputs are recorded and constancy of temperature is monitored. Insulation resistance is often verified during the drift test. For *gage-factor-change* verification the temperature should be stabilized at the specified level (e.g., one or both temperature extremes) before the specified value of strain is obtained in the specimen by loading.

Thermal hysteresis can be determined from outputs recorded, with no loading applied, while the temperature is cycled in an increasing and decreasing direction one or more (to obtain average values) times over the operating temperature range.

The *strain limit at temperature extremes* is measured on a qualification basis by repeating the strain limit test (see 13.5.2) at one or both limits of the operating temperature range.

Temperature-gradient error (*transient temperature response*) tests are performed, also on a qualification basis, by mounting the gages on a thin rectangular coupon, which remains unloaded during the test, and by exposing them to a step change in radiant heat by turning on a heat lamp to heat the specimen at an approximately known rate while outputs and specimen temperature are continuously recorded. The test can be repeated at different heating rates each of which is given by the voltage at which the lamp is made to operate.

13.5.4 Other environmental tests

Tests for effects of other specified environments, such as nuclear radiation, strong magnetic fields, ambient pressure (altitude), and illumination (on semiconductor strain gages) are sometimes performed on a qualification basis, usually without loading the gage.

13.6 Related Measuring Devices

13.6.1 Fatigue-life gages

Foil grids of special design have been developed for use as fatigue-life sensors. They are bonded to a structure in the same manner as a strain gage. The fatigue experience of the structure, in cyclic or random loading, causes a permanent change in the resistance of the gage. The characteristics of the gage must be chosen so as to match the characteristics of the specimen material for the range of strains to which the gage is to be subjected.

Quantitative fatigue data can be obtained by first performing a fatigue-vs-resistance-change calibration on a gage mounted on a piece of the material on which it will be used at a point where cracking will eventually start. The specimen is then cycled in a testing machine until cracking occurs to determine the resistance change in the gage at the time the cracking starts in the specimen. The amount of resistance change of the gage in relation to the accumulated fatigue damage of the specimen will be governed by the magnitude of any fixed preload, the magnitude and type of alternating load, and the program of any random loading.

13.6.2 Crack-propagation gages

Foil grids consisting of a number of closely-spaced short parallel conductors connected to a bus strip at one end and provided with individual tabs (terminals) at the opposite end have been used to follow the progress of a crack through a test specimen. The grid is applied over the expected crack so that each of the fine parallel conductors is broken as the crack propagates under it. The extent of the crack can then be determined, at any time, by checking continuity to the bus strip at each terminal or by connecting a resistance network to all terminals so that each broken conductor causes an appropriate change of the overall network resistance.

BIBLIOGRAPHY

1. Tomlinson, H., "On the Increase of Resistance to the Passage of an Electric Current Produced on Certain Wires by Stretching," *Proceedings of the Royal Society*, London, Vol. 26, pp. 401–410, 1877.

2. Hetenyi, M. (Ed.), *Handbook of Experimental Stress Analysis.* New York: John Wiley & Sons, Inc., 1950.

3. Yarnell, J., "Resistance Strain Gages: Their Construction and Use," *Electronic Engineering*, London, 1951.

4. Koch, J. J., Boiten, R. G., Biermasz, A. L., and Roszbach, G. P., "Strain Gauges— Theory and Application," Eindhoven, The Netherlands: Philips Technical Library, 1952.

5. Murray, W. M. and Stein, P. K., "Strain Gage Techniques," Lectures and Laboratory Exercises, Cambridge, Massachusetts: Massachusetts Institute of Technology, 1955.

6. Perry, C. and Lissner, H., *The Strain Gage Primer*. New York: McGraw-Hill Book Company, 1955.

7. Smith, C. S., "Piezoresistive Effect in Germanium and Silicon," *Physics Review*, Vol. 94, p. 42, April 1954.

8. "Characteristics and Applications of Resistance Strain Gages," *N.B.S. Circular No. 528*, Washington, D.C.: National Bureau of Standards, 1954.

9. Zelbstein, U., *Technique et Utilization des Jauges de Constrainte*. Paris: Dunod, 1956.

10. Mason, W. P. and Thurston, R. N., "Use of Piezoresistive Materials in the Measurement of Displacement, Force and Torque," *Journal of the Acoustical Society of America*, Vol. 29, p. 1096, 1957.

11. Turichin, A. M. and Novitskiy, P. V., "Provolochnyye Preobrazovateli i ikh Tekhnicheskoye Primeneniye" (Wire-Resistance Strain Gages and their Applications), Moscow: Gosudarstvennoye Energeticheskoye Izdatel'stvo, 1957.

12. Fink, K. and Rohrbach, C., *Handbuch der Spannungs und Dehnungsmessung* (Handbook of Stress and Strain Measurement). Duesseldorf, Germany: VDI Verlag, 1958.

13. Stein, P. K., "Pressure Effects on Strain Gages," *Strain Gage Readings*, Vol. I, No. 1, p. 43, 1958.

14. Stewart, R. J., "Strain Gages in Supersonic Aircraft," *Electronic Industries*, March 1958.

15. Stein, P. K., "Semiconductor Strain Gages: Gage Factors in the Hundreds," *Strain Gage Readings*, Vol. II, No. 2, p. 39, June–July 1959.

16. Stein, P. K., "Some Notes on Thermoelectric Effects in Strain Gage Circuits," *Strain Gage Readings*, Vol. II, No. 3, p. 17, 1959.

17. Bloss, R. L., "Characteristics of Resistance Strain Gages," *Preprint No. 90-NY60*, Pittsburgh: Instrument Society of America, September 1960.

18. Glagovskiy, B. A., Rakhman, B. M., and Zubkov, V. M., "Electric Resistance Tensometer" (Ti-Al alloy foil gage), U.S.S.R. Patent No. 153138, May 1960.

19. Hollander, L. E., Vick, G. L., and Diesel, T. J., "The Piezoresistive Effect and Its Applications," *Review of Scientific Instruments*, March 1960.

20. Padgett, E. D. and Wright, W. V., "Silicon Piezoresistive Devices," *Preprint No. NY60-42*, Pittsburgh: Instrument Society of America, 1960.

21. Stein, P. K., "The Frequency Response of Strain Gages," *Strain Gage Readings*, Vol. III, No. 2, p. 9, 1960.

22. Sanchez, J. C., "Semiconductor Strain Gages: A State of the Art Summary," *Strain Gage Readings*, Vol. IV, No. 4, October-November 1961.

23. Talmo, R. E., "Semiconductor Strain Gages Offer High Sensitivity," *Electronics*, February 1961.

24. Beyer, F. R. and Smith, J. O., "Evaluation of High Temperature Strain Gages for Enrico Fermi Reactor Vessel," *Applied Mechanics*, Vol. 2, No. 3, p. 81, March 1962.

25. Dean, M., III (Ed.), *Semiconductor and Conventional Strain Gages*. New York: Academic Press, 1962.

26. Leszynski, S. W., "The Development of Flame Sprayed Sensors," *I.S.A. Journal*, July 1962.

27. Sanchez, J. C., "Semiconductor Strain Gages—Standard Evaluation Production Testing Practices," *Preprint No. 11.2.62*, Pittsburgh: Instrument Society of America, 1962.

28. Wu, C. T., "Transverse Sensitivity of Bonded Foil Gages," *Experimental Mechanics*, Vol. 2, No. 11, p. 338, November 1962.

29. "Strain Gage Handbook," *Bulletin 4311A*, Waltham, Massachusetts: BLH Electronics, 1962.

30. Parker, R. L. and Krinsky, A., "Electrical Resistance-Strain Characteristics of Thin Evaporated Metal Films," *Journal of Applied Physics*, Vol. 34, No. 9, 1963.

31. Wnuk, S. P., Jr., "Development and Evaluation of Strain Gages for Cryogenics," *Preprint No. 38.2.63*, Pittsburgh: Instrument Society of America, 1963.

32. Allnut, R. B. and Palermo, P. M., "Strain Gage Measurements on Submarine Structures," *Preprint No. 16.11-2-64*, Pittsburgh: Instrument Society of America, 1964.

33. Bray, A. and Plassa, M., "The Strain Sensitivity of Ge and Cr-Si Thin Films Deposited Under Vacuum," *Paper 21-IT-193, Acta IMEKO 1964*, IMEKO, Budapest, Hungary, 1964.

34. Kemény, T., "Die Untersuchung von Praezisions-Mess-Streifen ohne Traegerfolie, sowie Kompensation des Kriechens für Praezisions-Gewichts-, Kraft- und Druckmessung" (Examination of Precision Strain Gages without Carrier Base and of Compensation for Creep in Precision Weight, Force and Pressure Measurement), *Paper 21-HU-143, Acta IMEKO 1964*, pp. 57–75, IMEKO, Budapest, Hungary, 1964.

35. Scott, I. G., "Temperature Coefficient Determination of Resistance Strain Gages," *Structures and Materials Note 290*, Melbourne: Department of Supply, Australian Defence Scientific Service, Aeronautical Research Laboratories, July 1964.

36. Shapiro, B. and Tolotta, S., "Encapsulated Strain Gage for Use in Steam Environment," *Preprint No. 16.11-4-64*, Pittsburgh: Instrument Society of America, 1964.

37. Weymouth, L. J., "Strain Gage Application by Flame Techniques," *Preprint No. 16.5-1-64*, Pittsburgh: Instrument Society of America, 1964.

38. "Strain-Gages, Bonded Resistance," *National Aerospace Standard NAS942*, Washington, D.C.: Aerospace Industries Association of America, Inc., 1964.

39. Bloss, R. L., Trumbo, J. T., and Melton, C. H., "Methods for Determining the Performance Characteristics of Resistance Strain Gages at Elevated Temperatures," *NBS-R-9019*, Washington, D.C.: National Bureau of Standards, December 1965.

40. Chironis, N. P., "Changes in Strain Gages" (Fatigue Life Gages), *Product Engineering*, Vol. 36, No. 25, pp. 83–88, December 1965.

41. Davis, H. J. and Horn, L., "Notes on the Use of Semiconductor Strain Gages," *Report TR-1285*, Washington, D.C.: Harry Diamond Laboratories, April 1965.

42. Dorsey, J., "N-Type Self-Compensating Strain Gages," *Experimental Mechanics*, Vol. 5, No. 9, pp. 27A–38A, September 1965.

43. Melton, C. J. and Bloss, R. L., "Four Methods of Determining Temperature Sensitivity of Strain Gages at High Temperatures," *ISA Journal*, pp. 69–74, October 1965.

44. Bradley, C. D., "Strain Indicator for Semiconductor Strain Gages," *ISA Journal*, pp. 55–57, January 1966.

45. Dearinger, J. A., Gesund, H., and Pincus, G., "Instrumentation for Measuring Strains and Deflections in Buildings," *Preprint No. 16.1-4-66*, Pittsburgh: Instrument Society of America, 1966.

46. Dykes, B. C., "Strain Measurement at Temperatures from 500 to 1800°F," *Preprint No. 16.9-1-66*, Pittsburgh: Instrument Society of America, 1966.

47. Johansson, J. W., "Strain Gages for Cryogenic Use," *Instruments and Control Systems*, Vol. 39, No. 1, pp. 116–118, January 1966.

48. Tatnall, F. G., *Tatnall on Testing*, Metals Park, Ohio: American Society for Metals, 1966.

49. Telinde, J. C., "Investigation of Strain Gages at Cryogenic Temperature," *Douglas Paper No. 3835*, Huntington Beach, California: Douglas MSD, 1966. (Presented to Society for Experimental Stress Analysis 1966 Annual Meeting.)

50. "Fatigue Life Gage Applications Manual," Romulus, Michigan: Micro-Measurements, Inc., 1966.

51. Onnen, O., "The Determination of Stress-Patterns Around Strain Gages and Their Influence on Measuring Accuracy," *Paper BRD-215, Acta IMEKO 1967*, IMEKO, Budapest, Hungary, 1967.

14 Temperature

14.1 Basic Concepts

14.1.1 Basic definitions

The **temperature** of a body is its thermal state considered with reference to its power of communicating heat to other bodies; it is a measure of the mean kinetic energy of the molecules of a substance; it is the potential of heat flow.

Heat is energy in transfer, due to temperature differences, between a system and its surroundings or between two systems, substances, or bodies. It has also been defined as the energy contained in a sample of matter comprising potential energy resulting from interatomic forces as well as kinetic energy associated with random motion of molecules in the sample.

Heat transfer is the transfer of heat energy by one or more of the following methods:

1. **Conduction**—by diffusion through solid material or through stagnant fluids (liquids or gases).
2. **Convection**—by the movement of a fluid (between two points).
3. **Radiation**—by electromagnetic waves.

581

Thermal equilibrium is a condition of a system and its surroundings (or two or more systems, substances, or bodies) when no temperature difference exists between them (no more heat transfer occurs between them).

The **ice point** (273.15°K) is the temperature at which ice is in equilibrium with air-saturated water at a pressure of 1 atmosphere.

The **steam point** (100°C) is the temperature at which steam is in equilibrium with pure water at a pressure of one standard atmosphere.

The **triple point of water** (273.16°K) is the temperature at which the solid, liquid, and vapor states of water are all in equilibrium. This occurs at the pressure of 4.6 mm Hg.

14.1.2 Related laws

Boyle's law
In a given quantity of gas, $pV = C$ (constant) if the temperature is held constant. (p = pressure, V = volume.)

Charles' law
In a given quantity of gas, $\dfrac{T}{V} = C$ (constant) if the pressure is held constant (T = absolute temperature), and $\dfrac{T}{P} = C$ (constant) if the volume is held constant.

Ideal-gas law

$$\frac{pv}{T} = R = \frac{p}{\rho T}$$

where: p = pressure
v = specific volume
T = absolute temperature
R = gas constant
ρ = density

Fourier's law (for conducted heat)

$$Q = -kA\frac{dt}{dL}$$

where: Q = heat transferred across a surface of A square feet (in Btu/hr)
t = temperature (in °F)
L = thickness (in inches) of homogeneous wall across which heat is conducted
k = *thermal conductivity* (characteristic of wall material)
$\dfrac{dt}{dL}$ = *temperature gradient*—rate of change of temperature with thickness of wall

Stefan-Boltzmann law (radiant heat)

$$Q_T = \sigma AT^4$$

where: Q_T = total heat radiated from an ideal blackbody surface (total hemispherical emission in all wavelengths)

σ = Stefan-Boltzmann constant

A = area of emitting surface

T = absolute temperature of emitting surface

Units of Q_T	Units of A	Units of T	Value of σ
watts (W)	in.^{2}	°R	3.49×10^{-12}
watts (W)	cm^2	°K	5.67×10^{-12}
ergs/second	cm^2	°K	5.67×10^{-5}
Btu/hour	ft^2	°R	0.172×10^{-8}

Heat transferred by radiation (British units)

$$Q_r = 0.173A\left[\left(\frac{T_1}{100}\right)^4 - \left(\frac{T_2}{100}\right)^4\right]F_e F_A$$

where: Q_r = radiant heat transferred (Btu/hr)

A = area of surface receiving radiation (ft^2)

T_1 = absolute temperature of radiation source (°R)

T_2 = absolute temperature of radiation receiver (°R)

F_e = emissivity factor (allows for nonblackbody characteristics of source)

F_A = angle factor (allows for relative position and geometry of source and receiver)

14.1.3 Units of measurement

Temperature is expressed in **degrees Celsius** (°C), **degrees Fahrenheit** (°F), **degrees Kelvin** (°K), or **degrees Rankine** (°R) depending on the temperature scale used. The International System of Units (SI) recognizes only the degree Kelvin. The °K is one of the six basic units of the SI. It is the unit of temperature in the thermodynamic scale defined by assigning the temperature value 273.16°K to the triple point of water.

Temperature scales were arbitrarily established and later confirmed by international agreement:

	Celsius	*Fahrenheit*	*Kelvin*	*Rankine*
Steam point	100 °C	212 °F	373.15 °K	671.67°R
Ice point	0 °C	32 °F	273.15 °K	491.67°R
Absolute zero	-273.15 °C	-459.67 °F	0 °K	0 °R
Scale conversion	°C = (°F $-$ 32) $\times \frac{5}{9}$		°F = $\frac{9}{5} \times$ °C + 32	
	°K = °C + 273.15		°R = °F + 459.67	

A convenient scale conversion table is shown in Table 14-1, located at the end of the chapter. Note that integers in both °F and °C columns facilitate interpolation.

Absolute temperature is expressed in °K or °R. The Kelvin scale is known as the *thermodynamic scale*.

Heat flux is expressed in **British thermal units per square foot per hour** $(Btu/ft^2/hr)$, **ergs per square centimeter per second** $(erg/cm^2/s)$, **joules per square meter per second** $(J/m^2/s)$, **watts per square inch** $(W/in.^2)$, **watts per square centimeter** (W/cm^2), or **calories per square centimeter per second** $(cal/cm^2/s)$.

14.2 Sensing Elements

Most temperature sensing elements, notably thermoelectric and resistive elements, also perform the transduction in temperature transducers. Such sensing/transduction elements are here referred to simply as "sensing elements."

14.2.1 Thermoelectric sensing elements

14.2.1.1 Thermoelectric effect. Thermoelectric sensing elements are used in *thermocouple circuits*. A basic thermocouple circuit, as shown in Figure 14-1, consists of a pair of wires of different metals joined together at one end (*sensing junction*) and terminated at their other end by terminals (*reference junction*) maintained at an equal and known temperature (*reference temperature*). A load resistance constituted by the input impedance of signal-conditioning or readout equipment completes the circuit. When a temperature difference exists between the sensing junction and the reference junction, an electromotive force is produced which causes current to flow through the circuit. This *thermoelectric effect*, caused essentially by contact potentials at the junctions, was discovered by Thomas J. Seebeck (1770–1831), a German physicist, and is also known as the *Seebeck effect*.

Each of the wires between the sensing junction and the reference junction and

Fig. 14-1.   Basic thermocouple circuit.

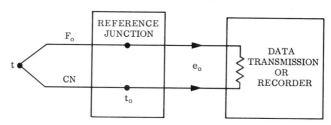

$$e_o = At + \tfrac{1}{2} Bt^2 + \tfrac{1}{3} Ct^3$$
$$t = \text{JUNCTION TEMPERATURE AT MEASUREMENT POINT}$$
$$A, B \,\&\, C = \text{CONSTANTS DEPENDING ON THE MATERIAL OF}$$
$$\text{WHICH THE THERMOCOUPLE IS MADE}$$

$$\text{FOR SMALL TEMPERATURE SPANS}$$
$$e_o = St$$
$$S = A + Bt + Ct^2$$

any intermediate terminals or connection in the wire must all be made of the same material. The connecting leads from the reference junction to the load resistance are usually copper wires. They must be copper whenever the associated wiring in the signal-conditioning or readout circuits is made of copper so as to avoid junctions of different metals in the portion of the thermocouple circuit located within this equipment.

Related phenomena are known as the Peltier and Thomson effects: When a current flows across a junction of two dissimilar conductors, heat (unrelated to normal I^2R heating) is absorbed or liberated dependent upon the direction of current flow (*Peltier effect*). When a current flows through a wire along which a temperature gradient exists, heat (unrelated to normal I^2R heating) is absorbed or liberated in the wire (*Thomson effect*).

14.2.1.2 Thermal electro-motive force. The magnitude of the thermoelectric potential produced depends on the wire materials selected and on the temperature difference between the two junctions. Figure 14-2 shows the thermal emf generated by various thermocouples. The values shown are based on a reference temperature of 0°C.

Tables showing thermal emf vs temperature, based on standard quality wire materials, have been developed for a variety of thermocouples by the National Bureau of Standards, the Instrument Society of America, and several manufacturers. Most tables are based on a reference temperature of 0°C; however, similar tables can be derived for other reference temperatures.

14.2.1.3 Thermocouple materials. The most commonly used thermocouples are Chromel-Alumel†, iron-Constantan‡, copper-Constantan, Chromel-Constantan, and platinum-platinum/rhodium alloy. A thermocouple of 90% platinum and 10% rhodium alloy to platinum (Pt-10Rh/Pt) is used to define the International Temperature Scale from 630.5 to 1063°C but is usable to temperatures up to 1482°C (2700°F). Copper-Constantan is generally used for temperature measurements to +371°C (700°F), iron-Constantan to +760°C (1400°F), Chromel-Alumel and Chromel-Constantan to +1260°C (2300°F), and platinum-platinum/13% rhodium to 1482°C (2700°F). The recently developed tungsten-rhenium thermocouples are useful for temperature measurements to +2760°C (5000°F). A gold/cobalt-copper combination shows good characteristics in the temperature range below −190°C.

The recommended upper temperature limit for thermocouples decreases with wire size. The limit for Chromel-Constantan, for example, decreases from 2300°F

---

†The words "Chromel" and "Alumel" are registered trade names of Hoskins Mfg. Co., Detroit, Mich. Chromel P is an alloy of 90% nickel, 10% chromium; Alumel is composed of 95% nickel, the remaining 5% containing aluminum, silicon, and manganese.

‡An alloy of 55% copper and 45% nickel.

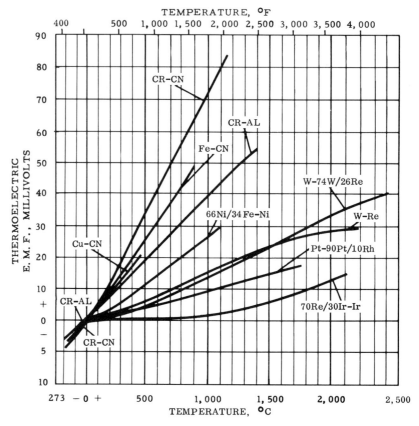

Fig. 14-2. Thermocouple output-vs-temperature characteristics.

Fig. 14-3. Thermopile, schematic diagram.

THERMOPILE CIRCUIT

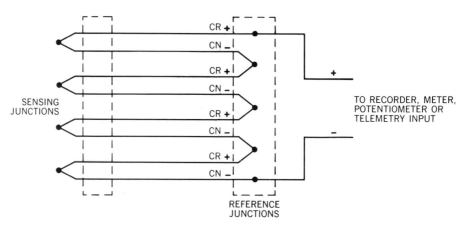

for a No. 8 AWG wire to 1600°F for a No. 28 AWG wire. Limits shown are primarily intended for enclosed-element ("protected") thermocouples. They are somewhat lower for exposed sensing junctions.

14.2.1.4 The **thermopile** is a combination of several thermocouples of the same materials connected in series. Figure 14-3 shows the schematic of a typical Chromel-Constantan (CR/CN) thermopile. The output of a thermopile is equal to the output from each thermocouple multiplied by the number of thermocouples in the thermopile assembly. All reference junctions must be at the same temperature. The polarities of emf shown in the figure are based on a temperature higher at the sensing junctions than at the reference junctions.

14.2.2 Resistive sensing elements

Resistive temperature-sensing elements are conductors and semiconductors which change their resistance as the temperature changes. Sensing elements used in *resistance thermometry* include primarily metal wires (platinum, nickel, and nickel alloys) and thermistors. Deposited metal films, germanium and silicon crystals, and carbon resistors are also used.

Values of *temperature coefficients of resistance* (α) and *resistivity* (ρ) for various metals and carbon are shown in Table 14-2. A negative coefficient signifies that the

Table 14-2   RESISTANCE PROPERTIES OF PURE, ANNEALED METALS
AND CARBON (SELECTED FROM VARIOUS SOURCES)

Material	Temperature coefficient of resistance (α) over 0–100°C range (ohms per ohm per °C)	Resistivity at 20°C (microhm-cm)
Aluminum	0.0042	2.69
Carbon	−0.0005	3500 (@ 0°C)
Constantan (55% Cu, 45% Ni)	±0.00002	49.0
Copper	0.0043	1.673
Gold	0.0039	2.3
Indium	0.0047	8.4 (@ 0°C)
Iron	0.0065	9.71
Manganin (86% Cu, 12% Mn, 2% Ni)	−0.00002	43.0
Nickel	0.00681	6.844
Nichrome (60% Ni, 16% Cr, 24% Fe)	0.0002	109.0
Palladium	0.00377	10.8
Platinum	0.00392	9.81 (@ 0°C)
Silver	0.0041	1.63
Tungsten	0.0046	5.5

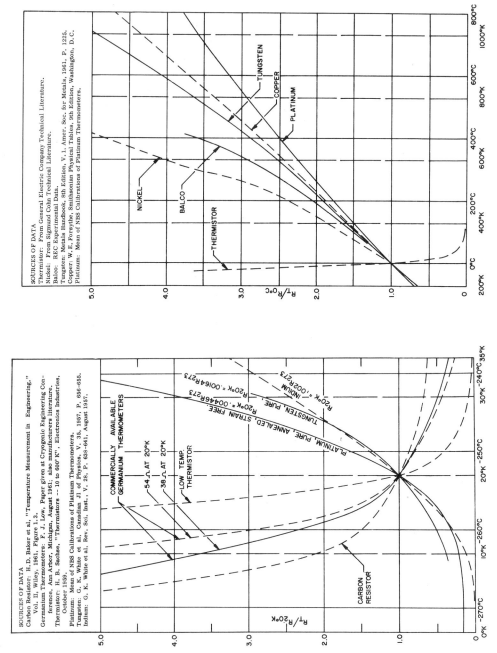

Fig. 14-4. Relative resistance-vs-temperature characteristics of resistive sensing elements (courtesy of Rosemount Engineering Co.).

SOURCES OF DATA
Thermistor: From General Electric Company Technical Literature.
Nickel: From Sigmund Cohn Electrical Literature.
Balco: REC Experimental Data.
Tungsten: Metals Handbook, 8th Edition, V. 1, Amer. Soc. for Metals, 1961, P. 1225.
Copper: W. E. Forsythe, Smithsonian Physical Tables, 9th Edition, Washington, D.C.
Platinum: Mean of NBS Calibrations of Platinum Thermometers.

SOURCES OF DATA
Carbon Resistor: H.D. Baker et al, "Temperature Measurement in Engineering,"
    Vol. II, Wiley, 1961, Figure 1.3.
Germanium Thermometers: F. J. Low, Paper given at Cryogenic Engineering Con-
    ference, Ann Arbor, Michigan, August 1961; also manufacturers literature.
Thermistor: H. B. Sachse, "Thermistors -- 10 to 600°K", Electronics Industries,
    October 1959.
Platinum: Mean of NBS Calibrations of Platinum Thermometers.
Tungsten: G. K. White et al, Canadian Jl of Physics, V. 35, 1957, P. 656-655.
Indium: G. K. White et al, Rev. Sci. Inst., V. 28, P. 638-641, August 1957.

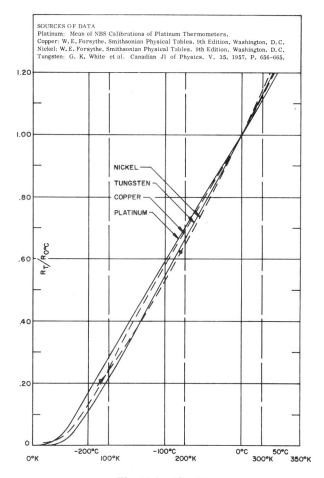

SOURCES OF DATA
Platinum: Mean of NBS Calibrations of Platinum Thermometers.
Copper: W. E. Forsythe, Smithsonian Physical Tables, 9th Edition, Washington, D. C.
Nickel: W. E. Forsythe, Smithsonian Physical Tables, 9th Edition, Washington, D. C.
Tungsten: G. K. White et al. Canadian Jl of Physics, V. 35, 1957, P. 656-665.

NICKEL
TUNGSTEN
COPPER
PLATINUM

Fig. 14-4. (Con't)

resistance will increase with decreasing temperature; this occurs typically in thermistors, carbon, and germanium, and in silicon at temperatures below $-50°C$. A high α is desirable in a temperature-sensing element so that a substantial easy-to-measure resistance change occurs even when the range of temperatures to be measured is narrow. Another desirable characteristic is high resistivity (resistance unit area per unit length) to reduce the amount of material required for use in practical resistive elements.

Figure 14-4 shows the variation of resistance with temperature for several commonly used materials.

An approximate relationship for the resistance-vs-temperature characteristics of conductors in the temperature range near $0°C$ can be calculated from the equation

$$R_t = R_0(1 + \alpha t)$$

where: R_t = resistance at temperature t (°C)

R_0 = resistance at 0°C

α = temperature (or *thermal*) coefficient of resistance

14.2.2.1 Conductors.

Platinum-Wire Elements. A winding of pure, strain-free, annealed platinum wire is widely used in transducers for precise temperature measurements. An element of this type is used to define the International Temperature Scale between the limits of −183°C and +630°C, and it is possible that this range will be extended in the future by international agreement. For practical applications this type of element can be used from −265 to +1050°C.

The resistance-vs-temperature relationship of a platinum-wire element is given by the *Callendar-VanDusen equation*

$$R_t = R_0 + R_0\alpha[t - \delta(0.01t - 1)(0.01t) - \beta(0.01t - 1)(0.01t)^3]$$

where: R_t = resistance at temperature t (°C)

R_0 = resistance at 0°C (ice-point resistance)

α, δ, β are constants; typical values are:

$\alpha = 0.003925$ (temperature coefficient of resistance near 0°C)

$\delta = 1.49$

$\beta = 0.11$ (if t is negative)

$\beta = 0$ (if t is positive)

This equation is applicable over the range −183°C to +630°C. Values of decreasing accuracy are obtained when these limits are exceeded on either side of the applicable range.

Platinum-wire elements are used in transducers for most laboratory work, for missile and space instrumentation, and for industrial measurements when special requirements for accuracy exist.

Nickel-Wire Elements. Elements wound from pure nickel wire or from nickel-alloy wire such as "Balco" are used in the medium temperature range (−100 to +300°C). They are less costly than platinum-wire elements and are relatively easier to manufacture. They are commonly used in transducers intended for industrial and aeronautical applications.

Tungsten-Wire Elements show promising performance over a wide temperature range. Their resistance-vs-temperature characteristics are not yet as well established as those of platinum wire. They compare in cost with platinum elements but are more difficul to manufacture, primarily because of the high temperatures required for their annealing and their brittleness resulting from this necessary process.

Copper-Wire Elements. Elements wound from copper wire were frequently used in the past. Although their resistance-vs-temperature characteristics show an unusually good linearity, their low resistivity requires the use of impractically long wire lengths. The development of the other wire elements has essentially made copper elements obsolete.

Metal-Film Elements. Metallic films deposited on insulating material, such as platinum on ceramic, have been used successfully for temperature measurement. The resistance-vs-temperature relationship of such sensing elements is empirical; i.e., it must be determined from measured points since no theoretical relationship has been established.

14.2.2.2 Semiconductors.

Thermistors are semiconductors whose resistance varies rapidly with temperature. The semiconductor materials, usually sintered mixtures of sulfides, selenides, or oxides of metals such as nickel, manganese, cobalt, copper, iron, and uranium, are formed into small glass-enclosed beads, disks, or rods. They have high resistivities and high negative temperature coefficients of resistance. Their resistance-vs-temperature characteristics are not linear, and it is difficult to maintain narrow resistance tolerances during their manufacture. Long-term drifts noted in their operation have been ascribed to diffusion of impurity ions resulting from the passage of the electric measuring current. They are normally used over a temperature range of -75 to $+250°C$. Special thermistors have been developed for temperature measurements in the range below $24°K$.

Germanium Crystals. This semiconductor sensing element, developed primarily for temperature measurements in the 1 to $35°K$ region, is a single crystal of germanium with controlled impurities. The repeatability of germanium thermometers has been reported as within $0.001°C$ near the helium boiling point and within $0.05°C$ near the hydrogen boiling point. Their resistance-vs-temperature characteristics show a nonlinearity similar to that of thermistors. The effects of magnetic fields on germanium crystals will require further study. A high degree of calibration interchangeability between elements is difficult to obtain in production.

Silicon Crystals. Small crystals of pure silicon with controlled impurities ("doping") show promising characteristics as semiconductor resistive temperature sensors. They differ from thermistors and germanium elements in that their temperature coefficient of resistance is positive above $-50°C$ and their resistance-vs-temperature characteristics are much more linear, particularly over their most usable range, $-50°C$ to $+250°C$. Below $-50°C$ their temperature coefficient of resistance becomes negative and the slope of their resistance-vs-temperature characteristics increases sharply. These characteristics can be changed by the use of different doping methods. Silicon elements have been used for liquid-hydrogen-temperature measurements.

Carbon Resistors. Commercially available carbon resistors have been used as temperature sensing elements in the region below 60°K, where their behavior is similar to that of semiconductors described above. Their high resistivity and negative temperature coefficient of resistance assures a relatively high resistance in the lower portion of their range. The resistance-vs-temperature relationship follows a simple empirical equation. Some of the disadvantages encountered in their use are their sensitivity to pressure variations when used in pressurized ducts or tanks, long-term drifts in their resistance, short-term resistance drifts after cycling between cryogenic and room temperatures, and a relatively long response time.

Gallium Arsenide p-n junction devices (diodes), which have a relatively low piezoresistive coefficient, have been designed for use in ranges from −270 to +300°C. They have also been reported usable around the boiling point of liquid helium.

14.2.3 Oscillating-crystal sensing elements

The sensitivity of oscillating quartz crystals to temperature has long been accepted in the field of communications engineering. This phenomenon is utilized for temperature measurements in the range −40 to +230°C in quartz crystals specially designed to have a highly linear temperature coefficient of frequency. The thin disc is sliced at an orientation known to provide such a linear coefficient from high-quality synthetic single-crystal quartz. When connected into an oscillator circuit and excited at its third overtone resonance (approximately 28 MHz), it can exhibit a frequency-vs-temperature slope of 1 kHz/°C. The output of the crystal-controlled oscillator is mixed with that of a reference oscillator so that a beat frequency is obtained which can be displayed on a frequency counter.

14.3 Design and Operation

14.3.1 Thermoelectric temperature transducers

Thermoelectric temperature transducers are commonly referred to as *thermocouples* even when they contain only the sensing junction, since their eventual connection to the reference junction and load resistance upon their installation (end use) can be taken for granted.

14.3.1.1 Thermocouples and extension wire. Thermocouples have been used for temperature measurement so extensively and for such a long time that standards were deemed necessary for some of the most commonly used types. Most countries have one or more national standards for thermocouples. In the U.S.A., standards were developed by technical societies and industrial organizations as well as by the Department of Defense. Military standards cover iron-Constantan, copper-Constantan, and Chromel-Alumel materials. Somewhat different requirements are set forth in civilian standards. U.S.A. Standard C96.1-1964, "Temperature Measurement Thermocouples," is intended to replace all other civilian standards covering the same subjects. Instrument Society of America (ISA) Recommended Practices RP1.1–1.7 (1959), "Thermocouples and Thermocouple Extension Wires," served as basis for the U.S.A. Standard.

Existing standards differentiate between thermocouple wires which form the sensing junction (measuring junction) and thermocouple extension wires, which can be connected to the thermocouple wires for longer cable runs without introducing thermoelectric potentials at these connections. Because some of their characteristics are less critical, extension wires can be made at less cost than thermocouple wires. When the appropriate extension wire is properly connected to a thermocouple, the reference junction is in effect transferred to the other end of the extension wires.

Accepted standards also discourage the use of the names of materials, especially proprietary names, to designate a thermocouple, because materials of different names can result in a thermocouple of equal characteristics. Such names as Chromel, Constantan, and Alumel are proprietary names, trade names registered by individual wire manufacturers. The same alloy often exists under a different name when made by a different manufacturer. In order to resolve such discrepancies a number of wire-type letter designations have been established. Copper-Constantan thermocouples, for example, are designated as "Type T" thermocouples. The matching extension wire for this material combination is "Type TX" extension wire.

Table 14-3 shows standard letter designations and color codes for commonly used thermocouples and extension wires, in addition to suggested abbreviations for materials (when letter designations are not used, e.g., on wiring diagrams). Polarities shown apply when sensing-junction temperature is higher than reference-junction temperature. Tracer colors in overall protective covering, denoting type and class of wire in military specifications, are not shown. Output values are shown in Table 14-4, which is located at the end of the chapter.

Normally used thermocouple-wire gages are: 8, 14, 20, 24, and 28 AWG (American Wire Gage) for Types J, K, and E; 14, 20, 24, and 28 AWG for Type T; and 24 AWG only for Types R and S. Extension wire gages are usually limited to 14, 16, and 20 AWG.

Other often used thermocouple-wire materials, not covered by U.S.A. Standard C96.1-1964 and not shown in Table 14-2, include platinum-30% rhodium/platinum-6% rhodium, iridium-40% rhodium/iridium, tungsten/iridium, tungsten/rhenium, tungsten/tungsten-26% rhenium, tungsten-5% rhenium/tungsten-26% rhenium, tungsten-3% rhenium/molybdenum, and certain other combinations intended for measurement of very high or cryogenic temperatures.

Table 14-3 Thermocouple Materials, Type Letters, and Color Codes for Duplex Insulated Thermocouples and Extension Wires

Typical conductor material +	Typical conductor material −	Material abbreviation	USA std. type	USA std. color code overall	USA std. +cond.	USA std. −cond.	MIL-STD-687 overall	MIL-STD-687 +cond.	MIL-STD-687 −cond.	Color used by
THERMOCOUPLES										
Iron	Constantan	Fe/CN	J	brown	white	red	red / lt. blue	grey / black	red / yellow	Army, Navy / Air Force
Copper	Constantan	Cu/CN	T	brown	blue	red	white / black	blue / red	red / yellow	Army, Navy / Air Force
Chromel	Alumel	CR/AL	K	brown	yellow	red	yellow / white	yellow / white	red / green	Army, Navy / Air Force
Chromel	Constantan	CR/CN	E	brown	purple	red				
90% Platinum 10% Rhodium	Platinum	Pt-10Rh/Pt	S							
87% Platinum 13% Rhodium	Platinum	Pt-13Rh/Pt	R							
EXTENSION WIRES										
Iron	Constantan	Fe/CN	JX	black	white	red	red / lt. blue	grey / black	red / yellow	Army, Navy / Air Force
Copper	Constantan	Cu/CN	TX	blue	blue	red	white / black	blue / red	red / yellow	Army, Navy / Air Force
Chromel	Alumel	CR/AL	KX	yellow	yellow	red	yellow / white	yellow / white	red / green	Army, Navy / Air Force
Chromel	Constantan	CR/CN	EX	purple	purple	red				
Copper	Alloy 11 (1)	—	SX (2)	green	black	red				
Iron	Alloy 125 (1)	—	WX (3)	white	green	red				

Notes: (1) Typical material as used by Thermo Electric Co., Inc.
(2) Used with type R and S thermocouples as alternate extension wire.
(3) Used with type K thermocouples as alternate extension wire.

594

14.3.1.2 Wire thermocouples. A basic thermocouple is made from two wires of thermocouple materials by stripping the insulation from the wires at the end of the cable and forming a junction between the two bare wire tips. Better service is usually obtained from wire thermocouples made from two-conductor (duplex) shielded or unshielded thermocouple cable, with solid or stranded conductors (depending on mechanical flexibility desired) or from swaged, metal-sheathed, ceramic-insulated thermocouple cable which exists in many sizes and with various conductor, insulator, and sheath materials. As the wire size decreases, the time constant will become shorter; however, ruggedness will also be lowered.

Junctions can be prepared by silver soldering, brazing, or welding. Since the melting point of the solder limits the usable temperature of the thermocouple and since both soldering and brazing add another metal as well as organic fluxes to the junction, the usually preferred method is welding. *Butt-welded* junctions are prepared by bending the two wire ends until they butt against each other and then welding the two ends together.

Cross-wire junctions are made by bending the two wire ends at an angle less than 90 degrees, laying one wire over the other, welding them together at this crossing, and then cutting excess wire material off the end. In *coiled-wire* junctions one wire is kept straight, and the other is wound tightly over the straight wire end. This type, as well as *twisted-wire* junctions, are usually silver soldered or brazed, but welding can be used with some material combinations.

Grounded enclosed ("protected") junctions frequently used in thermocouple probes are made in several ways: by brazing or welding wire of one thermocouple material to the inside of a probe tip of the other material, by joining both thermocouple wires together and to the inside of the tip, or by welding the two wire ends separately, but in close proximity to each other, to the inside of the probe tip. Exposed sensing junctions are frequently grounded when they are attached to a conducting surface whose temperature is to be measured.

14.3.1.3 Foil thermocouples (Figure 14-5) are designed for surface temperature measurements where thin flat sensing junctions are required. Each half of the symmetrical foil pattern is made from a different thermocouple material. A butt-type junction is formed at the point where the two foil segments meet. A foil thickness of 0.0005 in. or less is commonly used. Two design versions exist: a free-foil style with removable carrier base and a matrix type with the foil embedded in very thin plastic material. Either type can be bonded to flat or curved conducting or nonconducting surfaces. Related micro-miniature thin-film thermocouples have also been produced.

14.3.1.4 Immersion probes. Thermocouple probes are a frequently used adaptation of wire thermocouples. The sensing junction can be *exposed*, as in the probe

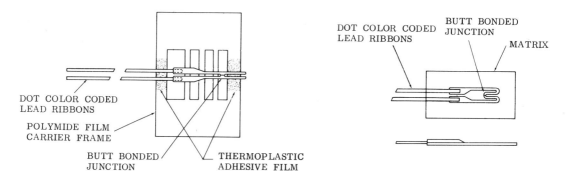

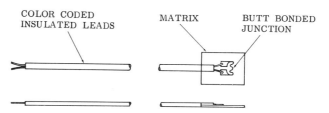

Fig. 14-5. Typical foil thermocouples (courtesy of RDF Corp.).

shown in Figure 14-6, or *enclosed*. Enclosed junctions can be either grounded or isolated. Semienclosed *stagnation* junctions admit the measured fluid to the junction while shielding it against fluid impact under rapid flow conditions. Essential portions of a typical thermocouple probe are the junction, the probe sheath, or *well*, the *head* with a possible mounting thread, the *seal* in immersion probes, which can be integral or of the *gland* type (as by use of a compression nut with sealing gland), and the electrical connector or the unshielded, shielded, or metal-sheathed connecting cable originating at the head.

Thermocouple probes are available in numerous sizes and configurations, most of which are custom-designed for a given application. Very thin wells with sharply pointed tips simulating the needle of a hypodermic syringe permit thermoelectric sensing of intravenous blood temperature. Probes with externally threaded wells, sealed to withstand high pressures, can be mounted in industrial presses, extrusion dies, and nozzles. Spring-loaded thermocouples, sometimes equipped with bayonet-lock heads, are used on internal components such as bearings. Probes with threaded bosses and provisions for a compression seal (O-ring or gasket) measure temperatures of liquids and gases in ducts, pipes, tanks, and other vessels.

14.3.1.5 Special configurations. Other special versions of thermocouple-type sensors include washers and pipe clamps with integral sensing junctions and connecting cable or leads and junctions embedded in very small threaded stainless-steel inserts for use in thin walls.

14.3.1.6 Differential thermocouples are useful when a measurement of only the temperature difference between two points must be made. Better system accuracy can be obtained for such relatively narrow-range measurements. In a differential thermocouple circuit a second sensing junction replaces the reference junction of known temperature. A wire of one material (e.g., iron) leads from a terminal block to the

Fig. 14-6. Short-stemmed thermocouple probe with exposed element and internal high-pressure seal (courtesy of Fenwal Electronics).

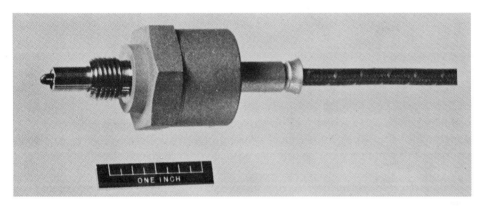

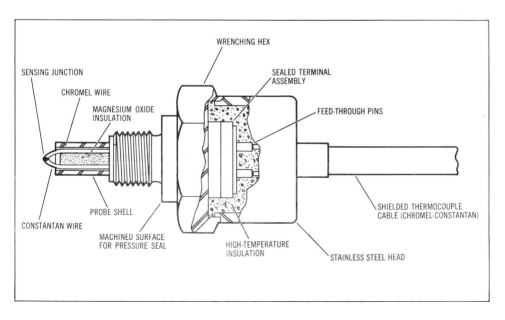

first sensing junction made with a wire of the other material (e.g., Constantan). The second wire, in turn, leads to the second sensing junction made with a wire of the first material (iron, in the example given). This third wire leads to the other terminal of the terminal block. The two terminals must be at the same temperature if further connections from them are to be made with copper wire. However, the temperature of the terminals need not be known. The approximate temperature of at least one of the two points of measurement should be known so that the applicable portion of the emf-vs-temperature curve can be determined.

14.3.2 Resistive temperature transducers (wire element)

14.3.2.1 Immersion probes (probe-type transducers). The major portions of a typical probe-type temperature transducer are shown in Figure 14-7. The *sensing element* is near the tip of the (perforated) probe shield, shell, or *sheath*. The housing, or *head*, can be filled with packing material, such as insulating fibers, which provide protection to the internal wires and trimming resistor against shock, vibration, and ambient-temperature fluctuations. The trimming resistor is sometimes used in series with the sensing element to adjust the total transducer resistance to a desired value. A precision-machined mounting thread is an integral part of the head. An internal *seal* located near the root of the mounting threads enables the probe to operate in high-pressure ducts or vessels without rupture or leakage. Figure 14-8 shows the components of a somewhat differently designed dual-exposed-element temperature probe intended for use in high-pressure gas bottles. The protecting tube is attached over the probe sheath during shipping, storage, and handling only.

Fig. 14-7. Probe-type transducer with exposed wire element (courtesy of Rosemount Engineering Co.).

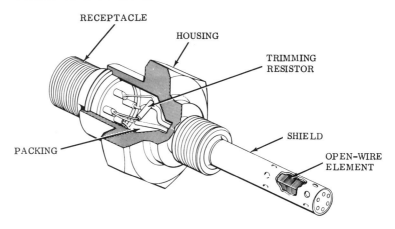

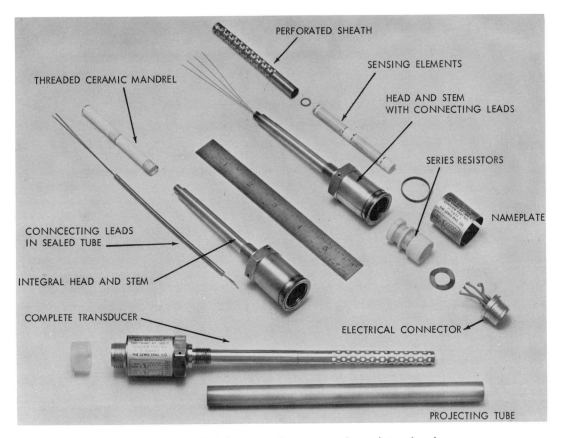

Fig. 14-8. Components of dual-element probe-type transducer; internal seal operates to 12,000 psi (courtesy of Lewis Engineering Co.).

A different probe design is shown in Figure 14-9. The sheath is completely enclosed to protect the sensing element from corrosive fluids. The receptacle pin insert, in the form of a hermetically sealed header, completes the seal on the internal portion of the transducer.

An enclosed-element industrial transducer is shown in Figure 14-10. The head is a waterproof junction box with provisions for a conduit coupling.

Probe Sensing Elements. The nickel or platinum wire element of the probe shown in Figure 14-9 is wound over a flat insulating *mandrel* sandwiched between two mica sheets for electrical insulation. Thermal conductor springs provide good thermal contact between element and sheath; they also act as shock absorbers. This sensing element has a time constant of approximately 2 s in agitated water. It is frequently used for aircraft and industrial measurements.

ALL METAL HEADER
WITH GLASS COMPRESSION
TERMINAL INSERTS

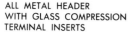

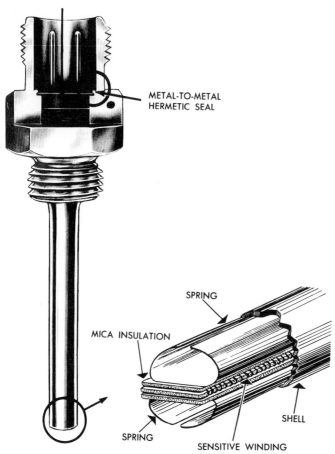

METAL-TO-METAL
HERMETIC SEAL

SPRING

MICA INSULATION

SPRING

SHELL

SENSITIVE WINDING

Fig. 14-9. Probe-type transducer with enclosed wire element (courtesy of Thomas A. Edison Industries).

Fig. 14-10. Industrial resistive immersion probe (courtesy of Thomas A. Edison Industries).

Platinum-wire elements were used primarily for laboratory applications, such as in reference-standard platinum "resistance thermometers," until missile and space-vehicle measurements established requirements for accurate, wide-range, fast-response, rugged temperature transducers. Basic design characteristics of this new family of transducers include full support of the wire to protect against mechanical shock and vibration, pure platinum wire ranging in thickness from 0.5 to 3 mils, thermal expansion matching of mandrel to wire, minimizing of strain in the wire during the winding process, and annealing after completion of the winding.

A group of platinum wire elements is shown in Figure 14-11. The ceramic *coated element* is constructed by coating a platinum tube with ceramic, winding the resistance wire over the coated tube, then coating the finished winding with ceramic. The ceramic material must be carefully chosen to match the thermal expansion of the platinum wire to avoid thermal stresses in the element when the temperatures vary. The ceramic material is fired at a temperature high enough to assure annealing of the winding. This type of element can be wound with very thin wire so that an ice-point resistance up to 6200 ohms can be obtained with a winding only 0.8 in. long. Although the element is electrically insulated by its ceramic coating, the heat barrier formed by the

Fig. 14-11. Typical sensing elements for probe-type resistive transducers (courtesy of Rosemount Engineering Co.).

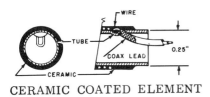

CERAMIC COATED ELEMENT

MINIATURE CERAMIC
COATED ELEMENT

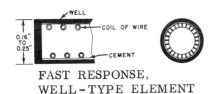

FAST RESPONSE,
WELL-TYPE ELEMENT

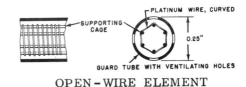

OPEN-WIRE ELEMENT

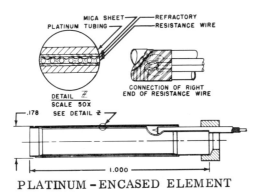

PLATINUM-ENCASED ELEMENT

coating is small enough to yield a time constant of approximately 200 ms in agitated (3 ft/s) water. Dual elements, electrically isolated from each other, can be prepared by interlacing the two windings (as in a bifilar winding) on the same mandrel.

The miniature ceramic coated element also has a time constant of about 200 ms in agitated water but can dissipate less power and is therefore more subject to self-heating effects on its resistance.

The well-type element is a completely *enclosed element*. The wire is not wound on a mandrel but is formed into a helix which is inserted into the sheath. A thin coating of cement on the inside of the sheath wall fixed the element in place and provides the necessary electrical insulation. Since only a fraction of the wire diameter is held by the cement, the winding remains strain-free even when strain occurs in the sheath. A time constant of approximately 200 ms in agitated (3 ft/s) water can be realized.

The open-wire element, which is also shown on the transducer in Figure 14-7, is an *exposed element*. The wire itself comes in contact with the measured fluid without any intermediate heat barrier. The wire is wound loosely over a supporting cage of thin platinum rods and coated with ceramic at the point of wire contact to provide fixing of the element and electrical insulation between wire and rod. The element has a time constant of less than 80 ms in water moving at 3 ft/s, and approximately 500 ms in air moving at 50 ft/s.

The platinum-encased element is another version of a fast-response enclosed element. The element is wound between the walls of a hollow cylinder. The measured fluid can circulate through the tube as well as around its external surface. Other successful element styles include exposed windings on finely threaded ceramic mandrels (as used in the transducer of Figure 14-8) and coated elements utilizing sintered Teflon instead of ceramic. Two or more elements can be wound on one threaded mandrel if two simultaneous but mutually isolated outputs are to be provided by one transducer.

A different type of element design is used in the temperature transducer of Figure 14-12, which is intended primarily for industrial applications. The sensing element consists of several small coils of pure platinum wire connected in series. Each coil is inserted into a hole within a ceramic rod and is bonded to the inside of the hole over a small portion of its circumference. Connecting leads are attached, and the rod is sealed. This construction results in a virtually strain-free sensing element. The completed rod is ground for a tight fit within the probe sheath so that good heat transfer is obtained and damaging "rattling" in a vibration environment is avoided.

Elements similar to those described can be wound with tungsten wire. Nickel and nickel-alloy wire elements are usually of less sophisticated construction so that low production costs can be maintained.

Probe-Design Characteristics. The stem inside the sheath sometimes supports the sensing element and normally contains the connecting leads of nickel, platinum, or other precious-metal wire. These leads are always thicker than the sensing-element wire so that a change in their resistance is negligible compared to the resistance change in the sensing wire. Good connecting-lead design emphasizes minimizing of any thermoelectric potential which could be generated if the connections of the two parallel leads are at different temperatures and act as thermocouples.

The leads are frequently sealed in ceramic which fills the portion of the stem near the probe's head. A high-pressure ceramic-to-metal seal is made either in the mounting thread area or throughout the supporting stem. Care is taken to avoid the possibility of leakage of the measured fluid, such as high-pressure gas, along the leads within the sealing area and outwards into the head.

The head of the transducer is the portion which remains external to the boss when the probe is installed. One piece of stainless-steel-bar stock is often machined to provide head, mounting, sealing surface, and a portion of the stem. Electrical connections on some transducers are made by means of a length of shielded high-temperature cable, moisture-sealed at its exit from the head. A more common configuration, however, uses a hermetically sealed receptacle. Industrial connecting heads are usually mounted to a thread at the lead-exit side of the transducer (see Figure 14-12).

Some probes now in use contain a Wheatstone bridge or other signal-conditioning network in their head. Others incorporate a trimming (series) resistor (see Figure 14-7) so selected that predetermined resistance values can be obtained at the end points of the transducer's nominal range.

14.3.2.2 Surface-temperature transducers. A large variety of wire-element transducer designs are used to measure surface temperatures. Named after the method of

Fig. 14-12. Probe-type resistive temperature transducer with multiple-coil sensing element (courtesy of Rosemount Engineering Co.).

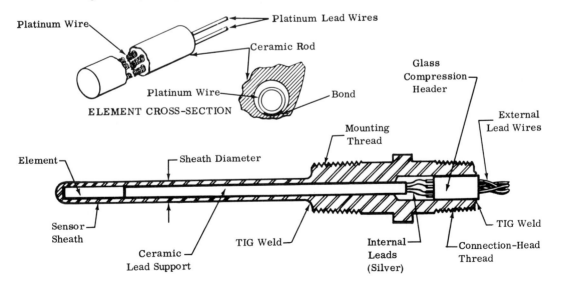

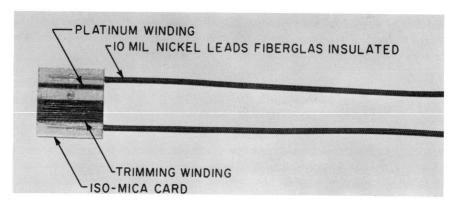

Fig. 14-13. Surface-temperature transducer with winding over insulator card (courtesy of Trans-Sonics, Inc.).

their installation, they include weldable, cementable, surface-transferable, pottable, clamp-on, and bolt-on (or washer) types.

The construction of a typical cementable surface-temperature transducer is shown in Figure 14-13. The sensing wire, which can be platinum or nickel, is wound around a flat insulating mandrel ("card"). The trimming winding is frequently omitted from the transducer itself and is replaced by a fixed or adjustable resistor external to the transducer, such as in the remotely located bridge network. The bare portions of the connecting leads are fastened to the card and act as terminals for the winding. The card, with winding and leads, is sandwiched between two cover sheets of thin phenolic- or epoxy-impregnated paper, cloth, or glass-base material, sealed around the edges of the card. Careful sealing is particularly necessary around the lead exits from the card.

The same basic card assembly can be used in *weldable* transducers in which the plastic cover sheets are replaced by metal foil. Insulating material is packed between the card and the inside surfaces of the metal sheets. Sealing around the edge of the metal-card-metal sandwich is accomplished by seam welding.

Fig. 14-14. Surface-temperature transducer with coiled element (courtesy of Rosemount Engineering Co.).

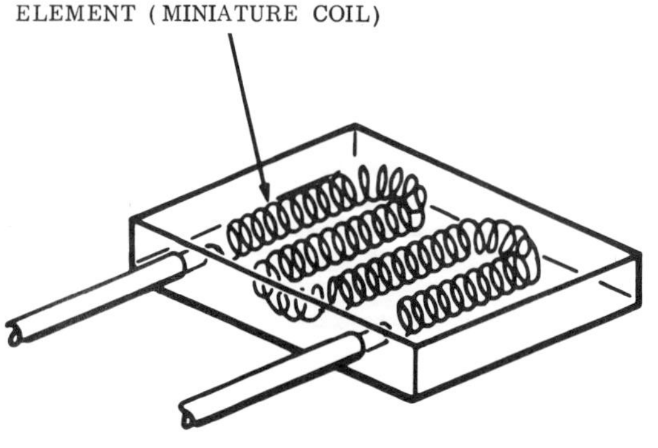

ELEMENT (MINIATURE COIL)

The surface-temperature transducer shown in Figure 14-14 has an entirely different construction. The coiled-wire sensing element is bonded to the inside of its small case at the bottom portion of its coil circumference only. Insulating material fills the inside of the case and prevents coil motion. The solid case construction allows a high lead pull-out strength. Transducers of this type have been designed for various mounting methods.

Bolt-on transducers are designed in the form of resistance wire spools sandwiched between two thin washers, sealed around their internal and external edges, and provided with connecting leads. Other bolt-on designs contain sealed cards wholly or partially surrounded by a sheetmetal piece with holes in its protruding edges.

Two-hole or four-hole bolt-on transducers can be preformed to fit over the curvature of a pipe of given outside diameter. Since they can then be clamped to the pipe, they are referred to as *clamp-on* type transducers. Cementable and weldable types can be preformed in a similar manner.

A *surface-transferable* transducer is a bare resistance wire or ribbon laid in zigzag fashion so that an overall square or rectangular outline is obtained. This element is stored between two plastic sheets and removed only for direct installation of the surface to be measured. The surface must first be coated with a thin layer of adhesive cement. The cement also provides electrical insulation if the surface is conductive. Another thin layer of cement is applied over the installed wire element. The cement must then be cured at a temperature and by a process specified by its manufacturer.

14.3.3 Metal-film-element resistive transducers

Various design approaches have been tried in the development of resistive transducers utilizing a metal film deposited on an insulator instead of a wire element wound on an insulating mandrel. The main advantages of a film element are its lower production cost and its relatively greater ruggedness compared to partially unsupported wire elements. Good results have been obtained with transducers using a platinum film less than one microinch in thickness deposited on a ceramic disk, cylinder, or wafer. In probe versions the film is deposited on the edge of a ceramic disk supported by the stiff connecting leads. A film of insulating material, such as Teflon®, can be applied over the sensing film. Surface sensing types have the film deposited on the edge or on one side of a ceramic disk approximately $\frac{1}{4}$ in. in diameter and 0.05 in. thick or on a $\frac{1}{8} \times \frac{1}{4}$ in. wafer as thin as 0.006 in. Weldable stainless-steel covers can be provided for surface sensors.

Usable ranges lie between the limits of -430 and $+800°F$. Ice-point resistance values between 300 and 8000 ohms can be obtained. Time constants of insulated-element transducers in agitated water range from 100 ms to over 1 s depending on insulation thickness, wafer mass, and heat conduction through supporting leads. Uninsulated elements yield time constants starting below 50 msec. The temperature coefficient of resistance of the film is between 30% and 80% of that of pure annealed platinum wire.

14.3.4 Transducers with semiconductor elements

14.3.4.1 Thermistors are characterized by their small size, usually high ice-point resistance, high negative temperature coefficient of resistance, fast time constant (associated with very small beads), and moderate cost. Another important characteristic is their limitation to use with small currents only (usually less than 100 μA).

Larger currents cause self-heating in the thermistor with resulting rapid lowering of its resistance. Careful control of excitation voltage to assure low current through the thermistor is essential to its use as temperature transducer. Thermistor resistance in the absence of self-heating has been referred to as *cold resistance*.

The basic resistance-vs-temperature characteristic of a thermistor at near-zero power is expressed by

$$\frac{R_T}{R_{T_0}} = e^{\beta(1/T - 1/T_0)}$$

where: R_T = cold resistance at measured absolute temperature T

R_{T_0} = known cold resistance at known absolute temperature T_0, usually stated at the ice point or at 298.15°K (25°C)

β = material constant (normally shown in °K)

(e = 2.718; base of natural logarithm)

Typical values of β are between 3000°K and 4500°K, as determined from resistance measurements at the ice point and at 50°C. Cold-resistance values for various types of thermistors range from 500 ohms to over 10 megohms at 25°C. Time constants, especially for bead thermistors, are quite short and are often specified as applicable in still air, where typical values are about 1 to 2 s. Time constants in agitated water (3 ft/s) are less than 50 ms. Dissipation constants (power required to raise thermistor temperature 1°C due to self-heating) vary between 0.1 and 2 mW.

Glass-coated *bead thermistors* are available in diameters from 0.01 to 0.09 in. Leads of platinum-iridium or "Dumet" material, varying in diameter from 0.001 to 0.012 in. can start at the same end or at opposite ends of the bead. Bead thermistors are frequently used as temperature transducers with no support or mountings other than their leads, particularly when very fast response is required.

Disk- and washer type thermistors have longer time constants than beads and are not in common use for temperature measurement.

Thermistor probes utilize various bead thermistors. The typical probes shown in Figure 14-15 contain a glass rod with a bead in its thickened tip. Exposed- and enclosed-element probes are available, with or without mounting thread or flange. Although usable measuring ranges lie within the limit of −100 and +600°F, they are more commonly used in the −30 to +300°F range. A special group of thermistors has been developed for use in the cryogenic temperature ranges. The time constant of an exposed-element probe is approximately 30 s in still air. Longer time constants prevail for enclosed elements.

Because of their relatively large resistance change over narrow temperature ranges, thermistor-type temperature transducers are frequently used in biomedical, marine-science, atmospheric-science, and other research measuring systems. Figure 14-16 illustrates an example of such a transducer used for measurement of temperature under water, in tanks, and in wells, where a permanently installed transducer is not required.

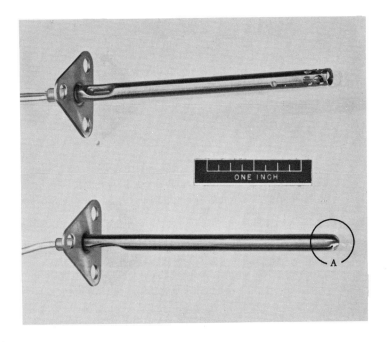

ONE INCH

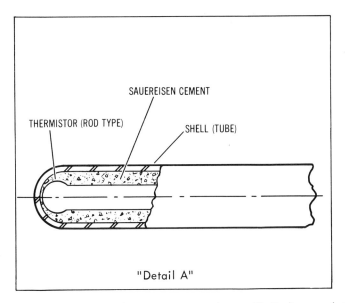

SAUEREISEN CEMENT

THERMISTOR (ROD TYPE)

SHELL (TUBE)

"Detail A"

Fig. 14-15. Flange-mounted temperature transducer with (top) exposed and (bottom) enclosed thermistor sensing element (courtesy of Fenwal Electronics).

Fig. 14-16. "Plumb bob" thermistor-type temperature transducer (courtesy of Atkins Technical Inc.).

14.3.4.2 Germanium-element transducers, as designed for cryogenic temperature measurements, use a doped germanium crystal approximately 0.1 in. thick at its center and 0.3 in. long. Dopants include arsenic, gallium, and (in slight amounts) antimony. Slightly thicker portions at each end of a typical crystal design support the connecting leads at their contact points. Four-wire crystals are available for use with precision readout equipment; current and voltage contacts are kept separate.

Two-wire crystals of the same configuration are used in conjunction with simpler measurement circuits. Transducer configurations include coated crystals, crystals encapsulated and hermetically sealed in metal covers, and immersion probes.

14.3.4.3 Silicon-element transducers have been designed as small, very thin ($0.09 \times 0.04 \times 0.0005$ in.) wafers, with leads at opposite ends, primarily for surface applications. They also exist in immersion-probe configurations. The resistance-vs-temperature curve is similar to that of metal wire between about 550°F and a point lying between 0°F and −100°F, depending on the manner of doping the single-crystal silicon material. Below this point the curve resembles that of a thermistor. Associated asymmetrical Wheatstone-bridge circuits can be used to compensate for the somewhat nonlinear curve. The current through the transducer must be kept very low. Due to the small mass-to-sensing-surface ratio a time constant of the bare sensor as low as 150 ms in still air can be realized.

14.3.4.4 Carbon resistors have a negative temperature coefficient which increases in magnitude sharply below about 10°K. Commercial carbon-composition resistors have been used experimentally in the cryogenic temperature range and have yielded satisfactory data in the liquid-hydrogen and liquid-helium regions. Wide practical applications of carbon resistors are severely limited by the low magnitude of the temperature coefficient of resistance above about 100°K where most temperature measurements are normally made. Carbon resistors have also been shown sensitive to variations in their ambient pressure.

14.3.4.5 Gallium-arsenide junction diodes have been developed for cryogenic and low-temperature measurements. When a constant forward current is maintained through a GaAs diode, the forward voltage varies almost linearly with temperature over the range 2 to 70°K and 100 to 300°K, although the slope of the voltage-vs-temperature curve at constant current is not the same for the two ranges. This characteristic offers certain advantages over the nonlinear characteristics of other semiconductor temperature transducer materials below 100°K. The relatively small size of the diode makes short time constants not too difficult to achieve.

14.3.5 Quartz-crystal temperature transducers

Transducers using oscillating crystals whose frequency changes with temperature can be adapted very easily for use with digital readout equipment. The frequency output obtained by electronically mixing the transducer-oscillator signal with a reference-oscillator signal is converted with negligible error into a number of counts per unit time which can be indicated on a digital display device either in °F or in °C depending on conversion-circuit constants and reference-oscillator frequency chosen. The crystal sensing element is contained in the tip of an approximately 8-in. long probe connected by coaxial cable to the electronics unit which contains power supply, oscillators, signal-conditioning, and display equipment. The transducer is best used in conjunction with the electronics unit supplied by the same manufacturer. Typical characteristics of the complete transducer system include a temperature range of −40 to +230°C, time constant in agitated (2 ft/s) water of about one second, terminal linearity (with end-points at 0 and 200°C) within ±0.15°C indicated even when the reference line extends to −40 and +230°C, and very small self-heating error.

14.4 Performance and Applications

14.4.1 Specification characteristics

The number and types of design and performance characteristics which should be considered for the specifications of temperature transducers vary widely for transducer designs of different complexity.

Wire Thermocouples—with sensing junction, connecting hardware, and associated circuits completed by the user—are specified by their material, wire gauge, number of conductors in cable, insulation and coverings, and accuracy ("standard" or "special"). Standard designations have been established for these characteristics and should be used where feasible. Manufacturers' standards exist for numerous types of metal-sheathed, ceramic-insulated thermocouple cable, and cable characteristics can be selected from such tabulated data.

Bead Thermistors intended to be assembled and connected by the user in a custom-designed package or installation can be specified by their bead covering and dimensions, lead configuration, dimensions and materials, cold resistance (with tolerances) at a specified temperature (usually 25°C), usable or intended temperature range, and either the material constant β or a curve of resistance vs temperature in

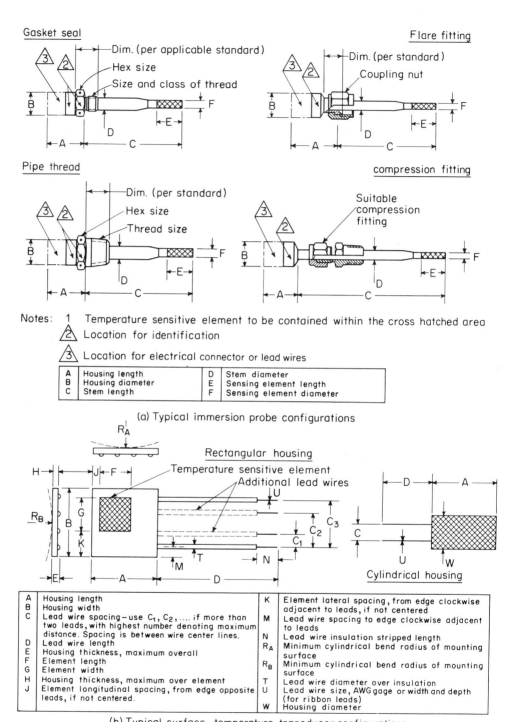

Gasket seal

3 ⚠ 2 ⚠ — Dim. (per applicable standard)
— Hex size
— Size and class of thread

Flare fitting

3 ⚠ 2 ⚠ — Dim. (per standard)
— Coupling nut

Pipe thread

3 ⚠ 2 ⚠ — Dim. (per standard)
— Hex size
— Thread size

compression fitting

Suitable compression fitting
3 ⚠ 2 ⚠

Notes: 1   Temperature sensitive element to be contained within the cross hatched area

2 ⚠   Location for identification

3 ⚠   Location for electrical connector or lead wires

A	Housing length	D	Stem diameter
B	Housing diameter	E	Sensing element length
C	Stem length	F	Sensing element diameter

(a) Typical immersion probe configurations

Rectangular housing

Temperature sensitive element
Additional lead wires

Cylindrical housing

A	Housing length	K	Element lateral spacing, from edge clockwise adjacent to leads, if not centered
B	Housing width	M	Lead wire spacing to edge clockwise adjacent to leads
C	Lead wire spacing—use C_1, C_2, ... if more than two leads, with highest number denoting maximum distance. Spacing is between wire center lines.	N	Lead wire insulation stripped length
D	Lead wire length	R_A	Minimum cylindrical bend radius of mounting surface
E	Housing thickness, maximum overall		
F	Element length	R_B	Minimum cylindrical bend radius of mounting surface
G	Element width		
H	Housing thickness, maximum over element	T	Lead wire diameter over insulation
J	Element longitudinal spacing, from edge opposite leads, if not centered.	U	Lead wire size, AWG gage or width and depth (for ribbon leads)
		W	Housing diameter

(b) Typical surface–temperature transducer configurations

Fig. 14-17. Dimensional specifications for temperature transducers.

610

graphical or tabular form. Additionally, the time constant and dissipation constant in still air can be specified.

Similar considerations apply to other types of temperature transducers when specified, essentially, as a sensing element. A larger number of characteristics have to be considered when a complete ("packaged") surface-temperature transducer needs to be specified. An even greater number of characteristics can be applicable to completely packaged immersion-probe type transducers. These characteristics are described below in more detail.

14.4.1.1 Mechanical design characteristics. After determining whether a probe (immersion) or surface-temperature transducer is to be specified, several essential design characteristics should be considered. Configuration and dimensions have to be chosen with extreme care. Typical examples for both immersion and surface types are illustrated in Figure 14-17. Mounting provisions and methods and their effect on configuration should be determined. The stem length of an immersion probe intended for installation in a pipe or duct should be so chosen as to place the center of the sensing element at a radial position located $0.72r$ (for turbulent flow) or $0.58r$ (for laminar flow) from the centerline of the duct, where r is the duct radius. Special modifications to the configuration, such as a stagnation fitting or a spring-loading arrangement, should be detailed. Allowable or actual weight should be included in the specification.

Measured fluids should be listed and associated limitations on sensing-element protection, stem materials, and housing of a surface transducer defined. The maximum transverse flow of the measured fluid should be specified for immersion probes whenever it can be established. Head (housing) material is usually determined on the basis of its compatibility with the stem material of a probe and with the ambient atmosphere. Special requirements may have to be shown for assembly methods, finish, nature of stem, and housing seal. The operating and proof pressure and a burst-pressure rating are always shown for an immersion probe. An allowable leakage rate can be stated as applicable at the specified proof pressure. Lead pull-out strength (maximum force not causing lead separation) and preformed or maximum bend radius are important characteristics of surface-temperature transducers.

The type (thermoelectric or resistive), material (e.g., nickel-wire, platinum-film, silicon, thermistor, Type E thermocouple) of the sensing element, and its exposure (exposed, enclosed, coated) for an immersion probe or mounting method for a surface transducer should also be stated. The identification markings or nameplate of the transducer should include such information as nomenclature, e.g., "Transducer, Temperature, Resistive, Platinum-Wire, Exposed-Element" or "Cementable Germanium Resistive Temperature Transducer." Other nameplate information includes part and serial number, manufacturer's name, identification of external electrical connections, and at least one operating characteristic deemed essential, such as nominal range or resistance at a stated temperature.

The external electrical connections (terminal block, leads, receptacle, etc.)

should be specified in detail. Consideration should be given to any separable parts to be furnished with the transducer, such as a "thermowell" (a removable probe sheath used in the transducer installation), a throw-away protecting tube, a disposable substrate (carrier) for surface-transferable transducers, certain mounting hardware, or a mating electrical connector.

14.4.1.2 Electrical design characteristics. Internal and external connections are best represented by a simple schematic which also indicates any grounding requirements. Typical schematics usable for this purpose are illustrated in Figure 14-18. It should be noted that the letters "P" (positive) and "N" (negative) are suffixed to letter-type thermocouple identifications. These illustrations as well as those of Figure 14-17 are based on draft preparatory efforts by I.S.A. Committee SP37.4 (Specifications and Tests for Resistive Temperature Transducers).

Insulation resistance or breakdown-voltage rating are normally shown for all types of transducers having an ungrounded sensing element. Additional characteristics of resistive transducers include nominal and maximum excitation current or excitation voltage and power for any integral conditioning circuitry and the resistance of any internal resistive components other than the sensing element. This may even apply to the resistance of internal leads where critical to the application.

The a-c reactance of a wirewound sensing element may have to be considered when the excitation current in the transducer's application is pulsed, a common occurrence in time-multiplexed telemetry subsystems. If this reactance can cause excess distortion in the pulse shape, an appropriate limit may have to be included in the specification.

14.4.1.3 Performance characteristics comprise range, output, and accuracy and response characteristics at room atmospheric pressure in the absence of shock and vibration and at stated measured-fluid or other measured temperatures (the measurand) and at environmental temperatures (other than the measurand).

Range is usually specified as a nominal temperature range over which all other performance characteristics apply. It is frequently selected as a range narrower than that given by the transducer's capability but required by the intended application of the transducer and limited by the end-point settings of associated signal-conditioning and readout equipment.

Maximum and Minimum Temperature are overload conditions beyond the specified range limits to which the transducer can be exposed without incurring damage or subsequent performance changes beyond stated tolerances.

Output can be stated as nominal full-scale output over the transducer's span between its range limits), e.g., "Full-Scale Output: 36 mV (with reference junction at 0°C)" or "Full-Scale Output: 250 ohms" or "Full-Scale Output: 210 kHz."

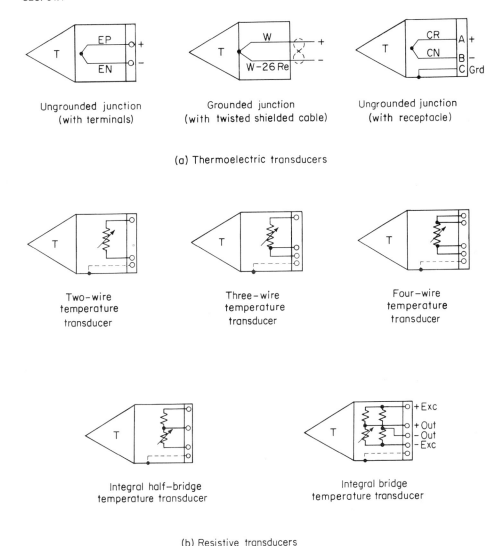

(a) Thermoelectric transducers

(b) Resistive transducers

Fig. 14-18. Typical internal schematic diagrams of temperature transducers.

Since most temperature transducers have a nonlinear output-vs-temperature relationship, the variation of output over the range is best described by a table listing output values at several temperatures including the range limits. The curve established by such a table can be referred to as "theoretical curve" or as *reference curve*. This curve is often based partly or wholly on empirical data.

A somewhat less practical but equally valid manner of stating output is to show one output value at one temperature ("reference output") and a graph or equation, including values for any required constants, describing the relative output-vs-temperature characteristics over the range. The actual reference curve can then be calculated on the basis of reference output and relative characteristics.

Repeatability is usually the only static accuracy characteristic specified for temperature transducers. Linearity is rarely specified since the reference curve of a large majority of temperature transducers is inherently nonlinear. Hysteresis and friction error are virtually nonexistent in temperature transducers. Repeatability, the maximum difference in repeated output readings from each other, can be expressed in percent of full-scale output but is usually stated in output units or output-equivalent temperature units, e.g., "Repeatability: within 0.01°C" or "Repeatability: within 1.0 ohm" or "Repeatability: within 0.05 mV." Repeatability has also been expressed in percent of reading (percent of reading-equivalent temperature correlated on the basis of a reference curve), e.g., "Repeatability: within 0.5% of reading." The tolerance specification can be followed by a statement of the period of time over which it is considered applicable.

Stability, when specified, refers essentially to repeatability over a much longer period of time than specified for repeatability. Hence, tolerances are always larger for stability than for repeatability.

Calibration Interchangeability is the maximum deviation over the specified range of the calibration curve of any one transducer from a reference curve established for a group of transducers. A reference curve is often established for a certain transducer part number. Calibration interchangeability then applies to all transducers identified by this part number. It is normally expressed in the same terms as repeatability but with bipolar tolerances, e.g., "Calibration Interchangeability: within ±2.0 ohms." Calibration interchangeability tolerances can be specified as larger in some portions of the range than in others.

Thermoelectric Potentials can be generated within a resistive temperature transducer. Internal electrical connections between wires of different metals can be so affected by a temperature gradient along any of these wires that a thermal emf is created. A limit on thermoelectric potential is usually specified as a maximum voltage to be generated within the transducer when a specific temperature difference exists between the sensing element and the external electrical connections of the transducer, e.g., "Thermoelectric Potential: 0.5 mV, maximum, when a temperature difference of 100°C exists between ends of cable and the sensing element."

Self-heating (I^2R heating) in resistive elements can cause measurement errors, particularly when semiconductor elements are used. Self-heating is dependent on the power dissipation of the element as established primarily by element resistance, current through the element, and heat transfer to the measured fluid or surface. As an example, self-heating and resultant measurement errors in a given sensing element, under identical current flow, will be much more severe if the measured fluid is relatively still air than when it is a cold, rapidly flowing liquid.

Self-heating error can be specified as the difference in output readings obtained when output is first measured with the nominal excitation current and then with a fraction (e.g., one-fifth) of the nominal excitation current.

Conduction Error is caused by heat conduction between the sensing element and the mounting of the transducer. It occurs primarily in immersion probes, especially when the stem is short. Conduction error can be specified as the maximum difference in output readings when the transducer is first immersed in an agitated temperature bath up to just below the electrical connections and when it is immersed up to the mounting flange or thread only while the head is artificially cooled or heated so that it attains the specified *head temperature*. The latter corresponds to "environmental temperature" in specifications of transducers for most measurands other than temperature. The difference between sensing-element and head temperatures should be chosen as large as it is estimated to occur in the end-use application of the transducer.

Mounting Error tolerances should be specified for surface temperature transducers This error is introduced mostly by strain in the transducer after it is welded, cemented, or otherwise attached to a surface. It has been referred to as "strain effects" or "strain error." It can be specified as the maximum difference in output readings before and after mounting the transducer to a sample surface by a specified mounting method.

An *Error Band* can be specified to combine certain errors and static accuracy characteristics. It establishes convenient acceptance criteria for temperature transducers. Accuracy requirements for thermoelectric transducers are customarily specified as "limits of error," or error-band. These limits of error include calibration interchangeability and repeatability. They are usually expressed in percent of reading for higher temperatures and in degrees Fahrenheit or Celsius for lower temperatures as the maximum (allowable) deviation from a tabulated reference curve.

Error bands for resistive temperature transducers include calibration interchangeability, repeatability, and, usually, self-heating error at nominal excitation unless a different excitation current is made applicable to the error-band specification. They are commonly expressed in ohms as a maximum (allowable) deviation from the specified reference curve. Tolerances may not be equal over all portions of the range and should be expressed in percent of output reading when the reference curve is very nonlinear.

Time Constant is the most important dynamic performance characteristic of a temperature transducer. Its specification, e.g., "Time Constant: 100 ms, maximum," is meaningless unless accompanied by a statement of the two limits of the step change in temperature and the type and flow rate of the measured fluid at both limits, e.g., "from still air at 25 $\pm 5°C$ to distilled water at 80 $\pm 2°C$ moving at 1 m per s."

Response Time is sometimes specified as 98% or 99% response time under the same test conditions as applicable to time constant.

Recovery Error is sometimes stated for transducers intended to be used at measured-fluid flow velocities in excess of Mach 0.1, primarily when the flow is transverse to a probe-type transducer. Recovery error is the error in total temperature (equivalent to transducer output indicated) caused by the assumption of a unity recovery factor. The recovery factor is the proportion of kinetic energy converted to heat expressed as the ratio of the difference between recovery temperature and absolute temperature to the difference between total temperature and absolute temperature.

14.4.1.4 Environmental characteristics are usually specified only for nonoperating environmental conditions; i.e., the performance of the transducer as temperature-measuring device is not monitored during application of the environment, and only post-environmental performance is verified. The effects of vibration are frequently specified only as absence of noise and electrical discontinuities. Acceleration effects on long-stemmed immersion probes can be specified in a similar manner. Post-vibration, post-acceleration, and post-shock performance can be specified as deviations from the reference curve. Neither of these are normally shown for surface-temperature transducers.

Among other environmental conditions which may need to be specified are long-term storage temperatures, thermal shock during storage, handling, and transportation, humidity, explosive gas mixtures and other atmospheric constituents, contaminants such as salt and soot (when not shown as measured fluid), and nuclear radiation.

14.4.2 Considerations for selection

Selection of any one type of temperature transducer can involve a relatively complex matrix of considerations. It is necessary to select a transducer design whose sensing element will attain the temperature of the measured fluid or solid in the time available for making the measurement. The output of a temperature transducer at any time is merely a measure of the temperature of its sensing element. It is often difficult to assure that this is, indeed, the same as the temperature that is intended to be measured.

For most applications a limited number of selection criteria are of paramount importance: the nature and characteristics of the measured fluid or solid, the measuring range, the time constant, and the type of associated signal-conditioning circuitry and readout equipment available or intended to be used. The relation of these criteria to selection of an optimum transducer design is further explained below.

14.4.2.1 Measured materials. Liquids and gases can be corrosive or noncorrosive; stagnant or moving at slow, medium, or high velocity; contained in pipes, ducts, tanks, vessels, or cavities or free.

The temperature of contained fluids is normally measured with a probe-type transducer whenever an opening can be made in the wall suitable for probe immersion and attachment. When such an opening is not feasible, the outside wall temperature can be measured with a transducer thermally insulated from the ambient atmosphere and from sources of heat radiation. Such external measurements are inherently less accurate than measurements of the fluid itself; however, the accuracy may be adequate for the purpose of the measurement. All exposed materials of an immersion probe must be compatible with any measured fluid that may come in contact with them. Electrically conducting fluids make the use of an exposed-element resistive transducer inadvisable. The probe design should provide sufficient structural strength to withstand transverse flow of the highest velocity known or estimated to occur. It should also be designed to minimize errors due to frictional heating of the transducer at fast flow. The stem of the probe should be long enough to minimize conduction error. It should also be of the correct length needed to place the element at a point in a duct where the mean temperature of the fluid will be measured. Special probe-tip fittings such as a stagnation housing may be required for measurements in contained fluids, but these are more often used in free fluids.

Probe-type transducers are also frequently used in free fluids moving at medium or high velocity. Stagnant free fluids can be measured with exposed resistive elements including bead thermistors or fine-wire thermocouples enclosed only in a housing designed to shield the transducer from mechanical impact and from heat radiation.

Solid materials can be metallic or nonmetallic, of varying thickness and cross-sectional configuration, and exposed to various ambient environmental conditions. An ideal temperature transducer would be made entirely of the same material as the solid and would not affect its configuration in any manner. A good practical transducer introduces as little foreign material into the solid as possible and affects its configuration as little as possible. This rule applies also to any electrical connections, housings, and mounting materials.

Most temperature measurements are made at, rather than below, the surface of the solid. Measurements in solids can be made with small embeddable transducers or with threaded inserts made preferably from the same material as the solid, except for the sensing element itself. Thin-wire thermocouples are usually suitable for such applications.

Surface-temperature measurements require very thin transducers whose housings, if any, should be of a material at least similar to that of the surface. A number of different types of weldable, cementable, and surface-transferable resistive transducers have been designed with an overall thickness, including the housing, of less than one millimeter. Foil thermocouples and thin-wire thermocouples are useful for surface measurements. Metal-sheathed ceramic-insulated thermocouple wire has been made with an overall diameter of less than 0.3 mm. When extension wires of larger diameter must be used for the sake of ruggedness, the connection between transducer wires and extension wires should be as remote as feasible from the transducer location to minimize changes to the configuration of the surface in the immediate vicinity of the transducer.

14.4.2.2 Measuring range. The range over which temperature is to be measured in any given application can limit the number of usable or optimum transducer designs. Platinum-wire resistive transducers, for example, can be used when the measuring range is any portion of the overall usable range from -200 to $+1000°C$. Chromel-Constantan thermocouples have approximately the same overall usable range. However, if a narrow portion of this is specified as the measuring range, e.g, -200 to $-180°C$, the resistive transducer would be preferable to the thermocouple. The former can be designed to provide a resistance change of about 100 ohms for the $20°C$ span, whereas the change in the emf from the thermocouple for the same span would be only about 0.6 mV. Low-error signal-conditioning and readout equipment (for a $20°C$ full-scale indication) is easier to devise for a 100 ohm change than for a 0.6 mV change.

Narrow ranges are often specified for continuously monitored processes or systems operations. Transducer ranges are sometimes widened slightly to include a fixed point (ice point, boiling point of a liquified gas, etc.). The inclusion of a fixed point as a convenient calibration check point within the range facilitates performance verification.

Some types of transducers are intended solely or primarily for narrow-range measurements of cryogenic temperatures below $-200°C$, e.g., carbon and germanium resistive transducers. Others are used when wide-range high temperatures are to be measured, e.g., Pt-10Rh/Pt thermocouples for wide portions of the range 0 to $1700°C$ or thermocouples of the tungsten-rhenium family for wide portions of the range 0 to $2760°C$.

Acceptability of large nonlinearity in transducer characteristics over the range, or a portion of the range, may also have to be considered. A platinum-wire-element transducer, for example, is usable between -200 and $-265°F$ if its sharply decreasing sensitivity below $-250°C$ presents no obstacle to the measurement system.

14.4.2.3 Time constant. When the temperature of a liquid in a storage tank must be measured once a day, the time constant of the transducer is not important. However, when temperature must be measured in a shock tube during an experiment, time constant becomes the most essential criterion for transducer selection. The time constant of a transducer sensing element is proportional to the ratio of its heat capacity to the heat transfer between element and measured material. For a given step change in temperature of a given fluid moving at a given velocity, the time constant of different transducer designs will usually be different. Relatively short time constants will be obtained with small bead thermistors and fine-wire thermocouples, whereas probe-type thermocouples with enclosed ungrounded sensing junctions installed in a separate thermowell will have relatively long time constants.

Time constants are often stated for a condition where the step change in temperature terminates in the temperature of a moving liquid, primarily because properly controlled tests for such a condition are convenient. Such statements are useful for comparison purposes when a number of different transducer designs intended for use in liquids are tested under the same conditions. Time constants, however, are most critical in gas-temperature measurements. The accurate prediction of a time constant, even in a known gas moving at a known velocity, from knowledge of a time constant in moving liquid involves complex calculations and, usually, assumptions which are often invalid. Coarse correlations have been attempted with some success for certain designs. Time constants in gases are best determined by tests simulating the conditions of the transducer application as closely as possible.

14.4.2.4 Associated circuits, primarily signal-conditioning circuits, often affect the selection of a temperature transducer and are important factors for its specification. For example, when a temperature measurement must be added to an established complex measuring system containing signal-conditioning equipment for thermocouples, the choice of a transducer for the new measurement is usually limited to the thermoelectric types unless the use of a resistive transducer offers indisputable advantages. Or, when a new measuring system is devised requiring all its transducers to furnish a 0 to 250 mV output, the cost of a resistive transducer with a bridge network is often less than that of a thermocouple with a reference junction, special wiring materials, and a differential amplifier.

A variety of different circuits is used with temperature transducers. Factors influencing their design include transducer type, full-scale transducer output, distance between point of measurement and either readout equipment or telemetry transmitter, the multiplexing and submultiplexing requirements in telemetry systems, presence of electric noise, environmental conditions, and accuracy needs.

Thermoelectric transducers always require a reference junction maintained at a known temperature. In the simplest temperature measuring system—a wire thermocouple and a readout device (voltmeter or voltage potentiometer)—the connection terminals on the device constitute the reference junction. The two terminals can be assumed to be at the same temperature. The device can be assumed to be stabilized at room temperature, which can be measured so that readings can be corrected, or whose variations are simply neglected for coarse-accuracy measurements when temperature-emf tables referenced to 25°C are used.

The classic thermoelectric temperature-measuring system employs a reference junction maintained at the ice point, as provided by an ice bath (see 14.5.1). Reference junctions in thermostatically controlled ovens have also been frequently used. Telemetry systems often employ a reference junction containing a resistive temperature transducer in intimate thermal contact with the isothermal block or other mass ("heatsink") used for the reference junction. One reference-junction block may be used for two or more reference junctions. The temperature of the reference junctions can be monitored continuously, and appropriate corrections can be made to temperature readings.

The resistive transducer at the reference junction is also used in *reference-junction compensation* (Figure 14-19). The compensation circuit is initially adjusted for a desired *simulated reference temperature*, e.g., 0°C, 100°C, or −196°C. When the temperature of the reference junction changes due to variations in ambient temperature, the output of the resistive temperature transducer is used to create an error

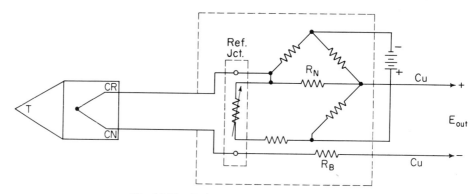

Fig. 14-19.   Reference-junction compensation.

Fig. 14-20.   Typical measuring circuits for resistive temperature transducers.

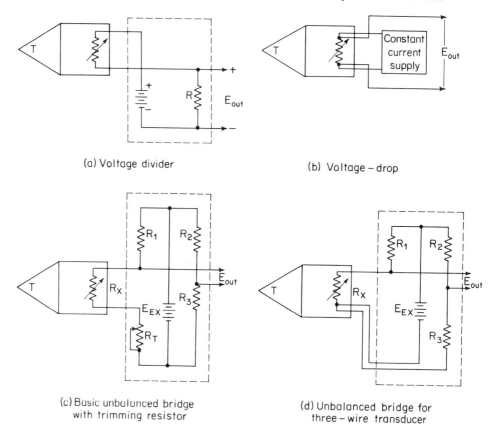

(a) Voltage divider

(b) Voltage – drop

(c) Basic unbalanced bridge
with trimming resistor

(d) Unbalanced bridge for
three – wire transducer

signal which causes a compensation voltage to be inserted in series with the thermocouple loop. With properly selected circuit constants in the compensation network, a simulated reference temperature can usually be maintained within $\pm 0.5°C$ when the actual reference-junction temperature varies between -100 and $+200°C$.

The compensation voltage provided by the network is zero when the actual and simulated reference temperatures are equal. The output of the compensator, E_o, is zero when the sensing-junction temperature equals the simulated reference temperature. In the typical circuit illustrated, the value of resistor R_N is related to the nonlinearity of the resistive temperature transducer and resistor R_B balances its leg of the thermocouple loop for the equivalent resistance of the compensation network (including the power supply) in the other leg.

Signal-conditioning circuits associated with resistive transducers include the voltage-divider and voltage-drop circuits shown in Figure 14-20 and various bridge networks only two of which are illustrated. The circuit of Figure 14-20(a) has been used primarily with thermistors. Different ratios of transducer resistance to fixed resistance (R) and requirements for either transducer or output grounding will introduce simple variations to the voltage-divider circuit. The voltage-drop circuit [Figure 14-20(b)], with E_o indicated by a voltmeter having a high input impedance, is sometimes used when lead-resistance effects are considered critical.

The unbalanced-bridge circuit shown in Figure 14-20(c) is typical for circuits used in many telemetry systems. The trimming resistor, R_T, connected in series with the transducer, R_X, is adjusted to balance the bridge at a transducer resistance corresponding to the desired lower end point of the temperature range to be measured.

The output voltage of the bridge, E_o, can be calculated for any combination of resistance values and excitation voltage, E_{EX}, by

$$E_o = \left(\frac{R_1}{R_1 + R_X + R_T} - \frac{R_2}{R_2 + R_3} \right) E_{EX}$$

A bridge circuit to which the transducer is connected as shown in Figure 14-20(d) is sometimes used to compensate for changes in the resistance of long connecting wires between transducer and bridge. Of the two wires of equal resistance connected to the same point at the transducer, one is placed in the excitation branch, the other in the opposite leg of the bridge. The third wire (to the other transducer terminal) is, of course, in the same bridge leg as the transducer. Hence, wire-resistance changes in the transducer leg are compensated by equal changes in the R_3 leg when the resistance of R_3 is approximately equal to that of the transducer.

More complex bridge circuits have been devised to compensate further for lead resistances, to provide adjustments for zero and slope of the output, and to linearize the output when transducer characteristics are nonlinear.

14.4.2.5 The application examples below illustrate some of the selection and usage criteria applied to typical temperature measurement problems. Only one of several possible solutions is shown for each problem.

1. Temperatures at various spots on a portable radio transmitter chassis are to be measured to determine excessive heating of parts. Desired accuracy is $\pm 5°F$, range $+70$ to $+200°F$, time constant up to 2 min. Appropriate thermocouple readout equipment, calibrated for iron-Constantan, is available.

Unshielded, insulated, No. 22 gage iron-Constantan (Type J) thermocouple wires are selected, and twisted-wire junctions are prepared and silver soldered. The junctions are soft soldered or metal-filled-epoxy cemented while being pressed against the surfaces to be measured. The leads are routed away from other components or surfaces and connected to the readout equipment a few feet away. Readings are taken after a specified warm-up time has elapsed.

2. The temperature of a portion of stainless-steel missile-tank skin is to be telemetered during flight. Desired accuracy is $\pm 30°F$, range 0 to $800°F$, required time constant up to 2 s. A compensated reference junction is available for use with Chromel-Constantan thermocouples; it can be set for a $0°F$ simulated reference temperature. An associated differential amplifier is available to amplify the approximately 30 mV output to the 5 V required as telemetry input.

A stainless-steel sheathed, 0.040-in. diameter, magnesium-oxide insulated thermocouple cable with No. 36 gage Chromel and Constantan conductors is selected. The sheath and insulation are stripped to within one inch of the wire ends. The conductors are individually spot welded, two spots $\frac{1}{2}$ in. apart on each wire, to the tank skin in close proximity to each other so that the junction is completed by the skin surface itself, and both welds can be considered to be at the same temperature. The sheathed cable is held to the skin surface by means of spot-welded tabs. The compensated reference junction is installed in a protected area, as close as possible to the measurement location in order to minimize lead losses in the thin thermocouple wires.

3. The temperature in the wall of a pump housing is to be measured while the pump is operating. Required accuracy is $\pm 10°F$, range $+40$ to $+300°F$, time constant up to 5 s. Readout equipment for copper-Constantan thermocouples is available, installed in a temperature-controlled area 50 ft from the pump.

A short, threaded, enclosed-element thermoelectric insert probe with No. 18 AWG copper-Constantan (Type T) wires is selected. The probe is fastened into a tapped hole to a depth of about half the wall thickness and is connected to the readout equipment with No. 18 AWG shielded, Type TX extension cable. The terminals of the readout equipment constitute the reference junction.

4. The temperature of a mildly corrosive and conductive liquid is to be measured as the liquid enters a pump. Required accuracy is $\pm 8°F$, range 0 to $300°F$, time constant up to 3 s, pressure up to 100 psig, required proof-pressure rating up to 150 psig, flow rate between 50 and 80 gpm, no temperature overloads expected, duct inside diameter at measurement point is 8 in. A tapped mounting boss is provided. An automatic resistance bridge intended for use with a wire-element transducer is available.

An immersion probe with an enclosed nickel-wire element is used. The base resistance (resistance at the lower range limit) for the desired range is 108 ohms, the ΔR for 300°F is 150 ohms. The bridge can be adjusted to indicate the proper temperature for these resistance characteristics. The transducer is provided with a mounting thread and sealing surface. The sensing element has a length of 2 in., starting from the tip. The stainless-steel stem has a diameter of 0.25 in., and stem length is chosen as 3.5 in. to the sealing surface, which is one inch above the inside of the duct wall when installed.

5. A medical temperature sensor is required for continuous monitoring of rectal temperature during an operation. Required accuracy is ± 0.1°F, range $+96$ to $+104$°F time constant up to 5 s.

A narrow, plastic-stem probe containing a bead thermistor at its tip is selected. The plastic material can withstand sterilization in alcohol. The thermistor has a high ΔR over the desired range. A low-voltage-excited resistance bridge is so designed that it is capable of adjustment for full-scale indication over the 8°F range. A microammeter with dial indications in terms of temperature is used for readout.

6. The temperature of liquid oxygen flowing through a high pressure duct is to be telemetered. Required accuracy is ± 0.5°F, range -275 to -300°F, required time constant less than 0.8 s; proof pressure of 1200 psig is required; flow rate is 60 to 150 gpm; duct inside diameter 6 in.; distance from duct inside wall to sealing surface of provided mounting boss is 0.8 in.; the bridge provided furnishes a proper voltage to the telemeter when the ΔR is 100 ohms, with a base resistance (for bridge null) of 500 ohms; average current through the element is 1.2 mA, possible overload current 20 mA. Measurement is to be taken in the presence of vibration of $25\,g$ rms amplitude at frequencies up to 2000 Hz.

A platinum-wire, coated-element, probe-type transducer is selected. Sensing element length, from tip, is 1.1 in. using closely-wound 0.8 mil wire. Resistance over the desired range varies from 370 to 470 ohms, requiring a 130 ohm trimming resistor in series with the element in the head of the transducer. A precision-machined mounting thread and a sealing surface for use with a metal O-ring are provided. Stem length, tip to sealing surface, is chosen as 3 in. A hermetically sealed receptacle is provided on the head for easy cable connection. A high-pressure ceramic-to-metal seal is used between element support and inside of head. Platinum leads are connected between element and connector to reduce a possible thermoelectric potential.

14.5 Calibration and Tests

14.5.1 Calibration equipment

The temperature sources used for calibration are fixed- and variable-temperature baths as well as melting- and freezing-point standards. Reference thermometers or thermocouples are used to measure the bath temperature accurately. Voltage potentiometers are used to obtain thermocouple readings, whereas precision resistance bridges are used for measurements on resistive transducers.

Fixed points are transition points at which a material changes its state (solid, liquid, gas) at a precisely known temperature. Frequently used fixed points are the boiling points of nitrogen, oxygen, and sulfur, the sublimation point (solid-gaseous) of carbon dioxide, the freezing points of water (ice point), tin, zinc, and antimony, and the melting point of silver and gold. The boiling points of helium and hydrogen have relatively more uncertainty but make useful fixed points in the low cryogenic range. The temperature of a *fixed-point bath* is usually monitored with a platinum resistance thermometer, the calibration of which has been certified by the cognizant government standards agency—in the U.S.A., the National Bureau of Standards. Dewar flasks, or *dewars*, similar in construction to vacuum jugs, are used for fixed-point baths.

Freezing-point standards, used to determine freezing points of metals, are usually not monitored by a reference thermometer or reference thermocouple. Instead, the pure metal samples themselves are certified.

Since the ice point (0°C) is basic to most temperature calibrations, the *ice bath* is the most commonly used fixed-point bath. An equilibrium suspension of crushed ice in distilled water is contained in a dewar.

The triple point of water ($+0.01$°C) can be measured accurately and with relative ease by using a *triple-point cell*. This measurement is now preferred over the ice point if a very accurate determination is required.

A liquid-nitrogen bath calibration setup is shown in Figure 14-21. The liquid nitrogen in the dewar is stirred by an *agitator*, a small propeller attached to a long shaft driven by an electric motor. The control knob over the motor is used to adjust the motor speed and, in turn, the *agitation rate*. A reference thermometer (platinum resistance), with its characteristic long stem, is clamped relatively high over the dewar but with its sensing element immersed well below the liquid surface in proximity to the platinum-wire-element transducer which is being calibrated. The reference thermometer is connected to a servo-operated, direct-reading resistance bridge ("automatic Mueller bridge"). The transducer under test (with white cable on the illustration) is connected to a precision resistance bridge ("guarded Wheatstone bridge").

Temperature baths are electrically heated double-wall tanks containing a bath medium appropriate for a given temperature range. The tank is set in a housing or console and is equipped with an agitator for stirring. A cooling tube may be coiled around the inner tank to permit rapid reduction of the bath temperature. Commonly used bath media include a mixture of dry ice and trichlorethylene, acetone, or methanol (*dry-ice-mixture bath*) for temperatures between -100°F and $+32$°F, acetone or methanol for temperatures between $+32$ and $+100$°F, silicone oil (such as Dow Corning No. 550) for the range 70°F to 500°F (*oil bath*), and eutectic salts (e.g., sodium and potassium nitrates and nitrides) for temperatures between 400°F and 1300°F (*salt bath*).

The bath temperature is controlled by very accurate thermostatic or proportional-

Fig. 14-21. Liquid-nitrogen bath calibration setup for temperature transducers.

control circuitry so that its stability within narrow temperature limits can be relied upon.

A calibration setup utilizing an oil bath is shown in Figure 14-22. Precision bridges of different types are used to measure the resistance of the transducer under test (white cable) and of the reference thermometer (black cable).

Calibration of temperature transducers in the cryogenic range (below −290°F) can also be performed on a point-to-point basis by use of a *cryostat*, a dewar of special design which permits gradual controlled warming of a liquified gas, used as bath medium, by heater control and controllable pressurization.

Electrically heated muffle ovens, kilns, and other furnaces are used primarily for the calibration of thermocouples at high temperatures. Optical pyrometers are employed for calibrations above 1063°C (gold point).

14.5.2 Calibration methods

14.5.2.1 Measured points. Most practical temperature transducers are characterized by an output-vs-temperature characteristic which follows a reference curve. A calibration based on a substantial number of measured points is, therefore, necessary only in the case of those transducers for which a predictable curve cannot be

Fig. 14-22.   Oil-bath calibration setup for temperature transducers.

established with adequate accuracy. Thermocouples and metal-wire-element trans-ducers are calibrated by measuring their output at only a few (but not less than two) temperature points and then arriving at their output-vs-temperature characteristics over the desired or nominal range by calculation or interpolation. The measured points should be selected fixed points if possible. Temperature points in variable-tem-perature baths are less accurate but frequently more convenient to obtain. They are acceptable for calibrations of transducers not used as transfer or reference standards when combined, if possible, with at least one fixed-point measurement.

14.5.2.2 Measurement error reduction. Close proximity of the sensing elements of the transducer under test and the reference standard is essential, as is the certainty of the absence of any temperature gradient between the two sensing elements. Ther-mocouple junctions should be brought into physical contact with the junction of the reference thermocouple whenever possible. A heat sink (e.g., copper block) is some-times used to assure equal temperature of two transducers in physical contact with it. Resistive transducers of the probe type should be very close to the reference resist-ance thermometer while immersed in a temperature bath. Both should also be kept away from the dewar wall.

Surface sensors can be calibrated against a reference thermocouple or resistance thermometer; in either case it is important to minimize any heat barrier between test unit and standard. Agitation of the bath medium is always required during calibration, even when cryogenic liquids are used. In the latter case the same medium, but in a gaseous state, can be introduced into the bottom of the dewar. The resulting bubbling can be used in lieu of mechanical agitation.

Depth of immersion in the bath is critical, particularly in the case of probes with short stems, so that *conduction errors* are minimized. When immersion depth is questioned, the use of the transducer in its actual application must be considered, and operating conditions, as they affect conduction error, may have to be simulated during calibration. Alternately, the conduction error can be determined separately.

A group of thermoelectric transducers made from the same reel of thermocouple wire can be calibrated on a sampling basis by cutting a length of wire from each end of the reel, forming a junction on each of the two thermocouple samples, and calibrating them against a reference thermocouple. The other transducers in the group can then be assumed to have characteristics sufficiently close to those of the two test specimens that no further calibration is required when a brief functional test on each transducer indicates that it operates. Output measurement at one temperature usually suffices as such a functional test (Figure 14-23).

14.5.2.3 Calculations. Most thermocouple calibrations are considered adequate when the maximum deviation of their output-vs-temperature characteristics from those shown on applicable standard tables can be established from a number of measured points.

The calibration of a platinum-wire element resistive transducer is calculated by use of the *Callendar-VanDusen Equation* after the constants α, β, and δ are determined. Only α is always based on measured points for each transducer. The constants β and δ may be determined as typical for a group of transducers of the same design and with elements made from the same reel of platinum wire.

Calculated calibration points of platinum-wire-element transducers intended for measurements in the cryogenic region (below $-183°C$) can be obtained by the *Corruccini Method*, named after R. J. Corruccini who developed the method at the National Bureau of Standards and proved the adequacy of its accuracy for applications by the industry. This method is based on the measurement of two or three fixed points, their correlation with equivalent points determined for a number of N.B.S.-calibrated reference thermometers, and interpolation to obtain intermediate calculated points.

Transducers with wire elements other than pure annealed platinum can be calibrated in a similar manner except that the Callendar-VanDusen Equation is not applicable. Their reference curves are linear over a narrow range and can be extended by empirical means to cover larger ranges.

Metal-film elements follow an empirical relationship established by a series of measured points.

Semiconductor-element transducers follow exponential curves more or less closely over limited temperature ranges and several measured points are necessary to establish a correlation between actual and theoretical behavior. Since the value

VENDOR'S MODEL NO.	TEST FACILITY		PART NO.
VENDOR	**THERMOELECTRIC TEMPERATURE TRANSDUCER**		SERIAL NO.
	INDIVIDUAL ACCEPTANCE TEST & CALIBRATION		RANGE

1. VISUAL: MECHANICAL ☐ FINISH ☐ NAME PLATE ☐ RECEPTACLE ☐

2. INSULATION RESISTANCE _____ MEGOHMS AT _____ VDC   3. HEAD TEMPERATURE ☐

4. PROOF PRESSURE _____ PSIA FOR _____ MINUTES   LEAKAGE _____ CC/MIN., ALLOWED _____ CC/MIN.

5. OPERATING PRESSURE: _____ PSIA @ _____ °F   FOR _____ MINUTES.

              RESULT: _____

6. SENSING WIRE SAMPLE CALIBRATION:

TEMP. (°F)	THEOR. OUTPUT (MV)	MEAS'D OUTPUT SAMPLE NO.1 (MV)	MEAS'D OUTPUT SAMPLE NO.2 (MV)	ERROR (°F) SAMPLE NO. 1	ERROR (°F) SAMPLE NO. 2	ALLOWED ERROR (°F)

7. EXTENSION WIRE SAMPLE TEST: MAX. ERROR: SAMPLE NO. 1 ____ °F,   SAMPLE NO. 2 ____ °F,   ALLOWED ____ °F ☐ N/A

WIRE SAMPLES REPRESENT SERIAL NOS.:

8. CALIBRATION

TEMPERATURE _____ °F   THEOR. OUTPUT _____ MV   MEAS'D OUTPUT _____ MV

           ERROR _____ °F   ALLOWED ERROR _____ °F

EQUIPMENT USED	DEFECTS NOTED OR COMMENTS

BY: _____   DATE _____

Fig. 14-23. Typical form for acceptance test and calibration of probe-type thermoelectric temperature transducer.

of the current passing through the element is critical, it should be shown on the calibration record.

14.5.3 Performance verification

14.5.3.1 Range, output, and error limits are obtained from the transducer's calibration record. A properly prepared calibration record shows not only the measured and calculated output-vs-temperature values, but also the deviations of each of these points (errors) or the maximum deviation of any point along the entire calibration curve (error band) from a specified or guaranteed reference ("theoretical") curve. The record should also indicate whether any specified error tolerance has been exceeded.

14.5.3.2 Repeatability is determined by two or more successive calibrations. One-point or two-point calibrations may suffice for this purpose.

14.5.3.3 An insulation resistance test is performed only when the transducer contains an ungrounded sensing element. If the transducer has a metallic case (cover, stem, head, or mounting), the insulation resistance is measured between the external electrical connections to both sides of the sensing element (connected together) and the case. Surface sensors without metallic cases, such as cementable transducers, can be clamped firmly to a metallic surface and the insulation resistance measured between this surface and both of the sensing-element leads connected together.

14.5.3.4 A proof-pressure test is performed on immersion probes when an internal seal is required. The probe is mounted into a narrow cylinder equipped with a pressure gage. A gas such as helium is admitted into the cylinder through a valve until the gage indicates the required proof pressure. The valve is closed and the gage is observed for a period of time so that any pressure drop, indicating a leakage, can be noticed. If the internal volume of the cylinder is known, and if the temperature of the gas is determined by connecting the transducer under test to appropriate readout equipment, the equivalent volume of gas escaping per unit time (*leakage rate*) can be calculated. Use of a helium leak detector is an alternate method. A hydraulic booster can be used to pressurize the gas in the test cylinder if proof pressure higher than that available from a gas bottle is required. A liquid pressurizing medium should be used, if at all, only when it is known that the transducer will never be required to provide a seal against gas pressure. If the temperature of the pressurizing medium is expected to affect the performance of the transducer's internal seal, the medium can be heated or cooled before it enters the cylinder.

14.5.3.5 Time constant and response time tests are performed by applying a step change of temperature to the sensing element of a transducer and observing the time required for the transducer's output to reach 63%, and a second specified level such as 99%, of the final value. The test always requires readout on an oscilloscope or on a high-speed strip-chart recorder. The associated amplifier or resistance bridge or other network should be so adjusted that a deflection of several inches will be obtained between the initial and final values of the transducer output. The test setup can include a fast-acting switch which is actuated when the transducer first comes in contact with the final fluid, and which is connected to the readout equipment so that a marker pulse is obtained.

The magnitude of the step change of temperature is normally selected to lie within the range of the transducer. Several methods can be used to provide the step change. The essential characteristic of any test setup is the minimization of transit time between immersion in the initial and final fluids. The initial fluid used most frequently is relatively still air at room temperature near the temperature bath containing the agitated final fluid which can be liquid nitrogen, liquid oxygen, a dry-ice mixture, distilled water, mineral oil, or silicone oil.

To obtain meaningful readings the velocity of the final liquid relative to the transducer should be established when still air is used as the initial fluid. A useful setup consists of a circular open pan, containing the temperature-controlled test liquid, which is rotated by a variable-speed motor. The transducer to be tested is immersed at a predetermined radius from the center of rotation when the pan rotates at the desired rate. This makes it possible to obtain a specific flow velocity (e.g., 1 m per s) at the point of immersion.

Time constants can also be obtained in heated air moving at a known mass-flow rate by rapidly changing the exposure of the transducer, by means of a two-way valve, from cold air to hot air at the same flow rates or by suddenly introducing heated air into an air stream in a duct in which the transducer has been installed.

Transducers with ranges solely in the cryogenic region can be tested by rapid transfer between two cold liquids. Use of room temperature air as final fluid is not recommended because evaporative cooling of the sensing element can introduce large errors. Transducers with exposed elements should not be immersed into a bath containing solids, such as ice crystals, which may damage the sensing element.

14.5.3.6 The thermoelectric potential test can be performed on a resistive transducer by connecting it to a thermocouple potentiometer while the sensing element is immersed in a bath maintained at a temperature near the transducer's upper-range end point and measuring the potential developed. Such potentials are normally less than one millivolt.

14.5.3.7 A head-temperature test can be used to determine the approximate conduction error of the transducer and effects of high or low ambient temperatures on the head. Actual use conditions should be simulated as closely as possible. An acceptable method is installation of the transducer from the inside of a horizontal duct section so that the stem protrudes from the duct. The duct is mounted over a temperature bath with just enough space between rim of bath and duct bottom to accommodate the agitator shaft (unless agitation is from the bottom of the bath). While the element is fully immersed in the bath, heated or cooled air at the desired temperature is blown through the duct and the transducer output is monitored.

14.5.3.8 Vibration and acceleration tests are frequently required for probe-type transducers used in aerospace applications or on heavy rotating machinery. The transducer is installed into a fixture by using its mounting thread or flange. The fixture is then installed on the test equipment (shaker or centrifuge), and the output is monitored for discontinuities or indications of shorting between portions of the sensing elements or between internal leads while the equipment is operated through the required vibration or acceleration program.

14.5.3.9 Additional tests for surface temperature transducers are often performed, especially on cementable and weldable types, to determine errors due to strains introduced by bending over a mandrel of specified diameter, by bonding with a specific cement and subsequent drying of this cement, or by installation on a test specimen on which tension or compression tests can be performed. An installation of the transducer on an appropriate sample surface must be made in accordance with specified methods.

Another important test is a verification of lead pull-out strength. This can be accomplished by clamping the transducer securely in a vertical position and carefully attaching a weight equivalent to the specified (minimum) pull-out force to each lead.

14.5.3.10 A visual inspection of those visible mechanical characteristics considered important normally precedes any acceptance or qualification test. Dimensional checks, scrutiny of welded, brazed, soldered, and threaded joints, nameplate verification, and examination of cable, connector, or other external electrical connections are usually included.

14.5.3.11 Additional environmental tests may be dictated by the expected environment of the temperature transducer's end-use application. Performance may have to be verified, e.g., by a one-point calibration at a convenient fixed point, after exposure of the transducer to mechanical shock and nuclear radiation. Verification tests during and after exposure to long-term immersion in a specified liquid, specified humidity levels, salt concentrations, or explosive gas mixtures in the ambient atmosphere and to variations in ambient pressure may also be necessary.

14.6 Related Measuring Devices

14.6.1 Measuring instruments for temperature extremes

Various methods have been developed to measure temperatures of hot gases, e.g., in arc discharges and plasmas, ranging from over 4000°K to over 100,000°K. One method uses the variation in conductivity of an ionized gas. Others are based on analysis of plasma emission spectra in the X-ray region, on intermittent-contact measurements, on the temperature dependence of thermal noise in a resistor, on controlled cooling of the hot gas, and on the measurement of sound velocities in a gaseous medium (see below).

A *magnetic thermometer* has been developed for measurements below 1°K. It operates on the basis of the variation in susceptibility with temperature of a paramagnetic salt placed between a primary and a secondary winding.

The temperature dependence of the dielectric constant of polycrystalline barium titanate has been employed in a capacitive temperature transducer for measurements between 2°K and room temperature.

An *acoustical thermometer* has been used to measure low gas temperatures by determination of the velocity of sound in the gas. The velocity of sound travelling through an ideal gas is directly proportional to the square root of the absolute temperature of the gas. This technique has been found applicable to the temperature measurement of helium gas near the helium boiling point. Similar acoustical thermometers have been applied by several researchers to measurements of high temperatures (3000 to 6000°K).

14.6.2 Radiation pyrometers

Radiation pyrometers are widely used for noncontacting temperature measurements between 700°C and about 3500°C. *Pyrometry* has been defined as the art of measuring high temperatures. The term is derived from the ancient Greek word for fire, *pyr* or *pyros*. In modern technical literature it is gradually being replaced by the general term "thermometry." Radiation pyrometers (radiation thermometers) measure thermal radiation and provide an output in terms of temperature or one which is convertible to temperature. They exist in three basic types: total-radiation, monochromatic-radiation, and radiation-ratio pyrometers. They share one external characteristic, a sighting device which is aimed at the object whose temperature is to be measured.

14.6.2.1 Total-radiation pyrometers are designed to respond to incident radiation at all wavelengths to the best extent possible. The sensing element is a blackened surface to which a thermoelectric or resistive temperature transducer is attached. A sensing element with a thermistor, metal-film, or metal-wire transducer is often referred to as a *bolometer*. Thermoelectric transducers comprise thermocouples and thermopiles. The temperature of the measured object is determined from the temperature rise of the sensing element on the basis of the Stefan-Boltzmann Law for total radiation (see 14.1.2) with suitable corrections for emissivity of the measured surface and the relative position and geometry of the surface as well as that of the pyrometer and its sensing-element configuration. Thermal radiant energy from a solid increases with temperature. The wavelength at which maximum radiation occurs decreases with increasing temperature. The latter phenomenon is utilized in the operation of the monochromatic and ratio pyrometers ("optical pyrometers").

14.6.2.2 The monochromatic radiation pyrometer or *brightness pyrometer* measures temperature as a function of incident radiation within a narrow band of wavelengths (*brightness temperature*). The center wavelength is typically 2.3 microns.

The "disappearing filament" pyrometer is one of the earliest versions of the brightness pyrometer. It cannot be classified as a transducer since it requires a human operator who varies and monitors the current through the tungsten ribbon filament of a lamp ("standard lamp") until the filament matches the image of the measured surface in brightness. The temperature of the surface is indicated by the lamp current. This instrument is a commonly used calibration device for contacting as well as noncontacting temperature transducers in the range 800 to 2300°C. It should be noted that the International Practical Temperature Scale (IPTS) above the gold point (1063°C) is defined on the basis of thermal radiation. Automatic versions of this pyrometer alternately image the measured surface and the standard lamp on a photoelectric sensor, e.g., by use of a rotating sector disc or by a vibrating mirror. The output due to unequal brightness is used as error signal in the lamp-filament control loop. The brightness of the lamp is automatically varied until a null condition is reached with equal brightness. The measured temperature is then determined from the lamp-filament current.

Infrared brightness pyrometers employ an internal blackbody-cavity radiation source as temperature reference instead of the tungsten lamp (which emits in the visible light spectrum). Two optical filters are commonly used to provide the necessary narrow bandwidth characteristic, one acting as a low-pass, the other as a high-pass filter to incident radiation wavelengths. A rotating sector disc alternates incidence on the sensing element (typically a thermistor bolometer) from reference and measured object. A rotating or oscillating shutter (optical "chopper") is often employed to provide an a-c rather than a d-c output from the sensing element. The sector disc itself may be employed for this purpose provided that it rotates at a sufficiently high speed. Relatively weak a-c signals are more conveniently handled by associated amplifiers and other conditioning circuitry.

14.6.2.3 The radiation ratio pyrometer or *two-color radiation pyrometer* measures temperature as a function of the ratio of incident radiation at one wavelength to that at a different wavelength. The two wavelengths are either in the visible spectrum (e.g., 0.53 and 0.62 microns) or in the infrared spectrum (e.g., 2.04 and 2.64 microns). The operation of this instrument is based on Wien's Law for the spectral energy distribution of blackbody radiation at relatively short wavelengths. The ratio of the energies at two wavelengths is a function of temperature and of the ratio of emissivities (of the source, at each of the two wavelengths) rather than of temperature and absolute emissivity (as is the case in the monochromatic pyrometers). This facilitates obtaining equally good instrument performance for high, low, and even varying emissivities of the measured surface, as long as their ratio at the two wavelengths is constant. In the ratio pyrometer the radiant energy is separately measured at the two wavelengths and their ratio is computed by integral circuitry. The output voltage of the pyrometer is convertible to temperature.

14.6.3 Heat-flux transducers

Two basic types of transducers are used to provide outputs in response to heat flux for direct measurements of heat transfer. The *calorimeter* is a device for measuring the change of heat content of an object or system; it responds to heat convection as well as radiation. The *radiometer* is intended to respond only to radiant heat flux.

14.6.3.1 Calorimeters. A typical *membrane calorimeter* or "Gardon Gauge" consists of a blackened Constantan membrane, rim-supported by an annular copper heat sink. A thin copper wire is butt-welded to the center of the membrane so that a differential thermocouple is formed whose first junction is at the center of the membrane and whose second junction is between the membrane rim and the copper heat sink. Heat flux applied to the membrane causes a temperature difference between the two junctions as heat absorbed by the membrane is transferred radially to the heat sink. The output of the differential thermocouple is proportional to the energy absorbed. Membrane calorimeters can be water-cooled internally when heat flux is to be measured over longer periods of time.

The *slug calorimeter* differs in construction from the membrane calorimeter in that a metal mass ("slug") is attached to the inside surface of the blackened membrane at its center. The slug can be made of copper, nickel, or stainless steel. A complete thermocouple sensing junction is attached to the slug, inside the calorimeter. A second junction can be formed since the membrane and heat sink are of different metals. Most of the heat absorbed will be retained by the slug. When only a slug thermocouple is used, the heat flux can be determined from the recorded temperature history of the slug. When two junctions are used to form a differential thermocouple, its output is used as a measure of heat flux on the basis of the temperature difference between the heat receiver (slug) and the heat sink.

14.6.3.2 Radiometers are commonly constructed by mounting a "window," typically of quartz or synthetic sapphire, over a membrane calorimeter so as to isolate it from convective heat flux. In the absence of any convective heating, a calorimeter will, of course, measure purely radiant heat flux. Low-flux-level calorimeters using thin-film blackened resistive or thermopile sensing elements have been constructed to operate under such conditions. When equipped with a window, such calorimeters operate as low-level radiometers in the presence of convective heating. Coarse radiant-heat-flux measurements can be obtained by using two calorimeters, one with the usual blackened surface, the other with a highly reflective (e.g., gold-plated and buffed) surface intended to respond to convective heating only and monitoring the difference in output between the two devices.

Infrared radiometers have been constructed in a manner similar to that described for infrared brightness pyrometers (14.6.2.2). These employ a radiation-selecting device such as an oscillating mirror to beam either incident radiation or calibration-source radiation through an optical filter onto a sensing element (e.g., a thermopile).

BIBLIOGRAPHY

1. Hoge, H. J. and Brickwedde, F. G., "Establishment of a Temperature Scale for the Calibration of Thermometers between 14 K and 83 K," *N.B.S. Research Paper RP1188*, Washington, D.C.: National Bureau of Standards, 1939.

2. *Temperature—Its Measurement and Control in Science and Industry.* Vol. I, New York: Reinhold Publishing Corp., 1941.

3. Griffiths, E., *Methods of Measuring Temperature.* London: Charles Griffin & Co., Ltd., 1947.

4. Faires, V. M., *Applied Thermodynamics.* New York: The Macmillan Company, 1948.

5. Weber, R. L. (Ed.), *Heat and Temperature Measurement.* Englewood Cliffs, N.J.: Prentice-Hall, Inc., 1950.

6. Brown, A. I. and Marco, S. M., *Introduction to Heat Transfer.* (2nd ed.) New York: McGraw-Hill Book Company, 1951.

7. Campbell, C. N., *Modern Pyrometry.* New York: Chemical Publishing Co., 1951.

8. Royds, R., *Measurement and Control of Temperatures in Industry.* London: Constable & Co., Ltd., 1951.

9. Lindorf, H., *Technische Temperaturmessungen (Technical Temperature Measurements).* Essen, W. Germany: W. Girardet, 1952.

10. Baker, H. D., Ryder, E. A., and Baker, N. H., *Temperature Measurement in Engineering.* Vols. I and II, New York: John Wiley & Sons, Inc., 1953.

11. McAdams, W. H., *Heat Transmission,* (3rd ed.). New York: McGraw-Hill Book Company, 1954.

12. Bauer, W. and Buschner, R., "Beitrag zur Messung der Lufttemperatur mit verschiedenen Formen des Strahlungsschutzes" (Report on Measurement of the Air Temperature with Various Types of Radiation Shielding), *German Weather Service Reports*, Vol. 3, No. 19, pp. 2–12, 1955. (Translation available from American Meteorological Society, Boston, Massachusetts.)

13. Lee, J. F. and Sears, F. W., *Thermodynamics*. Reading, Massachusetts: Addison-Wesley Publishing Co., 1955.

14. Shenker, H., Lauritzen, J. I., Jr., and Corruccini, R. J., "Reference Tables for Thermocouples," *N. B. S. Circular 561*, Washington, D.C.: Government Printing Office, 1955.

15. *Temperature—Its Measurement and Control in Science and Industry*. Vol. II, New York: Reinhold Publishing Corp., 1955.

16. "Preparation and Use of Chromel-Alumel Thermocouples for Turbojet Engines," *SAE Standard AIR46*, New York: Society of Automotive Engineers, 1956.

17. Corruccini, R. J., "Calibration of Platinum Resistance Thermometers," *N.B.S. Laboratory Note, File No. 57–25*, Boulder, Colorado: National Bureau of Standards, 1957.

18. Giedt, W. H., *Principles of Engineering Heat Transfer*. Princeton, N.J.: D. Van Nostrand Co. Inc., 1957.

19. Zemansky, M. W., *Heat and Thermodynamics*. (4th ed.) New York: McGraw-Hill Book Company, 1957.

20. Giedt, W. H., "Temperature Measurement in Solids," *Product Engineering*, Vol. 29, No. 29, 1958.

21. Knudsen, K. J., "Resistance Determination of Resistance Thermometers at Integral Temperatures," *Paper No. CP58-103*, New York: American Institute of Electrical Engineers (IEEE), 1958.

22. Werner, F. D., "Total Temperature Measurements," *Paper No. 58-AV-17*, New York: American Society of Mechanical Engineers, 1958.

23. Carslaw, H. S. and Jaeger, J. C., *Conduction of Heat in Solids*. (2nd ed.) New York: Oxford University Press, Inc., 1959.

24. Eckert, E. R. G. and Drake, R. N., Jr., *Heat and Mass Transfer*. New York: McGraw-Hill Book Company, 1959.

25. "Thermocouples and Thermocouple Extension Wires," *I.S.A. Recommended Practice RP1.1–1.7*, Pittsburgh: Instrument Society of America, 1959.

26. Saunders, R. N., "Methods of Measuring Temperature Sensor Time Constant and Self Heating," *Bulletin 7619*, Minneapolis: Rosemount Engineering Co., 1959.

27. Corruccini, R. J., "Interpolation of Platinum Resistance Thermometers, 20 to 373.15°K," *Review of Scientific Instruments*, Vol. 31, pp. 637–640, 1960.

28. DeLeo, R. V. and Werner, F. D., "Temperature Sensing from Aircraft with Immersion Sensors," *Preprint N.Y. 60-91*, Pittsburgh: Instrument Society of America, 1960.

29. Harrison, T. R., *Radiation Pyrometry and Its Underlying Principles of Radiant Heat Transfer*. New York: John Wiley & Sons, Inc., 1960.

30. Werner, F. D., "Time Constant and Self-Heating Effect for Temperature Probes in Moving Fluids," *Bulletin 106017*, Minneapolis: Rosemount Engineering Co., 1960.

31. Aronson, M. H. (Ed.), *Temperature Measurement and Control Handbook*. Pittsburgh: Instruments Publishing Co., Inc., 1961.

32. Bird, F. F. and Jackson, W. E., "Accuracy in Resistance Thermometer Measurement," *Preprint 157-LA-61*, Pittsburgh: Instrument Society of America, 1961.

33. Elliott, R. D., "Dynamic Behavior of Missile Skin Temperature Transducers," *Preprint 160-LA-61*, Pittsburgh: Instrument Society of America, 1961.

34. Halpern, C. and Moffat, R. J., "Bibliography of Temperature Measurement, January 1953 to June 1960," *N.B.S. Monograph 27*, Washington, D.C.: National Bureau of Standards, 1961.

35. Moffat, R. J., "Gas Temperature Measurement," *Report GMR-329*, Warren, Michigan: General Motors Research Laboratories, 1961.

36. Shearer, C. R., "Semiconductor Temperature Sensors," *Preprint No. 159-LA-61*, Pittsburgh: Instrument Society of America, 1961.

37. Stimson, H. F., "International Practical Temperature Scale of 1948 Text Revision of 1960," *N.B.S. Monograph 37*, Washington, D.C.: National Bureau of Standards, 1961.

38. "Temperature Measurement," *ASME Power Test Codes, Supplement on Instruments and Apparatus*, Part 3, New York: American Society of Mechanical Engineers, 1961.

39. Blakemore, J. S. "Germanium for Low-Temp Resistance Thermometry," *Instruments and Control Systems*, Vol. 35, pp. 94–95, May 1962.

40. Brooks, E. J., Kramer, W. C., and McGowan, R. D., "High-Temperature Sensors for Borax-V Boiling Fuel Rods," *Report ANL-6636*, Argonne, Illinois: Argonne National Laboratory, 1962.

41. Caldwell, F. R., "Thermocouple Materials," *N.B.S. Monograph 40*, Washington, D.C.: National Bureau of Standards, 1962.

42. Kostkowski, H. J. and Lee, R. D., "Theory and Methods of Optical Pyrometry," *N.B.S. Monograph 41*, Washington, D.C.: National Bureau of Standards, 1962.

43. Swindells, J. F., "Calibration of Temperature Measuring Instruments at the National Bureau of Standards," *N.B.S. Miscellaneous Publication 248*, Washington, D.C.: National Bureau of Standards, 1962.

44. Thomas, D. B., "A Furnace for Thermocouple Calibrations to 2000°C," *N.B.S. Journal of Research*, Vol. 66C, No. 3, July–September 1962.

45. *Temperature—Its Measurement and Control in Science and Industry.* Vol. III, Parts 1, 2, and 3, New York: Reinhold Publishing Corp., 1962.

46. "Wire, Thermocouple, Identification Marking and Color Code of," *Military Standard MIL-STD-687*, Washington, D.C.: U.S. Government Printing Office, January 1962.

47. Benedict, R. P., "Temperature Measurement in Moving Fluids," *Electro-Technology*, Vol. 72, No. 4, pp. 56–61, October 1963.

48. Hsu, S. T., *Engineering Heat Transfer.* Princeton, N.J.: D. Van Nostrand, Inc., 1963.

49. Nanigian, J., "Temperature Measurements and Heat Transfer Calculations in Rocket Nozzle Throats and Exit Cones," *Preprint No. 29.3.63*, Pittsburgh: Instrument Society of America, 1963.

50. Scagnetti, M. and Crabol, J., "Sondes Thermométriques à Film de Platine à Réponse Rapide" (Fast-Response Platinum-Film Temperature Probes), *La Recherche Aérospatiale*, No. 97, November–December 1963.

51. Stempel, F. C. and Rall, D. L., "Applications and Advancements in the Field of Direct Heat Transfer Measurements," *Preprint No. 8.1.63*, Pittsburgh: Instrument Society of America, 1963.

52. "Resistance Thermometers," *SAMA Standard RC5*, (2nd ed.) Chicago: Scientific Apparatus Makers Association, 1963.

53. "Standard Temperature-Electromotive Force (EMF) Tables for Thermocouples,"

ASTM Standard E230–63, Philadelphia: American Society for Testing and Materials, 1963.

54. "Thermistor Definitions and Test Methods," *EIA Standard RS-275*, Washington, D.C.: Electronic Industries Association, 1963.

55. "Thermocouple Thermometers (Pyrometers)," *SAMA Standard RC8*, (2nd ed.) Chicago: Scientific Apparatus Makers Association, 1963.

56. Bachmann, R. C., Chambers, J. T., and Foley, D. D., "Investigations of Heat-Flux Measurements with Calorimeters," *Preprint No. 16.19-3-64*, Pittsburgh: Instrument Society of America, 1964.

57. Barber, C. R., "Temperature Scale Measurement Techniques for the Range 10–90°K," *Paper 23-UK-246, Acta IMEKO 1964*, Budapest: IMEKO, 1964.

58. Clark, R. B., "Calibration and Stability of W/WRe Thermocouples to 2760°C (5000°F)," *Preprint No. 16.10-4-64*, Pittsburgh: Instrument Society of America, 1964.

59. Haas, A., "Ambient Temperature Compensated Non-Conducting Surface Thermometer," *Acta IMEKO 1964*, pp. 339–351, Budapest: IMEKO, 1964.

60. Hammond, D. L., Adams, C. A., and Schmidt, P., "A Linear Quartz Crystal Temperature Sensing Element," *Preprint No. 11.2-3-64*, Pittsburgh: Instrument Society of America, 1964.

61. Kandyba, V. V., "The Method and Instrument for Measuring Temperatures of Flames, Gas Flows and Plasmas," *Acta IMEKO 1964*, pp. 359–367, Budapest: IMEKO, 1964.

62. Perls, T. A., "Radiative Heat Transfer Measurements at High Intensities," *Acta IMEKO 1964*, pp. 431–445, Budapest: IMEKO, 1964.

63. Rall, D. L. and Stempel, F. C., "A Discussion of the Standardized Procedure for Calibrating Heat Flux Transducers," *Preprint No. 16.7-1-64*, Pittsburgh: Instrument Society of America, 1964.

64. Ruskin, J. M., Jr., "Thermistors as Temperature Transducers," *Data Systems Engineering*, Vol. 19, No. 2, pp. 24–27, February 1964.

65. "Standard Method for Calibration of Thermocouples by Comparison Techniques," *ASTM Standard E220-64*, Philadelphia: American Society for Testing and Materials, 1964.

66. "Industrial Platinum Resistance Thermometer Elements," *British Standard BS1904, 1952, Amended 1954*, London: British Standards Institute, 1964.

67. "Tentative Specification for Metal-Sheathed, Corrosion-Resistant Thermocouples for Nuclear Service," *ASTM Tentative Standard E235-64T*, Philadelphia: American Society for Testing and Materials, 1964.

68. "Temperature Measurement Thermocouples," *U.S.A. Standard C96.1-1964*, New York: U.S.A. Standards Institute, 1964.

69. Bedford, R. E., "Reference Tables for Platinum-40% Rhodium/Platinum-20% Rhodium Thermocouples," *Review of Scientific Instruments*, Vol. 36, No. 11, pp. 1571–1580, November 1965.

70. Chao, G. T. Y., Leslie, J. C., and Mancus, H. V., "A Direct Measuring Radiation Calorimeter for Determining Propellant Gas Emissivity," *Paper No. 65-358*, New York: American Institute of Aeronautics and Astronautics, 1965.

71. Chandon, H. C., Hopping, D. E., and Takeda, S., "A Standardized Temperature Measuring System with Constant Current Excitation and Platinum Element Transducers," *Preprint No. 9.1-1-65*, Pittsburgh: Instrument Society of America, 1965.

72. Goodall, D. H. J., "The Measurement of Temperature in the Range 3°K to 80°K Using Carbon Resistors," *Report CLM-R-47*, Culham, England: UKAEA, Research Group, 1965.

73. Montgomery, P. W. and Lowery, R. L., "Jet Temperature by IR Pyrometry," *I.S.A. Journal*, Vol. 12, No. 4, pp. 61–64, April 1965.

74. Plumb, H. H. and Cataland, G., "Acoustical Thermometer," *Science*, Vol. 50, pp. 155–159, October 1965.

75. Svet, D. Y., "Thermal Radiation," New York: Consultants Bureau, 1965.

76. Walker, B. E., Jr., Ewing, C. T., and Miller, R. R., "Instability of Refractory Metal Thermocouples," *N.R.L. Report 6231*, Washington, D.C.: U.S. Naval Research Laboratory, 1965.

77. "Method of Measuring the Temperature of Petroleum and Petroleum Products," *API STD 2543*, New York: American Petroleum Institute, 1965.

78. Chandon, H. C., Hopping, D. E., and Takeda, S., "Cryogenic Measurement with Standardized RTT's," *I.S.A. Journal*, Vol. 13, No. 1, pp. 47–50, January 1966.

79. Frazine, D. F., "The Design and Construction of Thin Film Radiation Thermopiles," *Report AEDC-TR-66-38*, Arnold Air Force Station, Tennessee: Arnold Engineering Development Center, USAF-AFSC, 1966.

80. Lapina, E. A., "Pirometri Ieluchenija" (Radiation Pyrometry), Moscow: PRIBORPROM, 1966.

81. Lovejoy, D. R., "Measure Radiation above the Gold Point," *I.S.A. Journal*, Vol. 13, No. 2, pp. 55–59, February 1966.

82. Nanigian, J., "Ribbon Thermocouples in the 3000 to 5000°F Range," *Instruments and Control Systems*, Vol. 39, No. 5, pp. 93–99, May 1966.

83. Pfeifer, G. D. and Mikk, G., "A Comparison and Evaluation of Several Radiant Energy Methods for Calibrating Heat Flux Sensors," *Preprint No. 16.15-2-66*, Pittsburgh: Instrument Society of America, 1966.

84. Sparks, L. L. and Powell, R. L., "Final Report on Thermometry Project," *N.B.S. Report 9249*, Boulder, Colorado: National Bureau of Standards, 1966.

85. Tartakovskij, D. F. and Fayans, A. Kh., "Termoelektricheskaja Pirometrija" (Thermoelectric Pyrometry), Moscow: PRIBORPROM, 1966.

86. Hoffman, D., "Dynamics of Mineral Insulated Thermoelements," *IMEKO-IV Paper DDR-160*, Budapest: IMEKO, 1967.

87. Lorenz, G., "Hochtemperaturmessung mit thermischem Rauschen," (High-temperature Measurement by Thermal Noise), *IMEKO-IV Paper No. DDR-130*, Budapest: IMEKO, 1967.

88. Rall, D. L. and Hornbaker, D. R., "A Practical Guide to Temperature Measurement," *Preprint No. P12-1-PHYMMID-67*, Pittsburgh: Instrument Society of America, 1967.

Table 14-1 Temperature Scale Conversion

Celsius	Fahrenheit	Kelvin	Rankine		Celsius	Fahrenheit	Kelvin	Rankine
Absolute Zero					−227.2	−377.0	45.9	82.7
−273.15	−459.67	0	0		−226.0	−374.8	47.1	84.9
					−225.0	−373.0	48.1	86.7
−273.1	−459.7	−0.	0.0		−224.0	−371.2	49.1	88.5
−272.0	−457.6	1.1	2.1		−223.3	−370.0	49.8	89.7
−270.0	−454.0	3.1	5.7		−222.0	−367.6	51.1	92.1
					−221.1	−366.0	52.0	93.7
Boiling Point of Helium					−220.0	−364.0	53.1	95.7
−268.9	−452.0	4.3	7.7		−218.9	−362.0	54.3	97.7
					−218.0	−360.4	55.1	99.3
−268.3	−451.0	4.8	8.7		−217.2	−359.0	55.9	100.7
−268.0	−450.4	5.1	9.3		−216.0	−356.8	57.1	102.9
−267.2	−449.0	5.9	10.7		−215.0	−355.0	58.1	104.7
−266.0	−446.8	7.1	12.9		−214.0	−353.2	59.1	106.5
−265.0	−445.0	8.2	14.7		−213.3	−352.0	59.8	107.7
−264.0	−443.2	9.1	16.5		−212.0	−349.6	61.1	110.1
−263.3	−442.0	9.8	17.7		−211.1	−348.0	62.0	111.7
−262.0	−439.6	11.1	20.1		−210.0	−346.0	63.1	113.7
−261.1	−438.0	12.0	21.7		−208.9	−344.0	64.3	115.7
−260.0	−436.0	13.1	23.7		−208.0	−342.4	65.1	117.3
					−207.2	−341.0	65.9	118.7
−258.9	−434.0	14.3	25.7		−206.0	−338.8	67.1	120.9
−258.0	−432.4	15.1	27.3		−205.0	−337.0	68.1	122.7
−257.2	−431.0	15.9	28.7		−204.0	−335.2	69.1	124.5
−256.0	−428.8	17.1	30.9		−203.3	−334.0	69.8	125.7
−255.0	−427.0	18.1	32.7		−202.0	−331.6	71.1	128.1
−254.0	−425.2	19.1	34.5		−201.1	−330.0	72.0	129.7
−253.3	−424.0	19.8	35.7		−200.0	−328.0	73.1	131.7
					−198.9	−326.0	74.3	133.7
Boiling Point of Hydrogen					−198.0	−324.4	75.1	135.3
−252.8	−423.0	20.3	36.6		−197.2	−323.0	75.9	136.7
					−196.0	−320.8	77.1	138.9
−252.2	−422.0	20.9	37.7					
−252.0	−421.6	21.1	38.1		**Boiling Point of Nitrogen**			
−251.1	−420.0	22.0	39.7		−195.8	−320.4	77.3	139.2
−250.0	−418.0	23.1	41.7					
−248.9	−416.0	24.3	43.7		−195.0	−319.0	78.1	140.7
−248.0	−414.4	25.1	45.3		−194.0	−317.2	79.1	142.5
−247.2	−413.0	25.9	46.7		−193.3	−316.0	79.8	143.7
−246.0	−410.8	27.1	48.9		−192.0	−313.6	81.1	146.1
−245.0	−409.0	28.1	50.7		−191.1	−312.0	82.0	147.7
−244.0	−407.2	29.1	52.5		−190.0	−310.0	83.1	149.7
−243.3	−406.0	29.8	53.7					
−242.0	−403.6	31.1	56.1		**Triple Point of Argon**			
−241.1	−402.0	32.0	57.7		−189.3	−308.8	83.8	150.9
−240.0	−400.0	33.1	59.7					
−238.9	−398.0	34.3	61.7		−188.9	−308.0	84.3	151.7
−238.0	−396.4	35.1	63.3		−188.3	−307.0	84.8	152.7
−237.2	−395.0	35.9	64.7		−188.0	−306.4	85.1	153.3
−236.0	−392.8	37.1	66.9		−187.2	−305.0	85.9	154.7
−235.0	−391.0	38.1	68.7		−186.0	−302.8	87.1	156.9
−234.0	−389.2	39.1	70.5		−185.0	−301.0	88.1	158.7
−233.3	−388.0	39.8	71.7		−184.0	−299.2	89.1	160.5
−232.0	−385.6	41.1	74.1		−183.3	−298.0	89.8	161.7
−231.1	−384.0	42.0	75.7		−183.0	−297.4	90.2	162.3
−230.0	−382.0	43.1	77.7					
−228.9	−380.0	44.3	79.7		**Boiling Point of Oxygen**			
−228.0	−378.4	45.1	81.3		−182.97	−297.36	90.18	162.31

Table 14-1 TEMPERATURE SCALE CONVERSION (*continued*)

°C	°F	°K	°R		°C	°F	°K	°R
−182.2	−296.0	90.9	163.7		−100.0	−148.0	173.1	311.7
−181.1	−294.0	92.0	165.7		−98.9	−146.0	174.3	313.7
−180.0	−292.0	93.1	167.7		−96.0	−140.8	177.1	318.9
−178.9	−290.0	94.3	169.7		−95.0	−139.0	178.1	320.7
−178.0	−288.4	95.1	171.3		−92.0	−133.6	181.1	326.1
−177.2	−287.0	95.9	172.7		−91.1	−132.0	182.0	327.7
−176.0	−284.8	97.1	174.9		−88.0	−126.4	185.1	333.3
−175.0	−283.0	98.1	176.7		−87.2	−125.0	185.9	334.7
−174.0	−281.2	99.1	178.5		−84.0	−119.2	189.1	340.5
−173.3	−280.0	99.8	179.7		−83.3	−118.0	189.8	341.7
					−80.0	−112.0	193.1	347.7
−172.0	−277.6	101.1	182.1		−78.9	−110.0	194.3	349.7
−171.1	−276.0	102.0	183.7					
−170.0	−274.0	103.1	185.7		Sublimation Point of Carbon Dioxide			
−168.9	−272.0	104.3	187.7		−78.5	−109.3	194.6	350.4
−168.0	−270.4	105.1	189.3					
−167.2	−269.0	105.9	190.7		−77.8	−108.0	195.4	351.7
−166.0	−266.8	107.1	192.9		−76.0	−104.8	197.1	354.9
−165.0	−265.0	108.1	194.7		−75.0	−103.0	198.1	356.7
−164.0	−263.2	109.1	196.5		−72.0	−97.6	201.1	362.1
−163.3	−262.0	109.8	197.7		−71.1	−96.0	202.0	363.7
					−68.0	−90.4	205.1	369.3
−162.0	−259.6	111.1	200.1		−67.2	−89.0	205.9	370.7
−161.1	−258.0	112.0	201.7		−64.0	−83.2	209.1	376.5
−160.0	−256.0	113.1	203.7		−63.3	−82.0	209.8	377.7
−158.9	−254.0	114.3	205.7		−60.0	−76.0	213.1	383.7
−158.0	−252.4	115.1	207.3					
−157.2	−251.0	115.9	208.7		−58.9	−74.0	214.3	385.7
−156.0	−248.8	117.1	210.9		−56.0	−68.8	217.1	390.9
−155.0	−247.0	118.1	212.7		−55.0	−67.0	218.1	392.7
−154.0	−245.2	119.1	214.5		−52.0	−61.6	221.1	398.1
−153.3	−244.0	119.8	215.7		−51.1	−60.0	222.0	399.7
					−48.0	−54.4	225.1	405.3
−152.0	−241.6	121.1	218.1		−44.0	−47.2	229.1	412.5
−151.1	−240.0	122.0	219.7		−43.3	−46.0	229.8	413.7
−150.0	−238.0	123.1	221.7		−40.0	−40.0	233.1	419.7
−148.9	−236.0	124.3	223.7		−38.9	−38.0	234.3	421.7
−148.0	−234.4	125.1	225.3		−36.0	−32.8	237.1	426.9
−147.2	−233.0	125.9	226.7		−35.0	−31.0	238.1	428.7
−144.0	−227.2	129.1	232.5					
−143.3	−226.0	129.8	233.7		−32.0	−25.6	241.1	434.1
−140.0	−220.0	133.1	239.7		−31.1	−24.0	242.0	435.7
−138.9	−218.0	134.3	241.7		−28.0	−18.4	245.1	441.3
−136.0	−212.8	137.1	246.9		−27.2	−17.0	245.9	442.7
−135.0	−211.0	138.1	248.7		−24.0	−11.2	249.1	448.5
−132.0	−205.6	141.1	254.1		−23.3	−10.0	249.8	449.7
−131.1	−204.0	142.0	255.7		−20.0	−4.0	253.1	455.7
					−16.0	3.2	257.1	462.9
−128.0	−198.4	145.1	261.3		−15.6	4.0	257.6	463.7
−127.2	−197.0	145.9	262.7		−12.0	10.4	261.1	470.1
−124.0	−191.2	149.1	268.5		−11.7	11.0	261.5	470.7
−123.3	−190.0	149.8	269.7		−8.0	17.6	265.1	477.3
−120.0	−184.0	153.1	275.7		−7.8	18.0	265.4	477.7
−118.9	−182.0	154.3	277.7		−4.0	24.8	269.1	484.5
−116.0	−176.8	157.1	282.9		−3.9	25.0	269.3	484.7
−115.0	−175.0	158.1	284.7					
−112.0	−169.6	161.1	290.1		Ice Point			
−111.1	−168.0	162.0	291.7		0.0	32.0	273.15	491.67
−108.0	−162.4	165.1	297.3		Triple Point of Water			
−107.2	−161.0	165.9	298.7		0.01	32.02	273.16	491.69
−104.0	−155.2	169.1	304.5					
−103.3	−154.0	169.8	305.7					

Table 14-1 Temperature Scale Conversion (*continued*)

°C	°F	°K	°R	°C	°F	°K	°R
4.0	39.2	277.1	498.9	124.0	255.2	397.1	714.9
4.4	40.0	277.6	499.7	124.4	256.0	397.6	715.7
8.0	46.4	281.1	506.1	128.0	262.4	401.1	722.1
8.3	47.0	281.5	506.7	128.3	263.0	401.5	722.7
12.0	53.6	285.1	513.3	132.0	269.6	405.1	729.3
12.2	54.0	285.4	513.7	132.2	270.0	405.4	729.7
16.0	60.8	289.1	520.5	136.0	276.8	409.1	736.5
16.1	61.0	289.3	520.7	136.1	277.0	409.3	736.7
20.0	68.0	293.1	527.7	140.0	284.0	413.1	743.7
24.0	75.2	297.1	534.9	144.0	291.2	417.1	750.9
24.4	76.0	297.6	535.7	144.4	292.0	417.6	751.7
28.0	82.4	301.1	542.1	148.0	298.4	421.1	758.1
28.3	83.0	301.5	542.7	148.3	299.0	421.5	758.7
32.0	89.6	305.1	549.3	152.0	305.6	425.1	765.3
32.2	90.0	305.4	549.7	152.2	306.0	425.4	765.7
36.0	96.8	309.1	556.5	156.0	312.8	429.1	772.5
36.1	97.0	309.3	556.7	156.1	313.0	429.3	772.7
40.0	104.0	313.1	563.7	160.0	320.0	433.1	779.7
44.0	111.2	317.1	570.9	164.0	327.2	437.1	786.9
44.4	112.0	317.6	571.7	164.4	328.0	437.6	787.7
48.0	118.4	321.1	578.1	168.0	334.4	441.1	794.1
48.3	119.0	321.5	578.7	168.3	335.0	441.5	794.7
52.0	125.6	325.1	585.3	172.0	341.6	445.1	801.3
52.2	126.0	325.4	585.7	172.2	342.0	445.4	801.7
56.0	132.8	329.1	592.5	176.0	348.8	449.1	808.5
56.1	133.0	329.3	592.7	176.1	349.0	449.3	808.7
60.0	140.0	333.1	599.7	180.0	356.0	453.1	815.7
64.0	147.2	337.1	606.9	184.0	363.2	457.1	822.9
64.4	148.0	337.6	607.7	184.4	364.0	457.6	823.7
68.0	154.4	341.1	614.1	188.0	370.4	461.1	830.1
68.3	155.0	341.5	614.7	188.3	371.0	461.5	830.7
72.0	161.6	345.1	621.3	192.0	377.6	465.1	837.3
72.2	162.0	345.4	621.7	192.2	378.0	465.4	837.7
76.0	168.8	349.1	628.5	196.0	384.8	469.1	844.5
76.1	169.0	349.3	628.7	196.1	385.0	469.3	844.7
80.0	176.0	353.1	635.7	200.0	392.0	473.1	851.7
84.0	183.2	357.1	642.9	204.0	399.2	477.1	858.9
84.4	184.0	357.6	643.7	204.4	400.0	477.6	859.7
88.0	190.4	361.1	650.1	208.0	406.4	481.1	866.1
88.3	191.0	361.5	650.7	208.3	407.0	481.5	866.7
92.0	197.6	365.1	657.3	212.0	413.6	485.1	873.3
92.2	198.0	365.4	657.7	212.2	414.0	485.4	873.7
96.0	204.8	369.1	664.5	216.0	420.8	489.1	880.5
96.1	205.0	369.3	664.7	216.1	421.0	489.3	880.7
				220.0	428.0	493.1	887.7
Steam Point				224.0	435.2	497.1	894.9
100.0	212.0	373.1	671.7	224.4	436.0	497.6	895.7
				228.0	442.4	501.1	902.1
104.0	219.2	377.1	678.9	228.3	443.0	501.5	902.7
104.4	220.0	377.6	679.7				
108.0	226.4	381.1	686.1	**Freezing Point of Tin**			
108.3	227.0	381.5	686.7	231.91	449.44	505.06	909.11
112.0	233.6	385.1	693.3	232.2	450.0	505.4	909.7
112.2	234.0	385.4	693.7	232.0	449.6	505.1	909.3
116.0	240.8	389.1	700.5	232.2	450.0	505.4	909.7
116.1	241.0	389.3	700.7	236.0	456.8	509.1	916.5
120.0	248.0	393.1	707.7	236.1	457.0	509.3	916.7
				240.0	464.0	513.1	923.7

642

Table 14-1 TEMPERATURE SCALE CONVERSION (*continued*)

°C	°F	°K	°R	°C	°F	°K	°R
244.0	471.2	517.1	930.9	368.3	695.0	641.5	1155.
244.4	472.0	517.6	931.7	372.0	701.6	645.1	1161.
248.0	478.4	521.1	938.1	372.2	702.0	645.4	1162.
248.3	479.0	521.5	938.7	376.0	708.8	649.1	1168.
				380.0	716.0	653.1	1176.
252.0	485.6	525.1	945.3				
252.2	486.0	525.4	945.7	384.0	723.2	657.1	1183.
256.0	492.8	529.1	952.5	384.4	724.0	657.6	1184.
256.1	493.0	529.3	952.7	388.0	730.4	661.1	1190.
260.0	500.0	533.1	959.7	388.3	731.0	661.5	1191.
264.0	507.2	537.1	966.9	392.0	737.6	665.1	1197.
264.4	508.0	537.6	967.7	392.2	738.0	665.4	1198.
268.0	514.4	541.1	974.1	396.0	744.8	669.1	1204.
268.3	515.0	541.5	974.7	396.1	745.0	669.3	1205.
272.0	521.6	545.1	981.3	400.0	752.0	673.1	1212.
				404.0	759.2	677.1	1219.
272.2	522.0	545.4	981.7				
276.0	528.8	549.1	988.5	404.4	760.0	677.6	1220.
276.1	529.0	549.3	989.	408.0	766.4	681.1	1226.
280.0	536.0	553.1	996.	408.3	767.0	681.5	1227.
284.0	543.2	557.1	1003.	412.0	773.6	685.1	1233.
284.4	544.0	557.6	1004.	412.2	774.0	685.4	1234.
288.0	550.4	561.1	1010.	416.0	780.8	689.1	1240.
288.3	551.0	561.5	1011.	416.1	781.0	689.3	1241.
292.0	557.6	565.1	1017.				
292.2	558.0	565.4	1018.	Freezing Point of Zinc			
296.0	564.8	569.1	1024.	419.5	787.1	692.6	1247.
296.1	565.0	569.3	1025.				
300.0	572.0	573.1	1032.	420.0	788.0	693.1	1248.
304.0	579.2	577.1	1039.	424.0	795.2	697.1	1255.
304.4	580.0	577.6	1040.	424.4	796.0	697.6	1256.
308.0	586.4	581.1	1046.	428.0	802.4	701.1	1262.
308.3	587.0	581.5	1047.	428.3	803.0	701.5	1263.
312.0	593.6	585.1	1053.	432.0	809.6	705.1	1269.
				432.2	810.0	705.4	1270.
312.2	594.0	585.4	1054.	436.0	816.8	709.1	1276.
316.0	600.8	589.1	1060.	436.1	817.0	709.3	1277.
316.1	601.0	589.3	1061.	440.0	824.0	713.1	1284.
320.0	608.0	593.1	1068.	444.0	831.2	717.1	1291.
324.0	615.2	597.1	1075.	444.4	832.0	717.6	1292.
324.4	616.0	597.6	1076.				
328.0	622.4	601.1	1082.	Boiling Point of Sulfur			
328.3	623.0	601.5	1083.	444.6	832.3	717.7	1292.
332.0	629.6	605.1	1089.				
332.2	630.0	605.4	1090.	445.0	833.0	718.1	1293.
				448.0	838.4	721.1	1298.
336.0	636.8	609.1	1096.	448.3	839.0	721.5	1299.
336.1	637.0	609.3	1097.	452.0	845.6	725.1	1305.
340.0	644.0	613.1	1104.	452.2	846.0	725.4	1306.
344.0	651.2	617.1	1111.	456.0	852.8	729.1	1312.
344.4	652.0	617.6	1112.	456.1	853.0	729.3	1313.
348.0	658.4	621.1	1118.	460.0	860.0	733.1	1320.
348.3	659.0	621.5	1119.	464.0	867.2	737.1	1327.
352.0	665.6	625.1	1125.	464.4	868.0	737.6	1328.
352.2	666.0	625.4	1126.				
356.0	672.8	629.1	1132.	468.0	874.4	741.1	1334.
				468.3	875.0	741.5	1335.
356.1	673.0	629.3	1133.	472.0	881.6	745.1	1341.
360.0	680.0	633.1	1140.	472.2	882.0	745.4	1342.
364.0	687.2	637.1	1147.	476.0	888.8	749.1	1348.
364.4	688.0	637.6	1148.				
368.0	694.4	641.1	1154.				

643

Table 14-1 Temperature Scale Conversion (*continued*)

°C	°F	°K	°R		°C	°F	°K	°R
476.1	889.0	749.3	1349.		612.0	1134.	885.1	1593.
480.0	896.0	753.1	1356.		612.2	1134.	885.4	1594.
484.0	903.2	757.1	1363.		616.0	1141.	889.1	1600.
484.4	904.0	757.6	1364.		616.1	1141.	889.3	1601.
488.0	910.4	761.1	1370.		620.0	1148.	893.1	1608.
488.3	911.0	761.5	1371.		624.0	1155.	897.1	1615.
492.0	917.6	765.1	1377.		624.4	1156.	897.6	1616.
492.2	918.0	765.4	1378.		628.0	1162.	901.1	1622.
496.0	924.8	769.1	1384.					
496.1	925.0	769.3	1385.		Freezing Point of Antimony			
					630.5	1166.9	903.65	1626.6
500.0	932.0	773.1	1392.					
504.0	939.2	777.1	1399.		632.2	1170.	905.4	1630.
504.4	940.0	777.6	1400.		636.0	1177.	909.1	1636.
508.0	946.4	781.1	1406.		636.1	1177.	909.3	1637.
508.3	947.0	781.5	1407.		640.0	1184.	913.1	1644.
512.0	953.6	785.1	1413.		644.0	1191.	917.1	1651.
512.2	954.0	785.4	1414.		644.4	1192.	917.6	1652.
516.0	960.8	789.1	1420.		648.0	1198.	921.1	1658.
516.1	961.0	789.3	1421.		648.3	1199.	921.5	1659.
520.0	968.0	793.1	1428.		652.0	1206.	925.1	1665.
					652.2	1206.	925.4	1666.
524.0	975.2	797.1	1435.					
524.4	976.0	797.6	1436.		656.0	1213.	929.1	1672.
528.0	982.4	801.1	1442.		656.1	1213.	929.3	1673.
528.3	983.0	801.5	1443.		660.0	1220.	933.1	1680.
532.0	989.6	805.1	1449.		664.0	1227.	937.1	1687.
532.2	990.0	805.4	1450.		664.4	1228.	937.6	1688.
536.0	996.8	809.1	1456.		668.0	1234.	941.1	1694.
536.1	997.0	809.3	1457.		668.3	1235.	941.5	1695.
540.0	1004.	813.1	1464.		672.0	1242.	945.1	1701.
544.0	1011.	817.1	1471.		672.2	1242.	945.4	1702.
					676.0	1249.	949.1	1708.
544.4	1012.	817.6	1472.					
548.0	1018.	821.1	1478.		676.1	1249.	949.3	1709.
548.3	1019.	821.5	1479.		680.0	1256.	953.1	1716.
552.0	1026.	825.1	1485.		684.0	1263.	957.1	1723.
552.2	1026.	825.4	1486.		684.4	1264.	957.6	1724.
556.0	1033.	829.1	1492.		688.0	1270.	961.1	1730.
556.1	1033.	829.3	1493.		688.3	1271.	961.5	1731.
560.0	1040.	833.1	1500.		692.0	1278.	965.1	1737.
564.0	1047.	837.1	1507.		692.2	1278.	965.4	1738.
564.4	1048.	837.6	1508.		696.0	1285.	969.1	1744.
					696.1	1285.	969.3	1745.
568.0	1054.	841.1	1514.					
568.3	1055.	841.5	1515.		700.0	1292.	973.1	1752.
572.2	1062.	845.4	1522.		704.0	1299.	977.1	1759.
576.0	1063.	849.1	1528.		704.4	1300.	977.6	1760.
576.1	1069.	849.3	1529.		708.0	1306.	981.1	1766.
580.0	1076.	853.1	1536.		708.3	1307.	981.5	1767.
584.0	1083.	857.1	1543.		712.0	1314.	985.1	1773.
584.4	1084.	857.6	1544.		712.2	1314.	985.4	1774.
588.0	1090.	861.1	1550.		716.0	1321.	989.1	1780.
588.3	1091.	861.5	1551.		716.1	1321.	989.3	1781.
					720.0	1328.	993.1	1788.
592.0	1098.	865.1	1557.					
592.2	1098.	865.4	1558.		724.0	1335.	997.1	1795.
596.0	1105.	869.1	1564.		724.4	1336.	997.6	1796.
600.0	1112.	873.1	1572.		728.0	1342.	1001.	1802.
604.0	1119.	877.1	1579.		728.3	1343.	1002.	1803.
604.4	1120.	877.6	1580.		732.0	1350.	1005.	1809.
608.0	1126.	881.1	1586.		732.2	1350.	1005.	1810.
608.3	1127.	881.5	1587.					

Table 14-1 TEMPERATURE SCALE CONVERSION (*continued*)

°C	°F	°K	°R		°C	°F	°K	°R
736.0	1357.	1009.	1816.		868.0	1594.	1141.	2054.
736.1	1357.	1009.	1817.		868.3	1595.	1142.	2055.
740.0	1364.	1013.	1824.		872.0	1602.	1145.	2061.
744.0	1371.	1017.	1831.		872.2	1602.	1145.	2062.
744.4	1372.	1018.	1832.		876.0	1609.	1149.	2068.
748.0	1378.	1021.	1838.		876.1	1609.	1149.	2069.
748.3	1379.	1022.	1839.		880.0	1616.	1153.	2076.
752.0	1386.	1025.	1845.		884.0	1623.	1157.	2083.
752.2	1386.	1025.	1846.		884.4	1624.	1158.	2084.
756.0	1393.	1029.	1852.		888.0	1630.	1161.	2090.
756.1	1393.	1029.	1853.		888.3	1631.	1162.	2091.
760.0	1400.	1033.	1860.		892.0	1638.	1165.	2097.
764.0	1407.	1037.	1867.		892.2	1638.	1165.	2098.
764.4	1408.	1038.	1868.		896.0	1645.	1169.	2104.
768.0	1414.	1041.	1874.		896.1	1645.	1169.	2105.
768.3	1415.	1042.	1875.		900.0	1652.	1173.	2112.
772.0	1422.	1045.	1881.		904.0	1659.	1177.	2119.
776.0	1429.	1049.	1888.		904.4	1660.	1178.	2120.
776.1	1429.	1049.	1889.		908.0	1666.	1181.	2126.
780.0	1436.	1053.	1896.		908.3	1667.	1182.	2127.
784.0	1443.	1057.	1903.		912.0	1674.	1185.	2133.
784.4	1444.	1058.	1904.		912.2	1674.	1185.	2134.
788.0	1450.	1061.	1910.		916.0	1681.	1189.	2140.
788.3	1451.	1062.	1911.		916.1	1681.	1189.	2141.
792.0	1458.	1065.	1917.		920.0	1688.	1193.	2148.
792.2	1458.	1065.	1918.		924.0	1695.	1197.	2155.
796.0	1465.	1069.	1924.		924.4	1696.	1198.	2156.
796.1	1465.	1069.	1925.		928.0	1702.	1201.	2162.
800.0	1472.	1073.	1932.		928.3	1703.	1202.	2163.
804.0	1479.	1077.	1939.		932.0	1710.	1205.	2169.
804.4	1480.	1078.	1940.		932.2	1710.	1205.	2170.
808.0	1486.	1081.	1946.		936.0	1717.	1209.	2176.
808.3	1487.	1082.	1947.		936.1	1717.	1209.	2177.
812.0	1494.	1085.	1953.		940.0	1724.	1213.	2184.
812.2	1494.	1085.	1954.		944.0	1731.	1217.	2191.
816.0	1501.	1089.	1960.		944.4	1732.	1218.	2192.
816.1	1501.	1089.	1961.		948.0	1738.	1221.	2198.
820.0	1508.	1093.	1968.		948.3	1739.	1222.	2199.
824.0	1515.	1097.	1975.		952.0	1746.	1225.	2205.
824.4	1516.	1098.	1976.		952.2	1746.	1225.	2206.
828.0	1522.	1101.	1982.		956.0	1753.	1229.	2212.
828.3	1523.	1102.	1983.		956.1	1753.	1229.	2213.
832.0	1530.	1105.	1989.		960.0	1760.	1233.	2220.
832.2	1530.	1105.	1990.					
836.0	1537.	1109.	1996.		Melting Point of Silver			
836.1	1537.	1109.	1997.		960.8	1761.	1234.	2221.
840.0	1544.	1113.	2004.					
844.0	1551.	1117.	2011.		961.1	1762.	1234.	2222.
844.4	1552.	1118.	2012.		964.0	1767.	1237.	2227.
848.0	1558.	1121.	2018.		964.4	1768.	1238.	2228.
848.3	1559.	1122.	2019.		968.0	1774.	1241.	2234.
852.0	1566.	1125.	2025.		968.3	1775.	1242.	2235.
852.2	1566.	1125.	2026.		972.2	1782.	1245.	2242.
856.0	1573.	1129.	2032.		976.0	1789.	1249.	2248.
856.1	1573.	1129.	2033.		976.1	1789.	1249.	2249.
860.0	1580.	1133.	2040.		980.0	1796.	1253.	2256.
864.0	1587.	1137.	2047.		984.0	1803.	1257.	2263.
864.4	1588.	1138.	2048.		984.4	1804.	1258.	2264.
					988.0	1810.	1261.	2270.

Table 14-1 TEMPERATURE SCALE CONVERSION (continued)

°C	°F	°K	°R
988.3	1811.	1267.	2271.
992.0	1818.	1265.	2277.
992.2	1818.	1265.	2278.
996.0	1825.	1269.	2284.
996.1	1825.	1269.	2285.
1000.	1832.	1273.	2292.
1004.	1839.	1277.	2299.
1008.	1846.	1281.	2306.
1012.	1854.	1285.	2313.
1016.	1861.	1289.	2320.
1020.	1868.	1293.	2328.
1024.	1875.	1297.	2335.
1028.	1882.	1301.	2342.
1032.	1890.	1305.	2349.
1036.	1897.	1309.	2356.
1040.	1904.	1313.	2364.
1044.	1911.	1317.	2371.
1048.	1918.	1321.	2378.
1052.	1926.	1325.	2385.
1056.	1933.	1329.	2392.
1060.	1940.	1333.	2400.

Melting Point of Gold

°C	°F	°K	°R
1063.	1945.	1336.	2405.

°C	°F	°K	°R
1064.	1947.	1337.	2407.
1068.	1954.	1341.	2414.
1072.	1962.	1345.	2421.
1076.	1969.	1349.	2428.
1080.	1976.	1353.	2436.
1084.	1983.	1357.	2443.
1088.	1990.	1361.	2450.
1092.	1998.	1365.	2457.
1096.	2005.	1369.	2464.
1100.	2012.	1373.	2472.
1110.	2030.	1383.	2490.
1120.	2048.	1393.	2508.
1130.	2066.	1403.	2526.
1140.	2084.	1413.	2544.
1150.	2102.	1423.	2562.
1160.	2120.	1433.	2580.
1170.	2138.	1443.	2598.
1180.	2156.	1453.	2616.
1190.	2174.	1463.	2634.
1200.	2192.	1473.	2652.
1210.	2210.	1483.	2670.
1220.	2228.	1493.	2688.
1230.	2246.	1503.	2706.
1240.	2264.	1513.	2724.
1250.	2282.	1523.	2742.
1260.	2300.	1533.	2760.
1270.	2318.	1543.	2778.
1280.	2336.	1553.	2796.
1290.	2354.	1563.	2814.
1300.	2372.	1573.	2832.
1310.	2390.	1583.	2850.
1320.	2408.	1593.	2868.
1330.	2426.	1603.	2886.

°C	°F	°K	°R
1340.	2444.	1613.	2904.
1350.	2462.	1623.	2922.
1360.	2480.	1633.	2940.
1370.	2498.	1643.	2958.
1380.	2516.	1653.	2976.
1390.	2534.	1663.	2994.
1400.	2552.	1673.	3012.
1410.	2570.	1683.	3030.
1420.	2588.	1693.	3048.
1430.	2606.	1703.	3066.
1440.	2624.	1713.	3084.
1450.	2642.	1723.	3102.
1460.	2660.	1733.	3120.
1470.	2678.	1743.	3138.
1480.	2696.	1753.	3156.
1490.	2714.	1763.	3174.
1500.	2732.	1773.	3192.
1510.	2750.	1783.	3210.
1520.	2768.	1793.	3228.
1530.	2786.	1803.	3246.
1540.	2804.	1813.	3264.
1550.	2822.	1823.	3282.
1560.	2840.	1833.	3300.
1570.	2858.	1843.	3318.
1580.	2876.	1853.	3336.
1590.	2894.	1863.	3354.
1600.	2912.	1873.	3372.
1610.	2930.	1883.	3390.
1620.	2948.	1893.	3408.
1630.	2966.	1903.	3426.
1640.	2984.	1913.	3444.
1650.	3002.	1923.	3462.
1660.	3020.	1933.	3480.
1670.	3038.	1943.	3498.
1680.	3056.	1953.	3516.
1690.	3074.	1963.	3534.
1700.	3092.	1973.	3552.
1710.	3110.	1983.	3570.
1720.	3128.	1993.	3588.
1730.	3146.	2003.	3606.
1740.	3164.	2013.	3624.
1750.	3182.	2023.	3642.
1760.	3200.	2033.	3660.

Melting Point of Platinum

°C	°F	°K	°R
1769.	3216.	1972.	3676.

°C	°F	°K	°R
1770.	3218.	2043.	3678.
1780.	3236.	2053.	3696.
1790.	3254.	2063.	3714.
1800.	3272.	2073.	3732.
1810.	3290.	2083.	3750.
1820.	3308.	2093.	3768.
1830.	3326.	2103.	3786.
1840.	3344.	2113.	3804.
1850.	3362.	2123.	3822.
1860.	3380.	2133.	3840.
1870.	3398.	2143.	3858.
1880.	3416.	2153.	3876.

Table 14-1 Temperature Scale Conversion (*continued*)

°C	°F	°K	°R	°C	°F	°K	°R
1890.	3434.	2163.	3894.	2470.	4478.	2743.	4938.
1900.	3452.	2173.	3912.	2480.	4496.	2753.	4956.
1910.	3470.	2183.	3930.	2490.	4514.	2763.	4974.
1920.	3488.	2193.	3948.	2500.	4532.	2773.	4992.
1930.	3506.	2203.	3966.	2510.	4550.	2783.	5010.
1940.	3524.	2213.	3984.	2520.	4568.	2793.	5028.
1950.	3542.	2223.	4002.	2530.	4586.	2803.	5046.
1960.	3560.	2233.	4020.	2540.	4604.	2813.	5064.
1970.	3578.	2243.	4038.	2550.	4622.	2823.	5082.
1980.	3596.	2253.	4056.	2560.	4640.	2833.	5100.
1990.	3614.	2263.	4074.	2570.	4658.	2843.	5118.
2000.	3632.	2273.	4092.	2580.	4676.	2853.	5136.
2010.	3650.	2283.	4110.	2590.	4694.	2863.	5154.
2020.	3668.	2293.	4128.	2600.	4712.	2873.	5172.
2030.	3686.	2303.	4146.	2610.	4730.	2883.	5190.
2040.	3704.	2313.	4164.	2620.	4748.	2893.	5208.
2050.	3722.	2323.	4182.	2630.	4766.	2903.	5226.
2060.	3740.	2333.	4200.	2640.	4784.	2913.	5244.
2070.	3758.	2343.	4218.	2650.	4802.	2923.	5262.
2080.	3776.	2353.	4236.	2660.	4820.	2933.	5280.
2090.	3794.	2363.	4254.	2670.	4838.	2943.	5298.
2100.	3812.	2373.	4272.	2680.	4856.	2953.	5316.
2110.	3830.	2383.	4290.	2690.	4874.	2963.	5334.
2120.	3848.	2393.	4308.	2700.	4892.	2973.	5352.
2130.	3866.	2403.	4326.	2710.	4910.	2983.	5370.
2140.	3884.	2413.	4344.	2720.	4928.	2993.	5388.
2150.	3902.	2423.	4362.	2730.	4946.	3003.	5406.
2160.	3920.	2433.	4380.	2740.	4964.	3013.	5424.
2170.	3938.	2443.	4398.	2750.	4982.	3023.	5442.
2180.	3956.	2453.	4416.	2760.	5000.	3033.	5460.
2190.	3974.	2463.	4434.	2770.	5018.	3043.	5478.
2200.	3992.	2473.	4452.	2780.	5036.	3053.	5496.
2210.	4010.	2483.	4470.	2790.	5054.	3063.	5514.
2220.	4028.	2493.	4488.	2800.	5072.	3073.	5532.
2230.	4046.	2503.	4506.	2810.	5090.	3083.	5550.
2240.	4064.	2513.	4524.	2820.	5108.	3093.	5568.
2250.	4082.	2523.	4542.	2830.	5126.	3103.	5586.
2260.	4100.	2533.	4560.	2840.	5144.	3113.	5604.
2270.	4118.	2543.	4578.	2850.	5162.	3123.	5622.
2280.	4136.	2553.	4596.	2860.	5180.	3133.	5640.
2290.	4154.	2563.	4614.	2870.	5198.	3143.	5658.
2300.	4172.	2573.	4632.	2880.	5216.	3153.	5676.
2310.	4190.	2583.	4650.	2890.	5234.	3163.	5694.
2320.	4208.	2593.	4668.	2900.	5252.	3173.	5712.
2330.	4226.	2603.	4686.	2910.	5270.	3183.	5730.
2340.	4244.	2613.	4704.	2920.	5288.	3193.	5748.
2350.	4262.	2623.	4722.	2930.	5306.	3203.	5766.
2360.	4280.	2633.	4740.	2940.	5324.	3213.	5784.
2370.	4298.	2643.	4758.	2950.	5342.	3223.	5802.
2380.	4316.	2653.	4776.	2960.	5360.	3233.	5820.
2390.	4334.	2663.	4794.	2970.	5378.	3243.	5338.
2400.	4352.	2673.	4812.	2980.	5396.	3253.	5856.
2410.	4370.	2683.	4830.	2990.	5414.	3263.	5874.
2420.	4388.	2693.	4848.	3000.	5432.	3273.	5892.
2430.	4406.	2703.	4866.	3010.	5450.	3283.	5910.
2440.	4424.	2713.	4884.	3020.	5468.	3293.	5928.
2450.	4442.	2723.	4902.	3030.	5486.	3303.	5946.
2460.	4460.	2733.	4920.	3040.	5504.	3313.	5964.

Table 14-1 TEMPERATURE SCALE CONVERSION (*concluded*)

°C	°F	°K	°R
3050.	5522.	3323.	5982.
3060.	5540.	3333.	6000.
3070.	5558.	3343.	6018.
3080.	5576.	3353.	6036.
3090.	5594.	3363.	6054.
3100.	5612.	3373.	6072.
3110.	5630.	3383.	6090.
3120.	5648.	3393.	6108.
3130.	5666.	3403.	6126.
3140.	5684.	3413.	6144.
3150.	5702.	3423.	6162.
3160.	5720.	3433.	6180.
3170.	5738.	3443.	6198.
3180.	5756.	3453.	6216.
3190.	5774.	3463.	6234.
3200.	5792.	3473.	6252.
3210.	5810.	3483.	6270.
3220.	5828.	3493.	6288.
3230.	5846.	3503.	6306.
3240.	5864.	3513.	6324.
3250.	5882.	2523.	6342.
3260.	5900.	3533.	6360.
3270.	5918.	3543.	6478.
3280.	5936.	3553.	6396.
3290.	5954.	3563.	6414.
3300.	5972.	3573.	6432.
3310.	5990.	3583.	6450.
3320.	6008.	3593.	6468.
3330.	6026.	3603.	6486.
3340.	6044.	3613.	6504.
3350.	6062.	3623.	6522.
3360.	6080.	3633.	6540.
3370.	6098.	3643.	6558.

Melting Point of Tungsten

°C	°F	°K	°R
3380.	6116.	3653.	6576.

°C	°F	°K	°R
3390.	6134.	3663.	6594.
3400	6152.	3673.	6612.
3410.	6170.	3683.	6630.
3420.	6188.	3693.	6648.
3430.	6206.	3703.	6666.
3440.	6224.	3713.	6684.
3450.	6242.	3723.	6702.
3460.	6260.	3733.	6720.
3470.	6278.	3743.	6738.
3480.	6296.	3753.	6756.
3490.	6314.	3763.	6774.
3500.	6332.	3773.	6792.
3510.	6350.	3783.	6810.

**Platinum Versus Platinum-10-Percent Rhodium Thermocouples
(Electromotive Force in Absolute Millivolts. Temperatures
in Degrees C (Int. 1948). Reference Junctions at 0°C.)**

°C	0	1	2	3	4	5	6	7	8	9	10	°C
	Millivolts											
0	0.000	0.005	0.011	0.016	0.022	0.028	0.033	0.039	0.044	0.050	0.056	0
10	.056	.061	.067	.073	.078	.084	.090	.096	.102	.107	.113	10
20	.113	.119	.125	.131	.137	.143	.149	.155	.161	.167	.173	20
30	.173	.179	.185	.191	.198	.204	.210	.216	.222	.229	.235	30
40	.235	.241	.247	.254	.260	.266	.273	.279	.286	.292	.299	40
50	.299	.305	.312	.318	.325	.331	.338	.344	.351	.357	.364	50
60	.364	.371	.377	.384	.391	.397	.404	.411	.418	.425	.431	60
70	.431	.438	.445	.452	.459	.466	.473	.479	.486	.493	.500	70
80	.500	.507	.514	.521	.528	.535	.543	.550	.557	.564	.571	80
90	.571	.578	.585	.593	.600	.607	.614	.621	.629	.636	.643	90
100	.643	.651	.658	.665	.673	.680	.687	.694	.702	.709	.717	100
110	.717	.724	.732	.739	.747	.754	.762	.769	.777	.784	.792	110
120	.792	.800	.807	.815	.823	.830	.838	.845	.853	.861	.869	120
130	.869	.876	.884	.892	.900	.907	.915	.923	.931	.939	.946	130
140	.946	.954	.962	.970	.978	.986	.994	1.002	1.009	1.017	1.025	140
150	1.025	1.033	1.041	1.049	1.057	1.065	1.073	1.081	1.089	1.097	1.106	150
160	1.106	1.114	1.122	1.130	1.138	1.146	1.154	1.162	1.170	1.179	1.187	160
170	1.187	1.195	1.203	1.211	1.220	1.228	1.236	1.244	1.253	1.261	1.269	170
180	1.269	1.277	1.286	1.294	1.302	1.311	1.319	1.327	1.336	1.344	1.352	180
190	1.352	1.361	1.369	1.377	1.386	1.394	1.403	1.411	1.419	1.428	1.436	190
200	1.436	1.445	1.453	1.462	1.470	1.479	1.487	1.496	1.504	1.513	1.521	200
210	1.521	1.530	1.538	1.547	1.555	1.564	1.573	1.581	1.590	1.598	1.607	210
220	1.607	1.615	1.624	1.633	1.641	1.650	1.659	1.667	1.676	1.685	1.693	220
230	1.693	1.702	1.710	1.719	1.728	1.736	1.745	1.754	1.763	1.771	1.780	230
240	1.780	1.789	1.798	1.806	1.815	1.824	1.833	1.841	1.850	1.859	1.868	240
250	1.868	1.877	1.885	1.894	1.903	1.912	1.921	1.930	1.938	1.947	1.956	250
260	1.956	1.965	1.974	1.983	1.992	2.001	2.009	2.018	2.027	2.036	2.045	260
270	2.045	2.054	2.063	2.072	2.081	2.090	2.099	2.108	2.117	2.126	2.135	270
280	2.135	2.144	2.153	2.162	2.171	2.180	2.189	2.198	2.207	2.216	2.225	280
290	2.225	2.234	2.243	2.252	2.261	2.271	2.280	2.289	2.298	2.307	2.316	290
300	2.316	2.325	2.334	2.343	2.353	2.362	2.371	2.380	2.389	2.398	2.408	300
310	2.408	2.417	2.426	2.435	2.444	2.453	2.463	2.472	2.481	2.490	2.499	310
320	2.499	2.509	2.518	2.527	2.536	2.546	2.555	2.564	2.573	2.583	2.592	320
330	2.592	2.601	2.610	2.620	2.629	2.638	2.648	2.657	2.666	2.676	2.685	330
340	2.685	2.694	2.704	2.713	2.722	2.731	2.741	2.750	2.760	2.769	2.778	340
350	2.778	2.788	2.797	2.806	2.816	2.825	2.834	2.844	2.853	2.863	2.872	350
360	2.872	2.881	2.891	2.900	2.910	2.919	2.929	2.938	2.947	2.957	2.966	360
370	2.966	2.976	2.985	2.995	3.004	3.014	3.023	3.032	3.042	3.051	3.061	370
380	3.061	3.070	3.080	3.089	3.099	3.108	3.118	3.127	3.137	3.146	3.156	380
390	3.156	3.165	3.175	3.184	3.194	3.203	3.213	3.222	3.232	3.241	3.251	390
400	3.251	3.261	3.270	3.280	3.289	3.299	3.308	3.318	3.327	3.337	3.347	400
°C	0	1	2	3	4	5	6	7	8	9	10	°C

Table 14-4 Thermocouple Reference Tables (*continued*)

Platinum Versus Platinum-10-Percent Rhodium Thermocouples—Continued
(Electromotive Force in Absolute Millivolts. Temperatures
in Degrees C (Int. 1948). Reference Junctions at 0°C.)

°C	0	1	2	3	4	5	6	7	8	9	10	°C
						Millivolts						
400	3.251	3.261	3.270	3.280	3.289	3.299	3.308	3.318	3.327	3.337	3.347	400
410	3.347	3.356	3.366	3.375	3.385	3.394	3.404	3.414	3.423	3.433	3.442	410
420	3.442	3.452	3.462	3.471	3.481	3.490	3.500	3.510	3.519	3.529	3.539	420
430	3.539	3.548	3.558	3.567	3.577	3.587	3.596	3.606	3.616	3.625	3.635	430
440	3.635	3.645	3.654	3.664	3.674	3.683	3.693	3.703	3.712	3.722	3.732	440
450	3.732	3.741	3.751	3.761	3.771	3.780	3.790	3.800	3.809	3.819	3.829	450
460	3.829	3.839	3.848	3.858	3.868	3.878	3.887	3.897	3.907	3.917	3.926	460
470	3.926	3.936	3.946	3.956	3.965	3.975	3.985	3.995	4.004	4.014	4.024	470
480	4.024	4.034	4.044	4.053	4.063	4.073	4.083	4.093	4.103	4.112	4.122	480
490	4.122	4.132	4.142	4.152	4.162	4.171	4.181	4.191	4.201	4.211	4.221	490
500	4.221	4.230	4.240	4.250	4.260	4.270	4.280	4.290	4.300	4.310	4.319	500
510	4.319	4.329	4.339	4.349	4.359	4.369	4.379	4.389	4.399	4.409	4.419	510
520	4.419	4.428	4.438	4.448	4.458	4.468	4.478	4.488	4.498	4.508	4.518	520
530	4.518	4.528	4.538	4.548	4.558	4.568	4.578	4.588	4.598	4.608	4.618	530
540	4.618	4.628	4.638	4.648	4.658	4.668	4.678	4.688	4.698	4.708	4.718	540
550	4.718	4.728	4.738	4.748	4.758	4.768	4.778	4.788	4.798	4.808	4.818	550
560	4.818	4.828	4.839	4.849	4.859	4.869	4.879	4.889	4.899	4.909	4.919	560
570	4.919	4.929	4.939	4.950	4.960	4.970	4.980	4.990	5.000	5.010	5.020	570
580	5.020	5.031	5.041	5.051	5.061	5.071	5.081	5.091	5.102	5.112	5.122	580
590	5.122	5.132	5.142	5.152	5.163	5.173	5.183	5.193	5.203	5.214	5.224	590
600	5.224	5.234	5.244	5.254	5.265	5.275	5.285	5.295	5.306	5.316	5.326	600
610	5.326	5.336	5.346	5.357	5.367	5.377	5.388	5.398	5.408	5.418	5.429	610
620	5.429	5.439	5.449	5.459	5.470	5.480	5.490	5.501	5.511	5.521	5.532	620
630	5.532	5.542	5.552	5.563	5.573	5.583	5.593	5.604	5.614	5.624	5.635	630
640	5.635	5.645	5.655	5.666	5.676	5.686	5.697	5.707	5.717	5.728	5.738	640
650	5.738	5.748	5.759	5.769	5.779	5.790	5.800	5.811	5.821	5.831	5.842	650
660	5.842	5.852	5.862	5.873	5.883	5.894	5.904	5.914	5.925	5.935	5.946	660
670	5.946	5.956	5.967	5.977	5.987	5.998	6.008	6.019	6.029	6.040	6.050	670
680	6.050	6.060	6.071	6.081	6.092	6.102	6.113	6.123	6.134	6.144	6.155	680
690	6.155	6.165	6.176	6.186	6.197	6.207	6.218	6.228	6.239	6.249	6.260	690
700	6.260	6.270	6.281	6.291	6.302	6.312	6.323	6.333	6.344	6.355	6.365	700
710	6.365	6.376	6.386	6.397	6.407	6.418	6.429	6.439	6.450	6.460	6.471	710
720	6.471	6.481	6.492	6.503	6.513	6.524	6.534	6.545	6.556	6.566	6.577	720
730	6.577	6.588	6.598	6.609	6.619	6.630	6.641	6.651	6.662	6.673	6.683	730
740	6.683	6.694	6.705	6.715	6.726	6.737	6.747	6.758	6.769	6.779	6.790	740
750	6.790	6.801	6.811	6.822	6.833	6.844	6.854	6.865	6.876	6.886	6.897	750
760	6.897	6.908	6.919	6.929	6.940	6.951	6.962	6.972	6.983	6.994	7.005	760
770	7.005	7.015	7.026	7.037	7.047	7.058	7.069	7.080	7.091	7.102	7.112	770
780	7.112	7.123	7.134	7.145	7.156	7.166	7.177	7.188	7.199	7.210	7.220	780
790	7.220	7.231	7.242	7.253	7.264	7.275	7.286	7.296	7.307	7.318	7.329	790
800	7.329	7.340	7.351	7.362	7.372	7.383	7.394	7.405	7.416	7.427	7.438	800
°C	0	1	2	3	4	5	6	7	8	9	10	°C

Table 14-4 THERMOCOUPLE REFERENCE TABLES (*continued*)

Platinum Versus Platinum-10-Percent Rhodium Thermocouples—Continued
(Electromotive Force in Absolute Millivolts. Temperatures
in Degrees C (Int. 1948). Reference Junctions at 0°C.)

°C	0	1	2	3	4	5	6	7	8	9	10	°C
	Millivolts											
800	7.329	7.340	7.351	7.362	7.372	7.383	7.394	7.405	7.416	7.427	7.438	800
810	7.438	7.449	7.460	7.470	7.481	7.492	7.503	7.514	7.525	7.536	7.547	810
820	7.547	7.558	7.569	7.580	7.591	7.602	7.613	7.623	7.634	7.645	7.656	820
830	7.656	7.667	7.678	7.689	7.700	7.711	7.722	7.733	7.744	7.755	7.766	830
840	7.766	7.777	7.788	7.799	7.810	7.821	7.832	7.843	7.854	7.865	7.876	840
850	7.876	7.887	7.898	7.910	7.921	7.932	7.943	7.954	7.965	7.976	7.987	850
860	7.987	7.998	8.009	8.020	8.031	8.042	8.053	8.064	8.076	8.087	8.098	860
870	8.098	8.109	8.120	8.131	8.142	8.153	8.164	8.176	8.187	8.198	8.209	870
880	8.209	8.220	8.231	8.242	8.254	8.265	8.276	8.287	8.298	8.309	8.320	880
890	8.320	8.332	8.343	8.354	8.365	8.376	8.388	8.399	8.410	8.421	8.432	890
900	8.432	8.444	8.455	8.466	8.477	8.488	8.500	8.511	8.522	8.533	8.545	900
910	8.545	8.556	8.567	8.578	8.590	8.601	8.612	8.623	8.635	8.646	8.657	910
920	8.657	8.668	8.680	8.691	8.702	8.714	8.725	8.736	8.747	8.759	8.770	920
930	8.770	8.781	8.793	8.804	8.815	8.827	8.838	8.849	8.861	8.872	8.883	930
940	8.883	8.895	8.906	8.917	8.929	8.940	8.951	8.963	8.974	8.986	8.997	940
950	8.997	9.008	9.020	9.031	9.042	9.054	9.065	9.077	9.088	9.099	9.111	950
960	9.111	9.122	9.134	9.145	9.157	9.168	9.179	9.191	9.202	9.214	9.225	960
970	9.225	9.236	9.248	9.260	9.271	9.282	9.294	9.305	9.317	9.328	9.340	970
980	9.340	9.351	9.363	9.374	9.386	9.397	9.409	9.420	9.432	9.443	9.455	980
990	9.455	9.466	9.478	9.489	9.501	9.512	9.524	9.535	9.547	9.559	9.570	990
1000	9.570	9.582	9.593	9.605	9.616	9.628	9.639	9.651	9.663	9.674	9.686	1000
1010	9.686	9.697	9.709	9.720	9.732	9.744	9.755	9.767	9.779	9.790	9.802	1010
1020	9.802	9.813	9.825	9.837	9.848	9.860	9.871	9.883	9.895	9.906	9.918	1020
1030	9.918	9.930	9.941	9.953	9.965	9.976	9.988	10.000	10.011	10.023	10.035	1030
1040	10.035	10.046	10.058	10.070	10.082	10.093	10.105	10.117	10.128	10.140	10.152	1040
1050	10.152	10.163	10.175	10.187	10.199	10.210	10.222	10.234	10.246	10.257	10.269	1050
1060	10.269	10.281	10.293	10.304	10.316	10.328	10.340	10.351	10.363	10.375	10.387	1060
1070	10.387	10.399	10.410	10.422	10.434	10.446	10.458	10.469	10.481	10.493	10.505	1070
1080	10.505	10.517	10.528	10.540	10.552	10.564	10.576	10.587	10.599	10.611	10.623	1080
1090	10.623	10.635	10.647	10.658	10.670	10.682	10.694	10.706	10.718	10.729	10.741	1090
1100	10.741	10.753	10.765	10.777	10.789	10.801	10.812	10.824	10.836	10.848	10.860	1100
1110	10.860	10.872	10.884	10.896	10.907	10.919	10.931	10.943	10.955	10.967	10.979	1110
1120	10.979	10.991	11.003	11.014	11.026	11.038	11.050	11.062	11.074	11.086	11.098	1120
1130	11.098	11.110	11.122	11.133	11.145	11.157	11.169	11.181	11.193	11.205	11.217	1130
1140	11.217	11.229	11.241	11.253	11.265	11.277	11.289	11.300	11.312	11.324	11.336	1140
1150	11.336	11.348	11.360	11.372	11.384	11.396	11.408	11.420	11.432	11.444	11.456	1150
1160	11.456	11.468	11.480	11.492	11.504	11.516	11.528	11.540	11.552	11.564	11.575	1160
1170	11.575	11.587	11.599	11.611	11.623	11.635	11.647	11.659	11.671	11.683	11.695	1170
1180	11.695	11.707	11.719	11.731	11.743	11.755	11.767	11.779	11.791	11.803	11.815	1180
1190	11.815	11.827	11.839	11.851	11.863	11.875	11.887	11.899	11.911	11.923	11.935	1190
1200	11.935	11.947	11.959	11.971	11.983	11.995	12.007	12.019	12.031	12.043	12.055	1200
°C	0	1	2	3	4	5	6	7	8	9	10	°C

Table 14-4 THERMOCOUPLE REFERENCE TABLES (*continued*)

Platinum Versus Platinum-10-Percent Rhodium Thermocouples—Continued
(Electromotive Force in Absolute Millivolts. Temperatures
in Degrees C (Int. 1948). Reference Junctions at 0°C.)

°C	0	1	2	3	4	5	6	7	8	9	10	°C
					Millivolts							
1200	11.935	11.947	11.959	11.971	11.983	11.995	12.007	12.019	12.031	12.043	12.055	1200
1210	12.055	12.067	12.079	12.091	12.103	12.115	12.127	12.139	12.151	12.163	12.175	1210
1220	12.175	12.187	12.200	12.212	12.224	12.236	12.248	12.260	12.272	12.284	12.296	1220
1230	12.296	12.308	12.320	12.332	12.344	12.356	12.368	12.380	12.392	12.404	12.416	1230
1240	12.416	12.428	12.440	12.452	12.464	12.476	12.488	12.500	12.512	12.524	12.536	1240
1250	12.536	12.548	12.560	12.573	12.585	12.597	12.609	12.621	12.633	12.645	12.657	1250
1260	12.657	12.669	12.681	12.693	12.705	12.717	12.729	12.741	12.753	12.765	12.777	1260
1270	12.777	12.789	12.801	12.813	12.825	12.837	12.849	12.861	12.873	12.885	12.897	1270
1280	12.897	12.909	12.921	12.933	12.945	12.957	12.969	12.981	12.993	13.005	13.018	1280
1290	13.018	13.030	13.042	13.054	13.066	13.078	13.090	13.102	13.114	13.126	13.138	1290
1300	13.138	13.150	13.162	13.174	13.186	13.198	13.210	13.222	13.234	13.246	13.258	1300
1310	13.258	13.270	13.282	13.294	13.306	13.318	13.330	13.342	13.354	13.366	13.378	1310
1320	13.378	13.390	13.402	13.414	13.426	13.438	13.450	13.462	13.474	13.486	13.498	1320
1330	13.498	13.510	13.522	13.534	13.546	13.558	13.570	13.582	13.594	13.606	13.618	1330
1340	13.618	13.630	13.642	13.654	13.666	13.678	13.690	13.702	13.714	13.726	13.738	1340
1350	13.738	13.750	13.762	13.774	13.786	13.798	13.810	13.822	13.834	13.846	13.858	1350
1360	13.858	13.870	13.882	13.894	13.906	13.918	13.930	13.942	13.954	13.966	13.978	1360
1370	13.978	13.990	14.002	14.014	14.026	14.038	14.050	14.062	14.074	14.086	14.098	1370
1380	14.098	14.110	14.122	14.133	14.145	14.157	14.169	14.181	14.193	14.205	14.217	1380
1390	14.217	14.229	14.241	14.253	14.265	14.277	14.289	14.301	14.313	14.325	14.337	1390
1400	14.337	14.349	14.361	14.373	14.385	14.397	14.409	14.421	14.433	14.445	14.457	1400
1410	14.457	14.469	14.481	14.493	14.504	14.516	14.528	14.540	14.552	14.564	14.576	1410
1420	14.576	14.588	14.600	14.612	14.624	14.636	14.648	14.660	14.672	14.684	14.696	1420
1430	14.696	14.708	14.720	14.732	14.744	14.755	14.767	14.779	14.791	14.803	14.815	1430
1440	14.815	14.827	14.839	14.851	14.863	14.875	14.887	14.899	14.911	14.923	14.935	1440
1450	14.935	14.946	14.958	14.970	14.982	14.994	15.006	15.018	15.030	15.042	15.054	1450
1460	15.054	15.066	15.078	15.090	15.102	15.113	15.125	15.137	15.149	15.161	15.173	1460
1470	15.173	15.185	15.197	15.209	15.221	15.233	15.245	15.256	15.268	15.280	15.292	1470
1480	15.292	15.304	15.316	15.328	15.340	15.352	15.364	15.376	15.387	15.399	15.411	1480
1490	15.411	15.423	15.435	15.447	15.459	15.471	15.483	15.495	15.507	15.518	15.530	1490
1500	15.530	15.542	15.554	15.566	15.578	15.590	15.602	15.614	15.625	15.637	15.649	1500
1510	15.649	15.661	15.673	15.685	15.697	15.709	15.721	15.732	15.744	15.756	15.768	1510
1520	15.768	15.780	15.792	15.804	15.816	15.827	15.839	15.851	15.863	15.875	15.887	1520
1530	15.887	15.899	15.911	15.922	15.934	15.946	15.958	15.970	15.982	15.994	16.006	1530
1540	16.006	16.017	16.029	16.041	16.053	16.065	16.077	16.089	16.100	16.112	16.124	1540
1550	16.124	16.136	16.160	16.160	16.171	16.183	16.195	16.207	16.219	16.231	16.243	1550
1560	16.243	16.254	16.266	16.278	16.290	16.302	16.314	16.325	16.337	16.349	16.361	1560
1570	16.361	16.373	16.385	16.396	16.408	16.420	16.432	16.444	16.456	16.467	16.479	1570
1580	16.479	16.491	16.503	16.515	16.527	16.538	16.550	16.562	16.574	16.586	16.597	1580
1590	16.597	16.609	16.621	16.633	16.645	16.657	16.668	16.680	16.692	16.704	16.716	1590
1600	16.716	16.727	16.739	16.751	16.763	16.775	16.786	16.798	16.810	16.822	16.834	1600
°C	0	1	2	3	4	5	6	7	8	9	10	°C

Table 14-4 THERMOCOUPLE REFERENCE TABLES (*continued*)

Platinum Versus Platinum-10-Percent Rhodium Thermocouples—Concluded
(Electromotive Force in Absolute Millivolts. Temperatures
in Degrees C (Int. 1948). Reference Junctions at 0°C.)

°C	0	1	2	3	4	5	6	7	8	9	10	°C
	Millivolts											
1600	16.716	16.727	16.739	16.751	16.763	16.775	16.786	16.798	16.810	16.822	16.834	1600
1610	16.834	16.845	16.857	16.869	16.881	16.893	16.904	16.916	16.928	16.940	16.952	1610
1620	16.952	16.963	16.975	16.987	16.999	17.010	17.022	17.034	17.046	17.058	17.069	1620
1630	17.069	17.081	17.093	17.105	17.116	17.128	17.140	17.152	17.163	17.175	17.187	1630
1640	17.187	17.199	17.211	17.222	17.234	17.246	17.258	17.269	17.281	17.293	17.305	1640
1650	17.305	17.316	17.328	17.340	17.352	17.363	17.375	17.387	17.398	17.410	17.422	1650
1660	17.422	17.434	17.446	17.457	17.469	17.481	17.492	17.504	17.516	17.528	17.539	1660
1670	17.539	17.551	17.563	17.575	17.586	17.598	17.610	17.621	17.633	17.645	17.657	1670
1680	17.657	17.668	17.680	17.692	17.704	17.715	17.727	17.739	17.750	17.762	17.774	1680
1690	17.774	17.785	17.797	17.809	17.821	17.832	17.844	17.856	17.867	17.879	17.891	1690
1700	17.891	17.902	17.914	17.926	17.938	17.949	17.961	17.973	17.984	17.996	18.008	1700
1710	18.008	18.019	18.031	18.043	18.054	18.066	18.078	18.089	18.101	18.113	18.124	1710
1720	18.124	18.136	18.148	18.159	18.171	18.183	18.194	18.206	18.218	18.229	18.241	1720
1730	18.241	18.253	18.264	18.276	18.288	18.299	18.311	18.323	18.334	18.346	18.358	1730
1740	18.358	18.369	18.381	18.393	18.404	18.416	18.427	18.439	18.451	18.462	18.474	1740
1750	18.474	18.486	18.497	18.509	18.520	18.532	18.544	18.555	18.567	18.579	18.590	1750
1760	18.590	18.602	18.613	18.625	18.637	18.648	18.660	18.672	18.683	18.695	—	1760
°C	0	1	2	3	4	5	6	7	8	9	10	°C

Table 14-4 THERMOCOUPLE REFERENCE TABLES (*continued*)

**Platinum Versus Platinum-13-Percent Rhodium Thermocouples
(Electromotive Force in Absolute Millivolts. Temperatures
in Degrees C (Int. 1948). Reference Junctions at 0°C.)**

°C	0	1	2	3	4	5	6	7	8	9	10	°C
	Millivolts											
0	0.000	0.005	0.011	0.016	0.022	0.027	0.033	0.038	0.043	0.049	0.055	0
10	.055	.061	.066	.072	.078	.083	.089	.095	.101	.106	.112	10
20	.112	.118	.124	.130	.136	.142	.148	.154	.160	.166	.172	20
30	.172	.178	.184	.190	.196	.203	.209	.215	.221	.228	.234	30
40	.234	.240	.246	.252	.259	.265	.272	.278	.285	.291	.298	40
50	.298	.304	.311	.317	.324	.330	.337	.343	.350	.357	.363	50
60	.363	.370	.377	.383	.390	.397	.403	.410	.417	.424	.431	60
70	.431	.438	.445	.451	.458	.465	.472	.479	.486	.493	.500	70
80	.500	.507	.514	.521	.528	.536	.543	.550	.557	.565	.572	80
90	.572	.579	.586	.594	.601	.609	.616	.623	.631	.638	.645	90
100	.645	.653	.660	.668	.675	.683	.690	.698	.705	.713	.721	100
110	.721	.728	.736	.744	.752	.759	.767	.775	.782	.790	.798	110
120	.798	.805	.813	.821	.829	.837	.845	.853	.861	.869	.877	120
130	.877	.885	.893	.901	.909	.917	.925	.933	.941	.949	.957	130
140	.957	.966	.974	.982	.990	.998	1.006	1.014	1.022	1.031	1.039	140
150	1.039	1.047	1.055	1.063	1.072	1.080	1.088	1.096	1.104	1.112	1.121	150
160	1.121	1.129	1.138	1.146	1.154	1.163	1.171	1.179	1.188	1.196	1.205	160
170	1.205	1.213	1.222	1.231	1.239	1.247	1.256	1.265	1.273	1.282	1.290	170
180	1.290	1.298	1.307	1.316	1.324	1.333	1.342	1.351	1.359	1.368	1.377	180
190	1.377	1.386	1.395	1.403	1.412	1.420	1.429	1.438	1.447	1.456	1.465	190
200	1.465	1.473	1.482	1.491	1.500	1.509	1.517	1.526	1.535	1.544	1.553	200
210	1.553	1.562	1.571	1.580	1.589	1.598	1.607	1.616	1.625	1.634	1.643	210
220	1.643	1.652	1.661	1.670	1.679	1.688	1.697	1.706	1.715	1.725	1.734	220
230	1.734	1.743	1.752	1.761	1.770	1.779	1.788	1.798	1.807	1.816	1.826	230
240	1.826	1.835	1.844	1.853	1.863	1.872	1.881	1.890	1.900	1.909	1.918	240
250	1.918	1.928	1.937	1.946	1.956	1.965	1.974	1.984	1.993	2.002	2.012	250
260	2.012	2.021	2.031	2.040	2.050	2.059	2.068	2.078	2.087	2.097	2.107	260
270	2.107	2.116	2.126	2.135	2.145	2.154	2.164	2.173	2.183	2.192	2.202	270
280	2.202	2.211	2.221	2.231	2.240	2.250	2.259	2.269	2.279	2.288	2.298	280
290	2.298	2.308	2.317	2.327	2.337	2.346	2.356	2.366	2.375	2.385	2.395	290
300	2.395	2.405	2.415	2.424	2.434	2.444	2.454	2.464	2.473	2.483	2.493	300
310	2.493	2.503	2.513	2.522	2.532	2.542	2.552	2.562	2.572	2.581	2.591	310
320	2.591	2.601	2.611	2.621	2.631	2.641	2.650	2.660	2.670	2.680	2.690	320
330	2.690	2.700	2.710	2.720	2.730	2.740	2.750	2.760	2.770	2.780	2.790	330
340	2.790	2.800	2.810	2.820	2.830	2.840	2.850	2.860	2.870	2.880	2.890	340
350	2.890	2.900	2.910	2.920	2.930	2.940	2.950	2.961	2.971	2.981	2.991	350
360	2.991	3.001	3.011	3.021	3.031	3.041	3.051	3.062	3.072	3.082	3.092	360
370	3.092	3.102	3.112	3.122	3.133	3.143	3.153	3.163	3.173	3.183	3.194	370
380	3.194	3.204	3.214	3.224	3.234	3.245	3.255	3.265	3.276	3.286	3.296	380
390	3.296	3.306	3.317	3.327	3.337	3.347	3.358	3.368	3.378	3.389	3.399	390
400	3.399	3.409	3.420	3.430	3.440	3.451	3.461	3.471	3.481	3.492	3.502	400
°C	0	1	2	3	4	5	6	7	8	9	10	°C

Table 14-4 THERMOCOUPLE REFERENCE TABLES (*continued*)

Platinum Versus Platinum-13-Percent Rhodium Thermocouples—Continued
(Electromotive Force in Absolute Millivolts. Temperatures
in Degrees C (Int. 1948). Reference Junctions at 0°C.)

°C	0	1	2	3	4	5	6	7	8	9	10	°C
						Millivolts						
400	3.399	3.409	3.420	3.430	3.440	3.451	3.461	3.471	3.481	3.492	3.502	400
410	3.502	3.512	3.523	3.533	3.544	3.554	3.565	3.575	3.586	3.596	3.607	410
420	3.607	3.617	3.627	3.638	3.648	3.659	3.669	3.680	3.690	3.701	3.712	420
430	3.712	3.722	3.732	3.743	3.753	3.764	3.774	3.785	3.796	3.806	3.817	430
440	3.817	3.827	3.838	3.848	3.859	3.870	3.880	3.891	3.901	3.912	3.923	440
450	3.923	3.933	3.944	3.954	3.965	3.976	3.987	3.997	4.008	4.018	4.029	450
460	4.029	4.039	4.050	4.060	4.071	4.081	4.092	4.102	4.113	4.123	4.134	460
470	4.134	4.145	4.156	4.166	4.177	4.187	4.198	4.209	4.219	4.230	4.241	470
480	4.241	4.251	4.262	4.273	4.283	4.294	4.305	4.315	4.326	4.337	4.348	480
490	4.348	4.358	4.369	4.380	4.390	4.401	4.412	4.422	4.433	4.444	4.455	490
500	4.455	4.466	4.477	4.488	4.498	4.509	4.520	4.531	4.542	4.552	4.563	500
510	4.563	4.574	4.585	4.596	4.607	4.618	4.629	4.640	4.651	4.662	4.672	510
520	4.672	4.683	4.694	4.705	4.716	4.727	4.738	4.749	4.760	4.771	4.782	520
530	4.782	4.793	4.804	4.815	4.826	4.837	4.848	4.859	4.870	4.881	4.893	530
540	4.893	4.904	4.915	4.926	4.937	4.948	4.959	4.970	4.981	4.992	5.004	540
550	5.004	5.015	5.026	5.037	5.048	5.059	5.070	5.081	5.092	5.104	5.115	550
560	5.115	5.126	5.137	5.148	5.159	5.170	5.182	5.193	5.204	5.215	5.226	560
570	5.226	5.238	5.249	5.260	5.271	5.282	5.293	5.304	5.316	5.327	5.338	570
580	5.338	5.349	5.360	5.371	5.383	5.394	5.405	5.416	5.428	5.439	5.450	580
590	5.450	5.461	5.472	5.484	5.495	5.507	5.518	5.529	5.540	5.551	5.563	590
600	5.563	5.574	5.586	5.597	5.609	5.620	5.631	5.642	5.654	5.665	5.677	600
610	5.677	5.688	5.700	5.711	5.723	5.734	5.746	5.757	5.769	5.780	5.792	610
620	5.792	5.803	5.814	5.826	5.837	5.849	5.861	5.872	5.883	5.895	5.907	620
630	5.907	5.918	5.930	5.941	5.952	5.964	5.976	5.987	5.999	6.010	6.022	630
640	6.022	6.033	6.044	6.056	6.068	6.079	6.091	6.102	6.114	6.126	6.137	640
650	6.137	6.149	6.160	6.171	6.183	6.194	6.206	6.218	6.229	6.240	6.252	650
660	6.252	6.264	6.275	6.287	6.299	6.310	6.321	6.333	6.344	6.356	6.368	660
670	6.368	6.380	6.391	6.403	6.415	6.427	6.438	6.450	6.461	6.473	6.485	670
680	6.485	6.497	6.508	6.520	6.532	6.544	6.555	6.567	6.579	6.590	6.602	680
690	6.602	6.614	6.626	6.637	6.649	6.661	6.672	6.684	6.696	6.708	6.720	690
700	6.720	6.732	6.744	6.756	6.768	6.779	6.791	6.803	6.815	6.827	6.838	700
710	6.838	6.850	6.862	6.874	6.886	6.898	6.910	6.922	6.934	6.946	6.957	710
720	6.957	6.969	6.981	6.993	7.005	7.017	7.029	7.040	7.052	7.064	7.076	720
730	7.076	7.088	7.100	7.112	7.124	7.136	7.147	7.159	7.171	7.183	7.195	730
740	7.195	7.207	7.219	7.231	7.243	7.255	7.267	7.279	7.291	7.303	7.315	740
750	7.315	7.327	7.339	7.351	7.364	7.376	7.388	7.400	7.412	7.424	7.436	750
760	7.436	7.448	7.460	7.472	7.485	7.497	7.509	7.521	7.533	7.545	7.557	760
770	7.557	7.570	7.582	7.594	7.606	7.618	7.631	7.643	7.655	7.667	7.679	770
780	7.679	7.692	7.704	7.716	7.728	7.740	7.752	7.765	7.777	7.789	7.801	780
790	7.801	7.814	7.826	7.838	7.850	7.863	7.875	7.888	7.900	7.912	7.924	790
800	7.924	7.936	7.949	7.961	7.973	7.986	7.998	8.010	8.022	8.035	8.047	800
°C	0	1	2	3	4	5	6	7	8	9	10	°C

Table 14-4 Thermocouple Reference Tables (*continued*)

Platinum Versus Platinum-13-Percent Rhodium Thermocouples—Continued
(Electromotive Force in Absolute Millivolts. Temperatures in Degrees C (Int. 1948). Reference Junctions at 0°C.)

°C	0	1	2	3	4	5	6	7	8	9	10	°C
						Millivolts						
800	7.924	7.936	7.949	7.961	7.973	7.986	7.998	8.010	8.022	8.035	8.047	800
810	8.047	8.059	8.071	8.084	8.096	8.109	8.121	8.134	8.146	8.158	8.170	810
820	8.170	8.182	8.195	8.208	8.220	8.232	8.245	8.257	8.269	8.281	8.294	820
830	8.294	8.306	8.319	8.331	8.343	8.356	8.369	8.381	8.394	8.406	8.419	830
840	8.419	8.431	8.444	8.456	8.469	8.481	8.494	8.506	8.519	8.531	8.544	840
850	8.544	8.556	8.569	8.581	8.594	8.606	8.619	8.631	8.644	8.656	8.669	850
860	8.669	8.681	8.694	8.706	8.719	8.732	8.744	8.757	8.769	8.782	8.795	860
870	8.795	8.807	8.820	8.832	8.845	8.858	8.870	8.883	8.895	8.908	8.921	870
880	8.921	8.933	8.946	8.959	8.971	8.984	8.996	9.009	9.021	9.034	9.047	880
890	9.047	9.060	9.072	9.085	9.098	9.111	9.123	9.136	9.149	9.161	9.175	890
900	9.175	9.188	9.200	9.213	9.226	9.239	9.251	9.264	9.277	9.290	9.303	900
910	9.303	9.316	9.328	9.341	9.354	9.367	3.379	9.392	9.405	9.418	9.431	910
920	9.431	9.444	9.456	9.469	9.482	9.495	9.508	9.520	9.533	9.546	9.559	920
930	9.559	9.572	9.585	9.598	9.610	9.623	9.636	9.649	9.661	9.674	9.687	930
940	9.687	9.700	9.713	9.726	9.739	9.752	9.765	9.778	9.790	9.803	9.816	940
950	9.816	9.829	9.842	9.855	9.868	9.881	9.894	9.907	9.920	9.933	9.946	950
960	9.946	9.960	9.973	9.986	9.999	10.012	10.025	10.038	10.051	10.064	10.077	960
970	10.077	10.090	10.103	10.116	10.130	10.143	10.156	10.169	10.182	10.195	10.208	970
980	10.208	10.221	10.234	10.247	10.260	10.274	10.287	10.300	10.313	10.326	10.339	980
990	10.339	10.352	10.366	10.379	10.392	10.405	10.419	10.432	10.445	10.458	10.471	990
1000	10.471	10.484	10.497	10.510	10.523	10.537	10.550	10.563	10.576	10.589	10.603	1000
1010	10.603	10.616	10.629	10.642	10.655	10.669	10.682	10.695	10.709	10.722	10.735	1010
1020	10.735	10.748	10.761	10.775	10.788	10.801	10.815	10.828	10.841	10.855	10.869	1020
1030	10.869	10.882	10.895	10.909	10.922	10.936	10.949	10.963	10.976	10.989	11.003	1030
1040	11.003	11.016	11.030	11.043	11.057	11.070	11.084	11.097	11.111	11.124	11.138	1040
1050	11.138	11.151	11.165	11.178	11.191	11.205	11.219	11.232	11.246	11.259	11.273	1050
1060	11.273	11.286	11.300	11.313	11.327	11.340	11.354	11.367	11.381	11.394	11.408	1060
1070	11.408	11.421	11.435	11.449	11.463	11.476	11.490	11.504	11.517	11.531	11.544	1070
1080	11.544	11.558	11.571	11.585	11.599	11.613	11.626	11.640	11.654	11.667	11.681	1080
1090	11.681	11.694	11.708	11.722	11.736	11.749	11.763	11.776	11.790	11.803	11.817	1090
1100	11.817	11.830	11.844	11.858	11.871	11.885	11.899	11.913	11.926	11.940	11.954	1100
1110	11.954	11.967	11.981	11.994	12.008	12.022	12.035	12.049	12.063	12.077	12.090	1110
1120	12.090	12.104	12.118	12.131	12.145	12.159	12.173	12.186	12.200	12.214	12.227	1120
1130	12.227	12.241	12.254	12.268	12.282	12.296	12.310	12.323	12.337	12.351	12.365	1130
1140	12.365	12.378	12.392	12.406	12.420	12.434	12.447	12.461	12.475	12.489	12.503	1140
1150	12.503	12.516	12.530	12.544	12.558	12.572	12.585	12.599	12.613	12.627	12.641	1150
1160	12.641	12.654	12.668	12.682	12.696	12.710	12.723	12.737	12.751	12.765	12.779	1160
1170	12.779	12.792	12.806	12.820	12.834	12.848	12.861	12.875	12.889	12.903	12.917	1170
1180	12.917	12.931	12.944	12.958	12.972	12.986	13.000	13.014	13.028	13.042	13.055	1180
1190	13.055	13.069	13.083	13.097	13.111	13.125	13.139	13.152	13.166	13.180	13.193	1190
1200	13.193	13.207	13.221	13.235	13.249	13.263	13.277	13.291	13.305	13.319	13.332	1200
1210	13.332	13.346	13.360	13.374	13.388	13.402	13.416	13.429	13.443	13.457	13.471	1210
1220	13.471	13.485	13.499	13.513	13.526	13.540	13.554	13.568	13.582	13.596	13.610	1220
1230	13.610	13.624	13.638	13.652	13.666	13.679	13.693	13.707	13.721	13.735	13.749	1230
1240	13.749	13.763	13.777	13.791	13.805	13.818	13.832	13.846	13.860	13.874	13.888	1240
1250	13.888	13.902	13.916	13.930	13.943	13.957	13.971	13.985	13.999	14.013	14.027	1250
1260	14.027	14.041	14.055	14.069	14.082	14.096	14.110	14.124	14.138	14.152	14.165	1260
°C	0	1	2	3	4	5	6	7	8	9	10	°C

Table 14-4 THERMOCOUPLE REFERENCE TABLES (*continued*)

Platinum Versus Platinum-13-Percent Rhodium Thermocouples—Concluded
(Electromotive Force in Absolute Millivolts. Temperatures
in Degrees C (Int. 1948). Reference Junctions at 0°C.)

°C	0	1	2	3	4	5	6	7	8	9	10	°C
						Millivolts						
1260	14.027	14.041	14.055	14.069	14.082	14.096	14.110	14.124	14.138	14.152	14.165	1260
1270	14.165	14.179	14.193	14.207	14.221	14.235	14.249	14.263	14.277	14.291	14.304	1270
1280	15.304	14.318	14.332	14.346	14.360	14.374	14.388	14.402	14.416	14.430	14.443	1280
1290	14.443	14.457	14.471	14.485	14.499	14.513	14.527	14.541	14.555	14.569	14.582	1290
1300	14.582	14.596	14.610	14.624	14.638	14.652	14.666	14.680	14.694	14.707	14.721	1300
1310	14.721	14.735	14.749	14.763	14.777	14.791	14.804	14.818	14.832	14.846	14.860	1310
1320	14.860	14.874	14.888	14.901	14.915	14.929	14.943	14.957	14.971	14.985	14.999	1320
1330	14.999	15.013	15.026	15.040	15.054	15.068	15.082	15.096	15.110	15.124	15.138	1330
1340	15.138	15.151	15.165	15.179	15.193	15.207	15.221	15.234	15.248	15.262	15.276	1340
1350	15.276	15.290	15.304	15.318	15.331	15.345	15.359	15.373	15.387	15.401	15.415	1350
1360	15.415	15.429	15.443	15.456	15.470	15.484	15.498	15.512	15.526	15.540	15.553	1360
1370	15.553	15.567	15.581	15.595	15.609	15.623	15.637	15.651	15.665	15.679	15.692	1370
1380	15.692	15.706	15.720	15.734	15.748	15.761	15.775	15.789	15.803	15.817	15.831	1380
1390	15.831	15.845	15.859	15.873	15.886	15.900	15.914	15.928	15.942	15.956	15.969	1390
1400	15.969	15.983	15.997	16.011	16.025	16.039	16.053	16.067	16.081	16.095	16.108	1400
1410	16.108	16.122	16.136	16.150	16.164	16.178	16.192	16.206	16.219	16.233	16.247	1410
1420	16.247	16.261	16.275	16.289	16.303	16.317	16.330	16.344	16.358	16.372	16.386	1420
1430	16.386	16.400	16.414	16.427	16.441	16.455	16.469	16.483	16.497	16.511	16.524	1430
1440	16.524	16.538	16.552	16.566	16.580	16.594	16.608	16.621	16.635	16.649	16.663	1440
1450	16.663	16.677	16.691	16.705	16.719	16.733	16.746	16.760	16.774	16.788	16.802	1450
1460	16.802	16.816	16.830	16.844	16.858	16.872	16.885	16.899	16.913	16.927	16.940	1460
1470	16.940	16.954	16.968	16.982	16.996	17.010	17.024	17.037	17.051	17.065	17.079	1470
1480	17.079	17.092	17.106	17.120	17.134	17.148	17.161	17.175	17.189	17.203	17.217	1480
1490	17.217	17.230	17.244	17.258	17.272	17.286	17.299	17.313	17.327	17.341	17.355	1490
1500	17.355	17.368	17.382	17.396	17.410	17.424	17.437	17.451	17.465	17.479	17.493	1500
1510	17.493	17.506	17.520	17.534	17.547	17.561	17.575	17.589	17.603	17.617	17.631	1510
1520	17.631	17.644	17.658	17.672	17.686	17.699	17.713	17.726	17.740	17.754	17.768	1520
1530	17.768	17.781	17.795	17.809	17.823	17.837	17.850	17.864	17.878	17.892	17.906	1530
1540	17.906	17.919	17.933	17.947	17.960	17.974	17.988	18.002	18.016	18.029	18.043	1540
1550	18.043	18.056	18.070	18.084	18.098	18.111	18.125	18.139	18.152	18.166	18.179	1550
1560	18.179	18.193	18.207	18.220	18.234	18.248	18.261	18.275	18.289	18.303	18.316	1560
1570	18.316	18.330	18.344	18.357	18.371	18.385	18.399	18.412	18.426	18.440	18.453	1570
1580	18.453	18.467	18.481	18.494	18.508	18.522	18.536	18.549	18.563	18.576	18.590	1580
1590	18.590	18.604	18.618	18.631	18.645	18.659	18.672	18.686	18.700	18.714	18.727	1590
1600	18.727	18.741	18.754	18.768	18.782	18.796	18.810	18.823	18.836	18.850	18.864	1600
1610	18.864	18.878	18.891	18.905	18.919	18.932	18.946	18.960	18.973	18.987	19.001	1610
1620	19.001	19.014	19.028	19.042	19.056	19.069	19.083	19.096	19.110	19.124	19.137	1620
1630	19.137	19.150	19.164	19.178	19.191	19.205	19.219	19.232	19.246	19.260	19.273	1630
1640	19.273	19.287	19.300	19.314	19.328	19.341	19.355	19.369	19.382	19.396	19.409	1640
1650	19.409	19.423	19.437	19.450	19.464	19.477	19.491	19.504	19.518	19.531	19.545	1650
1660	19.545	19.559	19.573	19.586	19.600	19.614	19.627	19.641	19.654	19.668	19.682	1660
1670	19.682	19.695	19.709	19.722	19.736	19.750	19.763	19.777	19.790	19.804	19.818	1670
1680	19.818	19.831	19.845	19.859	19.873	19.886	19.900	19.913	19.927	19.940	19.954	1680
1690	19.954	19.967	19.981	19.994	20.008	20.022	20.035	20.049	20.062	20.076	20.090	1690
°C	0	1	2	3	4	5	6	7	8	9	10	°C

Table 14-4 THERMOCOUPLE REFERENCE TABLES (*continued*)

Chromel-Alumel Thermocouples
(Electromotive Force in Absolute Millivolts. Temperatures in Degrees C (Int. 1948). Reference Junctions at 0°C.)

°C	0	1	2	3	4	5	6	7	8	9	10	°C
	Millivolts											
−190	−5.60	−5.62	−5.63	−5.65	−5.67	−5.68	−5.70	−5.71	−5.73	−5.74	−5.75	−190
−180	−5.43	−5.45	−5.46	−5.48	−5.50	−5.52	−5.53	−5.55	−5.57	−5.58	−5.60	−180
−170	−5.24	−5.26	−5.28	−5.30	−5.32	−5.34	−5.35	−5.37	−5.39	−5.41	−5.43	−170
−160	−5.03	−5.05	−5.08	−5.10	−5.12	−5.14	−5.16	−5.18	−5.20	−5.22	−5.24	−160
−150	−4.81	−4.84	−4.86	−4.88	−4.90	−4.92	−4.95	−4.97	−4.99	−5.01	−5.03	−150
−140	−4.58	−4.60	−4.62	−4.65	−4.67	−4.70	−4.72	−4.74	−4.77	−4.79	−4.81	−140
−130	−4.32	−4.35	−4.37	−4.40	−4.42	−4.45	−4.48	−4.50	−4.52	−4.55	−4.58	−130
−120	−4.06	−4.08	−4.11	−4.14	−4.16	−4.19	−4.22	−4.24	−4.27	−4.30	−4.32	−120
−110	−3.78	−3.81	−3.84	−3.86	−3.89	−3.92	−3.95	−3.98	−4.00	−4.03	−4.06	−110
−100	−3.49	−3.52	−3.55	−3.58	−3.61	−3.64	−3.66	−3.69	−3.72	−3.75	−3.78	−100
−90	−3.19	−3.22	−3.25	−3.28	−3.31	−3.34	−3.37	−3.40	−3.43	−3.46	−3.49	−90
−80	−2.87	−2.90	−2.93	−2.96	−3.00	−3.03	−3.06	−3.09	−3.12	−3.16	−3.19	−80
−70	−2.54	−2.57	−2.61	−2.64	−2.67	−2.71	−2.74	−2.77	−2.80	−2.84	−2.87	−70
−60	−2.20	−2.24	−2.27	−2.30	−2.34	−2.37	−2.41	−2.44	−2.47	−2.51	−2.54	−60
−50	−1.86	−1.89	−1.93	−1.96	−2.00	−2.03	−2.07	−2.10	−2.13	−2.17	−2.20	−50
−40	−1.50	−1.54	−1.57	−1.61	−1.64	−1.68	−1.72	−1.75	−1.79	−1.82	−1.86	−40
−30	−1.14	−1.17	−1.21	−1.25	−1.28	−1.32	−1.36	−1.39	−1.43	−1.47	−1.50	−30
−20	−0.77	−0.80	−0.84	−0.88	−0.92	−0.95	−0.99	−1.03	−1.06	−1.10	−1.14	−20
−10	−0.39	−0.42	−0.46	−0.50	−0.54	−0.58	−0.62	−0.66	−0.69	−0.73	−0.77	−10
(−)0	−0.00	−0.04	−0.08	−0.12	−0.16	−0.19	−0.23	−0.27	−0.31	−0.35	−0.39	(−)0
(+)0	0.00	0.04	0.08	0.12	0.16	0.20	0.24	0.28	0.32	0.36	0.40	(+)0
10	0.40	0.44	0.48	0.52	0.56	0.60	0.64	0.68	0.72	0.76	0.80	10
20	0.80	0.84	0.88	0.92	0.96	1.00	1.04	1.08	1.12	1.16	1.20	20
30	1.20	1.24	1.28	1.32	1.36	1.40	1.44	1.49	1.53	1.57	1.61	30
40	1.61	1.65	1.69	1.73	1.77	1.81	1.85	1.90	1.94	1.98	2.02	40
50	2.02	2.06	2.10	2.14	2.18	2.23	2.27	2.31	2.35	2.39	2.43	50
60	2.43	2.47	2.51	2.56	2.60	2.64	2.68	2.72	2.76	2.80	2.85	60
70	2.85	2.89	2.93	2.97	3.01	3.05	3.10	3.14	3.18	3.22	3.26	70
80	3.26	3.30	3.35	3.39	3.43	3.47	3.51	3.56	3.60	3.64	3.68	80
90	3.68	3.72	3.76	3.81	3.85	3.89	3.93	3.97	4.01	4.06	4.10	90
100	4.10	4.14	4.18	4.22	4.26	4.31	4.35	4.39	4.43	4.47	4.51	100
110	4.51	4.55	4.60	4.64	4.68	4.72	4.76	4.80	4.84	4.88	4.92	110
120	4.92	4.96	5.01	5.05	5.09	5.13	5.17	5.21	5.25	5.29	5.33	120
130	5.33	5.37	5.41	5.45	5.49	5.53	5.57	5.61	5.65	5.69	5.73	130
140	5.73	5.77	5.81	5.85	5.89	5.93	5.97	6.01	6.05	6.09	6.13	140
150	6.13	6.17	6.21	6.25	6.29	6.33	6.37	6.41	6.45	6.49	6.53	150
160	6.53	6.57	6.61	6.65	6.69	6.73	6.77	6.81	6.85	6.89	6.93	160
170	6.93	6.97	7.01	7.05	7.09	7.13	7.17	7.21	7.25	7.29	7.33	170
180	7.33	7.37	7.41	7.45	7.49	7.53	7.57	7.61	7.65	7.69	7.73	180
190	7.73	7.77	7.81	7.85	7.89	7.93	7.97	8.01	8.05	8.09	8.13	190
200	8.13	8.17	8.21	8.25	8.29	8.33	8.37	8.41	8.46	8.50	8.54	200
°C	0	1	2	3	4	5	6	7	8	9	10	°C

Table 14-4 THERMOCOUPLE REFERENCE TABLES (*continued*)

Chromel-Alumel Thermocouples—Continued
(Electromotive Force in Absolute Millivolts. Temperatures
in Degrees C (Int. 1948). Reference Junctions at 0°C.)

°C	0	1	2	3	4	5	6	7	8	9	10	°C
	Millivolts											
200	8.13	8.17	8.21	8.25	8.29	8.33	8.37	8.41	8.46	8.50	8.54	200
210	8.54	8.58	8.62	8.66	8.70	8.74	8.78	8.82	8.86	8.90	8.94	210
220	8.94	8.98	9.02	9.06	9.10	9.14	9.18	9.22	9.26	9.30	9.34	220
230	9.34	9.38	9.42	9.46	9.50	9.54	9.59	9.63	9.67	9.71	9.75	230
240	9.75	9.79	9.83	9.87	9.91	9.95	9.99	10.03	10.07	10.11	10.16	240
250	10.16	10.20	10.24	10.28	10.32	10.36	10.40	10.44	10.48	10.52	10.57	250
260	10.57	10.61	10.65	10.69	10.73	10.77	10.81	10.85	10.89	10.93	10.98	260
270	10.98	11.02	11.06	11.10	11.14	11.18	11.22	11.26	11.30	11.34	11.39	270
280	11.39	11.43	11.47	11.51	11.55	11.59	11.63	11.67	11.72	11.76	11.80	280
290	11.80	11.84	11.88	11.92	11.96	12.01	12.05	12.09	12.13	12.17	12.21	290
300	12.21	12.25	12.29	12.34	12.38	12.42	12.46	12.50	12.54	12.58	12.63	300
310	12.63	12.67	12.71	12.75	12.79	12.83	12.88	12.92	12.96	13.00	13.04	310
320	13.04	13.08	13.12	13.17	13.21	13.25	13.29	13.33	13.37	13.42	13.46	320
330	13.46	13.50	13.54	13.58	13.62	13.67	13.71	13.75	13.79	13.83	13.88	330
340	13.88	13.92	13.96	14.00	14.04	14.09	14.13	14.17	14.21	14.25	14.29	340
350	14.29	14.34	14.38	14.42	14.46	14.50	14.55	14.59	14.63	14.67	14.71	350
360	14.71	14.76	14.80	14.84	14.88	14.92	14.97	15.01	15.05	15.09	15.13	360
370	15.13	15.18	15.22	15.26	15.30	15.34	15.39	15.43	15.47	15.51	15.55	370
380	15.55	15.60	15.64	15.68	15.72	15.76	15.81	15.85	15.89	15.93	15.98	380
390	15.98	16.02	16.06	16.10	16.14	16.19	16.23	16.27	16.31	16.36	16.40	390
400	16.40	16.44	16.48	16.52	16.57	16.61	16.65	16.69	16.74	16.78	16.82	400
410	16.82	16.86	16.91	16.95	16.99	17.03	17.07	17.12	17.16	17.20	17.24	410
420	17.24	17.29	17.33	17.37	17.41	17.46	17.50	17.54	17.58	17.62	17.67	420
430	17.67	17.71	17.75	17.79	17.84	17.88	17.92	17.96	18.01	18.05	18.09	430
440	18.09	18.13	18.17	18.22	18.26	18.30	18.34	18.39	18.43	18.47	18.51	440
450	18.51	18.56	18.60	18.64	18.68	18.73	18.77	18.81	18.85	18.90	18.94	450
460	18.94	18.98	19.02	19.07	19.11	19.15	19.19	19.24	19.28	19.32	19.36	460
470	19.36	19.41	19.45	19.49	19.54	19.58	19.62	19.66	19.71	19.75	19.79	470
480	19.79	19.84	19.88	19.92	19.96	20.01	20.05	20.09	20.13	20.18	20.22	480
490	20.22	20.26	20.31	20.35	20.39	20.43	20.48	20.52	20.56	20.60	20.65	490
500	20.65	20.69	20.73	20.77	20.82	20.86	20.90	20.94	20.99	21.03	21.07	500
510	21.07	21.11	21.16	21.20	21.24	21.28	21.32	21.37	21.41	21.45	21.50	510
520	21.50	21.54	21.58	21.63	21.67	21.71	21.75	21.80	21.84	21.88	21.92	520
530	21.92	21.97	22.01	22.05	22.09	22.14	22.18	22.22	22.26	22.31	22.35	530
540	22.35	22.39	22.43	22.48	22.52	22.56	22.61	22.65	22.69	22.73	22.78	540
550	22.78	22.82	22.86	22.90	22.95	22.99	23.03	23.07	23.12	23.16	23.20	550
560	23.20	23.25	23.29	23.33	23.38	23.42	23.46	23.50	23.54	23.59	23.63	560
570	23.63	23.67	23.72	23.76	23.80	23.84	23.89	23.93	23.97	24.01	24.06	570
580	24.06	24.10	24.14	24.18	24.23	24.27	24.31	24.36	24.40	24.44	24.49	580
590	24.49	24.53	24.57	24.61	24.65	24.70	24.74	24.78	24.83	24.87	24.91	590
600	24.91	24.95	25.00	25.04	25.08	25.12	25.17	25.21	25.25	25.29	25.34	600
°C	0	1	2	3	4	5	6	7	8	9	10	°C

Table 14-4 THERMOCOUPLE REFERENCE TABLES (*continued*)

Chromel-Alumel Thermocouples—Continued
(Electromotive Force in Absolute Millivolts. Temperatures
in Degrees C (Int. 1948). Reference Junctions at 0°C.)

°C	0	1	2	3	4	5	6	7	8	9	10	°C
						Millivolts						
600	24.91	24.95	25.00	25.04	25.08	25.12	25.17	25.21	25.25	25.29	25.34	600
610	25.34	25.38	25.42	25.47	25.51	25.55	25.59	25.64	25.68	25.72	25.76	610
620	25.76	25.81	25.85	25.89	25.93	25.98	26.02	26.06	26.10	26.15	26.19	620
630	26.19	26.23	26.27	26.32	26.36	26.40	26.44	26.48	26.53	26.57	26.61	630
640	26.61	26.65	26.70	26.74	26.78	26.82	26.86	26.91	26.95	26.99	27.03	640
650	27.03	27.07	27.12	27.16	27.20	27.24	27.28	27.33	27.37	27.41	27.45	650
660	27.45	27.49	27.54	27.58	27.62	27.66	27.71	27.75	27.79	27.83	27.87	660
670	27.87	27.92	27.96	28.00	28.04	28.08	28.13	28.17	28.21	28.25	28.29	670
680	28.29	28.34	28.38	28.42	28.46	28.50	28.55	28.59	28.63	28.67	28.72	680
690	28.72	28.76	28.80	28.84	28.88	28.93	28.97	29.01	29.05	29.10	29.14	690
700	29.14	29.18	29.22	29.26	29.30	29.35	29.39	29.43	29.47	29.52	29.56	700
710	29.56	29.60	29.64	29.68	29.72	29.77	29.81	29.85	29.89	29.93	29.97	710
720	29.97	30.02	30.06	30.10	30.14	30.18	30.23	30.27	30.31	30.35	30.39	720
730	30.39	30.44	30.48	30.52	30.56	30.60	30.65	30.69	30.73	30.77	30.81	730
740	30.81	30.85	30.90	30.94	30.98	31.02	31.06	31.10	31.15	31.19	31.23	740
750	31.23	31.27	31.31	31.35	31.40	31.44	31.48	31.52	31.56	31.60	31.65	750
760	31.65	31.69	31.73	31.77	31.81	31.85	31.90	31.94	31.98	32.02	32.06	760
770	32.06	32.10	32.15	32.19	32.23	32.27	32.31	32.35	32.39	32.43	32.48	770
780	32.48	32.52	32.56	32.60	32.64	32.68	32.72	32.76	32.81	32.85	32.89	780
790	32.89	32.93	32.97	33.01	33.05	33.09	33.13	33.18	33.22	33.26	33.30	790
800	33.30	33.34	33.38	33.42	33.46	33.50	33.54	33.59	33.63	33.67	33.71	800
810	33.71	33.75	33.79	33.83	33.87	33.91	33.95	33.99	34.04	34.08	34.12	810
820	34.12	34.16	34.20	34.24	34.28	34.32	34.36	34.40	34.44	34.48	34.53	820
830	34.53	34.57	34.61	34.65	34.69	34.73	34.77	34.81	34.85	34.89	34.93	830
840	34.93	34.97	35.02	35.06	35.10	35.14	35.18	35.22	35.26	35.30	35.34	840
850	35.34	35.38	35.42	35.46	35.50	35.54	35.58	35.63	35.67	35.71	35.75	850
860	35.75	35.79	35.83	35.87	35.91	35.95	35.99	36.03	36.07	36.11	36.15	860
870	36.15	36.19	36.23	36.27	36.31	36.35	36.39	36.43	36.47	36.51	36.55	870
880	36.55	36.59	36.63	36.67	36.72	36.76	36.80	36.84	36.88	36.92	36.96	880
890	39.96	37.00	37.04	37.08	37.12	37.16	37.20	37.24	37.28	37.32	37.36	890
900	37.36	37.40	37.44	37.48	37.52	37.56	37.60	37.64	37.68	37.72	37.76	900
910	37.76	37.80	37.84	37.88	37.92	37.96	38.00	38.04	38.08	38.12	38.16	910
920	38.16	38.20	38.24	38.28	38.32	38.36	38.40	38.44	38.48	38.52	38.56	920
930	38.56	38.60	38.64	38.68	38.72	38.76	38.80	38.84	38.88	38.92	38.95	930
940	38.95	38.99	39.03	39.07	39.11	39.15	39.19	39.23	39.27	39.31	39.35	940
950	39.35	39.39	39.43	39.47	39.51	39.55	39.59	39.63	39.67	39.71	39.75	950
960	39.75	39.79	39.83	39.86	39.90	39.94	39.98	40.02	40.06	40.10	40.14	960
970	40.14	40.18	40.22	40.26	40.30	40.34	40.38	40.41	40.45	40.49	40.53	970
980	40.53	40.57	40.61	40.65	40.69	40.73	40.77	40.81	40.85	40.89	40.92	980
990	40.92	40.96	41.00	41.04	41.08	41.12	41.16	41.20	41.24	41.28	41.31	990
1000	41.31	41.35	41.39	41.43	41.47	41.51	41.55	41.59	41.63	41.67	41.70	1000
°C	0	1	2	3	4	5	6	7	8	9	10	°C

Table 14-4 THERMOCOUPLE REFERENCE TABLES (*continued*)

Chromel-Alumel Thermocouples—Concluded
(Electromotive Force in Absolute Millivolts. Temperatures
in Degrees C (Int. 1948). Reference Junctions at 0°C.)

°C	0	1	2	3	4	5	6	7	8	9	10	°C
	Millivolts											
1000	41.31	41.35	41.39	41.43	41.47	41.51	41.55	41.59	41.63	41.67	41.70	1000
1010	41.70	41.74	41.78	41.82	41.86	41.90	41.94	41.98	42.02	42.05	42.09	1010
1020	42.09	42.13	42.17	42.21	42.25	42.29	42.33	42.36	42.40	42.44	42.48	1020
1030	42.48	42.52	42.56	42.60	42.63	42.67	42.71	42.75	42.79	42.83	42.87	1030
1040	42.87	42.90	42.94	42.98	43.02	43.06	43.10	43.14	43.17	43.21	43.25	1040
1050	43.25	43.29	43.33	43.37	43.41	43.44	43.48	43.52	43.56	43.60	43.63	1050
1060	43.63	43.67	43.71	43.75	43.79	43.83	43.87	43.90	43.94	43.98	44.02	1060
1070	44.02	44.06	44.10	44.13	44.17	44.21	44.25	44.29	44.33	44.36	44.40	1070
1080	44.40	44.44	44.48	44.52	44.55	44.59	44.63	44.67	44.71	44.74	44.78	1080
1090	44.78	44.82	44.86	44.90	44.93	44.97	45.01	45.05	45.09	45.12	45.16	1090
1100	45.16	45.20	45.24	45.27	45.31	45.35	45.39	45.43	45.46	45.50	45.54	1100
1110	45.54	45.58	45.62	45.65	45.69	45.73	45.77	45.80	45.84	45.88	45.92	1110
1120	45.92	45.96	45.99	46.03	46.07	46.11	46.14	46.18	46.22	46.26	46.29	1120
1130	46.29	46.33	46.37	46.41	46.44	46.48	46.52	46.56	46.59	46.63	46.67	1130
1140	46.67	46.70	46.74	46.78	46.82	46.85	46.89	46.93	46.97	47.00	47.04	1140
1150	47.04	47.08	47.12	47.15	47.19	47.23	47.26	47.30	47.34	47.38	47.41	1150
1160	47.41	47.45	47.49	47.52	47.56	47.60	47.63	47.67	47.71	47.75	47.78	1160
1170	47.78	47.82	47.86	47.89	47.93	47.97	48.00	48.04	48.08	48.12	48.15	1170
1180	48.15	48.19	48.23	48.26	48.30	48.34	48.37	48.41	48.45	48.48	48.52	1180
1190	48.52	48.56	48.59	48.63	48.67	48.70	48.74	48.78	48.81	48.85	48.89	1190
1200	48.89	48.92	48.96	49.00	49.03	49.07	49.11	49.14	49.18	49.22	49.25	1200
1210	49.25	49.29	49.32	49.36	49.40	49.43	49.47	49.51	49.54	49.58	49.62	1210
1220	49.62	49.65	49.69	49.72	49.76	49.80	49.83	49.87	49.90	49.94	49.98	1220
1230	49.98	50.01	50.05	50.08	50.12	50.16	50.19	50.23	50.26	50.30	50.34	1230
1240	50.34	50.37	50.41	50.44	50.48	50.52	50.55	50.59	50.62	50.66	50.69	1240
1250	50.69	50.73	50.77	50.80	50.84	50.87	50.91	50.94	50.98	51.02	51.05	1250
1260	51.05	51.09	51.12	51.16	51.19	51.23	51.27	51.30	51.34	51.37	51.41	1260
1270	51.41	51.44	51.48	51.51	51.55	51.58	51.62	51.66	51.69	51.73	51.76	1270
1280	51.76	51.80	51.83	51.87	51.90	51.94	51.97	52.01	52.04	52.08	52.11	1280
1290	52.11	52.15	52.18	52.22	52.25	52.29	52.32	52.36	52.39	52.43	52.46	1290
1300	52.46	52.50	52.53	52.57	52.60	52.64	52.67	52.71	52.74	52.78	52.81	1300
1310	52.81	52.85	52.88	52.92	52.95	52.99	53.02	53.06	53.09	53.13	53.16	1310
1320	53.16	53.20	53.23	53.27	53.30	53.34	53.37	53.41	53.44	53.47	53.51	1320
1330	53.51	53.54	53.58	53.61	53.65	53.68	53.72	53.75	53.79	53.82	53.85	1330
1340	53.85	53.89	53.92	53.96	53.99	54.03	54.06	54.10	54.13	54.16	54.20	1340
1350	54.20	54.23	54.27	54.30	54.34	54.37	54.40	54.44	54.47	54.51	54.54	1350
1360	54.54	54.57	54.61	54.64	54.68	54.71	54.74	54.78	54.81	54.85	54.88	1360
1370	54.88	54.91	—	—	—	—	—	—	—	—	—	1370
°C	0	1	2	3	4	5	6	7	8	9	10	°C

Table 14-4 THERMOCOUPLE REFERENCE TABLES (continued)

Iron-Constantan Thermocouples (Modified 1913)
(Electromotive Force in Absolute Millivolts. Temperatures in Degrees C (Int. 1948). Reference Junctions at 0°C.)

°C	0	1	2	3	4	5	6	7	8	9	10	°C
						Millivolts						
−190	−7.66	−7.69	−7.71	−7.73	−7.76	−7.78	—	—	—	—	—	−190
−180	−7.40	−7.43	−7.46	−7.49	−7.51	−7.54	−7.56	−7.59	−7.61	−7.64	−7.66	−180
−170	−7.12	−7.15	−7.18	−7.21	−7.24	−7.27	−7.30	−7.32	−7.35	−7.38	−7.40	−170
−160	−6.82	−6.85	−6.88	−6.91	−6.94	−6.97	−7.00	−7.03	−7.06	−7.09	−7.12	−160
−150	−6.50	−6.53	−6.56	−6.60	−6.63	−6.66	−6.69	−6.72	−6.76	−6.79	−6.82	−150
−140	−6.16	−6.19	−6.22	−6.26	−6.29	−6.33	−6.36	−6.40	−6.43	−6.46	−6.50	−140
−130	−5.80	−5.84	−5.87	−5.91	−5.94	−5.98	−6.01	−6.05	−6.08	−6.12	−6.16	−130
−120	−5.42	−5.46	−5.50	−5.54	−5.58	−5.61	−5.65	−5.69	−5.72	−5.76	−5.80	−120
−110	−5.03	−5.07	−5.11	−5.15	−5.19	−5.23	−5.27	−5.31	−5.35	−5.38	−5.42	−110
−100	−4.63	−4.67	−4.71	−4.75	−4.79	−4.83	−4.87	−4.91	−4.95	−4.99	−5.03	−100
−90	−4.21	−4.25	−4.30	−4.34	−4.38	−4.42	−4.46	−4.50	−4.55	−4.59	−4.63	−90
−80	−3.78	−3.82	−3.87	−3.91	−3.96	−4.00	−4.04	−4.08	−4.13	−4.17	−4.21	−80
−70	−3.34	−3.38	−3.43	−3.47	−3.52	−3.56	−3.60	−3.65	−3.69	−3.74	−3.78	−70
−60	−2.89	−2.94	−2.98	−3.03	−3.07	−3.12	−3.16	−3.21	−3.25	−3.30	−3.34	−60
−50	−2.43	−2.48	−2.52	−2.57	−2.62	−2.66	−2.71	−2.75	−2.80	−2.84	−2.89	−50
−40	−1.96	−2.01	−2.06	−2.10	−2.15	−2.20	−2.24	−2.29	−2.34	−2.38	−2.43	−40
−30	−1.48	−1.53	−1.58	−1.63	−1.67	−1.72	−1.77	−1.82	−1.87	−1.91	−1.96	−30
−20	−1.00	−1.04	−1.09	−1.14	−1.19	−1.24	−1.29	−1.34	−1.39	−1.43	−1.48	−20
−10	−0.50	−0.55	−0.60	−0.65	−0.70	−0.75	−0.80	−0.85	−0.90	−0.95	−1.00	−10
(−)0	0.00	−0.05	−0.10	−0.15	−0.20	−0.25	−0.30	−0.35	−0.40	−0.45	−0.50	(−)0
(+)0	0.00	0.05	0.10	0.15	0.20	0.25	0.30	0.35	0.40	0.45	0.50	(+)0
10	0.50	0.56	0.61	0.66	0.71	0.76	0.81	0.86	0.91	0.97	1.02	10
20	1.02	1.07	1.12	1.17	1.22	1.28	1.33	1.38	1.43	1.48	1.54	20
30	1.54	1.59	1.64	1.69	1.74	1.80	1.85	1.90	1.95	2.00	2.06	30
40	2.06	2.11	2.16	2.22	2.27	2.32	2.37	2.42	2.48	2.53	2.58	40
50	2.58	2.64	2.69	2.74	2.80	2.85	2.90	2.96	3.01	3.06	3.11	50
60	3.11	3.17	3.22	3.27	3.33	3.38	3.43	3.49	3.54	3.60	3.65	60
70	3.65	3.70	3.76	3.81	3.86	3.92	3.97	4.02	4.08	4.13	4.19	70
80	4.19	4.24	4.29	4.35	4.40	4.46	4.51	4.56	4.62	4.67	4.73	80
90	4.73	4.78	4.83	4.89	4.94	5.00	5.05	5.10	5.16	5.21	5.27	90
100	5.27	5.32	5.38	5.43	5.48	5.54	5.59	5.65	5.70	5.76	5.81	100
110	5.81	5.86	5.92	5.97	6.03	6.08	6.14	6.19	6.25	6.30	6.36	110
120	6.36	6.41	6.47	6.52	6.58	6.63	6.68	6.74	6.79	6.85	6.90	120
130	6.90	6.96	7.01	7.07	7.12	7.18	7.23	7.29	7.34	7.40	7.45	130
140	7.45	7.51	7.56	7.62	7.67	7.73	7.78	7.84	7.89	7.95	8.00	140
150	8.00	8.06	8.12	8.17	8.23	8.28	8.34	8.39	8.45	8.50	8.56	150
160	8.56	8.61	8.67	8.72	8.78	8.84	8.89	8.95	9.00	9.06	9.11	160
170	9.11	9.17	9.22	9.28	9.33	9.39	9.44	9.50	9.56	9.61	9.67	170
180	9.67	9.72	9.78	9.83	9.89	9.95	10.00	10.06	10.11	10.17	10.22	180
190	10.22	10.28	10.34	10.39	10.45	10.50	10.56	10.61	10.67	10.72	10.78	190
200	10.78	10.84	10.89	10.95	11.00	11.06	11.12	11.17	11.23	11.28	11.34	200
°C	0	1	2	3	4	5	6	7	8	9	10	°C

Table 14-4 THERMOCOUPLE REFERENCE TABLES (*continued*)

Iron-Constantan Thermocouples (Modified 1913)—Continued
(Electromotive Force in Absolute Millivolts. Temperatures in Degrees C (Int. 1948). Reference Junctions at 0°C.)

°C	0	1	2	3	4	5	6	7	8	9	10	°C
						Millivolts						
200	10.78	10.84	10.89	10.95	11.00	11.06	11.12	11.17	11.23	11.28	11.34	200
210	11.34	11.39	11.45	11.50	11.56	11.62	11.67	11.73	11.78	11.84	11.89	210
220	11.89	11.95	12.00	12.06	12.12	12.17	12.23	12.28	12.34	12.39	12.45	220
230	12.45	12.50	12.56	12.62	12.67	12.73	12.78	12.84	12.89	12.95	13.01	230
240	13.01	13.06	13.12	13.17	13.23	13.28	13.34	13.40	13.45	13.51	13.56	240
250	13.56	13.62	13.67	13.73	13.78	13.84	13.89	13.95	14.00	14.06	14.12	250
260	14.12	14.17	14.23	14.28	14.34	14.39	14.45	14.50	14.56	14.61	14.67	260
270	14.67	14.72	14.78	14.83	14.89	14.94	15.00	15.06	15.11	15.17	15.22	270
280	15.22	15.28	15.33	15.39	15.44	15.50	15.55	15.61	15.66	15.72	15.77	280
290	15.77	15.83	15.88	15.94	16.00	16.05	16.11	16.16	16.22	16.27	16.33	290
300	16.33	16.38	16.44	16.49	16.55	16.60	16.66	16.71	16.77	16.82	16.88	300
310	16.88	16.93	16.99	17.04	17.10	17.15	17.21	17.26	17.32	17.37	17.43	310
320	17.43	17.48	17.54	17.60	17.65	17.71	17.76	17.82	17.87	17.93	17.98	320
330	17.98	18.04	18.09	18.15	18.20	18.26	18.32	18.37	18.43	18.48	18.54	330
340	18.54	18.59	18.65	18.70	18.76	18.81	18.87	18.92	18.98	19.03	19.09	340
350	19.09	19.14	19.20	19.26	19.31	19.37	19.42	19.48	19.53	19.59	19.64	350
360	19.64	19.70	19.75	19.81	19.86	19.92	19.97	20.03	20.08	20.14	20.20	360
370	20.20	20.25	20.31	20.36	20.42	20.47	20.53	20.58	20.64	20.69	20.75	370
380	20.75	20.80	20.86	20.91	20.97	21.02	21.08	21.13	21.19	21.24	21.30	380
390	21.30	21.35	21.41	21.46	21.52	21.57	21.63	21.68	21.74	21.79	21.85	390
400	21.85	21.90	21.96	22.02	22.07	22.13	22.18	22.24	22.29	22.35	22.40	400
410	22.40	22.46	22.51	22.57	22.62	22.68	22.73	22.79	22.84	22.90	22.95	410
420	22.95	23.01	23.06	23.12	23.17	23.23	23.28	23.34	23.39	23.45	23.50	420
430	23.50	23.56	23.61	23.67	23.72	23.78	23.83	23.89	23.94	24.00	24.06	430
440	24.06	24.11	24.17	24.22	24.28	24.33	24.39	24.44	24.50	24.55	24.61	440
450	24.61	24.66	24.72	24.77	24.83	24.88	24.94	25.00	25.05	25.11	25.16	450
460	25.16	25.22	25.27	25.33	25.38	25.44	25.49	25.55	25.60	25.66	25.72	460
470	25.72	25.77	25.83	25.88	25.94	25.99	26.05	26.10	26.16	26.22	26.27	470
480	26.27	26.33	26.38	26.44	26.49	26.55	26.61	26.66	26.72	26.77	26.83	480
490	26.83	26.89	26.94	27.00	27.05	27.11	27.17	27.22	27.28	27.33	27.39	490
500	27.39	27.45	27.50	27.56	27.61	27.67	27.73	27.78	27.84	27.90	27.95	500
510	27.95	28.01	28.07	28.12	28.18	28.23	28.29	28.35	28.40	28.46	28.52	510
520	28.52	28.57	28.63	28.69	28.74	28.80	28.86	28.91	28.97	29.02	29.08	520
530	29.08	29.14	29.20	29.25	29.31	29.37	29.42	29.48	29.54	29.59	29.65	530
540	29.65	29.71	29.76	29.82	29.88	29.94	29.99	30.05	30.11	30.16	30.22	540
550	30.22	30.28	30.34	30.39	30.45	30.51	30.57	30.62	30.68	30.74	30.80	550
560	30.80	30.85	30.91	30.97	31.02	31.08	31.14	31.20	31.26	31.31	31.37	560
570	31.37	31.43	31.49	31.54	31.60	31.66	31.72	31.78	31.83	31.89	31.95	570
580	31.95	32.01	32.06	32.12	32.18	32.24	32.30	32.36	32.41	32.47	32.53	580
590	32.53	32.59	32.65	32.71	32.76	32.82	32.88	32.94	33.00	33.06	33.11	590
600	33.11	33.17	33.23	33.29	33.35	33.41	33.46	33.52	33.58	33.64	33.70	600
°C	0	1	2	3	4	5	6	7	8	9	10	°C

Table 14-4 THERMOCOUPLE REFERENCE TABLES *(continued)*

Iron-Constantan Thermocouples (Modified 1913)—Concluded
(Electromotive Force in Absolute Millivolts. Temperatures
in Degrees C (Int. 1948). Reference Junctions at 0°C.)

°C	0	1	2	3	4	5	6	7	8	9	10	°C
	Millivolts											
600	33.11	33.17	33.23	33.29	33.35	33.41	33.46	33.52	33.58	33.64	33.70	600
610	33.70	33.76	33.82	33.88	33.94	33.99	34.05	34.11	34.17	34.23	34.29	610
620	34.29	34.35	34.41	34.47	34.53	34.58	34.64	34.70	34.76	34.82	34.88	620
630	34.88	34.94	35.00	35.06	35.12	35.18	35.24	35.30	35.36	35.42	35.48	630
640	35.48	35.54	35.60	35.66	35.72	35.78	35.84	35.90	35.96	36.02	36.08	640
650	36.08	36.14	36.20	36.26	36.32	36.38	36.44	36.50	36.56	36.62	36.69	650
660	36.69	36.75	36.81	36.87	36.93	36.99	37.05	37.11	37.18	37.24	37.30	660
670	37.30	37.36	37.42	37.48	37.54	37.60	37.66	37.73	37.79	37.85	37.91	670
680	37.91	37.97	38.04	38.10	38.16	38.22	38.28	38.34	38.41	38.47	38.53	680
690	38.53	38.59	38.66	38.72	38.78	38.84	38.90	38.97	39.03	39.09	39.15	690
700	39.15	39.22	39.28	39.34	39.40	39.47	39.53	39.59	39.65	39.72	39.78	700
710	39.78	39.84	39.91	39.97	40.03	40.10	40.16	40.22	40.28	40.35	40.41	710
720	40.41	40.48	40.54	40.60	40.66	40.73	40.79	40.86	40.92	40.98	41.05	720
730	41.05	41.11	41.17	41.24	41.30	41.36	41.43	41.49	41.56	41.62	41.68	730
740	41.68	41.75	41.81	41.87	41.94	42.00	42.07	42.13	42.19	42.26	42.32	740
750	42.32	42.38	42.45	42.51	42.58	42.64	42.70	42.77	42.83	42.90	42.96	750
°C	0	1	2	3	4	5	6	7	8	9	10	°C

Table 14-4 THERMOCOUPLE REFERENCE TABLES (*continued*)

Copper-Constantan Thermocouples
(Electromotive Force in Absolute Millivolts. Temperatures
in Degrees C (Int. 1948). Reference Junctions at 0°C.)

°C	0	1	2	3	4	5	6	7	8	9	10	°C
						Millivolts						
−190	−5.379	−5.395	−5.411	—	—	—	—	—	—	—	—	−190
−180	−5.205	−5.223	−5.241	−5.258	−5.276	−5.294	−5.311	−5.328	−5.345	−5.362	−5.379	−180
−170	−5.018	−5.037	−5.056	−5.075	−5.094	−5.113	−5.132	−5.150	−5.169	−5.187	−5.205	−170
−160	−4.817	−4.838	−4.858	−4.878	−4.899	−4.919	−4.939	−4.959	−4.978	−4.998	−5.018	−160
−150	−4.603	−4.625	−4.647	−4.669	−4.690	−4.712	−4.733	−4.754	−4.775	−4.796	−4.817	−150
−140	−4.377	−4.400	−4.423	−4.446	−4.469	−4.492	−4.514	−4.537	−4.559	−4.581	−4.603	−140
−130	−4.138	−4.162	−4.187	−4.211	−4.235	−4.259	−4.283	−4.307	−4.330	−4.354	−4.377	−130
−120	−3.887	−3.912	−3.938	−3.964	−3.989	−4.014	−4.039	−4.064	−4.089	−4.114	−4.138	−120
−110	−3.624	−3.651	−3.678	−3.704	−3.730	−3.757	−3.783	−3.809	−3.835	−3.861	−3.887	−110
−100	−3.349	−3.377	−3.405	−3.432	−3.460	−3.488	−3.515	−3.542	−3.570	−3.597	−3.624	−100
−90	−3.062	−3.091	−3.120	−3.149	−3.178	−3.207	−3.235	−3.264	−3.292	−3.320	−3.349	−90
−80	−2.764	−2.794	−2.824	−2.854	−2.884	−2.914	−2.944	−2.974	−3.003	−3.033	−3.062	−80
−70	−2.455	−2.486	−2.518	−2.549	−2.580	−2.611	−2.642	−2.672	−2.703	−2.733	−2.764	−70
−60	−2.135	−2.167	−2.200	−2.232	−2.264	−2.296	−2.328	−2.360	−2.392	−2.423	−2.455	−60
−50	−1.804	−1.838	−1.871	−1.905	−1.938	−1.971	−2.004	−2.037	−2.070	−2.103	−2.135	−50
−40	−1.463	−1.498	−1.532	−1.567	−1.601	−1.635	−1.669	−1.703	−1.737	−1.771	−1.804	−40
−30	−1.112	−1.148	−1.183	−1.218	−1.254	−1.289	−1.324	−1.359	−1.394	−1.429	−1.463	−30
−20	−0.751	−0.788	−0.824	−0.860	−0.897	−0.933	−0.969	−1.005	−1.041	−1.076	−1.112	−20
−10	−0.380	−0.417	−0.455	−0.492	−0.530	−0.567	−0.604	−0.641	−0.678	−0.714	−0.751	−10
(−)0	0.000	−0.038	−0.077	−0.115	−0.153	−0.191	−0.229	−0.267	−0.305	−0.343	−0.380	(−)0
(+)0	0.000	0.038	0.077	0.116	0.154	0.193	0.232	0.271	0.311	0.350	0.389	(+)0
10	0.389	0.429	0.468	0.508	0.547	0.587	0.627	0.667	0.707	0.747	0.787	10
20	0.787	0.827	0.868	0.908	0.949	0.990	1.030	1.071	1.112	1.153	1.194	20
30	1.194	1.235	1.277	1.318	1.360	1.401	1.443	1.485	1.526	1.568	1.610	30
40	1.610	1.652	1.694	1.737	1.779	1.821	1.864	1.907	1.949	1.992	2.035	40
50	2.035	2.078	2.121	2.164	2.207	2.250	2.293	2.336	2.380	2.423	2.467	50
60	2.467	2.511	2.555	2.599	2.643	2.687	2.731	2.775	2.820	2.864	2.908	60
70	2.908	2.953	2.997	3.042	3.087	3.132	3.177	3.222	3.267	3.312	3.357	70
80	3.357	3.402	3.448	3.493	3.539	3.584	3.630	3.676	3.722	3.767	3.813	80
90	3.813	3.859	3.906	3.952	3.998	4.044	4.091	4.138	4.184	4.230	4.277	90
100	4.277	4.324	4.371	4.418	4.465	4.512	4.559	4.606	4.654	4.701	4.749	100
110	4.749	4.796	4.843	4.891	4.939	4.987	5.035	5.083	5.131	5.179	5.227	110
120	5.227	5.275	5.323	5.372	5.420	5.469	5.518	5.566	5.615	5.663	5.712	120
130	5.712	5.761	5.810	5.859	5.908	5.957	6.007	6.056	6.105	6.155	6.204	130
140	6.204	6.254	6.303	6.353	6.403	6.453	6.503	6.553	6.603	6.653	6.703	140
150	6.703	6.753	6.803	6.853	6.904	6.954	7.004	7.055	7.106	7.157	7.208	150
160	7.208	7.258	7.309	7.360	7.411	7.462	7.513	7.565	7.616	7.667	7.719	160
170	7.719	7.770	7.822	7.874	7.926	7.978	8.029	8.080	8.132	8.184	8.236	170
180	8.236	8.288	8.340	8.392	8.445	8.497	8.549	8.601	8.654	8.707	8.759	180
190	8.759	8.812	8.864	8.917	8.970	9.023	9.076	9.129	9.182	9.235	9.288	190
200	9.288	9.341	9.394	9.448	9.501	9.555	9.608	9.662	9.715	9.769	9.823	200
°C	0	1	2	3	4	5	6	7	8	9	10	°C

Table 14-4 THERMOCOUPLE REFERENCE TABLES (*continued*)

Copper-Constantan Thermocouples—Concluded
(Electromotive Force in Absolute Millivolts. Temperatures in Degrees C (Int. 1948). Reference Junctions at 0°C.)

°C	0	1	2	3	4	5	6	7	8	9	10	°C
						Millivolts						
200	9.288	9.341	9.394	9.448	9.501	9.555	9.608	9.662	9.715	9.769	9.823	200
210	9.823	9.877	9.931	9.985	10.039	10.093	10.147	10.201	10.255	10.309	10.363	210
220	10.363	10.417	10.471	10.526	10.580	10.635	10.689	10.744	10.799	10.854	10.909	220
230	10.909	10.963	11.018	11.073	11.128	11.183	11.238	11.293	11.348	11.403	11.459	230
240	11.459	11.514	11.569	11.624	11.680	11.735	11.791	11.847	11.903	11.959	12.015	240
250	12.015	12.071	12.126	12.182	12.238	12.294	12.350	12.406	12.462	12.518	12.575	250
260	12.575	12.631	12.688	12.744	12.800	12.857	12.913	12.970	13.027	13.083	13.140	260
270	13.140	13.197	13.254	13.311	13.368	13.425	13.482	13.539	13.596	13.653	13.710	270
280	13.710	13.768	13.825	13.882	13.939	13.997	14.055	14.112	14.170	14.227	14.285	280
290	14.285	14.343	14.400	14.458	14.515	14.573	14.631	14.689	14.747	14.805	14.864	290
300	14.864	14.922	14.980	15.038	15.096	15.155	15.213	15.271	15.330	15.388	15.447	300
310	15.447	15.506	15.564	15.623	15.681	15.740	15.799	15.858	15.917	15.976	16.035	310
320	16.035	16.094	16.153	16.212	16.271	16.330	16.389	16.449	16.508	16.567	16.626	320
330	16.626	16.685	16.745	16.804	16.864	16.924	16.983	17.043	17.102	17.162	17.222	330
340	17.222	17.281	17.341	17.401	17.461	17.521	17.581	17.641	17.701	17.761	17.821	340
350	17.821	17.881	17.941	18.002	18.062	18.123	18.183	18.243	18.304	18.364	18.425	350
360	18.425	18.485	18.546	18.607	18.667	18.727	18.788	18.849	18.910	18.971	19.032	360
370	19.032	19.093	19.154	19.215	19.276	19.337	19.398	19.459	19.520	19.581	19.642	370
380	19.642	19.704	19.765	19.827	19.888	19.949	20.011	20.072	20.134	20.195	20.257	380
390	20.257	20.318	20.380	20.442	20.504	20.565	20.627	20.688	20.750	20.812	20.874	390
°C	0	1	2	3	4	5	6	7	8	9	10	°C

Chromel-Constantan Thermocouples
(Electromotive Force in Absolute Millivolts. Temperatures in Degrees C (Int. 1948). Reference Junctions at 0°C.)

°C	0	10	20	30	40	50	60	70	80	90	100	°C
						Millivolts						
−100	−5.18	−5.62	−6.04	−6.44	−6.83	−7.20	−7.55	−7.87	−8.17	−8.45	−8.71	−100
(−)0	0.00	−0.58	−1.14	−1.70	−2.24	−2.77	−3.28	−3.78	−4.26	−4.73	−5.18	(−)0
(+)0	0.00	0.59	1.19	1.80	2.41	3.04	3.68	4.33	4.99	5.65	6.32	(+)0
100	6.32	7.00	7.69	8.38	9.08	9.79	10.51	11.23	11.95	12.68	13.42	100
200	13.42	14.17	14.92	15.67	16.42	17.18	17.95	18.72	19.49	20.26	21.04	200
300	21.04	21.82	22.60	23.39	24.18	24.97	25.76	26.56	27.35	28.15	28.95	300
400	28.95	29.75	30.55	31.36	32.16	32.96	33.77	34.58	35.39	36.20	37.01	400
500	37.01	37.82	38.62	39.43	40.24	41.05	41.86	42.67	43.48	44.29	45.10	500
600	45.10	45.91	46.72	47.53	48.33	49.13	49.93	50.73	51.54	52.34	53.14	600
700	53.14	53.94	54.74	55.53	56.33	57.12	57.92	58.71	59.50	60.29	61.08	700
800	61.08	61.86	62.65	63.43	64.21	64.99	65.77	66.54	67.31	68.08	68.85	800
900	68.85	69.62	70.39	71.15	71.92	72.68	73.44	74.20	74.95	75.70	76.45	900
°C	0	10	20	30	40	50	60	70	80	90	100	°C

Appendix

A

The International System of Units (SI) and Decimal Prefixes

A.1 Basic Units

Quantity	Unit	Symbol	Definition
Length	Meter	m	The length equal to 1,650,763.73 wavelengths in vacuum of the radiation corresponding to the transition between the energy levels $2p_{10}$ and $5d_5$ of the krypton-86 atom (XI CGPM† 1960).
Mass	Kilogram	kg	The mass of the international prototype kilogram in the custody of the Bureau International des Poids et Mesures (BIPM), Sèvres, near Paris (III CGPM, 1901).
Time interval	Second	s	The interval occupied by 9,192,631,770 cycles of the radiation corresponding to the ($F = 4$, $M_F = 0$)–($F = 3$, $M_F = 0$) transition of the cesium-133 atom when unperturbed by exterior fields (XII CGPM, 1964).
Electric current	Ampere	A	The constant current which, if maintained in two straight parallel conductors of infinite length, of

Quantity	Unit	Symbol	Definition
			negligible cross section, and placed at a distance of 1 meter apart in a vacuum, will produce a force between them equal to 2×10^{-7} newton per meter length (IX CGPM, 1948).
Temperature	Degree Kelvin	°K	The thermodynamic scale of temperature defined by means of the triple point of water as the fundamental fixed point, attributing to it the temperature 273.16 degrees Kelvin exactly; on this scale the temperature of the ice point is 273.15°K (0°C). Kelvin and Celsius temperature intervals are identical (X CGPM, 1954).
Luminance	Candela	cd	The magnitude of the candela is such that the luminance of the total radiator, at the temperature of solidification of platinum, is 60 candelas per square centimeter (Comité International des Poids et Mesures, 1964, Procès-Verbaux, 121; IX CGPM, 1948).

†CGPM—Conférence Général des Poids et Mesures

A.1.1 Conversion of length and mass units

$$1 \text{ m} = 39.37 \text{ inches}$$
$$1 \text{ kg} = 2.2046 \text{ pounds}$$
$$1 \text{ kg} = 35.27 \text{ ounces}$$
$$1 \text{ micron } (\mu) = 10^{-6} \text{ m}$$
$$1 \text{ metric ton} = 10^3 \text{ kg}$$

A.2 Supplementary Units

Quantity	Unit	Symbol	Typical Conversion (approx.)
Plane angle	radian	rad	1 rad = 57.296 deg.
Solid angle	steradian	sr	1 sr = 0.159 hemisphere

A.3 Derived Units (Having Names Based on Basic and Supplementary Units)

Quantity	Unit	Symbol	Typical Conversion (approx.)
Area	square meter	m^2	$1 \ m^2 = 1549.977 \ in.^2$
Volume	cubic meter	m^3	$1 \ m^3 = 35.3145 \ ft^3$,
			$1 \ liter = 10^{-3} \ m^3$
Velocity	meter per second	m/s	$1 \ m/s = 3.281 \ ft/s$
Acceleration	meter per second squared	m/s^2	$1 \ m/s^2 = 0.102 \ g$
Angular velocity	radian per second	rad/s	$1 \ rad/s = 9.5493 \ rpm$
Angular acceleration	radian per second squared	rad/s^2	
Density	kilogram per cubic meter	kg/m^3	$1 \ kg/m^3 = 0.99885 \ oz/ft^3$
Kinematic viscosity	square meter per second	m^2/s	$1 \ m^2/s = 10^4 \ stokes$
Luminance	candela per square meter	cd/m^2	
Magnetic field strength	ampere per meter	A/m	$10^3/4\pi \ A/m = 1 \ oersted$

A.4 Derived Units (Having Special Names)

Quantity	Unit	Symbol	Interrelation
Capacitance	farad	F	$A \cdot s/V$
Inductance	henry	H	$V \cdot s/A$
Electric charge	coulomb	C	$A \cdot s$
Voltage (emf)	volt	V	W/A
Electric field strength	volt per meter	V/m	
Resistance	ohm	Ω	V/A
Frequency	hertz	Hz	s^{-1}
Energy (work, heat)	joule	J	$N \cdot m$
Power	watt	W	J/s
Magnetic flux	weber	Wb	$V \cdot s$
Magnetic flux density	tesla	T	Wb/m^2
Luminous flux	lumen	lm	$cd \cdot sr$
Illumination	lux	lx	lm/m^2
Force	newton	N	$kg \cdot m/s^2$
Pressure	newton per square meter	N/m^2	
Pressure	pascal	Pa	N/m^2
Dynamic viscosity	newton-second per square meter	$N\text{-}s/m^2$	$N \cdot s/m^2$

A.5 Decimal Prefixes

Multiplier	Prefix	Symbol
10^{12}	tera	T
10^9	giga	G
10^6	mega	M
10^3	kilo	k
10^2	hecto	h
10	deka	da
10^{-1}	deci	d
10^{-2}	centi	c
10^{-3}	milli	m
10^{-6}	micro	μ
10^{-9}	nano	n
10^{-12}	pico	p
10^{-15}	femto	f
10^{-18}	atto	a

Suggested Graphical Symbols for Transducers in Electrical Schematics and Wiring Diagrams

B.1 Introduction

The basic graphical symbol for a measuring transducer is a square with an added equilateral triangle. The triangle symbolizes the sensing element. A letter in the triangle identifies the measurand. The square symbolizes the transduction element and transducer case in general. The specific transduction element, or electrical circuitry including a transduction element, is represented by a symbol within the square from which all required excitation and output connections and any optional additional connections lead to the side of the square opposite the triangle.

The relative dimensions of the complete transducer symbol are such that for any transduction element or electrical circuitry dimension shown as x the sides of the square and of the triangle are $2x$. Existing standards and conventions cover the representation of specific external electrical connections such as receptacles, terminals, and shielded or unshielded cables.

Compatibility with existing standard MIL-STD-15-1 and with symbols recommended by the International Electrotechnical Commission (IEC) has been attempted where feasible.

The basic transducer symbol and some of its adaptations are contained in the ISA S37 series of transducer standards (Instrument Society of America).

B.2 Transducer Measuring

B.2.1 General

The triangle should point towards the left.

A dotted line, indicating a mechanical link, can connect the sensing-element side of the square with a portion of the transduction element if required for clarity.

See paragraph 3 of this Appendix for special transducer symbols not using the general representation.

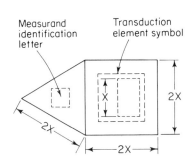
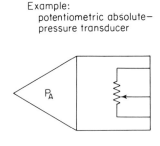

B2.2 Measurand identification letters

a	Acceleration (rectilinear)	h_A	Humidity (absolute)
α	Acceleration (angular)	h_{DP}	Dew point
D	Displacement (general)	L	Liquid level (general)
D_L	Displacement (rectilinear)	L_N	Liquid level (continuous)
D_A	Displacement (angular)	L_X	Liquid level (discrete)
Q	Flow rate (general)	P	Pressure (general)
Q_V	Flow rate (volumetric)	P_A	Pressure (absolute)
Q_M	Flow rate (mass)	P_G	Pressure (gage)
F	Force and Torque (general)	P_D	Pressure (differential)
F_T	Force (tension only)	P_{SPL}	Sound pressure
F_C	Force (compression only)	T	Temperature
F_{CT}	Force (tension and compression)	T_Q	Heat flux (total)
F_M	Torque	T_{QR}	Heat flux (radiant only)
h	Humidity (general)	v	Velocity and Speed (rectilinear)
h_R	Humidity (relative)	ω	Velocity and Speed (angular)

B.2.3 Transduction elements

B.2.3.1 Capacitive

General   Example of adaptation
                        (variable–differential capacitive)

B.2.3.2 Electromagnetic

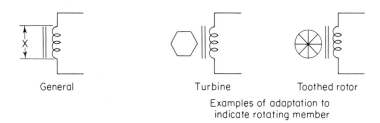

General   Turbine   Toothed rotor
                    Examples of adaptation to
                    indicate rotating member

B.2.3.3 Inductive

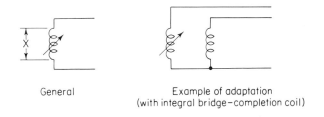

General   Example of adaptation
                    (with integral bridge–completion coil)

B.2.3.4 Piezoelectric

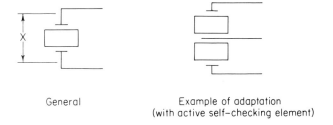

General   Example of adaptation
                    (with active self–checking element)

B.2.3.5 Potentiometric

General

Example of adaptation
(with center–tapped element)

B.2.3.6 Reluctive

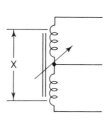

Inductance–bridge

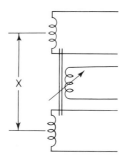

Differential–transformer
(example)

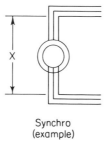

Synchro
(example)

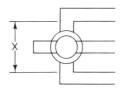

Resolver
(example)

B.2.3.7 Resistive

General

B.2.3.8 Strain-gage

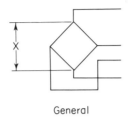

General

B.2.3.9 Thermoelectric

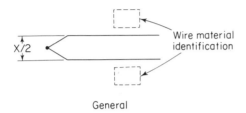

General

B.2.3.10 Any transduction element with integral excitation and signal conditioning

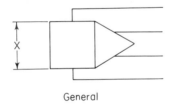

General

B.2.3.11 Any transduction element with null- or force-balance servo operation

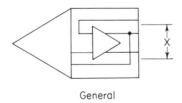

General

B.3 Special Symbols

These symbols can be used without the basic representation when applied to the specific measurand shown.

B.3.1 Photoconductive light sensor (or photoconductive transduction element for other measurands)

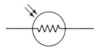

B.3.2 Photovoltaic light sensor (or photovoltaic transduction element for other measurands)

B.3.3 Photodiode light sensor (or photodiode transduction element for other measurands)

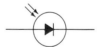

B.3.4 Resistive strain sensor (strain gage)

Note: Downward arrow can be used to indicate decreasing resistance.

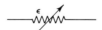

B.3.5 Wire-thermocouple (thermoelectric temperature transducer, wire-type) (see also B.2.3.9)

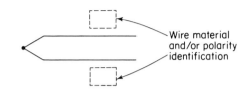

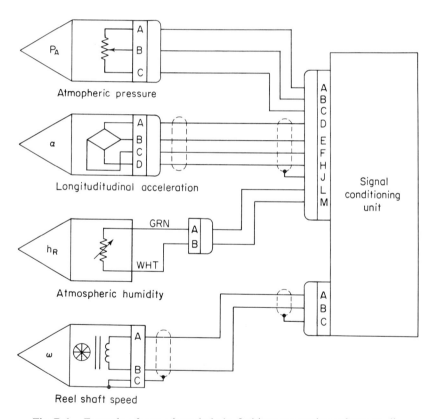

Fig. B-1. Example of use of symbols in fictitious measuring subsystem diagram.

Appendix C

Glossary of Terms Relating to Transducers Used in Measuring Systems

This Glossary is intended primarily for general reference purposes and, secondarily, for a rapid reference to some of the terminology used in this book. Terms related to each of the measurands are defined at the beginning of the appropriate chapter. A number of these terms are also shown in this Glossary, sometimes with slightly different wording. Mention of a term in this Glossary does not necessarily imply that it is to be considered a "preferred" term.

Although many of the terms can be used in transducer specifications, their use as referee material in procurement disputes and referencing of this Glossary in specifications is not recommended, since details necessary for such use have been omitted in most cases. For specification terminology a national or industrial standard should be used, notably ISA-S37.1, "Electrical Transducer Nomenclature and Terminology," which was developed by ISA Committee SP37 under the author's chairmanship.

A number of dictionaries, glossaries, standards, and other listings were consulted in the development of this terminology. Besides the various ISA RP37- and S37-series Standards, which reflect a consensus by experts of the required rewording of existing definitions, the best source was found to be "The International Dictionary of Physics and Electronics" (Second Edition, D. Van Nostrand Co., Inc., Princeton, 1961). Valuable information was also found in the Inter-Range Instrumentation

678

Group's "Glossary of Telemetry Transducer Terms" (1963) which is included in the U.S. Air Force "Telemetry Transducer Handbook" (Report WADD-TR-61-67, 1963) together with other useful definitions. Among additional source material searched were a number of IRE Standards (Institute of Electrical and Electronic Engineers), U.S.A. Standards for Electrical and Automatic-Control terms, and glossaries published by a number of transducer manufacturers.

Absolute humidity: The mass of water vapor present in a unit volume of air or other fluid.

Absolute pressure: The pressure measured relative to zero pressure.

Acceleration: A vector quantity equal to the time rate of change of velocity with respect to a reference system.

Acceleration error: The maximum difference, at any measurand value within the specified range, between output readings taken with and without the application of specified constant acceleration along specified axes.

Acceleration sensitivity: Acceleration error expressed in percent of full-scale output per g (or other unit of acceleration).

Accelerometer: An acceleration transducer.

Accuracy: The ratio of error to full-scale output or the ratio of error to output.

Active element (active arm, active leg): The Wheatstone-bridge arm in which a resistive transducer is connected.

Altitude: The vertical distance above a stated reference level. *Note:* Unless otherwise stated, this reference is mean sea level.

Ambient conditions: The conditions (pressure, temperature, etc.) of the medium surrounding the case of the transducer.

Ambient-pressure: The absolute pressure of the medium surrounding the transducer.

Ambient-pressure error: The maximum change in output, at any measurand value within the specified range, when the ambient pressure is changed between specified values.

Amplifying (transducer): A transducer with an integrally-packaged output amplifier.

Analog output: Transducer output which is a continuous function of the measurand (except as modified by the resolution of the transducer.)

Attitude: The relative orientation of a vehicle or object represented by its angles of inclination to three orthogonal reference axes.

Attitude error: The acceleration error due to the orientation of the transducer relative to the direction in which gravity acts upon the transducer.

Bellows: A mechanical element of generally cylindrical shape with cylindrical walls containing deep convolutions. *Note:* Used as pressure-sensing element, as seal between mechanically linked members, and in shaft couplings.

"Best straight line": A line midway between the two parallel straight lines closest together and enclosing all output-vs-measurand values on a calibration curve.

"Best straight line" with forced zero: The "best straight line" which meets the additional requirement of passing through the zero-measurand, zero-output point.

"Best straight line" with Y intercept: The "best straight line" which meets the additional requirement of passing through a point at zero measurand (i.e., on the y-axis of a calibration curve) and a specified value of transducer output.

Bondable (transducer): A transducer which is designed to be permanently mounted to a surface by means of adhesives.

Bonded strain gage: A resistive strain transducer which is permanently attached over the length and width of its active element.

Bourdon tube: A pressure-sensing element

consisting of a twisted or curved tube of noncircular cross section which tends to be straightened by the application of internal pressure.

Breakdown-voltage rating: The d-c or sinusoidal a-c voltage which can be applied across specified insulated portions of a transducer without causing arcing or conduction above a specified current value across the insulating material.

Bridge resistance: A general term for the resistive input or output impedance of a strain-gage type transducer (not containing integral active excitation- or output-conditioning circuitry).

Burst-pressure rating: The pressure which may be applied to the sensing element or the case (as specified) of a transducer without rupture of either the sensing element or transducer case as specified.

Calibration: A test during which known values of measurand are applied to the transducer and corresponding output readings are recorded.

Calibration curve: A graphical representation of the calibration record.

Calibration cycle: The application of known values of measurand and the recording of corresponding output readings over the full (or specified portion of the) range of a transducer in an ascending and descending direction.

Calibration record: A record (e.g., table or graph) of the measured relationship of the transducer output to the applied measurand over the transducer range.

Calibration traceability: The relation of a transducer calibration, through a specified step-by-step process, to an instrument or group of instruments calibrated by a governmental standards agency such as (in the U.S.A.) the National Bureau of Standards.

Calibration uncertainty: The maximum calculated error in the output values shown in a calibration record, due to causes not attributable to the transducer.

Capacitive transduction: Conversion of the measurand into a change of capacitance.

Capsule: A pressure-sensing element consisting of two metallic diaphragms joined around their peripheries.

Carrier: An a-c voltage having a frequency suitably high to be modulated by electrical signals representing measurand levels.

Case pressure: (See Burst-pressure rating, Proof pressure, or Reference pressure.)

Center of seismic mass: The point within an acceleration transducer where acceleration forces are considered to be summed.

Common-mode rejection: The ability of a differential amplifier to reject common-mode voltages, usually expressed as the attenuation of a common-mode voltage by the amplifier.

Common-mode voltage: A voltage which appears simultaneously at both input terminals of an amplifier with respect to a common reference point.

Compensated temperature range: The operating temperature range of a transducer incorporating temperature-compensation. (See Operating temperature range and Compensation.)

Compensation: Provision of a supplemental device, circuit, or special materials to counteract known sources of error.

Conduction error: The error in a temperature transducer due to heat conduction between the sensing element and the mounting of the transducer.

Conformance, Conformity: The closeness of a calibration curve to a specified curve. (See also Error Band.)

Continuous rating: The rating applicable to specified operation for a specified uninterrupted length of time.

Creep: A change in output occurring over a specific time period while the measurand and all environmental conditions are held constant.

Critical damping: Damping at the point of change between the underdamped and the overdamped conditions. (See Damping.)

Cross-axis acceleration: (See Transverse acceleration.)

Cross sensitivity, Cross-axis sensitivity: (See Transverse sensitivity.)

Damping: The energy dissipating characteristic which, together with natural frequency, determines the limit of frequency response and the response-time characteristics of a transducer. *Note:* In response to a step change of the measurand, an underdamped system oscillates about its final steady value before coming to rest at that value; an overdamped system comes to rest without overshoot; and a critically damped system is at the point of change between the underdamped and overdamped conditions.

Damping, magnetic: Damping effected by use of current induced in electrical conductors by changes in magnetic flux.

Damping, viscous: Damping effected by use of the viscosity of liquids or gases.

Damping ratio: The ratio of the degree of actual damping to the degree of damping required for critical damping.

Dark current: The current flowing in photoemissive or photoconductive light transducers in the absence of any incident photons or other radiation.

D-C output transducer: A transducer with integral demodulator, rectifier, or frequency integrator.

Dead band: Of a switch-type transducer, the portion of the range between switch actuation by increasing and decreasing measurand.

Dead volume: The total volume of the internal cavity between the sensing element and the most external portion of the pressure port of a pressure transducer, with room barometric pressure applied.

Definition: (See Resolution.)

Depth: (See Pressure depth.)

Detector: (See Transducer.)

Dew point: The temperature at which the saturation water vapor pressure is equal to the partial pressure of the water vapor in the atmosphere. *Note:* The relative humidity at the dew point is 100%.

Diaphragm: A sensing element consisting of a thin and usually circular plate which is deformed by pressure differential applied across the plate.

Dielectric strength: (See Breakdown voltage and Insulation resistance.)

Differential pressure: The difference in pressure between two points of measurement.

Digital output: Transducer output that represents the magnitude of the measurand in the form of a series of discrete quantities coded in a system of notation.

Directivity: The solid angle, or the angle in a specified plane, over which sound or radiant energy incident on a transducer is measured within specified tolerances in a specified band of measurand frequencies.

Discrete-increment output: A transducer output which represents the magnitude of the measurand in the form of discrete or quantized values.

Displacement: The change in position of a body or point with respect to a reference point.

Distance: The spatial separation between two objects or points.

Distortion: (See Harmonic content.)

Dithering: The application of intermittent or oscillatory forces just sufficient to minimize static friction within the transducer.

Double amplitude: The peak-to-peak value.

Drift: An undesired change in output over a period of time, which change is not a function of the measurand. (See also Stability.)

Dual-output transducer: A transducer providing two separate and noninteracting outputs which are functions of the applied measurand.

Dynamic calibration: A calibration during which the measurand varies with time in a specified manner and the output is recorded as a function of time.

Dynamic characteristics: Those characteristics of a transducer which relate to its response to variations of the measurand with time.

Electromagnetic transduction: Conversion of the measurand into an output induced in a conductor by a change in magnetic flux, in the absence of excitation.

End points: The outputs at the specified upper and lower limits of the range.

End-point line: The straight line between the end points.

Environmental conditions: Specified external conditions (shock, vibration, temperature, etc.) to which a transducer may be exposed during shipping, storage, handling, and operation.

Environmental conditions, nonoperating: Environmental conditions after exposure to which a transducer must perform in some specified manner.

Environmental conditions, operating: Environmental conditions during exposure to which a transducer must perform in some specified manner.

Error: The algebraic difference between the indicated value and the true value of the measurand.

Error band: The band of maximum deviations of output values from a specified reference line or curve due to those causes attributable to the transducer.

Error curve: A graphical representation of errors obtained from a specified number of calibration cycles.

Excitation: The external electrical voltage and/or current applied to a transducer for its proper operation.

Field of view: The solid angle, or the angle in a specified plane, over which radiant energy incident on a transducer is measured within stated tolerances.

Flow rate: The time rate of motion of a fluid, expressed as fluid quantity per unit time.

Flowmeter: A flow-rate transducer.

Fluid: A gas or liquid.

Fluid damping: (See Viscous damping and Damping.)

Force: The vector quantity necessary to cause a change in momentum.

Frequency, natural: The frequency of free (not forced) oscillations of the sensing element of a fully assembled transducer. Also, that frequency of a sinusoidally applied measurand at which the transducer output lags the measurand by 90 degrees.

Frequency, resonant: The measurand frequency at which a transducer responds with maximum output amplitude.

Frequency, ringing: The frequency of the oscillatory transient occurring in the transducer output as a result of a step change in measurand.

Frequency-modulated output: An output in the form of frequency deviations from a center frequency, where the deviation is a function of the applied measurand.

Frequency output: An output in the form of frequency which varies as a function of the applied measurand.

Frequency response: The change with frequency of the output/measurand amplitude ratio (and of the phase difference between output and measurand) for a sinusoidally varying measurand applied to a transducer within a stated range of measurand frequencies.

Frequency response, calculated: The frequency response of a transducer calculated from its transient response, its mechanical properties, or its geometry, and so identified.

Friction error: The maximum change in output before and after minimizing friction within the transducer by dithering.

Friction-free calibration: Calibration under conditions minimizing the effect of static friction, obtained by dithering.

Friction-free error band: The error band applicable at room conditions and with friction within the transducer minimized by dithering.

Full-scale output: The algebraic difference between the end points. *Note:* Sometimes expressed as "±(half the algebraic difference)," e.g., "±2.5 V."

Gage factor: A measure of the ratio of the relative change of resistance to the relative change in length of a resistive strain transducer (strain gage).

Gage pressure: Pressure measured relative to ambient pressure.

Grease line: Liquid-filled tubing between a pressure transducer and a measuring point, frequently isolated at the measuring point from the measured fluid by a membrane.

Gyro (A contraction of **Gyroscope**): A transducer which makes use of a self-contained spatial directional reference.

Hall effect: The development of a transverse electric potential gradient in a current-carrying conductor or semiconductor upon the application of a magnetic field.

Harmonic content: The distortion in a transducer's sinusoidal output, in the form of harmonics other than the fundamental component.

Harmonic motion: A vibration whose instantaneous amplitude varies sinusoidally with time.

Head: The height of a liquid column at the base of which a given pressure would be developed.

Heat conduction: The transfer of heat energy by diffusion through solid material or through stagnant fluids.

Heat convection: The transfer of heat energy by the movement of a fluid between two points.

Heat flux: The time rate of flow of heat energy per unit area.

Heat radiation: The transfer of heat energy by electromagnetic waves.

Hypsometer: A pressure measuring device which measures the boiling or condensation temperature of a liquid in equilibrium with its vapor while the pressure of the vapor equals the pressure to be measured.

Hysteresis: The maximum difference in output at any measurand value within the transducer's range when the value is approached first with increasing and then with decreasing measurand. *Note:* Friction error is included with hysteresis unless dithering is specified.

Hysteresis, thermal: (See Thermal hysteresis.)

Illumination: The luminous flux density on a surface, expressed as luminous flux per unit area.

Impact pressure: The pressure in a moving fluid exerted parallel to the direction of flow due to the velocity of the flow.

Inductive transduction: Conversion of the measurand into a change of the self-inductance of a single coil.

Input: (See Excitation or Measurand.)

Input impedance: The impedance (presented to the excitation source) measured across the excitation terminals of a transducer.

Instability: (See Stability.)

Insulation resistance: The resistance measured between specified insulated portions of a transducer when a specified d-c voltage is applied.

Integrating transducer: A transducer whose output is a time integral function of the measurand.

Intermittent rating: The rating applicable to specified operation over a stated number of time intervals of specified duration.

Ionizing transduction: Conversion of the measurand into a change in ionization current, such as through a gas between two electrodes.

Jerk: The time rate of change of acceleration. Expressed in ft/s^3, cm/s^3, or *g*/s.

Leakage rate: The maximum rate at which a fluid at a specified pressure is permitted to leak through a specified sealed portion of a transducer.

Least-squares line: The straight line for which the sum of the squares of the residuals (deviations) is minimized.

Life, cycling: The specified minimum number of full-range excursions or specified partial-range excursions over which a transducer will operate as specified without changing its performance beyond specified tolerances.

Life, operating: The specified minimum length of time over which the specified continuous and intermittent rating of a transducer applies without change in transducer performance beyond specified tolerances.

Life, storage: The specified minimum length of time over which a transducer can be exposed to specified storage conditions without changing its performance beyond specified tolerances.

Linearity: The closeness of a calibration curve to a specified straight line.

Linearity, end-point: Linearity referred to a straight line between the end points (the end-point line).

Linearity, independent: Linearity referred to the "best straight line."

Linearity, independent, with forced zero: Linearity referred to the "best straight line" with forced zero. (See "Best straight line" with forced zero.)

Linearity, independent, with y intercept: Linearity referred to the "best straight line" with y intercept. (See "Best straight line" with y intercept.)

Linearity, least-squares: Linearity referred to a straight line for which the sum of the squares of the residuals is minimized (the least-squares line).

Linearity, terminal: Linearity referred to the terminal line. (See Terminal line.)

Linearity, theoretical-slope: Linearity referred to a straight line between the theoretical end points (the theoretical slope).

Line pressure: (See Reference pressure.)

Load impedance: The impedance presented to the output terminals of a transducer by the associated external circuitry.

Loading error: An error due to the effect of the load impedance on the transducer output.

Magnetoresistive effect: The change in the resistance of a conductor or semiconductor due to the application of a magnetic field.

Magnetostrictive effect: The change in length of a ferromagnetic material subjected to an increasing or decreasing longitudinal magnetic field (Joule Effect). The inverse Joule Effect is known as the Villari Effect.

Mass: The inertial property of a body.

Maximum (ambient) temperature: The value of the highest ambient temperature that a transducer can be exposed to, with or without excitation applied, without being damaged or subsequently showing a performance degradation beyond specified tolerances.

Maximum excitation: The maximum value of excitation voltage or current that can be applied to the transducer at room conditions without causing damage or performance degradation beyond specified tolerances.

Maximum fluid temperature: The value of the highest measured-fluid temperature that a transducer can be exposed to, with or without excitation applied, without being damaged or subsequently showing a performance degradation beyond specified tolerances.

Mean output curve: The curve through the mean values of output during any one calibration cycle or a different specified number of calibration cycles.

Measurand: A physical quantity, property, or condition which is measured.

Measured fluid: The fluid which comes in contact with the sensing element.

Mechanical impedance: The complex ratio of force to velocity during simple harmonic motion.

Moisture: The amount of liquid adsorbed or absorbed by a solid.

Motion: The change in position of a body or point with respect to a reference system.

Mounting error: The error resulting from

mechanical deformation of the transducer caused by mounting the transducer and making all measurand and electrical connections.

Natural frequency: (See Frequency, natural.)

Nonlinearity: (See Linearity.)

Nonoperating conditions: (See Environmental conditions, nonoperating.)

Nonrepeatability: (See Repeatability.)

Null: A condition (typically a condition of balance) which results in a minimum absolute value of output.

Operating conditions: (See Environmental conditions, operating.)

Operating temperature range: (See Temperature range, operating.)

Output: The electrical quantity produced by a transducer which is a function of the applied measurand.

Output impedance: The impedance across the output terminals of a transducer presented by the transducer to the associated external circuitry.

Output noise: The unwanted component (typically of broad frequency spectrum) of the output of a transducer.

Output regulation: The change in output due to a change in excitation.

Overload: The maximum magnitude of measurand that can be applied to a transducer without causing a change in performance beyond specified tolerance.

Overrange: (See Overload.)

Overshoot: The amount of output measured beyond the final steady output value in response to a step change in the measurand.

Partial pressure: The pressure exerted by one constituent of a mixture of gases.

Photoconductive transduction: Conversion of the measurand into a change in resistance or conductivity of a semiconductor material by a change in the amount of illumination incident upon the material.

Photoemissive transduction: Conversion of the measurand into a change of the emission of electrons due to a change in the incidence of photons on a photocathode.

Photovoltaic transduction: Conversion of the measurand into a change in the voltage generated when a junction between certain dissimilar materials is illuminated.

Piezoelectric transduction: Conversion of the measurand into a change in the electrostatic charge or voltage generated by certain materials when mechanically stressed.

Piezoresistive transduction: Conversion of the measurand into a change in the resistance of a conductor or semiconductor by a change in the mechanical stress applied to it.

Position: The spatial location of a body or point with respect to a specified reference point.

Potentiometric transduction: Conversion of the measurand into a voltage ratio by a change in the position of a movable contact on a resistance element across which excitation is applied.

Precision: (See Repeatability and Stability.)

Pressure: The force acting on a surface, measured as force per unit area. (See Absolute pressure, Differential pressure, Gage pressure, and Reference pressure.)

Pressure altitude: Altitude determined on the basis of its known relationship to pressure.

Pressure depth: Depth (usually below mean sea level) determined on the basis of its known relationship to pressure.

Primary element, Primary detector: (See Sensing element.)

Proof pressure: The maximum pressure which may be applied to the sensing element of a transducer without changing the transducer performance beyond specified tolerances.

Proximity: The spatial closeness between two objects or points.

Pyroelectric effect: The separation of electric charge in a crystal by heating.

Random vibration: Nonperiodic vibration described only in statistical terms, most commonly taken to mean vibration characterized by an amplitude distribution which essentially follows the normal error curve (Gaussian distribution) with the required amplitude and frequency limits stated.

Range: The measurand values over which a transducer is intended to measure, specified by their upper and lower limits.

Recovery time: The time interval, after a specified event (e.g., overload) after which a transducer again performs within its specified tolerances.

Reference pressure: The pressure relative to which a differential-pressure transducer measures pressure.

Reference-pressure error: The error resulting from variations of a differential-pressure transducer's reference pressure within the applicable reference pressure range.

Reference-pressure range: The range of reference pressures which can be applied without changing the differential-pressure transducer's performance beyond specified tolerances for reference pressure error, if any such error is allowed.

Relative humidity: The ratio of the water-vapor pressure actually present to the water vapor required for saturation at a given temperature, expressed in percent.

Reliability (of a transducer): A measure of the probability that a transducer will continue to perform within specified limits of error for a specified length of time under specified conditions.

Reluctive transduction: Conversion of the measurand into an a-c voltage change by a change in the reluctance path between two or more coils or separated portions of one coil when a-c excitation is applied to the coil(s).

Repeatability: The ability of a transducer to reproduce output readings when the same measurand value is applied to it consecutively, under the same conditions, and in the same direction.

Resistive transduction: Conversion of the measurand into a change of resistance.

Resolution: The magnitude of output step changes (expressed in percent of full-scale output) as the measurand is continuously varied over the range.

Resolution, average: The reciprocal of the total number of transducer output steps over the measuring range, multiplied by 100, and expressed in percent of full-scale output (or percent voltage ratio).

Resolution, maximum: The magnitude of the largest of all transducer output steps over the measuring range, expressed in percent of full-scale output (or percent voltage ratio).

Resonances: Amplified vibrations of transducer components, within narrow frequency bands, observable in the output as vibration is applied along specific transducer axes.

Resonant frequency: (See Frequency, resonant.)

Response time: The length of time required for the output of a transducer to rise to a specified percentage of its final value as a result of a step change of measurand.

Responsivity: The ratio of the output amplitude to the product of the effective radiation flux density (at a given wavelength) and the active area of the sensing surface of a light transducer.

Ringing period: The period of time during which the amplitude of output oscillations, excited by a step change in measurand, exceed the steady-state output value.

Riometer: An instrument which measures ionospheric absorption by comparing the noise power of extraterrestrial radiations with the noise power of a local source.

Ripple: The rms a-c component of a transducer's d-c output voltage, at a constant measurand level, expressed in percent of the average value of the full-scale output.

Rise time: The length of time for the output of a transducer to rise from a small specified percentage of its final value to a large specified percentage of its final value.

Room conditions: Ambient environmental conditions, which have been established on the basis of environments found in most work-areas as follows (by ISA S37.1):

1. Temperature: $25 \pm 10°C (77 \pm 18°F)$.
2. Relative humidity: 90% or less.
3. Barometric pressure: 26 to 32 in. Hg. (760 ± 80 mm Hg.)

Sealed-reference differential pressure: The pressure difference between an unknown pressure and the pressure of a gas in a sealed reference chamber integral with a transducer.

Self-generating: Providing an output signal without applied excitation. Examples are piezoelectric, electromagnetic, and thermoelectric transducers.

Self-heating: Internal heating resulting from electrical energy dissipated within the transducer.

Sensing element: That part of the transducer which responds directly to the measurand.

Sensitivity: The ratio of the change in transducer output to a change in the value of the measurand.

Sensitivity shift: A change in the slope of the calibration curve due to a change in sensitivity.

Sensor: (See Transducer.)

Servo-type transducer: A transducer in which the output of the transduction element is amplified and fed back so as to balance the forces applied to the sensing element or its displacements. The output is a function of the feedback signal.

Shock: A sudden nonperiodic or transient excitation of a mechanical system.

Shunt calibration: A test of a measuring system in which shunt-calibration resistors are used.

Shunt-calibration resistor: A resistor which, when placed across specified points of the electrical circuit of the transducer, will electrically simulate a specified percentage of the full-scale output of the transducer.

Sound pressure: The total instantaneous pressure at a given point in the presence of a sound wave minus the static pressure at that point.

Sound-pressure level: The ratio of the rms sound pressure to a specified reference pressure, expressed in db.

Source impedance: The impedance presented to the transducer's excitation terminals by the excitation source.

Span: The algebraic difference between the limits of the range.

Speed: A scalar quantity equal to the magnitude of the time rate of change of displacement.

Stability: The ability of a transducer to retain its repeatability (and other specified performance characteristics) for a relatively long period of time.

Stagnation pressure: The sum of the static pressure and the impact pressure.

Static calibration: A calibration performed under room conditions by application of the measurand to the transducer in discrete amplitude intervals.

Static error band: The error band applicable at room conditions and in the absence of any vibration, shock, or acceleration (unless one of these is the measurand).

Static pressure: The pressure of a fluid, exerted normal to the surface along which the fluid flows.

Stimulus: (See Measurand.)

Strain: The deformation of a solid resulting from a stress, measured as the ratio of the dimensional change to the total value of the dimension in which the change occurs.

Strain error: The error resulting from a strain imposed on a surface to which the transducer is mounted.

Strain gage: A resistive strain transducer.

Strain-gage transduction: Conversion of the measurand into a change of resistance due to strain usually in two or four arms of a Wheatstone bridge (strain-gage bridge).

Tachometer: An angular-speed transducer.

Temperature error: The maximum change in output, at any measurand value within the specified range, when the transducer temperature is changed from room temperature to specified temperature extremes.

Temperature-error band: The error band applicable over stated environmental temperature limits.

Temperature-gradient error: The transient deviation in output of a transducer at a given measurand value when the ambient temperature or the measured-fluid temperature changes at a specified rate between specified magnitudes.

Temperature range, compensated: The operating temperature range of a transducer which contains provisions for thermal compensation.

Temperature range, fluid: The range of temperatures of the measured fluid, when it is not the ambient fluid, within which operation of the transducer is intended.

Temperature range, operating: The range of ambient temperatures, given by their extremes, within which the transducer is intended to operate. *Note:* Within this range of ambient temperatures, all tolerances specified for thermal effects on performance are applicable.

Terminal line: A theoretical slope for which the theoretical end points are 0% and 100% of both measurand and output.

Theoretical curve: The specified relationship (table, graph, or equation) of the transducer output to the applied measurand over the range.

Theoretical end points: The specified points between which the theoretical curve is established and to which no end-point tolerances apply.

Theoretical slope: The straight line between the theoretical end points.

Thermal coefficient of resistance: The relative change in resistance of a conductor or semiconductor per unit change in temperature over a stated range of temperatures expressed in ohms per ohm per degree F or C.

Thermal hysteresis: The maximum difference in output, at a given measurand value within the specified range and at a given temperature, when this temperature is approached in the increasing and in the decreasing portion of a temperature cycle whose maximum temperature is substantially beyond the given temperature.

Thermal sensitivity shift: The sensitivity shift due to changes of the ambient temperature from room temperature to the limits of the operating temperature range.

Thermal zero shift: The zero shift due to changes of the ambient temperature from room temperature to the specified limits of the operating temperature range.

Thermocouple: A thermoelectric temperature transducer.

Thermoelectric transduction: Conversion of the measurand into a change in the emf generated by a temperature difference between the junctions of two selected dissimilar materials.

Thermowell: A pressure-tight tubular receptacle designed for the installation in it, by insertion and fastening, of the sensing-element portion of a temperature transducer.

Threshold: The smallest change in the measurand that will result in a measurable change in transducer output.

Time constant: The length of time required for the output of a transducer to rise to 63% of its final value as a result of a step change of measurand.

Torque: The moment of force.

Total pressure: (See Stagnation pressure.)

Transducer: A device which provides a usable output in response to a specified measurand.

Transduction element: The (electrical) portion of a transducer in which the output originates.

Transient response: The response of a transducer to a step change in measurand.

Transmitter: In telemetry, the apparatus which generates the carrier (usually at radio frequencies), modulates the carrier (in response to signals representative of measurements), and couples the modulated carrier to a radio or wire link.

Transverse acceleration: An acceleration perpendicular to the sensitive axis of the transducer.

Transverse sensitivity: The maximum sensitivity of a transducer to a specified value of transverse acceleration or other transverse measurand. Usually expressed in percent of the sensitivity in the sensitive axis.

Triboelectric effect: The separation of electric charges by friction between surfaces.

Turbine: A bladed rotor which turns at a speed nominally proportional to the volume rate of flow.

Unbonded strain gage: Strain-sensitive wire stretched and unsupported between ends.

Variable (*n.*): (See Measurand.)

Velocity: A vector quantity equal to the time rate of change of displacement with respect to a reference system.

Vibration: An oscillatory velocity, acceleration, or other mechanical quantity.

Vibration error: The maximum change in output, at any measurand value within the specified range, when vibration levels of specified amplitude and range of frequencies are applied to the transducer along specified axes.

Voltage ratio: For potentiometric transducers, the ratio of output voltage to excitation voltage, usually expressed in percent.

Warm-up period: The period of time, starting with the application of excitation to the transducer, required to assure that the transducer will perform within all specified tolerances.

Weight: The gravitational force of attraction.

Zero balance: The tolerances of zero-measurand output.

Zero-measurand output: The transducer output with rated excitation and zero measurand applied to the transducer.

Zero shift: A change in zero-measurand output over a specified period of time and at room conditions. *Note:* Zero shift is an error characterized by a parallel displacement of the entire calibration curve.

Index